BIOMEDICAL APPLICATIONS OF MAGNETIC PARTICLES

BIOMEDICAL APPLICATIONS OF MAGNETIC PARTICLES

Edited by
Jeffrey N. Anker and O. Thompson Mefford

CRC Press is an imprint of the
Taylor & Francis Group, an **informa** business

First edition published [2021]
by CRC Press
6000 Broken Sound Parkway NW, Suite 300, Boca Raton, FL 33487-2742

and by CRC Press
2 Park Square, Milton Park, Abingdon, Oxon, OX14 4RN

CRC Press is an imprint of Taylor & Francis Group, LLC

Library of Congress Cataloging-in-Publication Data

Names: Anker, Jeffrey N., editor. | Mefford, Thompson (O. Thompson), editor.
Title: Biomedical applications of magnetic particles / edited by Jeffrey N. Anker and Thompson Mefford.
Description: First edition. | Boca Raton : CRC Press, 2021. | Includes bibliographical references and index.
Identifiers: LCCN 2020033558 (print) | LCCN 2020033559 (ebook) | ISBN 9781439839683 (hardback) | ISBN 9781315117058 (ebook)
Subjects: LCSH: Magnetic materials. | Biomedical engineering. |Nanobiotechnology. | Nanoparticles.
Classification: LCC TK454.4.M3 .B56 2021 (print) | LCC TK454.4.M3 (ebook) | DDC 610.28/4--dc23
LC record available at https://lccn.loc.gov/2020033558
LC ebook record available at https://lccn.loc.gov/2020033559

ISBN: 978-1-439-83968-3 (hbk)
ISBN: 978-1-315-11705-8 (ebk)

Typeset in Times New Roman
by MPS Limited, Dehradun

Table of Contents

Foreword

A little over eight years ago, both of us (Jeff Anker and Thompson Mefford) arrived at Clemson University as idealistic new assistant professors. What connected us was a shared love of the "magic" of magnetic nanoparticles. These unique materials give us the ability to "see" into the body, remotely manipulate structures, and, ultimately, treat diseases. In this shared appreciation, we organized the first meeting on the "Frontiers in Biomagnetic Particles," which has continued every two years since (currently in its fifth incarnation). Following the first meeting, we discussed the need for a book to give to those new to the field. Essentially, we asked ourselves, "What do we wish we had known when first starting in this field?"

This inquiry lead to us to seek the experts in each of the various subfields to contribute to the content provided herein. In retrospect, we are enamored with the results from our friends and colleagues. Also, while preparing this book, we lost a friend and mentor, Dr. Steve Klaine. It is hoped that his contribution to this book may in a small part help instill his legacy.

While the material in this field is fascinating, the individuals contributing are the greatest resource.

We would like to thank all the authors for their contributions to this book. We would also like to acknowledge the contribution of Godfrey Kimball, who helped copy editing at Clemson University. We are also grateful to CRC Press for helping make this possible.

Jeff Anker and O. Thompson Mefford, co-Editors
Clemson, SC

Preface

When children play with magnets and magnetic materials, magic happens and that magic can linger on into adulthood. I remember 25 years ago, I was at a meeting with a group of magnetic particle collaborators. During a break, one of them took out a vial filled with iron filings, emptied it on a thin white cardboard, grabbed a small magnet and started to move the filings around from underneath the cardboard, turning them into rings, trees, and pillars. Each of us tried to generate ever more amazing artwork, and it was a full 15 minutes later that one of the people present reminded us to get back "to business."

Over the quarter century since then, my fascination with magnetic particles has not weakened. Today we work with ultrasmall magnetic nanoparticles and ferrofluids. Many people might not realize that humans did not invent them—they were synthesized in small single-cell organisms such as magnetotactic bacteria many millions of years ago. It is only in the last 20 years that we have learned how these organisms build magnetic nanoparticles in a highly organized and genetically predetermined way. Although we can come close to synthesizing these perfectly shaped and sized nanoparticles, we still have a lot to learn to catch up with nature.

This book is about many things—how to make magnetic nanoparticles and how to use them: as contrast agents, for magnetically targeted drug delivery, in biosensors, and for energy dispersive processes. In a field with such broad applications, researchers must not only be very interdisciplinary but also understand the basics. The basics explained in this book by experts include general magnetism, magnetic fields and forces, magnetic particle interactions, and the use of instruments to investigate single and multiple magnetic nanoparticles.

This is not the only book about magnetic nanoparticles, but it is special in that it brings together a key group of leading experts. The readable way they cover all the important subfields will help you appreciate the diversity of their research subjects, will demonstrate how much is still being investigated, and will highlight why they enjoy working on the fine details. They also show you where the field is going, and where future applications and uses for magnetic particles lie—key information for curious scientists and future collaborators.

I suggest that you wait for a rainy day, grab a magnet and your best magnetic nanoparticles, and check out how they influence each other. And then take this book and try to figure out what just happened in front of your eyes, by reading the basics as well as the applied chapters. I'm sure that you will read about new subjects, learn some new facts that will be valuable in your research, and come away with new ideas for your work. The same things the book did for me. Enjoy!

Urs Hafeli

Editor Bios

Jeffrey N. Anker is a Wallace R. Roy Distinguished Professor of Chemistry and BioEngineering at Clemson University. He earned his BS degree in applied physics at Yale University in 1998. He earned his doctorate at The University of Michigan in 2005, working for Professor Raoul Kopelman to develop magnetically modulated optical nanoprobes (MagMOONs) to measure chemical concentrations and mechanical properties of solutions. For this work, he was awarded a grand prize at the 2002 National Inventor's Hall of Fame Collegiate Inventor's Competition. From 2005 to 2008, Dr. Anker worked as an NIH National Science Research Award (NSRA) postdoctoral research fellow at Northwestern University under the guidance of Professor Richard Van Duyne. His postdoctoral research focused on developing real-time high-resolution plasmonic nanosensors. He joined the Clemson Chemistry Department in August 2008. Current research focuses on imaging and spectroscopy using magnetic, plasmonic, x-ray excited micro- and nanosensors, implantable sensors, orthopedic devices, and medical imaging. Along with Thompson Mefford, he founded the Frontiers in BioMagnetic Particles Meeting Series. Awards include: NSF CAREER award (2013); Clemson Faculty Collaboration Award (2014); Clemson University School of Health Research (CUSHR), and Greenville Health System Embedded Faculty Fellow (2018); University Research, Scholarship, and Artistic Achievement Award (USRAAA) (2018); and senior member of the National Academy of Inventors (2019).

O. Thompson Mefford is an Associate Professor in the Department of Materials Science Engineering where he holds a David and Mary Ann Bishop Dean's Professorship, along with an additional appointment in the Department of Bioengineering at Clemson University. He earned his BS degree in Polymer and Textile Chemistry at Clemson in 2003 and a PhD in Macromolecular Science and Engineering from Virginia Tech in 2007, where he worked on the development of treatments for retinal detachment using hydrophobic ferrofluids. Before returning back to Clemson, Dr. Mefford developed methods for the fabrication and functionalization of microfluidic devices as a Post-doctoral Researcher for The Ohio State University Department of Chemistry.

Dr. Mefford joined the faculty of the Department of Materials Science and Engineering at Clemson in the Fall of 2008. His research focuses on developing stable, polymer-iron oxide nanoparticle complexes and composites for biomedical applications. These applications include developing materials for magnetically modulated energy delivery, MRI contrast agents, and drug delivery systems. He currently organizes the Frontiers in Magnetic Particles Conference (www.magneticnanoparticle.com). In his free time, Dr. Mefford is found running, cycling, sailing, backpacking, and home brewing.

List of Contributors

Jeffrey N. Anker*
Chemistry and Bioengineering Department
Clemson University
Clemson, SC, USA

Mahaveer S. Bhojani
Department of Radiation Oncology
University of Michigan
Ann Arbor, MI, USA

Jordan T. Burbage
Department of Biological Sciences
Institute of Environmental Toxicology
Clemson University
Clemson, SC, USA

Mark A. Burns
Department of Chemical Engineering
University of Michigan
Ann Arbor, MI, USA

and

Department of Biomedical Engineering
University of Michigan
Ann Arbor, MI, USA

Sarah H. Cartmell
School of Biomaterials
Materials Science Centre
The University of Manchester
Manchester, UK

Matthew R. J. Carroll
School of Physics
The University of Western Australia
Crawley, WA, Australia

Adam J. Cole
Department of Pharmaceutical Sciences
University of Michigan
Ann Arbor, MI, USA

Allan E. David*
Department of Chemical Engineering
Auburn University
Auburn, AL, USA

Cindi L. Dennis*
Material Measurement Laboratory
National Institute of Standards and Technology (NIST)
Gaithersburg, Maryland, USA

Jon Dobson*
Departments of Biomedical Engineering and Materials Science & Engineering
University of Florida
Gainesville, Florida USA

and

Institute for Science & Technology in Medicine
Keele University
Keele, UK

Urs Hafeli
Department Pharmaceutics and Biopharmaceutics
The University of British Columbia
Vancouver, BC, Canada

Micheal J. House
School of Physics
The University of Western Australia
Crawley, WA, Australia

Alexander Hrin
Department of Chemistry
University of Michigan
Ann Arbor, MI, USA

Dale L. Huber*
Sandia National Laboratories
Center for Integrated Nanotechnologies (CINT)
Albuquerque, NM, USA

Paivo Kinnunen
Department of Chemistry
University of Michigan
Ann Arbor, MI, USA

and

Fibre and Particle Engineering Research Unit
University of Oulu
Oulu, Finland

* Corresponding author

Stephen J. Klaine
Department of Biological Sciences
Institute of Environmental Toxicology
Clemson University
Clemson, SC, USA

Raoul Kopelman*
Department of Chemistry
University of Michigan
Ann Arbor, MI, USA

and

Department of Biomedical Engineering
University of Michigan
Ann Arbor, MI, USA

Yunzi Li
Department of Chemical Engineering
University of Michigan
Ann Arbor, MI, USA

and

Biosciences Division
Oak Ridge National Laboratory
Oak Ridge, TN, USA

O. Thompson Mefford*
Department of Materials Science and Engineering
Clemson University
Clemson, SC, USA

Paul W. Millhouse
Department of Chemistry
Clemson University
Clemson, SC, USA

Timothy G. St Pierre*
Department of Physics,
The University of Western Australia
Crawley, WA, Australia

Carlos Rinaldi*
Department of Chemical Engineering
University of Florida
Gainesville, FL, USA

and

J. Crayton Pruitt Family Department of Biomedical Engineering and Department of Chemical Engineering
University of Florida
Gainesville, FL, USA

Steven L. Saville
Department of Materials Science and Engineering
Clemson University
Clemson, SC, USA

MythreyiUnni
Department of Chemical Engineering
University of Florida
Gainesville, FL, USA

Unaiza Uzair
Department of Chemistry
Clemson University
Clemson, SC, USA

Robert C. Woodward*
School of Physics
The University of Western Australia
Crawley, WA, Australia

Erika C. Vreeland
Sandia National Laboratories
Albuquerque, NM, USA

Benjamin Yellen*
Department of Mechanical Engineering and Materials Science
Center for Biologically Inspired Materials and Material Systems (CBIMMS)
Duke University
Durham, NC, USA

1 Introduction to Biomedical Applications of Magnetic Nanoparticles

Jeffrey N. Anker and O. Thompson Mefford

CONTENTS

1.1 PURPOSE

The goal of this chapter is to provide a broad overview of properties and biomedical applications of magnetic nanoparticles. Subsequent chapters will elucidate the fundamental magnetic nanoparticle physics and chemistry and explore important biomedical applications. This brief introduction aims to place this information found in the remainder of this book into a broader context.

1.2 BIOMEDICAL APPLICATIONS OF MAGNETIC PARTICLES

A grand challenge in biomedical sciences is to detect and control the location and time of biochemical and biomechanical processes. For example, in treating cancer, the challenge is to direct the therapeutic to the tumor while minimizing damage to normal tissue, which can in principle be facilitated by targeting the drug and/or triggering the drug release. Similarly, the ability to switch cellular signaling on and off and detect this activity is critical for understanding and ultimately controlling cellular activity. A variety of external triggers are available including light (Fomina, Sankaranarayanan, and Almutairi 2012, Deisseroth 2011), ultrasound (Hernot and Klibanov 2008, Deckers and Moonen 2010, Mitragotri 2005), chemical signals (e.g. pH [Huh et al. 2012] or enzymes [De La Rica, Aili, and Stevens 2012]), electrical current or voltage (Murdan 2003), temperature changes (Schmaljohann 2006), magnetic fields (Dobson 2008, Pankhurst et al. 2003), and combinations thereof. Magnetism is an especially useful handle for controlling and detecting stimuli-responsive materials because magnetic fields are easily generated at a distance (using either permanent magnets or electromagnets) and biological tissue is essentially transparent to these magnetic fields. This provides a unique advantage to magnetic sensing and stimulation compared to other methods.

Only a select group of materials displays large magnetic moments at room temperature and magnetic fields that are readily accessible (<2.5 T). These magnetic materials are: elemental iron, nickel, and cobalt; compounds made from these three elements (e.g. certain alloys, oxides, sulfides, carbides, and borides), and a few other materials (e.g. chromium dioxide, and gadolinium metal at $<20\,°C$, and manganese compounds including MnBi, MnSb, MnAs, MnB, Au_2MnAl, and Cu_2MnAl). Almost all other materials have very small magnetic moments at room temperature and their magnetic properties are apparent only under unusual conditions (e.g. nuclear magnetic resonance [NMR] spectroscopy detects magnetic properties of atomic nuclei using radiofrequency excitation, and largely diamagnetic biological samples can be levitated in very strong fields with large field gradients).

Among room temperature magnetic materials, only maghemite (gamma crystal phase of Fe_2O_3) is found naturally in cells. Specifically, maghemite is found in magnetotactic bacteria, which use it to orient and navigate and stay in nutrient-rich environments (Matsunaga and Okamura 2003, Blakemore 1975, Martel et al. 2009). Maghemite is also found in birds, fish, and other animals, where it is believed to serve as a compass to help them navigate (Kirschvink and Gould 1981) (although chemical and electroinductive mechanisms have also been proposed for sensing magnetic fields, termed "magnetoreception," in some organisms) (Lohmann 2010). In humans, a small amount has been found in the brain (Dobson 2002) as well as exogenous magnetic materials from welding fumes (Nakadate et al. 1998) and air pollution (Maher et al. 2016) or from ingestion of iron particles (e.g. in cereal) (Hoppe, Hulthén, and Hallberg 2006). Aside from these esoteric examples, cells and tissues do not contain maghemite and are essentially unaffected by applied magnetic fields and do not affect these fields. This means that any magnetic material added to a cell or tissue can be controlled and imaged separately from the tissue.

1.3 WHY NANOPARTICLES?

In general, nanoparticles become interesting and useful when properties emerge that are specific to their size and shape and/or their composition is chosen to have multiple properties that work together synergistically. Magnetic nanoparticles are quintessential examples of materials with emergent properties. The magnetic moment of magnetic particles arises because interactions between atoms in ferromagnetic or ferromagnetic crystal cause the spin from their valence electrons to align together providing very large magnetic moments compared to the same number of atoms with randomly aligned spins (e.g., in a dissolved iron salt). This increased net magnetic moment of the particle leads to dramatically enhanced alignment energies and quite different behavior in response to applied magnetic fields. As an analogy, it is far easier to throw a snowball than the same amount of water dispersed in a puddle (or worse yet, the same number of water molecules as vapor in the air) because in a snowball the water molecules are mechanically linked together so that the position of one depends on the position of its neighbors. Although this analogy should not be taken too literally (atomic spin is not the same as atomic position and the coupling mechanisms are different too), it illustrates how the behavior of previously independent atoms changes dramatically when they are coupled together.

For example, Figure 1.1 shows how the magnetic energy for a magnetite (Fe_3O_4) nanoparticle depends upon the diameter of the nanoparticle in a 10,000 Oe applied field ($\mu_0H = 1$ T in SI units). We

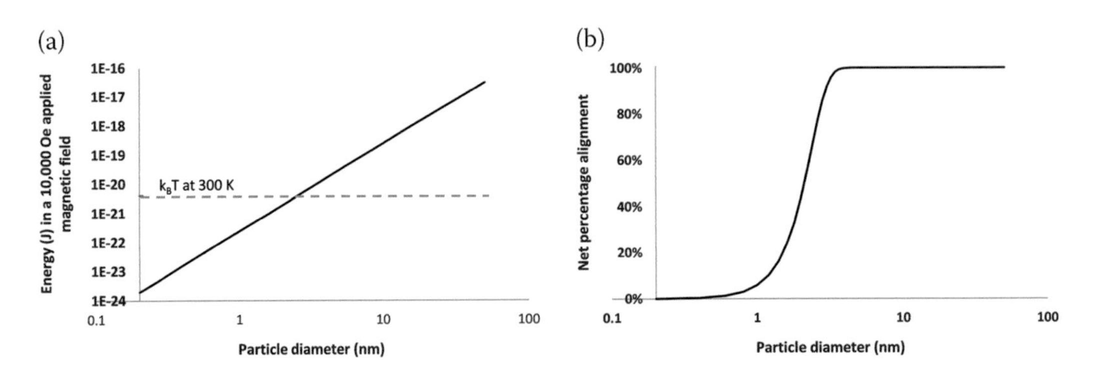

FIGURE 1.1 **(a)** Dependence of magnetic energy on particle diameter for a particle with magnetization equal to the saturation magnetization of magnetite (470 kA/m), **(b)** Net magnetic alignment as a function of diameter.

assumed that the particle has a saturation magnetization of 470 emu/cm^3 (or 470 kA/m in SI units) and calculated the energy as $E = \mu_0 M_S \cdot HV$, where E is the energy in Joules, M_S is the saturation magnetization in kA/m, H is the applied field (μ_0 H = 1T) and V is the particle volume in m^3 ($\frac{4}{3}\pi r^3$ for a sphere). The magnetic energy increases with the particle size, and becomes equal to thermal energy at 300 K ($k_B T$) when it is 2.6 nm in diameter (Figure 1.1a). Above this size, most of the moments will be aligned in the applied field; below this size the net alignment rapidly decreases. Although this is a crude calculation, which neglects effects from particle shape, crystalline anisotropy, and differences in magnetic properties of the surface of the particle, it shows the rapid change in magnetic energy and alignment with size, and illustrates the need for using magnetic particle that are at least several nanometers in diameter. Chapters 2 and 3 provide more details on the fundamentals of magnetism and their application to nanoparticle rotation and translation.

In general, the larger the particle is, the more force and torque can be applied to it with an external magnetic field. However, if the particle becomes too large, it is likely to form aggregates at reasonable concentrations due to magnetic interparticle forces (see Chapter 4). When the particle or aggregates are larger than a few microns, the materials interact with biological systems differently and are likely to block blood capillaries. Thus, magnetic nanoparticles in the 3–100 nm range (or materials/constructs made with them, such as magnetic nanoparticle-labelled cells or gels loaded with magnetic nanoparticles and drugs) are uniquely suited for biomedical applications. Figure 1.2 shows the size regime of biological materials. In general the 3–100 nm size range corresponds to proteins and viruses or intracellular organelles. The figure also shows the relevant ranges for behavior of magnetite nanoparticles (~3–25 nm superparamagnetic; 25–130 nm single domain, >130 nm multidomain). The cutoffs are not absolute and depend partly on the particle shape and composition as well as the time scale used to probe the particles. The figure also shows the range of

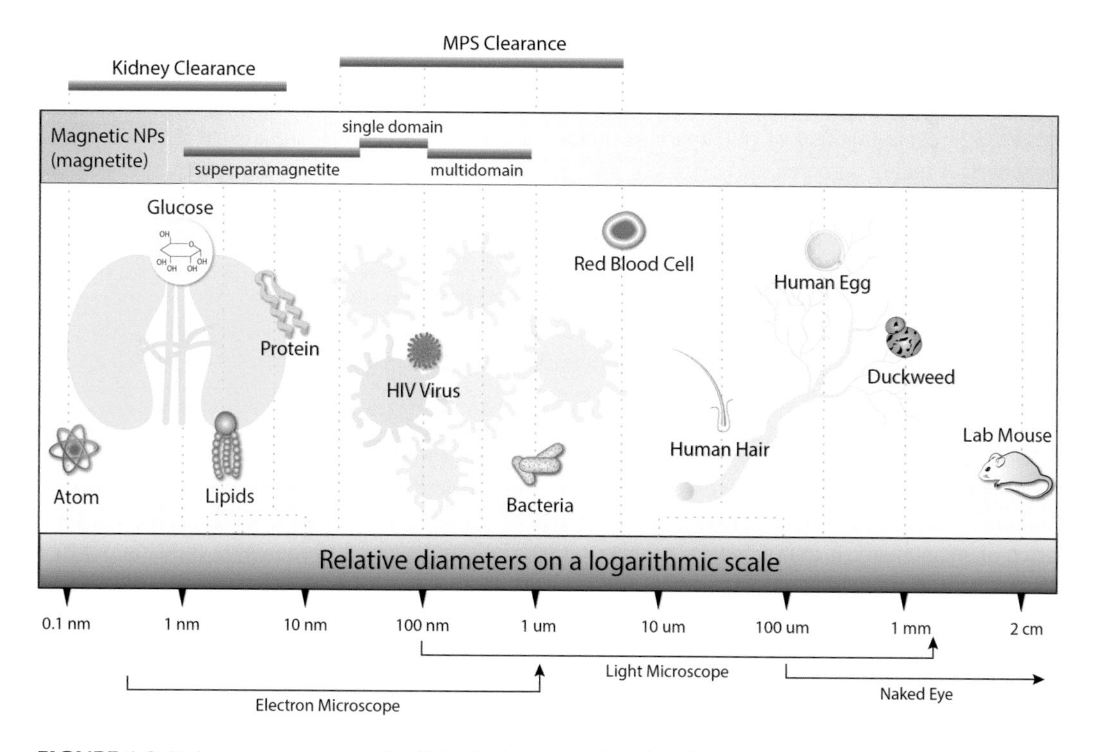

FIGURE 1.2 Relevant size regimes for biological samples (ranging from atoms to lab mouse) and magnetic nanoparticles. Rough ranges for clearance regimes and microscopy observation regimes are also shown.

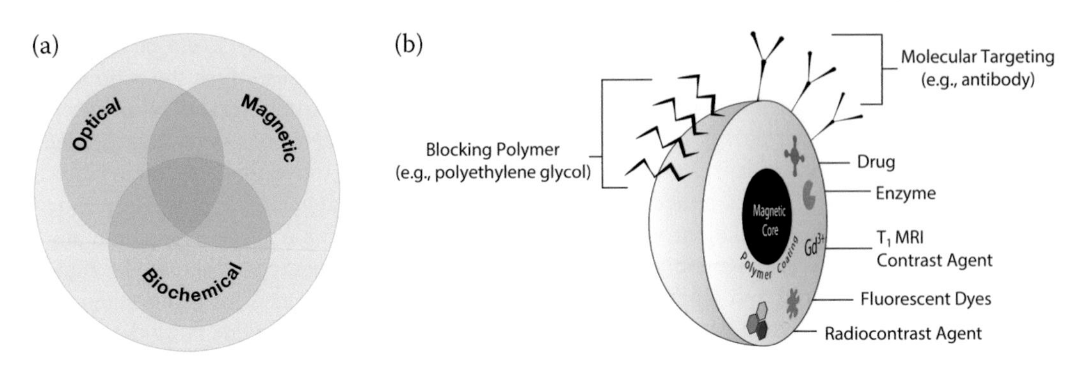

FIGURE 1.3 Multifunctional particle schematics. **(a)** Combining multiple properties into a single particle or group of particles, **(b)** Multiple possible components and functions integrated into the same particle.

particle diameters where particles are filtered by the kidneys and excreted in urine (<~6 nm), or taken up by the MPS immune system cells (10 nm to 8 μm), with >8 μm blocking capillaries.

For biological applications, one generally wants to control nanoparticle size, shape, composition, and surface properties for the application. Nanoparticles are well-suited for rational design because they have "real estate" in the core and surface, which can be filled with different materials/coatings to control their surface chemistry, drug content, and optical and magnetic properties. Figure 1.3 is a schematic showing how multiple components and functionalities can be loaded into a single vehicle. Although administering components separately is simpler and often appropriate (e.g. a chemotherapy drug can be used separately from a magnetic resonance imaging (MRI) agent to see the tumor), combining multiple components within a single particle often generates new capabilities (e.g. drug can be magnetically directed/retained at a specific location only if it is attached to a magnetic particle). In addition to bringing particles to a specific location with magnetic forces, torques can be used to separate target molecules, pull or twist on cellular receptors, aggregate together or pull apart, or modulate optical or chemical signals for particles with nonspherical magnetic cores and orientation-dependent optical or chemical properties.

1.4 SUMMARY OF SUBSEQUENT CHAPTERS

Following this introductory chapter, the book is divided into two sections. The first section is on fundamentals for working with magnetic particles. The second covers practical aspects of nanoparticle synthesis and biomedical applications. Each chapter is written by specialists in their field.

Section I Fundamentals for Working with Magnetic Particle Fundamentals

Chapter 2 A Conceptual Introduction to the Fundamentals of Magnetic Fields, Magnetic Materials, and Magnetic Particles for Biomedical Applications. This chapter describes the basics of magnetism. It explains how magnetic materialtions respond to magnetic fields and why this is size and shape dependent.

Chapter 3 Magnetic Forces and Torques: Separation, Tweezing, and Materials Assembly in Biology. This chapter builds upon Chapter 2's fundamental introduction and provides more details on how magnetic particles respond to external magnetic fields and field gradients, especially in viscous fluids.

Chapter 4 Colloidal Interactions of Magnetic Nanoparticles. This chapter discusses the properties of colloids. Several forces control interparticle separation. Under many conditions, particles can irreversibly aggregate, which can reduce repeatability of results and can also adversely affect many of the applications. For example, the ionic concentrations of biological media may result in aggregation, if the particle system is not properly engineered. The chapter discusses how surface coatings can improve stability under various conditions.

Chapter 5 Magnetic Characterization: Instruments and Methods. This chapter describes the techniques and instruments used to characterize magnetic nanoparticles, focusing especially on their magnetic properties.

Section II Practical Aspects of Nanoparticle Synthesis and Biomedical Applications

Chapter 6 Synthesis and Functionalization of Magnetic Particles. This chapter explains the theory and practice of how to synthesize and functionalize magnetic nanoparticles.

Chapter 7 Nanomagnetic Actuation: Controlling Cell Behavior with Magnetic Nanoparticles. Magnetic nanoparticles can be attached to specific cellular receptors and tissues in order to directly apply forces to study and control the resulting biochemical pathways. This chapter reviews the approaches and applications of nanomagnetic actuation.

Chapter 8 Magnetic nanoparticles: Challenges and opportunities in drug delivery. The goal in any drug delivery system is to maximize therapeutic effectiveness, while minimizing side effects. Magnetic particles are very interesting in this regard because they can be pulled to specific locations to increase local drug delivery and reduce systemic loads. This chapter covers both magnetic and physiological considerations for magnetic drug delivery.

Chapter 9 Magnetic Particle Biosensors. This chapter discusses how magnetic nanoparticles can be used in chemical sensors. Magnetic fields can selectively modulate or separate only the magnetic particles to improve sensitivity, speed, and/or specificity.

Chapter 10 Magnetic Contrast Imaging: Magnetic nanoparticles as probes in living systems. Magnetic nanoparticles are excellent contrast agents for imaging within the body because they provide a specific signal that can be detected through tissue. This chapter reviews magnetic nanoparticles and imaging techniques. It focuses on MRI, but also briefly describes magnetomotive optical coherence tomography, magnetic particle imaging (MPI), and magnetomotive ultrasound.

Chapter 11 Energy Dissipation by Magnetic Nanoparticles: Basic Principles for Biomedical Applications. Magnetic particles generate heat when exposed to alternating magnetic fields. This heating is potentially useful as a method of controlling drug release in gels and also killing cancer through local heating, either directly or in concert with other chemo or radiation therapies.

Chapter 12 Toxicology of Magnetic Nanoparticles. This chapter reviews the literature on magnetic nanoparticle toxicity, which is important for many biomedical applications.

REFERENCES

Blakemore, R. 1975. Magnetotactic bacteria. *Science* 190 (4212), 377–379.

De La Rica, R., D. Aili, and M. M. Stevens. 2012. Enzyme-responsive nanoparticles for drug release and diagnostics. *Advanced Drug Delivery Reviews* 64 (11), 967–978.

Deckers, R., and C. T. W. Moonen. 2010. Ultrasound triggered, image guided, local drug delivery. *Journal of Controlled Release* 148 (1), 25–33.

Deisseroth, K. 2011. Optogenetics. *Nature Methods* 8 (1), 26–29.

Dobson, J. 2002. Investigation of age-related variations in biogenic magnetite levels in the human hippocampus. *Experimental Brain Research* 144 (1), 122–126.

Dobson, J. 2008. Remote control of cellular behaviour with magnetic nanoparticles. *Nature Nanotechnology* 3 (3), 139–143.

Fomina, N., J. Sankaranarayanan, and A. Almutairi. 2012. Photochemical mechanisms of light-triggered release from nanocarriers. *Advanced Drug Delivery Reviews* 64 (11), 1005–1020.

Hernot, S., and A. L. Klibanov. 2008. Microbubbles in ultrasound-triggered drug and gene delivery. *Advanced Drug Delivery Reviews* 60 (10), 1153–1166.

Hoppe, M., L. Hulthén, and L. Hallberg. 2006. The relative bioavailability in humans of elemental iron powders for use in food fortification. *European Journal of Nutrition* 45 (1), 37–44.

Huh, K. M., H. C. Kang, Y. J. Lee, and Y. H. Bae. 2012. pH-sensitive polymers for drug delivery. *Macromolecular Research* 20 (3), 224–233.
Kirschvink, J. L., and J. L. Gould. 1981. Biogenic magnetite as a basis for magnetic field detection in animals. *Biosystems* 13 (3), 181–201.
Lohmann, K. J. 2010. Q&A: Animal behaviour: Magnetic-field perception. *Nature* 464 (7292), 1140–1142.
Maher, B. A., I. A. M. Ahmed, V. Karloukovski, D. A. MacLaren, P. G. Foulds, D. Allsop, D. M. A. Mann, R. Torres-Jardón, and L. Calderon-Garciduenas. 2016. Magnetite pollution nanoparticles in the human brain. *Proceedings of the National Academy of Sciences* 113 (39), 10797–10801.
Martel, S., M. Mohammadi, O. Felfoul, Z. Lu, and P. Pouponneau. 2009. Flagellated magnetotactic bacteria as controlled MRI-trackable propulsion and steering systems for medical nanorobots operating in the human microvasculature. *The International Journal of Robotics Research* 28 (4), 571–582.
Matsunaga, T., and Y. Okamura. 2003. Genes and proteins involved in bacterial magnetic particle formation. *Trends in Microbiology* 11 (11), 536–541.
Mitragotri, S. 2005. Healing sound: The use of ultrasound in drug delivery and other therapeutic applications. *Nature Reviews Drug Discovery* 4 (3), 255–260.
Murdan, S. 2003. Electro-responsive drug delivery from hydrogels. *Journal of Controlled Release* 92 (1), 1–17.
Nakadate, T., Y. Aizawa, T. Yagami, Y.-Q. Zheg, M. Kotani, and K. Ishiwata. 1998. Change in obstructive pulmonary function as a result of cumulative exposure to welding fumes as determined by magnetopneumography in Japanese arc welders. *Occupational and Environmental Medicine* 55 (10), 673–677.
Pankhurst, Q. A., J. Connolly, S. K. Jones, and J. Dobson. 2003. Applications of magnetic nanoparticles in biomedicine. *Journal of Physics D: Applied Physics* 36 (13), 167–181.
Schmaljohann, D. 2006. Thermo- and pH-responsive polymers in drug delivery. *Advanced Drug Delivery Reviews* 58 (15), 1655–1670.

Section I

Magnetic Particle Fundamentals

2 A Conceptual Introduction to the Fundamentals of Magnetic Fields, Magnetic Materials, and Magnetic Particles for Biomedical Applications

Timothy G. St Pierre

CONTENTS

2.1 INTRODUCTION

The chapter starts by introducing the concepts of magnetic fields and magnetic moments. These are the basic physical entities that are used in the measurement, description, and explanation of all magnetic phenomena. A description of a general experimental approach to measuring the magnetic properties of materials is given to illustrate the relationship between magnetic moments and magnetic fields. The major classes of magnetic behavior of materials that are observable with such an experimental approach are then considered. These classes of behavior include (1) materials that show increases in magnetic moment in the same direction as and linearly related to an applied magnetic field, (2) materials that show increases in magnetic moment in the opposite direction to and linearly related to an applied magnetic field, (3) materials that show non-linear relationships between their magnetic moment and the applied magnetic field, and (4) materials that exhibit magnetic moments that are dependent on the history of their magnetic field exposure (the phenomenon of magnetic hysteresis). The observed effects of varying the temperature of the materials on their magnetic behavior are also described.

After the reader is introduced to these surprisingly varied types of behavior, magnetic materials are considered from a microscopic perspective. The materials are considered to be comprised of atoms or molecules that behave as microscopic magnetic moments (in other words, particles that have a north and south magnetic pole). Many of the experimentally observed magnetic phenomena seen in materials can be explained or modeled by consideration of the magnetic interactions between these microscopic magnetic moments. By considering the different microscopic explanations for the different types of observed magnetic behavior in materials, the concepts of paramagnetism, ferromagnetism, antiferromagnetism, ferrimagnetism and diamagnetism are introduced.

A very large fraction of biomedical applications of magnetic materials involves small particles of magnetic materials in fluids. As such, this chapter covers an introduction to magnetic nanoparticles and the associated phenomenon of superparamagnetism. The chapter also introduces some of the factors that need to be considered when using magnetic fields to move magnetic particles through fluids. Finally, some examples of design considerations for magnetic particles for specific biomedical applications are used to illustrate the relevance of a basic understanding of magnetic materials to this field of research. These applications are covered in more detail in subsequent chapters.

2.2 MAGNETIC FIELDS AND MAGNETIC MOMENTS

Magnetic fields are generated by the movement of electric charges. A flow of electric charges is known as an electric current. The electric current flowing in a loop generates a magnetic field with a form known as a *dipole field* as illustrated in Figure 2.1a. This form of magnetic field is called a dipole field because it resembles the magnetic field that would be generated from two oppositely "charged" poles, or "magnetic charges", as shown in Figure 2.1b. The magnetic field lines flow from the north pole to the south pole. Figure 2.1b is, however, hypothetical, because magnetic monopoles have not been found in nature. Magnetic poles are always found in pairs, such as in the case of the familiar bar magnet, see Figure 2.1c. The naming convention for the poles of a magnet comes from the well-known phenomenon of the propensity of bar magnets to orient in the Earth's magnetic field when suspended on a pivot. One end of the magnet tends to seek the North Pole of the Earth and hence is termed a north-seeking pole while the other end of the magnet tends to seek the South Pole of the Earth and is hence termed a south-seeking pole. The terms *north-seeking pole* and *south-seeking pole* are usually shortened to *north pole* and *south pole* for convenience. At distances very close to the dipole source, the magnetic fields generated by an electric current loop, a pair of hypothetical magnetic charges, and a bar magnet are not identical (Figure 2.1a–c). However, if we are viewing the magnetic field of a dipole from a relatively large distance, or if the dipole is very small, the magnetic field will tend to look like that illustrated in Figure 2.1d and is similar for all dipoles. An electron is an example of a very

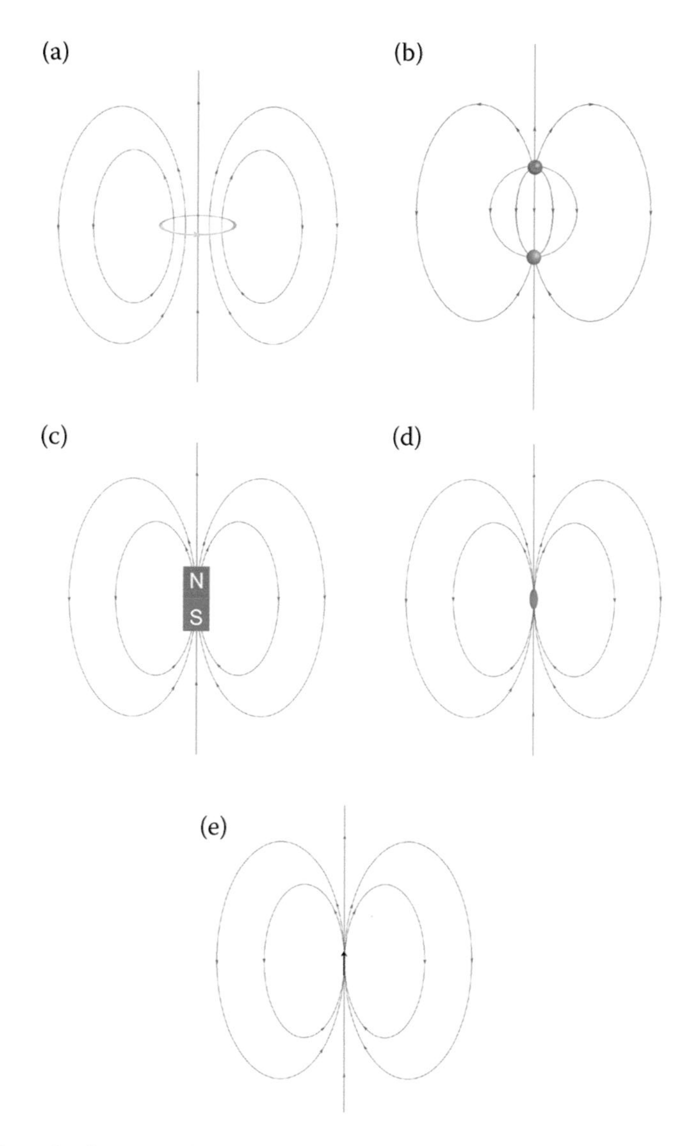

FIGURE 2.1 Schematic diagrams of **(a)** a magnetic dipole field generated by a loop of electric current, **(b)** a magnetic field generated from two oppositely "magnetically charged" magnetic poles, **(c)** a magnetic field generated by a bar magnet, **(d)** a magnetic field generated by a very small magnetic dipole (the far-field), **(e)** a magnetic dipole field with magnetic dipole represented as an arrow with the arrowhead at the north pole.

small magnetic dipole. In the original classical view of physics, the electron was thought to be a small spinning ball of electric charge, hence explaining the origin of its magnetic dipole nature. In fact, an electron is a point-like object and its properties are determined by quantum mechanics. Nevertheless, the spinning charged ball concept can be useful in many situations. Magnetic dipoles are often represented schematically by an arrow with the head of the arrow representing the north pole (see Figure 2.1e). The length of the arrow can indicate the strength of the dipole source (see Sections 2.3 and 2.5.5).

2.2.1 Magnetic Flux

The magnetic field lines shown in Figure 2.1a–e are known as lines of magnetic flux. The direction of the flux indicates the direction of the force that would be experienced by a north pole at a particular point in space. The density of flux lines is proportional to the magnitude of the force experienced by a north pole. Magnetic flux is usually represented by the symbol φ and the magnetic flux density, defined as the amount of flux passing through a unit area, A, is given by the symbol $\boldsymbol{B}$.

$$\boldsymbol{B} = \varphi / A. \tag{2.1}$$

We can imagine situations where the magnetic flux density is uniform in space (Figure 2.2a and b) or situations where the flux lines converge or diverge (Figure 2.2c). A magnetic field gradient is said to exist in situations where flux lines converge or diverge.

(a)

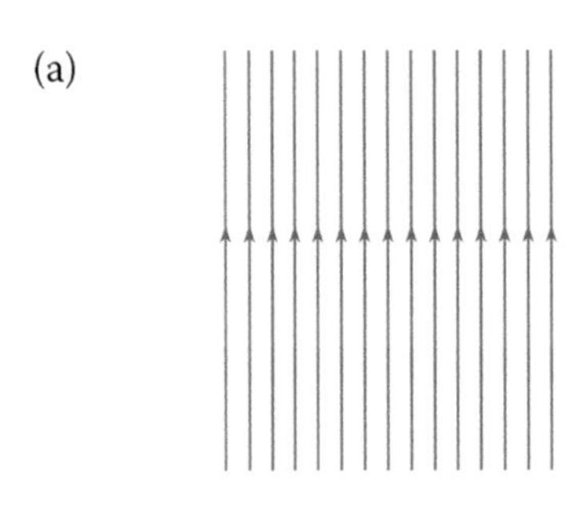

(b)

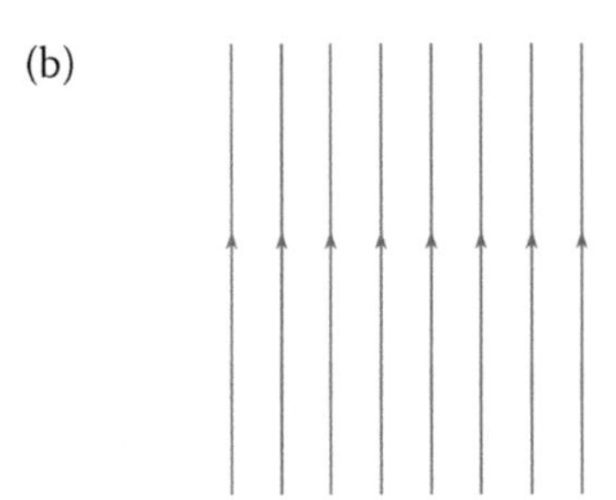

(c)

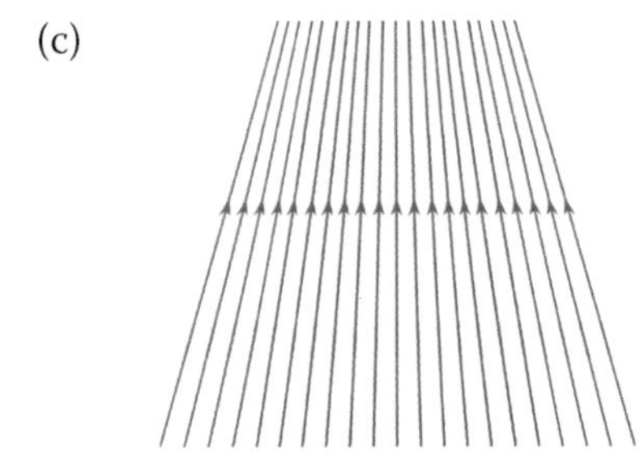

FIGURE 2.2 Schematic diagrams showing flux lines for a region of space with **(a)** uniform high flux density, **(b)** uniform low flux density, and **(c)** a magnetic field gradient.

2.2.2 Magnetic Moments

A magnetic dipole in a field with flux density, $\boldsymbol{B}$, experiences a torque, $\boldsymbol{\tau}$. The magnitude of the torque depends on $\boldsymbol{B}$ and the magnetic dipole moment, $\boldsymbol{m}$ (see Figure 2.3a and b).

$$\tau = mB\sin(\theta) \tag{2.2}$$

where θ is the angle between $\boldsymbol{m}$ and $\boldsymbol{B}$. The torque is manifested because, for a given flux density, south poles experience a force equal but opposite in direction to that experienced by an equivalent north pole (Figure 2.3a and b). It is worth noting that in a uniform magnetic flux density there is no net translational force on the dipole because the forces on the north and south pole balance each other. In a magnetic field gradient, however, one of the poles will be in a slightly higher flux density than the other and hence will experience a magnitude of force greater than the other (Figure 2.3c). Thus, in a magnetic field gradient, a dipole experiences a net translational force in the direction of the field gradient.

Note that both magnetic dipole moment and magnetic flux density are vector quantities. In other words, they have both magnitude and direction associated with them.

2.3 INTERACTION OF MAGNETIC MATERIALS WITH MAGNETIC FIELDS

A material with a net magnetic moment is said to be magnetized. Magnetization, $\boldsymbol{M}$, is the magnetic moment per unit volume within the material (although sometimes it is quoted in the form of magnetic moment per unit mass or per mole). From a microscopic perspective, the magnetization depends on the number density of magnetic dipoles within the material, the magnitude of the magnetic dipoles within the material, and the arrangement of the magnetic dipoles within the

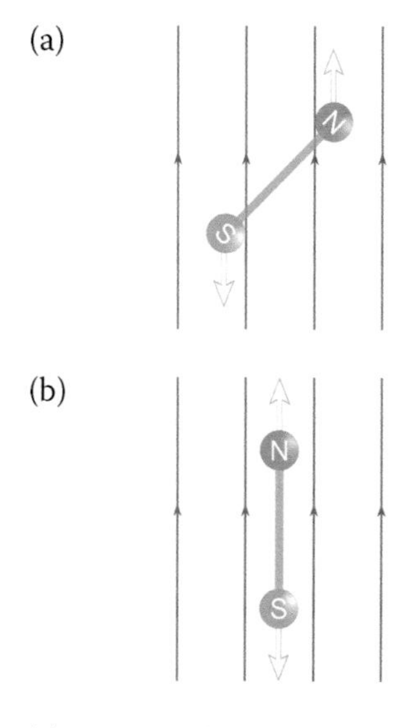

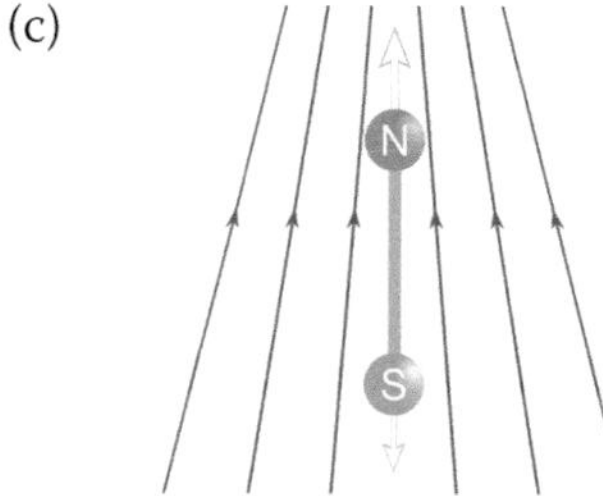

FIGURE 2.3 **(a)** The origin of the torque on a magnetic dipole in a region of space with uniform flux density arises from the equal but oppositely directed forces on the north and south pole of the dipole. **(b)** There is no net translational force on a dipole in uniform flux density. **(c)** A net translational force is experienced by a dipole in a magnetic field gradient because one pole experiences a greater magnitude force than the other.

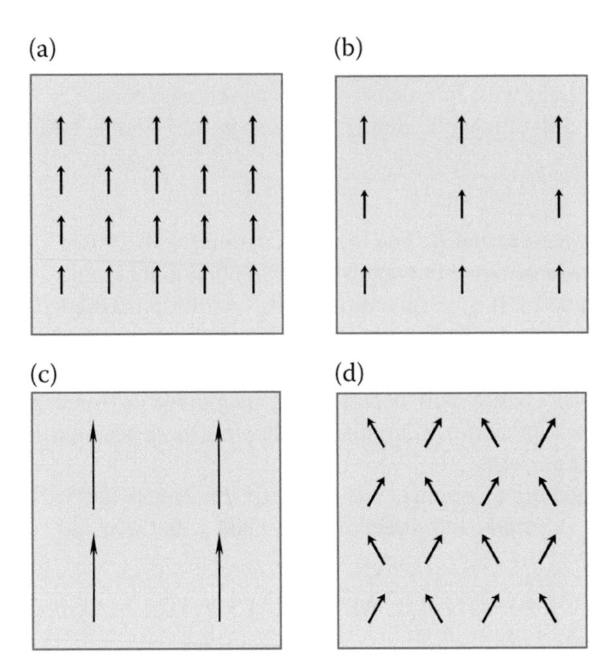

FIGURE 2.4 Schematic diagrams showing the arrangements of magnetic dipoles within magnetic materials. Material **(a)** has a high magnetization owing to a high density of aligned magnetic dipoles, while material **(b)** has a lower magnetization owing to the lower density of dipoles. While material **(c)** has a very low density of dipoles, the magnitude of the dipole moments is larger than the other materials. Material **(d)** has a high density of dipoles but the non-linear arrangement will result in a lower magnetization than a material with the same density of aligned dipoles.

material (Figure 2.4). The arrangement of the magnetic dipoles is important because their magnetic moments are vector quantities and as such their vector sum is dependent on the relative orientation of each dipole. Magnetization in materials arises mainly from unpaired electron spins and to a lesser extent from the orbital motion of electrons within the material.

2.3.1 Generating Magnetic Fields in the Laboratory

There are many ways to generate magnetic fields in the laboratory. To generate a uniform magnetic field, a solenoid is often used (Figure 2.5). An electric current, I, is passed through a conducting coil with several uniformly spaced turns. Assuming that the space inside the coil is a vacuum (or air), the magnitude of the magnetic flux density generated inside the coil is proportional to the electric current, I, and the number of turns per unit length, n, of the coil. For the case of a long thin solenoid, the flux density within the solenoid is parallel to the axis of the solenoid and is given by

$$B = \mu_0 I\, n = \mu_0 N\, I/L \tag{2.3}$$

where μ_0 is a universal constant called the permeability of free space, N is the total number of turns in the solenoid, and L is the length of the solenoid.

2.3.2 Sample of Material in a Magnetic Field

Now imagine inserting a sample of material into the coil (Figure 2.6a). Generally, the orbital and spin magnetic moments within atoms of the sample will respond to the applied magnetic flux in some way. If a material had no magnetic response, the flux lines would not be perturbed (Figure 2.6b). If, on the other hand, a strongly "magnetic" material such as iron or cobalt is inserted into the field, the sample will tend to concentrate the magnetic flux (Figure 2.6c). Most materials that are considered to be "non-magnetic" do in fact display a very weak response to a magnetic field known as *diamagnetism*. Diamagnetic materials such as water, protein, fat, and plastics tend to weakly repel and expel magnetic flux (Figure 2.6d).

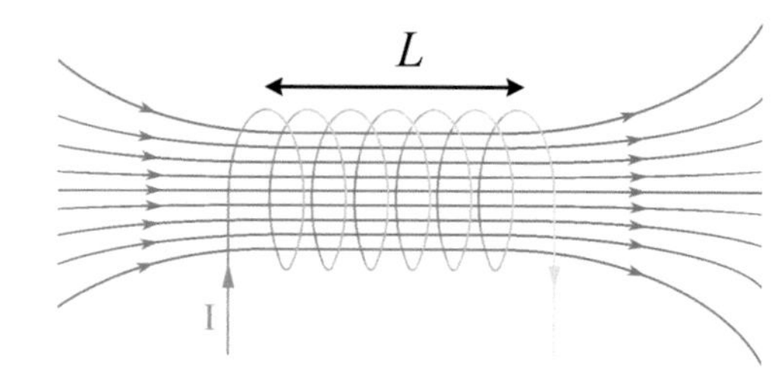

FIGURE 2.5 A solenoid is an elongated coil of conducting wire with *N* turns. A relatively uniform magnetic flux density is generated inside the solenoid of length L when an electric current, *I*, is passed through the wire.

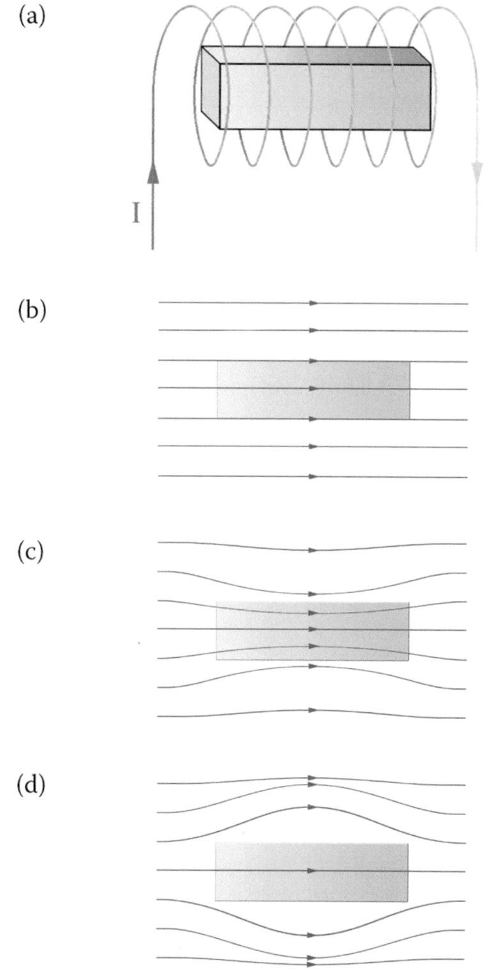

FIGURE 2.6 **(a)** Schematic diagram of a sample of material in a solenoid. **(b)** If the material shows no magnetization, the magnetic flux lines pass through the material unperturbed. **(c)** If a magnetic material such as metallic iron is placed in the solenoid, the material tends to concentrate the magnetic flux lines. **(d)** If a diamagnetic material such as plastic or water is placed in the solenoid, the magnetic flux lines are weakly repelled (the effect is exaggerated for clarity in this diagram).

2.3.3 The B-Field and the H-Field

Imagine that we have a sample of material inside a solenoid. We wish to predict the value of the magnetic flux density, $\boldsymbol{B}$, at some point inside the material. The value of $\boldsymbol{B}$ will depend on (1) the geometry of the solenoid and the electric current running through it, (2) the magnetic properties of the material, and (3) the geometry of the material. It is useful to introduce the concept of a vector field quantity called the magnetic field strength, $\boldsymbol{H}$. We can think of $\boldsymbol{H}$ as being a theoretical construct that is determined by the geometry of the electric current in the solenoid. We can then write down a relationship between the flux density, $\boldsymbol{B}$, the magnetization, $\boldsymbol{M}$, and the magnetic field strength, $\boldsymbol{H}$, at each point in space as follows:

$$\boldsymbol{B} = \mu_0(\boldsymbol{H} + \boldsymbol{M}). \tag{2.4}$$

As such, in the absence of any material in the solenoid, there would be no magnetization (because of an absence of magnetic dipoles) and so Equation 2.4 reduces to

$$\boldsymbol{B} = \mu_0\boldsymbol{H}. \tag{2.5}$$

2.4 MEASURING THE MAGNETIC MOMENT OF A SPECIMEN

Figure 2.7 shows a typical arrangement for measuring the magnetic moment of a specimen. The specimen is mounted on a device that enables the specimen to be moved through a coil of conducting wire (the inner coil in the figure). Very often in practice, the sample is repeatedly moved through the coil (such as in a vibrating sample magnetometer) (see Chapter 5 for additional details). The rate of change of magnetic flux through the coil as the specimen is moved generates a voltage across the terminals of the coil, which can be measured with a voltmeter. The voltage generated is proportional to the magnetic moment of the specimen. Such devices can be calibrated to enable measurement of the magnetic moment of specimens. The outer coil in Figure 2.7 represents a coil through which an electric current can be passed in order to generate a uniform magnetic field in the sample space. If the magnetic field generated is kept at a constant value, it will not affect the voltage across the terminals of the inner sensing coil, because only fields changing with time generate voltages in coils. In this way, the device illustrated in Figure 2.7 can be used to measure the magnetic moment of a sample as a function of the strength of a magnetic field applied by the outer coil. In this configuration, the device is known as a magnetic susceptometer. It enables us to observe how susceptible is the magnetic moment of a sample to an applied magnetic field. Many magnetic susceptometers are equipped to enable the temperature of the sample to be varied using a furnace or cryostat, thus enabling magnetic moments of samples to be measured as a function of both temperature and magnetic field strength. If the volume, V, of the sample is known, the magnetization can be deduced because it is given by

$$\boldsymbol{M} = \boldsymbol{m}/V. \tag{2.6}$$

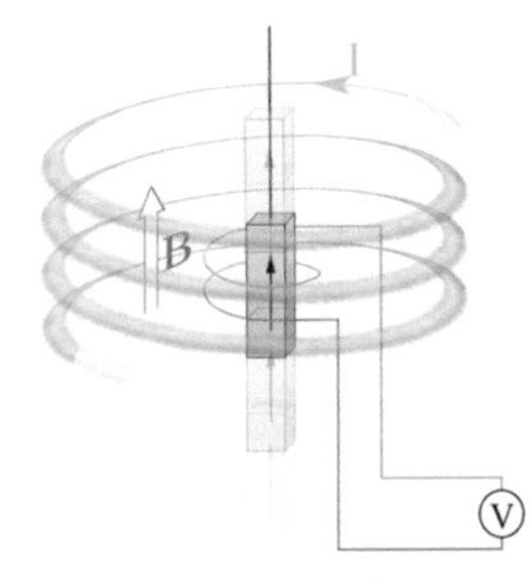

FIGURE 2.7 Schematic diagram of a magnetic susceptometer. The sample to be measured is moved through the inner sensing coil. The change in magnetic flux passing through the coil caused by the moving sample generates a voltage across the voltmeter, V. The magnitude of the voltage is related to the size of the magnetic moment on the sample and the speed with which it passes through the coil. The outer coil can be used to generate an applied magnetic flux density to the sample.

2.4.1 The Response of the Magnetization of a Material to Magnetic Field Strength, *H*

Generally the magnetization of a material changes in magnitude as the strength of the magnetic field, H, is varied (see Figure 2.8, for example). The magnitude of the response of the magnetization to the applied magnetic field is called the *magnetic susceptibility* of the material. The magnetic susceptibility, χ, is sometimes written as

$$\chi = M/H \tag{2.7}$$

but is sometimes written as the gradient of M vs H,

$$\chi = dM/dH. \tag{2.8}$$

Therefore, care needs to be taken when interpreting or reporting values of χ.

Diamagnetic materials, such as water or plastics, have a very weak negative response of their magnetization to an applied field, i.e. they have a small negative magnetic susceptibility. Materials that we usually think of as being "magnetic," such as iron or cobalt, have positive magnetic susceptibilities.

The magnetization, $\boldsymbol{M}$, of a material responds to the magnetic field strength, $\boldsymbol{H}$, in a variety of ways. The response depends on the type of material and the temperature of the material. The response can sometimes also depend on the previous history of the magnetic field strengths and directions applied to the material. The response of the magnetization to the applied magnetic field strength is not always linear, as depicted in Figure 2.8. The response is often non-linear, especially when the range of applied field strength is large or the temperature is low (e.g. when using cryogenic sample temperatures). Figure 2.9 shows some examples of non-linear responses of magnetization to applied magnetic field strength typical of a certain class of materials that do not exhibit magnetization in the absence of an applied magnetic field. Figure 2.9 shows how higher temperatures generally lead to a smaller but more linear response of M to H while lower temperatures lead to a stronger but more non-linear response. Note how the magnetization tends to saturate at high magnetic field strengths and low temperatures. Note also how the response of M to H is approximately linear if only small magnetic field strengths are applied.

For some materials, the low-field magnetic susceptibility is inversely proportional to temperature (Figure 2.10). This relationship is known as *Curie's law*. Curie's law can be written as

$$\chi = C/T \tag{2.9}$$

where C is a constant characteristic of the material, known as the Curie constant, and T is the absolute temperature measured in kelvins.

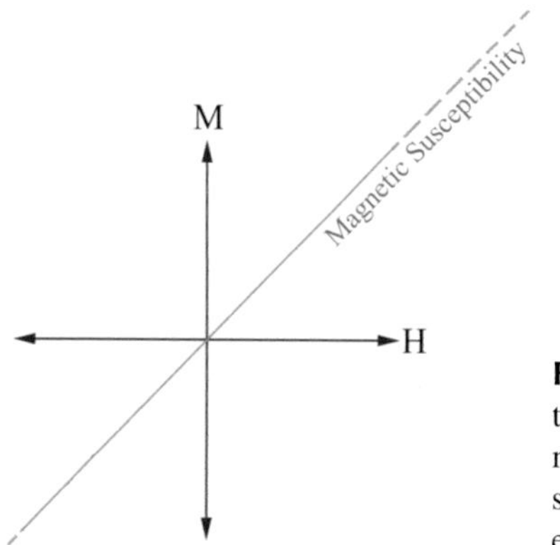

FIGURE 2.8 The response of the magnetization, M, of a sample to the magnetic field strength, H, applied to the sample is known as the magnetic susceptibility. Materials with high magnetic susceptibility show a steep gradient on the M vs H plot. Diamagnetic materials exhibit a very small negative gradient.

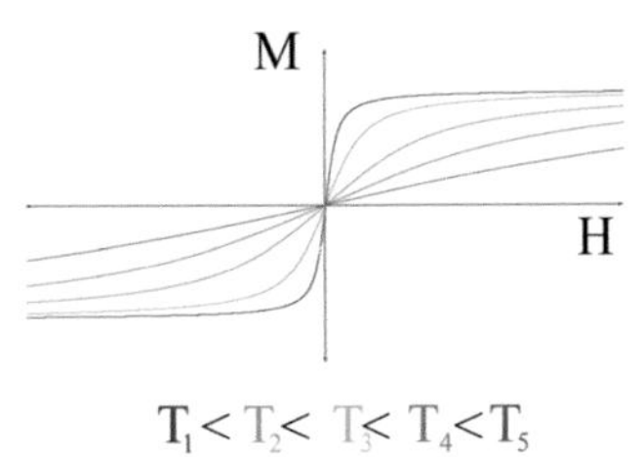

FIGURE 2.9 Many materials show a non-linear relationship between magnetization, M, and applied magnetic field strength, H. The relationship usually becomes weaker and more linear as the temperature of the material is increased.

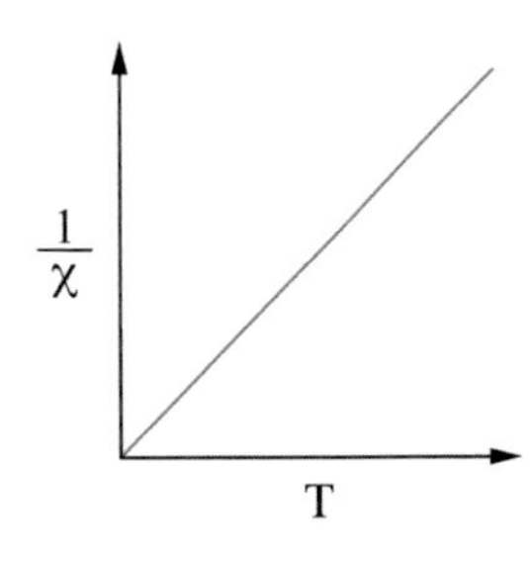

FIGURE 2.10 For some materials in low applied magnetic field strengths, the inverse of the magnetic susceptibility is proportional to the temperature. Such materials are said to obey Curie's law.

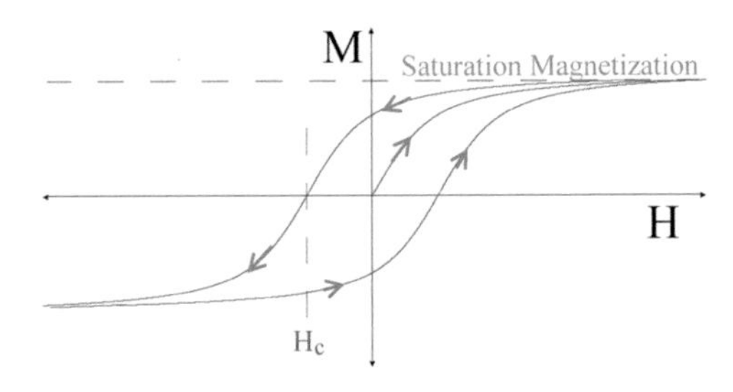

FIGURE 2.11 Magnetization (M) vs magnetic field strength (H) for a material that exhibits magnetic hysteresis. In this example, the material starts in the demagnetized state in zero field. The magnetic field strength is ramped up to a large positive field strength so that the magnetization saturates. The magnetic field strength is then ramped down to zero and subsequently increased in the negative direction so that eventually the magnetization saturates in the negative direction. H_c is the coercive field strength. The magnetization remaining after removal of the applied field is the remanent magnetization (M_r).

As mentioned previously, for some materials the response of M to H depends on the history of the magnetic field strengths that have been applied. Such materials exhibit the phenomenon of magnetic hysteresis (Figure 2.11). Figure 2.11 illustrates a history of the response of M to H for such a material. The magnetic material initially started with zero magnetization in zero applied magnetic field strength. As the field strength was increased, the magnetization initially increased linearly. As the field strength was increased further, the magnetization saturated. The magnetic field strength was then decreased. The figure shows the magnetization decreasing with decreasing applied field strength but not as rapidly as it had initially increased. Complete removal of the applied magnetic field did not return the magnetization to zero. The magnetization remaining after the application and removal of the magnetic field is known as the *remanent magnetization*, M_r. Figure 2.11 shows that the magnetic field was then increased in the opposite direction, i.e. H became more negative. As the magnetic field strength became more negative, the magnetization decreased. The magnetic field strength at which the magnetization is brought to zero value is known as the *coercive field strength*, H_c. Increasing the magnetic field strength further in the negative direction caused the magnetization to increase in the negative direction. In other words, once the magnetic field strength becomes more negative than the coercive field strength, the magnetization starts to increase in the direction of the applied magnetic

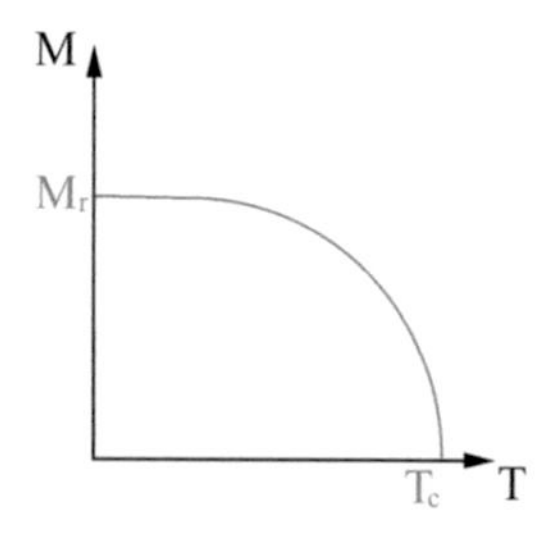

FIGURE 2.12 The remanent magnetization (M_r) of a material decreases with increasing temperature. The temperature at which the material becomes demagnetized is known as the Curie temperature (T_c).

field. Eventually the magnetization saturated in the direction of the reversed applied magnetic field. At this point, the situation is usually symmetrically identical to that of the saturated magnetization at high magnetic field strength in the opposite direction. Note that both the remanent magnetization and coercive field strength are dependent on the history of magnetic fields applied to the sample and are not unique characteristics of the material.

2.4.2 Permanent Magnets

Magnetic materials that retain a magnetization in the absence of an applied magnetic field are sometimes known as permanent magnets (as opposed to "temporary" magnets, which are magnetized only in the presence of an applied magnetic field and lose their magnetization when the applied field is reduced to zero). Permanent magnets must necessarily exhibit the phenomenon of magnetic hysteresis as shown in Figure 2.11. The magnetization of a permanent magnet in zero applied magnetic field is the remanent magnetization. Heating a magnetized specimen in zero applied field tends to reduce the magnetization (Figure 2.12). The temperature at which the remanent magnetization is reduced to zero is known as the Curie temperature (T_c). Heating a sample above its Curie temperature is a way of demagnetizing it. The process is known as thermal demagnetization.

2.5 A MICROSCOPIC PERSPECTIVE ON MAGNETIC MATERIALS

We will now revisit the experimentally observed magnetic behaviors that we have discussed previously and try to understand them from a microscopic point of view.

2.5.1 Diamagnetism

Most materials are diamagnetic. The phenomenon is subtle and is generally not observed in everyday life. When a magnetic field is applied to a diamagnetic material such as water, for example, the application of the field causes very minor changes in the orbitals of the electrons in the atoms and molecules. These changes result in very weak dipole moments being generated that oppose the applied magnetic field. This phenomenon is related to Lenz's law, which states that when a changing magnetic field is applied to a conductor, an electric current is induced in the conductor that generates a magnetic field that opposes the changing applied field. As such, diamagnetic materials become very weakly magnetized in the opposite direction to an applied magnetic field. The resulting magnetic susceptibility is independent of temperature.

2.5.2 Paramagnetism

Let us consider a hypothetical situation of a classical gas comprised of atoms or molecules each of which is a small magnetic dipole. It is assumed that there are no magnetic interactions between the molecules (or that the interactions are extremely weak and so can be ignored). In zero magnetic field, we would expect the gas to have zero magnetization because the thermal random buffeting of the molecules would mean that the vector sum of all the magnetic dipole moments would be zero (Figure 2.13a). On application of a magnetic field, however, we would expect some degree of orientation of the molecular magnetic dipoles because of the torque experienced by magnetic dipoles in an applied magnetic field (Figure 2.13b). Hence the gas will have attained a magnetization. Very strong applied magnetic fields will tend to result in a greater degree of alignment of the molecular dipoles and hence will increase the magnetization of the gas. It should be noted that once the dipoles are fully aligned, the magnetization can increase no further and hence is saturated. Heating the gas will tend to reduce the degree of alignment of the molecular dipoles through enhanced thermal buffeting and hence will decrease the magnetization of the gas. Because the energy of a magnetic dipole in a given magnetic flux density is related to its orientation (Figure 2.14), theoretical thermodynamic models have been developed to quantitatively describe these phenomena. The models use Boltzmann statistics together with the previously described physics of torques on magnetic dipoles (that do not interact with each other) in applied magnetic fields (Equation 2.2) to predict the behavior of magnetization as a function of applied field and temperature. The models are based either on classical physics (Langevin model), which describes a classical gas, or on quantum physics, which takes into account the quantum nature of atoms and molecules (Brillouin model). The models generate magnetization versus field strength curves that look very similar to those shown in Figure 2.9. The classical model generates an M vs H curve known as the Langevin function while the quantum model generates a curve known as the Brillouin function. They both look qualitatively similar to the eye. Both models obey Curie's law (Equation 2.9) at low applied field strengths. Materials that exhibit magnetic behavior that can be described by these models are known as paramagnetic materials and exhibit paramagnetism. Examples include ferrous sulfate crystals and

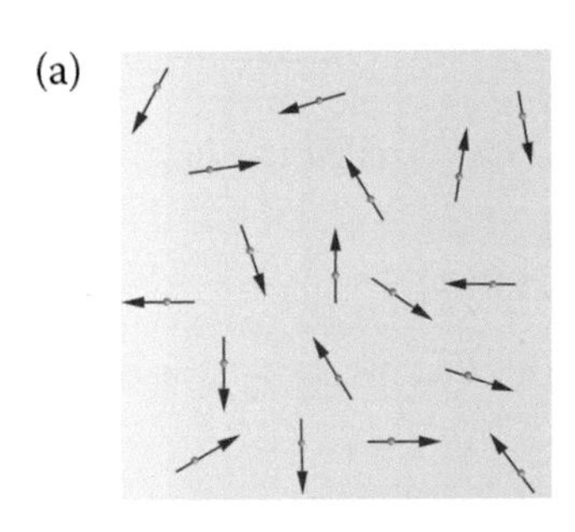

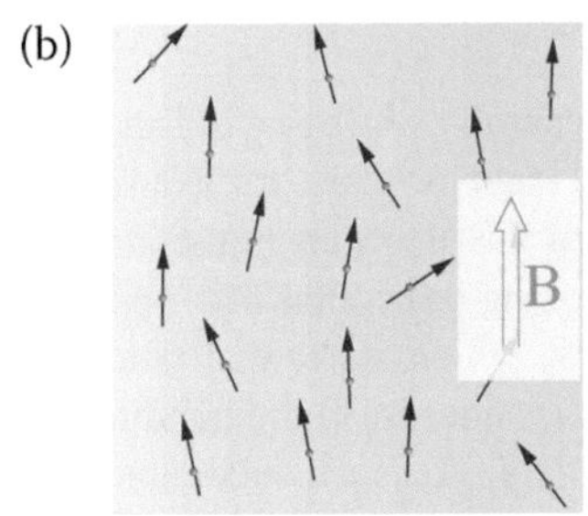

FIGURE 2.13 Schematic diagram showing **(a)** the randomly fluctuating magnetic moments of a classical paramagnetic gas in zero applied magnetic field, **(b)** the effect of applying a magnetic field so that the magnetic moments of the atoms experience a flux density B. In the magnetic field, the magnetic moments tend to wobble about the direction of B such that the gas has a net magnetization.

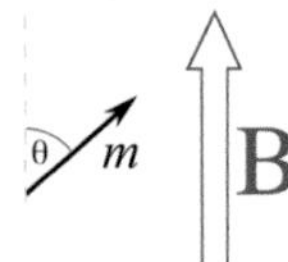

$E = -mB\cos\theta$

FIGURE 2.14 The free energy of a magnetic moment in a magnetic field with flux density B is dependent on the magnitude of the magnetic moment, the magnitude of the magnetic flux density, and the angle between the magnetic moment and flux density.

ionic solutions of magnetic atoms. Although these materials are not gases, they are characterized by having very weak interactions between their constituent atomic magnetic moments.

2.5.3 Ferromagnetism

Materials that retain a magnetization in zero applied magnetic field must have some internal interactions that result in a net alignment of the atomic magnetic moments (Figure 2.15). Classical interactions between magnetic dipoles would not result in parallel alignment of magnetic dipoles as shown in Figure 2.15. They would tend to favor some antiparallel alignments because of the tendency of like poles to repel each other and opposite poles to attract each other. The very strong interaction between magnetic dipoles in permanent magnet materials is quantum mechanical in origin and is known as the exchange interaction. Exchange interactions are very short-range interactions but are much stronger than classical dipole interactions. Materials in which the exchange interactions result in parallel alignment of atomic magnetic moments are known as ferromagnetic materials. Examples include metallic iron and cobalt.

Thermal energy can be used to overcome exchange interactions. The Curie temperature is the temperature above which thermal energy is sufficient to overcome the exchange interaction energy (Figure 2.16). The Curie temperature of a material can be used as an indicative measure of exchange interaction strength.

Ferromagnetic materials tend to form magnetic domains (Figure 2.17a). Each magnetic domain is magnetized in a different direction from its neighbors. The formation of magnetic domains in a

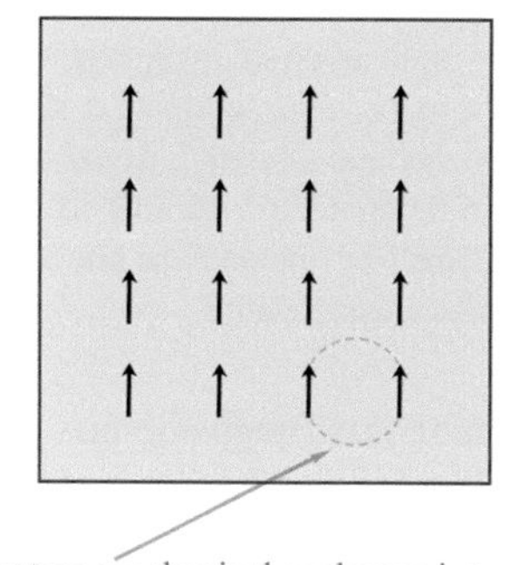

FIGURE 2.15 Schematic diagram showing the relative orientations of atomic magnetic moments within a ferromagnetic material. Quantum mechanical exchange interactions between neighboring atoms are responsible for the parallel alignment. Exchange interactions have a much higher energy associated with them than the energy of a dipole in a magnetic field. Hence, in most circumstances, applied fields are not able to overcome the effects of exchange interactions.

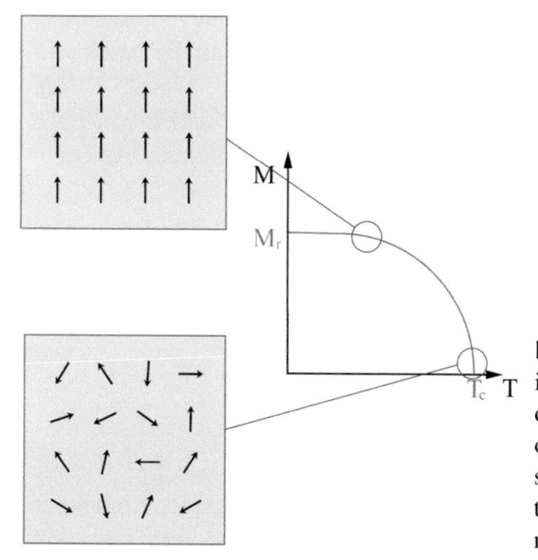

FIGURE 2.16 Heat can be used to overcome exchange interactions. Thermal energy can cause oscillations of the orientations of atomic magnetic moments relative to each other. Above the Curie temperature (T_c), thermal energy is sufficient to completely overcome the exchange interactions between atomic magnetic moments in a ferromagnetic material.

(a)

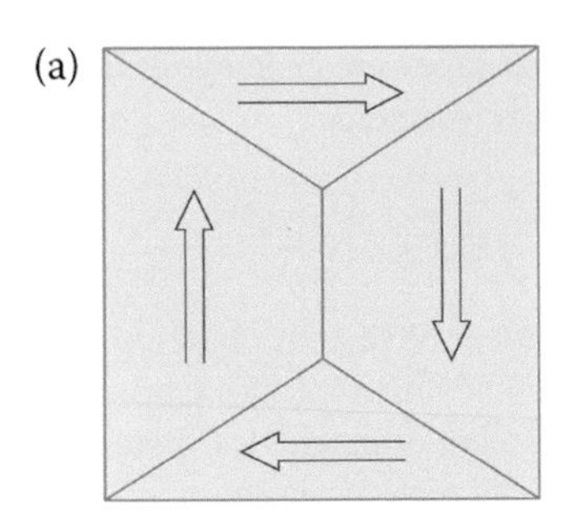

(b)

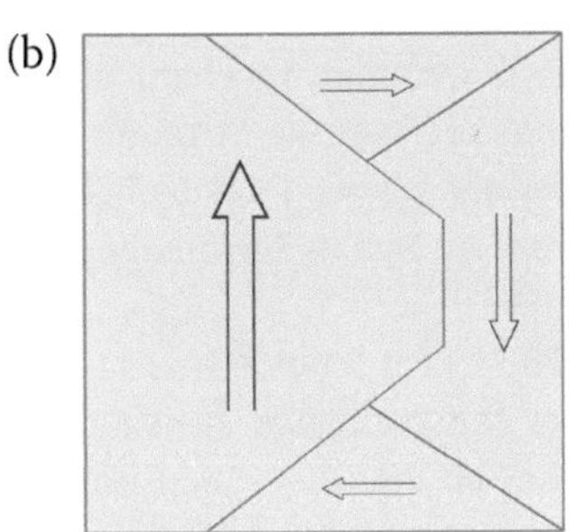

(c)

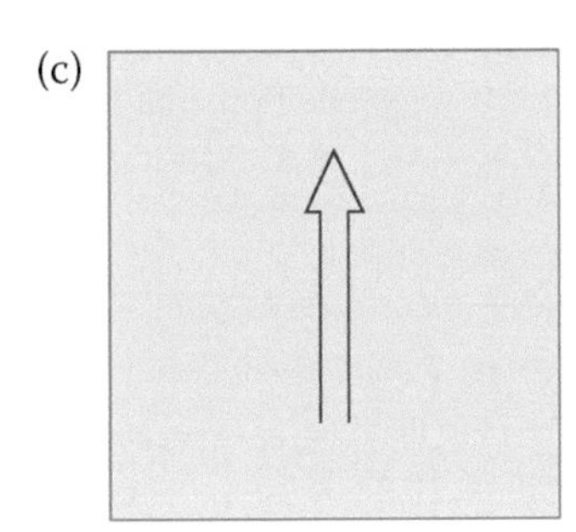

FIGURE 2.17 Schematic diagram of magnetic domains within a ferromagnetic material. In the demagnetizated state **(a)**, the vector sum of the magnetic moments of the individual magnetic domains is zero. If a magnetic field is applied to the material, domains with magnetizations in the same direction as the applied field tend to grow while those with opposed magnetizations shrink **(b)**. If very large fields are applied, it is possible to create a single magnetic domain **(c)**. In this state, the magnetization is saturated. Removal of the applied field will not necessarily result in the material returning to state **(a)**. The walls between domains may get "stuck" or "pinned" at imperfections in the crystal lattice resulting in remanent magnetization.

material minimizes the energy of the system by minimizing stray magnetic flux (magnetic flux that emanates from the material into the surrounding space). The magnetization of the domains relative to each other tends to be such that the magnetic flux owing to the magnetization is predominantly conducted within the material thus minimizing the energy of the system. Applying an external magnetic field to the material changes the domain structure. Domains with magnetizations in the direction of the applied magnetic field grow while those with magnetizations opposite in direction to the applied magnetic field shrink (Figure 2.17b). Application of very strong magnetic fields will eventually saturate the magnetization by creating a single magnetic domain (Figure 2.17c). Removing the field does not necessarily return the domain structure to its original state because the domain walls can get "stuck" or "pinned" at imperfections in the crystal structure. As such, magnetic hysteresis (Figure 2.11) is observed.

The boundary between two magnetic domains is called a domain wall. A domain wall has a finite thickness, which is typically of the order of 100 nm (Figure 2.18). If a particle of ferromagnetic material is smaller than the typical domain wall thickness for that material, then the particle will have a single magnetic domain only even in zero applied magnetic field. It will be too small to accommodate multiple magnetic domains (Figure 2.18).

2.5.4 Antiferromagnetism

In some materials, exchange interactions favor antiparallel alignment of atomic magnetic moments (Figure 2.19a). Antiferromagnetic materials are magnetically ordered but have zero remanent

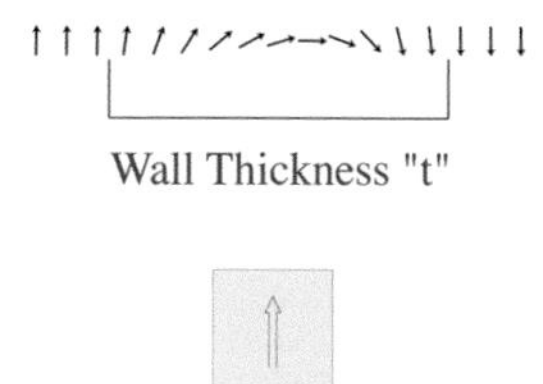

FIGURE 2.18 The boundary between two magnetic domains is called a domain wall. Domain walls have a finite thickness (t). The atomic magnetic moments exhibit a half twist in their orientation spread out across the domain wall. If a particle of ferromagnetic material is smaller than the characteristic domain wall thickness for that material, then it will consist of a single magnetic domain only. Domain wall thicknesses are typically of the order of 100 nm.

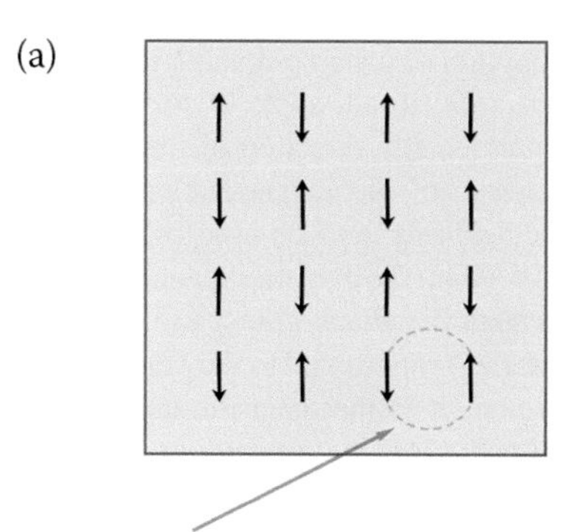

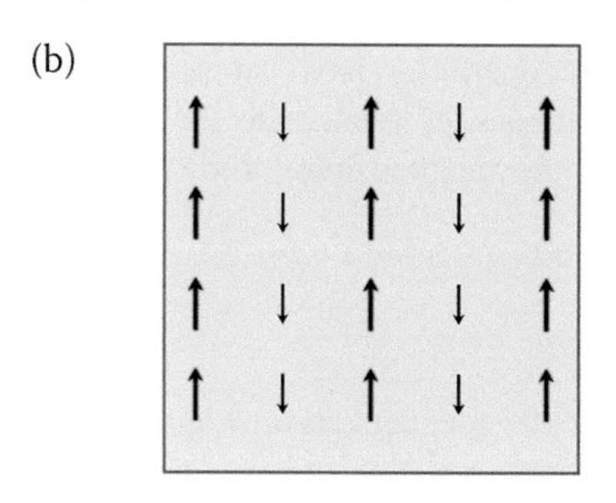

FIGURE 2.19 Schematic diagrams showing the relative orientations of atomic magnetic moments in **(a)** antiferromagnetic materials, and **(b)** ferrimagnetic materials.

magnetization because the antiparallel atomic magnetic moments vectorially add to zero. Antiferromagnets also have very low magnetic susceptibility because the exchange interactions between the atomic magnetic moments are very strong and so applied fields can have only a relatively small influence on the energy of the system. Many metal oxides are antiferromagnetic, such as hematite (α-Fe_2O_3), one of the iron oxides. As with ferromagnetic materials, thermal energy can be used to overcome exchange interactions in antiferromagnetic materials. Magnetic order is broken down above the Néel temperature (c.f. Curie temperature for ferromagnetic materials).

2.5.5 Ferrimagnetism

There is also a class of materials that exhibit antiferromagnetic exchange interactions but with different magnitude atomic magnetic moments on each antiparallel sublattice (Figure 2.19b). The imbalance of magnetic moments on the two sublattices results in a net magnetization for the material. As such, ferrimagnets behave much like ferromagnets. Examples of ferrimagnets include the iron oxides magnetite (Fe_3O_4) and maghemite (γ-Fe_2O_3).

2.6 SMALL-PARTICLE MAGNETISM

In this next section, we will consider the magnetic behavior of a magnetic particle small enough such that it comprises a single magnetic domain only. Such particles are typically less than 100 nm

(a)

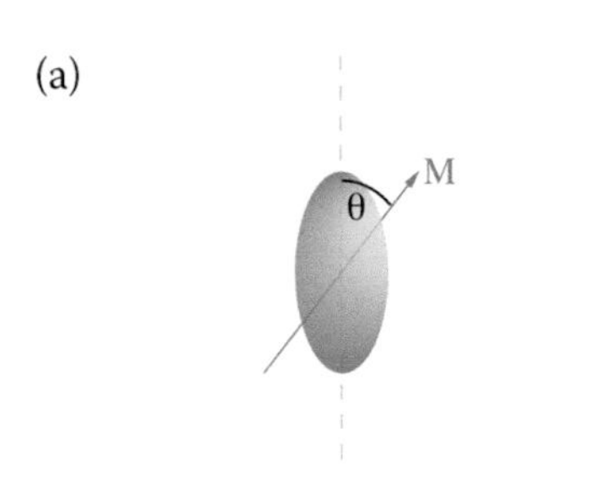

(b)

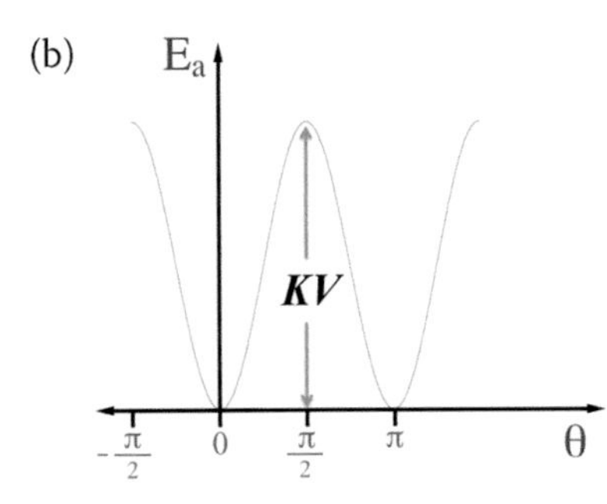

FIGURE 2.20 **(a)** Schematic diagram of a single-domain magnetic particle with its magnetization (M) at an angle θ to an axis corresponding to an "easy" or low-energy direction of magnetization. **(b)** A plot of the free energy of the magnetization of the particle in zero applied magnetic field as a function of its orientation with respect to the easy axis direction. To turn the magnetization from an easy axis direction through π radians (180°) to the other easy axis direction, an energy barrier proportional to the volume (V) of the particle has to be overcome. K is the magnetic anisotropy energy constant for the material and KV is the magnetic anisotropy energy barrier for the particle.

in size and are often described as nanoparticles. Magnetic nanoparticles have an associated magnetic anisotropy energy that favors magnetization along certain axes relative to the crystal lattice. Figure 2.20a shows a diagram of such a particle with the magnetization shown at an arbitrary angle, θ, relative to an easy axis of magnetization.

2.6.1 The Stoner–Wohlfarth Particle Model

In the case of uniaxial magnetic anisotropy, the free energy of the magnetization is given approximately by

$$E_a = KV \sin^2(\theta) \tag{2.10}$$

where V is the volume of the particle and K is a constant for the material known as the magnetic anisotropy constant. A plot of Equation 2.10 is shown in Figure 2.20b. The figure illustrates that there are two low-energy directions along which the particle can be magnetized. The two directions are antiparallel to each other. The two low-energy orientations of the magnetization are separated by an energy barrier of magnitude KV. A magnetization at 90° (or $\pi/2$ radians) to the easy axis is an unfavored high-energy direction. At low temperatures, the magnetic moment of the particle will be trapped in one of the energy wells (Figure 2.21a). Thermal energy is insufficient to buffet the magnetization over the energy barrier during the time of observation at low temperature. In this situation the magnetic moment of the particle is said to be *blocked*. At higher temperatures, thermal energy can buffet the magnetic moment of the particle over the energy barrier and the particle magnetic moment will be observed to fluctuate between the two energy wells many times during the time of observation (Figure 2.21b). In this situation the particle magnetic moment is said to be *unblocked*.

2.6.2 Magnetic Blocking Temperature

The magnetic blocking temperature, T_b, is the temperature below which the magnetic moment of a particle is blocked and above which it is unblocked. The blocking temperature depends on both the size of the particle and the timescale of observation. Larger particles

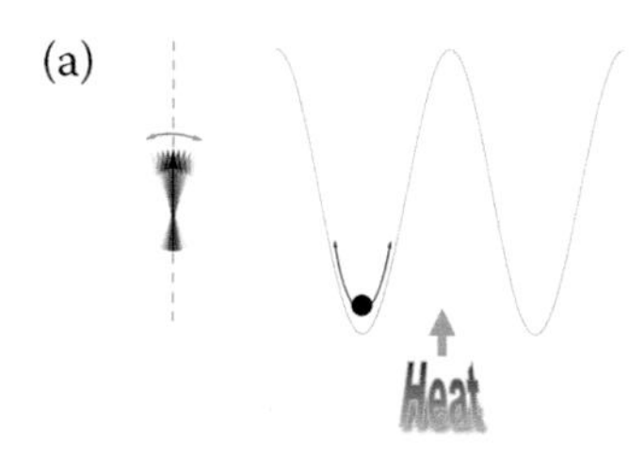

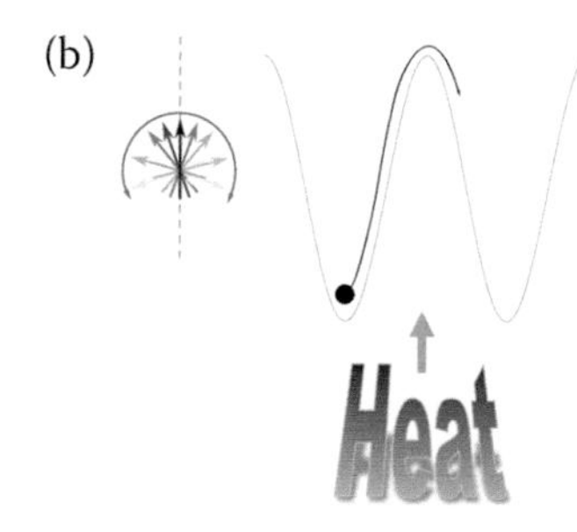

FIGURE 2.21 **(a)** At low temperatures the magnetization of a single-domain magnetic particle tends to become trapped in one of the magnetic anisotropy energy wells. Only small fluctuations of the magnetic moment of the particle around the easy axis of magnetization are possible. In this situation the magnetic particle is said to be blocked. **(b)** Increasing the temperature of the particle increases thermal buffeting of the magnetic moment. At sufficiently high temperatures, the magnetic moment can be observed to jump the energy barrier and fluctuate back and forth. In this situation the magnetic particle is said to be unblocked.

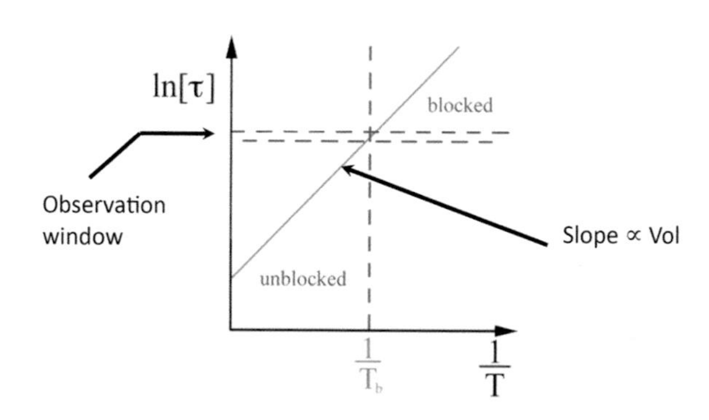

FIGURE 2.22 The characteristic time τ between jumps across the magnetic anisotropy energy barrier for a single-domain magnetic particle has a logarithmic relationship with the inverse of temperature, T. This behavior is a form of the Arrhenius law. For a given material, larger particles have steeper gradients for the relationship between $\ln(\tau)$ and $1/T$. Since any given experimental technique will tend to have a rather limited range of logarithmic observational timescales, particles are often observed to be either blocked or unblocked. In many situations where a particle size distribution is present, an experimental technique may indicate that the distribution is split into those particles that are blocked (larger volume particles) and those that are unblocked (smaller volume particles). Because of the logarithmic relationship between temperature T and τ, only a very small fraction of the particles in the distribution will exhibit complex transitional behavior between blocked and unblocked behaviors.

have higher blocking temperatures because they have higher magnetic anisotropy energy barriers (Equation 2.10). The longer the observation time, the more likely it is that the magnetic moment will be observed to flip from one energy well to the other. As such, longer observation times result in lower apparent blocking temperatures. There is a logarithmic relationship between the average time between magnetic moment flips, τ, and temperature, T (see Figure 2.22). Most experimental techniques have a limited window of observation timescales and as such particles tend to appear either blocked or unblocked. If an ensemble of particles with a range of particle sizes is being studied, it may appear to consist of two populations of particles, blocked and unblocked, if the particle size distribution spans the critical size for blocking at the given temperature and for the observation timescale being employed. It should be noted that particles that appear blocked with one experimental technique may appear unblocked with another experimental technique with a different characteristic observation time.

2.6.3 Effects of Applied Magnetic Field on a Single-Domain Particle

Figure 2.23 shows the effect on the free energy of magnetization of applying a magnetic field along the easy axis of magnetization. With zero magnetic field applied, the two energy wells are of equal depth. Above the magnetic blocking temperature, the magnetic moment will be observed to fluctuate rapidly between the two energy wells, spending equal time in each well. As such, the time-averaged magnetization will be zero because the time-averaged magnetization vector will be zero. As the magnetic field strength is increased from zero to H_1, one of the energy wells deepens while the other becomes more shallow. If the temperature is above the blocking temperature, we would observe the magnetic moment flipping between the two energy wells with the moment spending more time in the deep well than the shallow well. In this scenario, the time-averaged magnetization vector is non-zero and hence the particle is observed to have a finite magnetization. At higher magnetic field strengths, such as H_2, the energy well representing the direction of magnetization opposite to the field is completely eliminated.

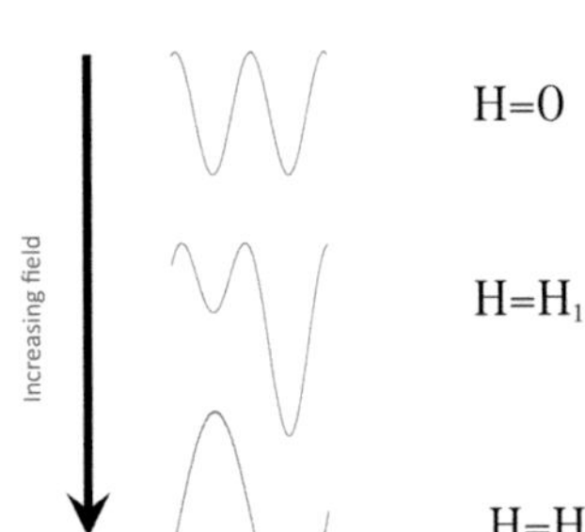

FIGURE 2.23 Effect of applying an external magnetic field along the direction of the easy axis of magnetization of a single-domain magnetic particle on the zero field double well potential. At high magnetic field strengths, the potential energy of the magnetic moment of the particle in the applied field dominates over the magnetic anisotropy energy.

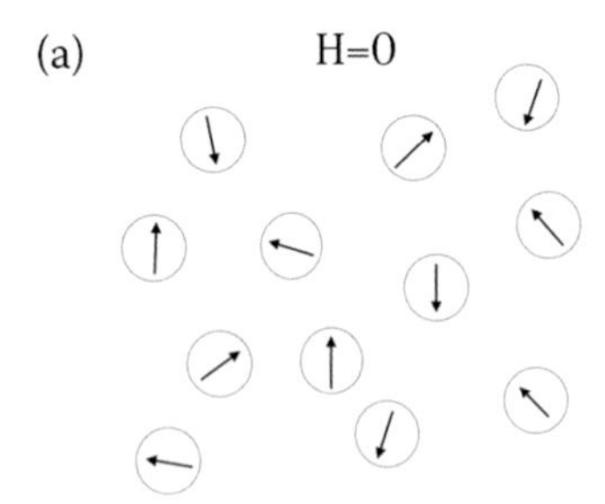

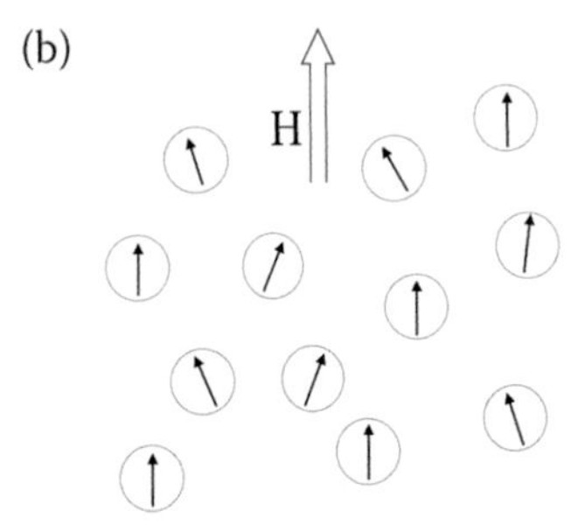

FIGURE 2.24 **(a)** In zero applied magnetic field, the magnetic moments of superparamagnetic single-domain particles fluctuate randomly resulting in zero net magnetization for the ensemble of particles. **(b)** Application of a magnetic field to an ensemble of superparamagnetic single-domain magnetic particles results in the magnetic moment of each particle fluctuating or wobbling about an average direction along the applied magnetic field direction. As such, the ensemble of particles exhibits a net magnetization in the direction of the applied field.

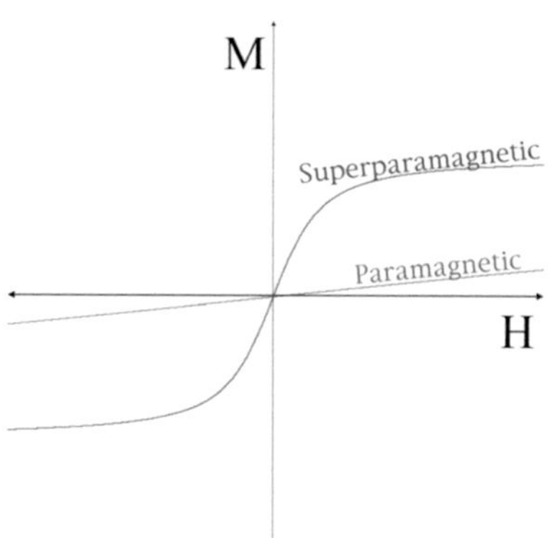

FIGURE 2.25 The magnetic moments on superparamagnetic single-domain magnetic particles behave in much the same way as classical magnetic moments in a gas. As such, the response of their magnetization to an applied magnetic field is similar to paramagnetic systems except that the magnetic moment per particle is much greater. Ensembles of both classical paramagnetic and superparamagnetic moments show M vs H behavior described by the Langevin function.

Figure 2.24 illustrates schematically the effect of applying a magnetic field to an ensemble of single magnetic domain particles above their blocking temperature. In zero field, the magnetic moments of the particles fluctuate randomly. In the applied field, H, the magnetic moments of the particles wobble about the direction of the applied field. Unblocked particles that respond to a magnetic field in this way are known as superparamagnetic particles. The term *superparamagnetism* is used to draw an analogy with Langevin's classical paramagnetism model of a gas of magnetic dipole moments (Figure 2.13). Instead of the small atomic or molecular magnetic moments associated with paramagnetism, superparamagnetism involves the much larger magnetic moments associated with single-domain magnetic particles.

The response of ideal superparamagnets to an applied magnetic field is described by the Langevin model (Figure 2.25) with the M vs H curve being a Langevin function. The behavior is qualitatively similar to that of paramagnets (Figure 2.9). At room temperature, superparamagnetic materials (based of ferromagnetic or ferrimagnetic structures) have a much greater magnetic susceptibility per atom than paramagnetic materials. Superparamagnets are often ideal for applications where a high magnetic susceptibility is required and/or where zero magnetic remanence is required.

2.7 MAGNETIC PARTICLES IN FLUIDS

Many of the clinical and biotechnological applications of magnetic particles discussed in subsequent chapters of this book involve suspensions of particles in fluids. Here we review some of the basic principles governing the behavior of magnetic particles in fluids. Many different forces are involved in determining the behavior of magnetic particles in fluids. These forces include the force of applied magnetic fields on the particles, viscous drag forces of the fluid on the particles, interparticle magnetic forces, interparticle electrostatic forces, interparticle entropic forces, and gravitational force.

A uniform magnetic field applied to a magnetic particle suspended in a fluid will have the effect of orienting the magnetic dipole of the particle but will not exert a translational force on the dipole (Figure 2.3). As described earlier, the reason for this effect is that the uniform field exerts equal but oppositely directed forces on the north and south poles of the particle. These forces can cause a torque on the particle but because they are balanced, no translational force results. If, however, the suspended particle is in a magnetic field gradient (Figure 2.3c), then one pole will be in a region of space with a higher flux density than the other pole and hence the force on one pole will be greater than that on the other. In this case, the imbalance of the two forces results in a net translational force, which can move the particle through the fluid. The translational magnetic force in the x-direction generated on a dipole with magnetic moment, m, that is oriented in the x-direction by a magnetic field gradient $\mathrm{d}B/\mathrm{d}x$ is

(a)

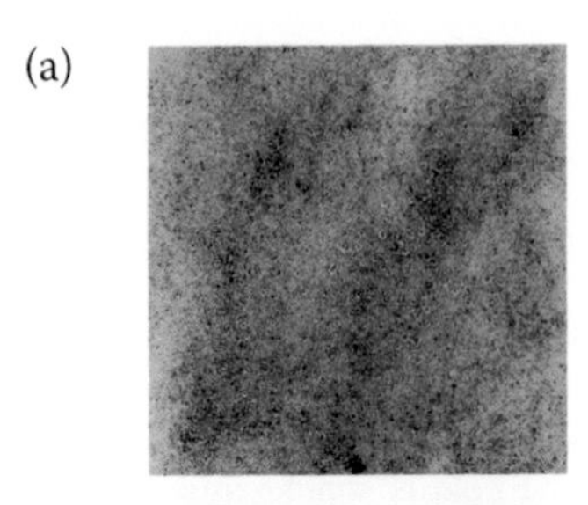

(b)

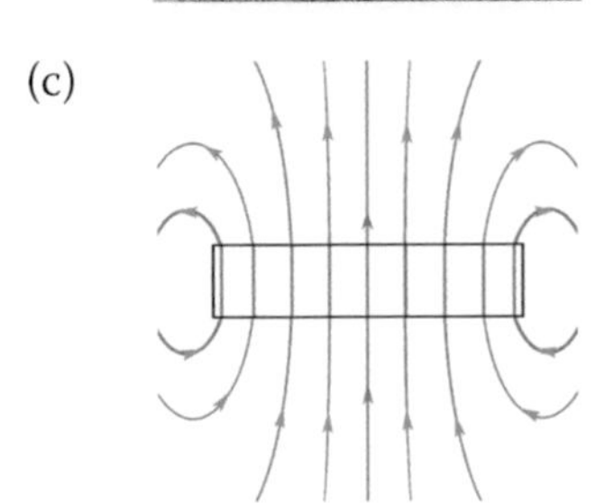

(c)

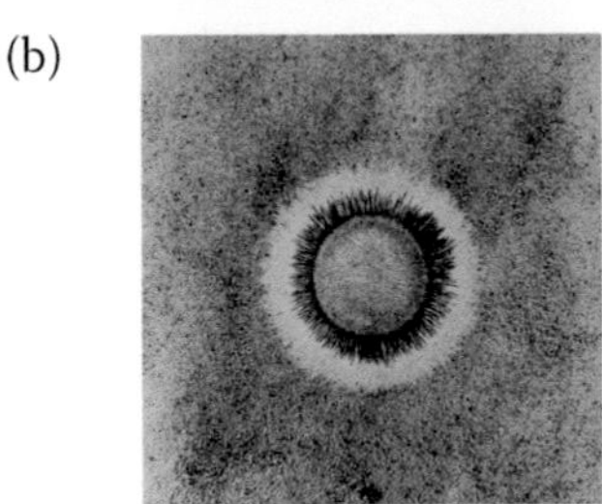

FIGURE 2.26 **(a)** Photograph looking down on some iron filings sprinkled on a piece of paper. **(b)** The north pole of a cylindrical disk-shaped permanent magnet has been slowly brought up to the underside of the paper. Note how the iron filings have tended to move toward the edges of the magnet rather than to the center of the face of the magnet. **(c)** Diagram looking at the cylindrical disk-shaped permanent magnet in the horizontal plane. Note that the magnetic field gradient is much stronger at the edges of the magnet. The magnetic flux density is also slightly higher at the edge of the magnet (for reasons outside the scope of this chapter).

$$F_{\mathrm{m}} = m \, \mathrm{d}B/\mathrm{d}x. \tag{2.11}$$

Magnetic field gradients are most commonly generated at the edges and corners of magnetized objects. This phenomenon can be easily observed in the classic schoolroom experiment involving the sprinkling of iron filings onto a piece of paper, below which is positioned a permanent magnet (Figure 2.26). The iron filings tend to make their way to the edges and corners of the magnet rather than to the center of the faces of the magnet. The smaller the radius of curvature of corners and edges, the greater is the associated magnetic field gradient. This principle is exploited in high field-gradient magnetic separators (Figure 2.27). The typical design of a magnetic separator comprises a column packed with fine wire or small beads with high magnetic susceptibility and low magnetic remanence. The fluid bearing the magnetic particles to be separated is passed through the column while an external magnetic field is applied. The external magnetic field magnetizes the fine wire or small beads in the column so that magnetic particles suspended in the fluid are attracted to the wire or beads by virtue of the strong magnetic field gradients associated with their small radii of curvature. Once the particle-bearing fluid has passed through the column, the external magnetic field can be removed to demagnetize the fine wire or small beads and hence release the captured particles. The captured particles, thus released, can then be retrieved by flushing the column with a fluid.

2.7.1 Reynolds Number

The Reynolds number of an object in a fluid is the ratio of the inertial forces to the viscous forces experienced by the object. Micron and sub-micron particles in water have very low Reynolds numbers. Under these circumstances, particles very quickly reach their terminal velocity on

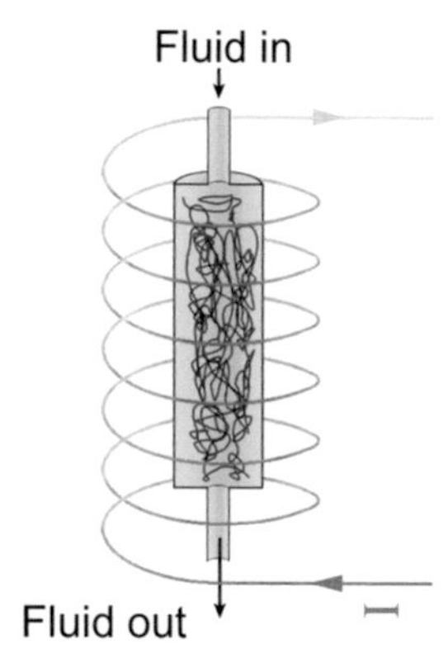

FIGURE 2.27 Schematic diagram of a magnetic separation column. The column contains very fine magnetizable wire with close to zero remanent magnetization. The electric current carrying coil can be used to generate an external magnetic field on the fine magnetizable wire in the column, thus magnetizing it. Magnetic particle bearing fluid can then be passed through the column. The external magnetic field magnetizes both the fine wire and the particles. The high magnetic field gradients around the fine wire result in attractive forces between the particles and wire, thus trapping the magnetic particles if the flow rate of fluid is sufficiently low. After passing the fluid through the column, the applied magnetic field can be removed to demagnetize the wire and release the magnetic particles. The magnetic particles can then be retrieved by flushing the column with an appropriate fluid.

application of a translational force. As such, the velocity of a low Reynolds number particle is proportional to the externally applied force. In the case of a low Reynolds number magnetic particle in a magnetic field gradient, the translational force increases with both the magnitude of the field gradient and the magnitude of the magnetic moment of the particle (see Equation 2.11). Because larger particles of a given magnetic material have larger magnetic moments, they will experience larger translational forces in a given magnetic field gradient. Stoke's law predicts that the speed attained by a particle moving through a viscous fluid is inversely proportional to its radius and proportional to the applied force. Because the magnetic moment on a spherical particle increases as the cube of the radius, larger particles of a given magnetic material move faster through a viscous fluid than do smaller particles of the same material. Hence, magnetic separation columns preferentially remove larger magnetic particles or aggregates of magnetic particles.

2.7.2 Interparticle Interactions

Similar to other colloidal systems, magnetic particles can be stabilized in a fluid suspension either through maintaining an electrostatic charge on each particle thus maintaining mutual electrostatic repulsion of particles or through suitable coating of polymeric chains, which induces mutual repulsion of particles because of entropic forces that arise when polymer chains from different particles interact in close proximity. Without these stabilizing influences, particles can aggregate due to attractive van der Waals interactions. In the case of magnetic particles, however, magnetic dipole interactions also have to be taken into account. When magnetic particles are magnetized, each particle induces magnetic field gradients in its immediate vicinity. These local magnetic field gradients induce attractive magnetic interactions between the particles. If the stabilizing forces are not sufficiently strong to overcome the attractive magnetic interactions, particles will start to aggregate or form chains which, when large enough, will cause sedimentation owing to gravitational forces.

Larger magnetic particles have larger magnetic moments and hence stronger interparticle interactions. Hence suspensions of larger magnetic particles tend to be less stable than suspensions of smaller particles. Nearly all magnetic particle suspensions will form aggregates or chains of particles when exposed to a magnetic field because of the induced magnetization in the particles. However, in some cases the aggregation is reversible upon removal of the applied magnetic field while in other cases the aggregation is irreversible. In the latter case, the mutual induction of magnetic dipole moments between the particles within an aggregate, even after removal of the applied field, is sufficient to overcome any repulsive electrostatic or entropic interactions. One of the useful characteristics of superparamagnetic particles is their high magnetic susceptibility. However, if aggregation of particles occurs, the low-field magnetic susceptibility may be reduced

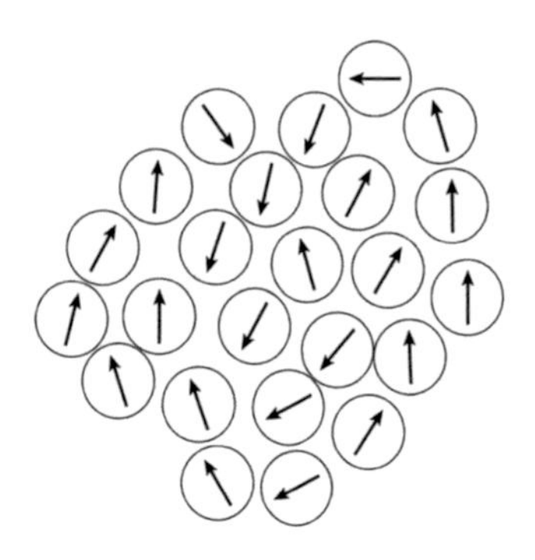

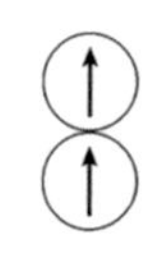

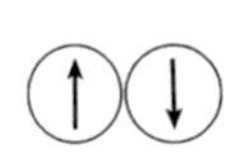

FIGURE 2.28 If single-domain magnetic particles aggregate, dipole interactions between the particles will tend to result in a lower low-field magnetic susceptibility of the particles. North poles are attracted to south poles and vice versa. To magnetize the ensemble, applied magnetic fields need to be strong enough to overcome the interparticle dipole interactions. Hence higher magnetic fields are required to achieve a given magnetization for aggregated particles relative to dispersed particles.

because the interparticle dipole interactions tend to reduce the overall magnetization of the aggregate (Figure 2.28).

2.8 PHYSICAL DESIGN CONSIDERATIONS FOR MAGNETIC PARTICLES FOR SPECIFIC APPLICATIONS

2.8.1 Magnetic Carriers

In many applications it is important for magnetic particles to be stable in suspension. In these applications it is generally desirable for the particles to have a high magnetic susceptibility and low remanent magnetization so that it is easy to "switch on" and "switch off" the magnetization of the particles using a modest externally applied magnetic field. In the switched-off state, the particles are less likely to aggregate because interparticle magnetic interactions are minimized. Single magnetic domain superparamagnetic particles are ideal for these applications (Figure 2.25).

2.8.2 Magnetic Microspheres

One of the disadvantages of dispersed superparamagnetic particles is that they have very low magnetophoretic mobility. In other words, it is difficult to induce significant translational motion in them using external magnetic field gradients because they are very small and hence have relatively small magnetic moments resulting in small magnetic forces relative to the viscous drag forces. For applications where efficient magnetic extraction of particles is required, magnetic microspheres can be used. A typical design of a magnetic microsphere is shown in Figure 2.29. Superparamagnetic nanoparticles are bound within a polymer microsphere so that the magnetic properties of the microsphere show all of the advantages of superparamagnetic particles (such as high magnetic susceptibility and low remanent magnetization) but have much greater magnetophoretic mobilities than individually dispersed superparamagnetic nanoparticles because the ratio of magnetic driving force to viscous drag force is increased.

2.8.3 Magnetic Hyperthermia

Several applications of magnetic particles as sources of heat for hyperthermia therapy have been proposed (Thiesen and Jordan 2008, Jordan et al. 1993). The idea is to locate the particles close to cancerous tissue and then apply an alternating magnetic field to induce heat in the particles. Other potential uses of alternating field induced heating of magnetic particles include the triggered release of drugs from polymer matrices surrounding magnetic nanoparticles (Kumar and Mohammad 2011). The alternating magnetic field will take the magnetization of particles around their M vs H hysteresis loops (Figure 2.11). The amount of heat generated per cycle is proportional to the area

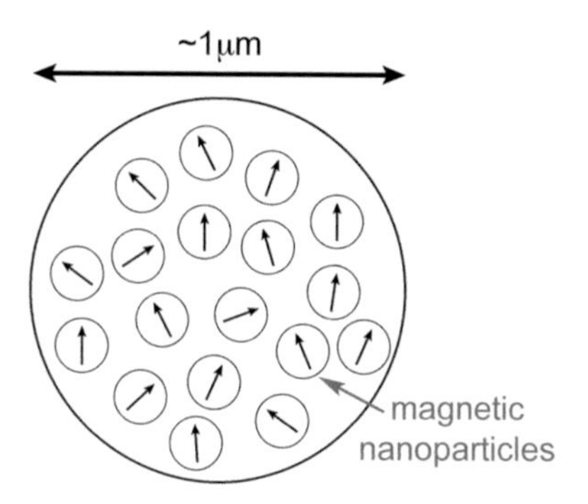

FIGURE 2.29 Magnetic microspheres: Micron-sized particles of ferromagnetic material would have large magnetic moments but would not have the advantage of superparamagnetism (i.e. zero remanent magnetization). Micron-sized particles of paramagnetic material would have very low magnetic susceptibilities. By embedding superparamagnetic particles into a micron-sized polymer sphere, relatively large magnetic susceptibilities with low magnetic remanence can be obtained. Such spheres will have minimal magnetic interactions in zero field thus reducing unwanted aggregation but will have reasonably high magnetophoretic mobility and hence can be moved through a fluid with externally applied magnetic field gradients.

within the hysteresis loop. The mechanisms that lead to hysteretic behavior in magnetic materials are manifold. For magnetic nanoparticles suspended in a fluid environment, the origins of hysteresis can be related to rotation of the magnetization vector relative to the crystal axes of the particle (sometimes referred to as Néel relaxation), the rotation of the particle itself (sometimes referred to as Brownian relaxation), or both. It should be kept in mind that the degree of hysteresis shown by a magnetic material depends on the rate at which the applied magnetic field changes i.e. on the frequency of the alternating magnetic field. Design of the equipment for application of the alternating magnetic fields has to take into account the conflicting requirements of a high number of cycles per second and a high magnetic field amplitude. Instruments designed to work at high frequency are not able to achieve as high an amplitude of magnetic field as those that work at lower frequencies. Hence, it is desirable for the magnetic particles to have high magnetic susceptibilities and low coercive fields (Figure 2.30). In this way, relatively large hysteresis loop areas can be swept out with small magnetic field strength amplitudes. For further details on magnetic hyperthermia, see Chapter 11.

2.8.4 Particles for Brownian Rotation Based Biosensors

Several applications of magnetic particles as biosensors have been proposed based on the change in hydrodynamic radius of the particles upon binding of biomolecules to their surfaces (Connolly and St Pierre 2001, Goodwill 2011). In these applications, the frequency dependent response of the particles to an applied alternating magnetic field is studied. Particles are required that are magnetically blocked and stable in suspension. At very low frequencies, the particles oscillate in a rotational fashion in phase with the applied field. At very high frequencies of applied field, the particles do not rotate because they are unable to keep up with the rapidly oscillating field. The frequency at which particles transition from being able to keep up with the oscillating field to not being able to rotate fast enough is determined by the hydrodynamic radius of the particle. For a given torque, larger particles rotate more slowly than smaller particles. Hence, there is a characteristic frequency determined by the hydrodynamic radius of the particle at which the magnetic moment of the particle rotates at 90° out of phase with the oscillating field. Measurement of this frequency enables sensitive measurement of the hydrodynamic radius of the particle and hence enables determination of whether or not biomolecules have bound to the surface.

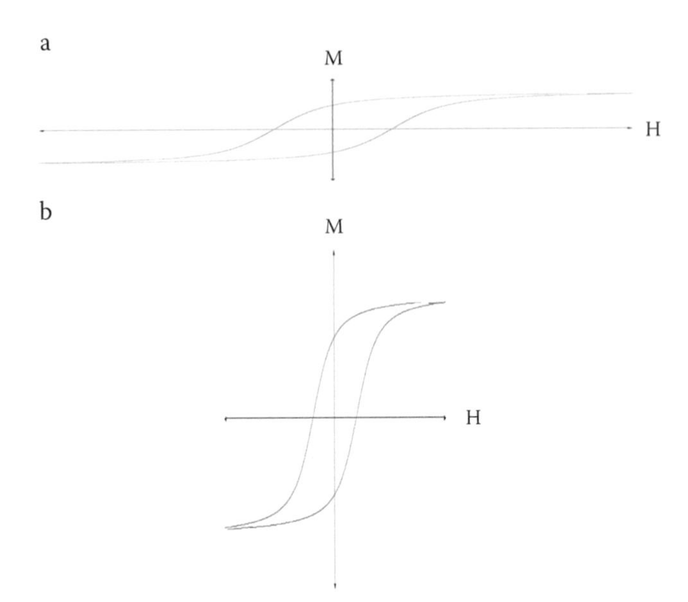

FIGURE 2.30 Heat is generated in a magnetic material when it is cycled around its *M* vs *H* hysteresis loop. The heat generated is proportional to the area of the hysteresis loop. Magnetic hyperthermia applications use high-frequency alternating magnetic fields with rather limited magnetic field amplitudes. Hence, materials with a low remanent magnetization to coercive field ratios as in **(a)** are not suitable. Materials with a high remanent magnetization to coercive field ratio as shown in **(b)** enable larger areas to be swept out with small amplitude oscillations of *H*.

2.9 SUMMARY

This chapter has provided a brief conceptual introduction to the magnetic properties of materials and particles used in the field of biomedical applications of magnetic particles. While most of the major relevant magnetic phenomena have been covered, it should be noted that the science of magnetic materials covers a very broad and complex range of phenomena that cannot be comprehensively covered in a chapter like this one. Many of the concepts introduced in this chapter have been models of ideal situations. Most real materials will show more complicated behavior. Nevertheless, the discussion of such behavior will almost always refer to the concepts introduced here. More comprehensive texts such as Cullity and Graham (2008) and Jiles (1998) are recommended as further reading in order to gain a deeper understanding of magnetic materials.

REFERENCES

Connolly, J., and T. G. St Pierre. 2001. Proposed biosensors based on time-dependent properties of magnetic fluids. *Journal of Magnetism and Magnetic Materials* 225 (1–2), 156–160.

Cullity, B. D., and C. D. Graham. 2008. *Introduction to Magnetic Materials*. 2nd ed. Piscataway, NJ: Wiley-IEEE Press.

Goodwill, P. W. 2011. Ferrohydrodynamic relaxometry for magnetic particle imaging. *Applied Physics Letters* 98 (26), 262502.

Jiles, D. 1998. *Introduction to Magnetism and Magnetic Materials*. 2nd ed. London: Chapman and Hall/CRC.

Jordan, A., P. Wust, H. Fählin, W. John, A. Hinz, and R. Felix. 1993. Inductive heating of ferrimagnetic particles and magnetic fluids: Physical evaluation of their potential for hyperthermia. *International Journal of Hyperthermia* 9 (1), 51–68. doi:10.3109/02656739309061478.

Kumar, C. S. S. R., and F. Mohammad. 2011. Magnetic nanomaterials for hyperthermia-based therapy and controlled drug delivery. *Advanced Drug Delivery Reviews* 63 (9), 789–808.

Thiesen, B., and A. Jordan. 2008. Clinical applications of magnetic nanoparticles for hyperthermia. *International Journal of Hyperthermia* 24 (6), 467–474. doi:10.1080/02656730802104757.

3 Magnetic Forces and Torques: Separation, Tweezing, and Materials Assembly in Biology

Benjamin Yellen

CONTENTS

3.1 INTRODUCTION

Magnetic separation is a vibrant field with a long history dating back to the 1970s and many applications have been developed and commercialized in minerals separation (Watson and Ellwood 1994, Liu and Friedlaender 1994, Kelland, Hiresaki, Friedlaender, and Takayasu, 1981), oil refinement (Jeong, Petrakis, Takayasu, and Friedlaender, 1982), and separation of mammalian cells (Hultgren, Tanase, Chen, and Reich, 2004, Zborowski, Moore, Williams, and Chalmers, 2002; Zborowski, Sun, Moore, Williams, and Chalmers, 1999, Nakamura et al. 2001, Thomas, Abraham, Otter, Blackmore, and Lansdorp, 1992), proteins (Bucak, Jones, Laibinis, and Hatton, 2003), bacteria (Bahaj, James, and Moeschler, 1996), and other biological materials (Dauer and Dunlop 1991, Takayasu, Duske, Ash, and Friedlaender, 1982). Work in the 1970s and 1980s was mostly concerned with large-scale magnetic separation for treatment of wastewater, oil, and mineral feeds; a comprehensive summary of these results can be found in several reviews (Setchell 1985, Oberteuffer 1974). More recently the focus has shifted to biological separation, and some of the recent results can be found in Yavuz, Prakash, Mayo, and Colvin (2009) and Friedman and Yellen (2005). Magnetic tweezing in biological systems also has a long history dating back to the work of Francis Crick (a Nobel Laureate for his work on DNA) who introduced the use of magnetic particles for testing the mechanical properties of the cytoplasm (Crick 1950, Crick and Hughes 1949). Since then, this approach has been adopted to study the mechanical properties of single molecules, such as proteins and DNA (Mosconi, Allemand, Bensimon, and Croquette, 2009b, Strick, Allemand, Bensimon, and Croquette, 1998, Smith, Finzi, and Bustamante, 1992), as well as the complex interplay between mechanical force and bio-chemical functions (Leuba et al. 2003, Thompson and Michael 2008, Gore et al. 2006, Ristic et al. 2005). Magnetic forces have also been used to assemble biological materials in controlled arrangements for applications ranging from microarrays (Barbee and Huang 2008, Lai et al. 2010, Tanase et al. 2005) to tissue engineering (Krebs et al. 2009, Ino, Ito, and Honda, 2007, Ito et al. 2005b, Tranquillo, Girton, Bromberek, Triebes, and Mooradian, 1996).

Magnetic fields offer many advantages for applying remote force to biomaterials as compared with other approaches based on electrical, optical, and acoustic fields. Perhaps the most important advantage is that most biomaterials and biofluids are transparent to magnetic fields (i.e. their magnetic permeability is roughly equivalent to vacuum), which allows magnetic fields to penetrate deeply into biological systems and living organisms without significant scattering and other shielding effects encountered by electrical, optical, and acoustic fields. The simplicity of magnetic instrumentation is another advantage of magnetic manipulation, which can consist of devices as simple as a permanent magnet or electromagnet, making these systems more accessible to biologists and medical practitioners. Finally, magnetic manipulation is biologically benign. Quasi-static magnetic fields do not lead to significant heating or electrochemical flows, such as those typically induced by optical and electrical fields. Most magnetic materials, in particular iron oxide, are relatively inert and are naturally occurring in many organisms (Bahaj et al. 1996). Thus, magnetic fields and forces can be applied over long periods without significantly interfering with normal biological processes. These advantages have made the use of magnetic materials and magnetic forces very popular in the fields of drug delivery (Yellen, Hovorka, and Friedman, 2005, Forbes, Yellen, Barbee, and Friedman, 2003), cell separation (Zborowski et al. 1999, Thomas et al. 1992), magnetic tweezing (Mosconi, Allemand, Bensimon, and Croquette, 2009a, Lionnet, Joubaud, Lavery, Bensimon, and Croquette, 2006), and sensing (Gaster et al. 2009, McNaughton, Kehbein, Anker, and Kopelman, 2006).

This review will summarize the recent and future trends concerning magnetic forces and torque in biological applications, with a particular focus on new ideas and applications. The review will begin with a discussion of the fundamentals of magnetic force and torque and a discussion of the general strategies for modeling magnetic materials with arbitrary shape. The equations for force and torque will be derived for several limiting cases, most notably spheres and spheroids, which are of practical interest to biological researchers. A basic discussion of fluid dynamics will also be provided to relate the dynamic response of magnetic materials under force application. These results will be used to frame the discussion of three independent applications of magnetic manipulation. The first topic will review magnetic separation systems and provide a comparison of conventional linear separation with more recent non-linear separation systems. The second topic will review the use of magnetic force and torque in biophysics, with a focus on magnetic tweezers for interrogation of single molecules, cellular, and sub-cellular systems. The last topic will focus on an application of magnetic force for assembling novel material structures, which have applications in microarrays and tissue engineering.

3.2 FUNDAMENTALS OF MAGNETIC FORCE, TORQUE, AND DYNAMICS

Calculating the interaction of magnetic materials with an external field typically requires a solution to a magnetostatic boundary value problem to determine the B and H fields everywhere in space. These fields are related through the constitutive relationship: $B = \mu_0(H + M)$, where M is the magnetization of the medium. A material's magnetization may have both a field-independent component, M_R, which is the remanent magnetization of the material (i.e. permanent magnetization resulting from intrinsic ferromagnetic properties), and a field-dependent component, $M = \chi H$ (i.e. paramagnetic properties). In weak fields, it is reasonable to assume that the magnetic susceptibility, χ, is a constant (i.e. field-independent), implying a linear constitutive relationship between the field and magnetization. In practice, the magnetic susceptibility, χ, depends on the material's composition and crystal structure as well as its shape, and is best expressed as a tensor for different crystallographic or geometric orientations. With regards to shape, χ is largest for prolate ellipsoids along its major axis. In strong fields, on the other hand, the magnetic susceptibility tends to follow a non-linear, field-dependent constitutive relationship (i.e. Langevin behavior), which reflects the process of saturating the magnetization in the macroscopic sense, or in the microscopic sense saturating the alignment of individual magnetic spins within the material.

The magnetostatic boundary value problem can be solved by various numerical approaches using conventional finite element and finite difference software packages (e.g. FEMLAB®, COMSOL®) and different iteration approaches (Jackson 1975). Analytical solutions are also possible when the material's geometry corresponds to a classical shape (e.g. sphere, needle, thin disk) (Stratton 1941). Due to the absence of free currents and the quasi-static assumption, the B fields are free from both divergence and curl, which allows for solution of the magnetic fields in terms of a more convenient magnetostatic scalar potential, φ, rather than the vector potential, A. The magnetostatic problem is solved by imposing two boundary conditions, requiring that: (1) the scalar potential is continuous and single valued on all boundaries, which is equivalent to stating that $(H_1 - H_2) \times \hat{n} = 0$ on interfaces where $\hat{n}$ is the unit vector of the surface normal, and (2) the normal component of the magnetic flux density is continuous across all boundaries, i.e. $(B_1 - B_2) \cdot \hat{n} = 0$. Once the scalar potential is computed everywhere, the H fields can be derived from $H = -\nabla\varphi$, and the B fields from the constitutive relationship, depending on the type of material and its history of field exposure (i.e. hysteresis).

3.2.1 Magnetic Force and Torque

For the sake of intuition, we can begin with a discussion of potential energy; however, it should be noted that magnetic force and torques are the experimentally measurable quantities, whereas the notion of potential energy is an abstraction used purely for convenience. Because isolated monopole moments do not exist in magnetic materials, the lowest order approximation is to consider the finite body as a distribution of dipoles. This overall dipole moment of the body amounts to integrating the magnetization (magnetic dipole density) over its entire volume and replacing the finite body with a point particle located at its geometric center. This analysis is similar to the treatment of point masses in the field of mechanics, which introduces much simplicity for dynamic analysis. The potential energy of a point dipole in an external field is given by:

$$U(\vec{r}, \theta) = -\vec{m}(\vec{r}) \cdot \vec{B}(\vec{r}) \tag{3.1}$$

where the position vector $\vec{r}$ denotes the location and orientation of the point dipole relative to an arbitrarily chosen origin and coordinate axes. By comparison with mechanics, the force and torque on the point dipole are the spatial and angular derivatives of the potential energy function, given by:

$$\vec{F} = -\nabla U \tag{3.2}$$

$$\vec{T} = \frac{dU}{d\theta}(\widehat{m} \times \hat{B}) \tag{3.3}$$

Conceptually, these relations imply two important features for a particle with positive magnetization: (1) a particle is forced to move into regions of increasing magnetic field where its potential energy can be minimized, and (2) a particle experiences a force couple that rotates its magnetization to align with the external field. Table 3.1 illustrates the typical forces and torque that may be experienced by particles commonly used in biomedical applications. Due to intrinsic

TABLE 3.1
Estimates of Force and Maximum Torque on Typical Particles with Different Properties and Exposed to Different External Fields

Magnetic Field (T)	Field Gradient (T/m)	Particle Radius (μm)	Susceptibility (χ)	Force (pN)	Maximum Torque (pN nm)
0.1	1000	1.0	1	~300	$\sim 3 \cdot 10^7$
0.1	0	1.0	1	0	$\sim 3 \cdot 10^7$
0.1	1000	0.1	1	0.3	$\sim 3 \cdot 10^4$
0.1	1000	0.1	10	3	$\sim 3 \cdot 10^5$

scaling relationship in Equations 3.1–3.3, the largest factor affecting force and torque is the diameter of the particle (scaling relationship is proportional to a^3) whereas the forces scale linearly with material properties, χ. Table 3.1 also illustrates that a particle experiences zero force in a uniform field, which is a consequence of the lack of magnetic monopole moments.

The point dipole approximation is reasonably good for a variety of particle shapes, and in some cases the solution is exact, such as the case of a linearly magnetizable sphere in a uniform field. Elongated particles (ellipsoids) will also behave like point dipoles, but with an effective magnetization that depends strongly on their aspect ratio. This problem has been solved previously, and the resulting effective dipole moment is presented without derivation following the notation of Kao (1961):

$$\vec{m} = \frac{\mu_1 - \mu_2}{\mu_2 - (\mu_2 - \mu_1)G} V\vec{H} \tag{3.4}$$

where the shape parameter, G, is given by:

$$G = \frac{ab^2}{2} \int_0^{\infty} \frac{ds}{(s + a^2)^{3/2}(s + b^2)} \tag{3.5}$$

where $2a$ and $2b$ are the lengths of the major and minor principal axes, respectively of a spheroidal object that is assumed to be aligned with its major principal axis along the external field direction. For spherical particles in which $a = b$, the shape-dependent parameter is $G = 1/3$, leading to the well-known expression for a linearly magnetizable spherical particle in a uniform magnetic field (Panofsky and Phillips 1955):

$$\vec{m} = 3\frac{\mu_1 - \mu_2}{\mu_1 + 2\mu_2} V\vec{H} = \bar{\mu}_{eff} V\vec{H} \tag{3.6}$$

This analysis implies that the magnetic moment of a strongly magnetizable spherical particle (μ_1) in a weakly magnetizable surrounding medium (μ_2), where $\mu_1 \gg \mu_2$ has a limiting value for its effective permeability of $\bar{\mu}_{eff} = 3$. A prolate spheroid can have a much larger effective permeability (due to a weaker de-magnetizing influence) that may exceed 10 for large aspect ratio needle-like particles. The opposite effect is true for an oblate spheroid, such as a disk-like object, which can have an effective permeability close to 1.

Once the effective dipole moment of the particle is calculated, the force can be computed from the particle's interaction with an external field gradient (Melcher 1981):

$$\vec{F} = \mu_0 (\vec{m} \cdot \nabla)\vec{H} \tag{3.7}$$

where m can have both ferromagnetic and paramagnetic characteristics, i.e. $m = m_R + \mu_{eff} VH$, and the field H is due to sources external to the particle.

The torque on spheroidal particles can also be computed using the effective dipole moment approach:

$$\vec{\tau} = \vec{m} \times \vec{B} \tag{3.8}$$

where like before the particle can experience torque as a result of its permanent dipole moment or its field-inducible dipole moment. For spherical particles, the only torque that can arise is from possession of a permanent dipole moment, because the field-inducible dipole moment is always collinear with the external field direction. For spheroidal particles, on the other hand, the instantaneous dipole moment does not always point in the same direction as the external field due to anisotropy in the de-magnetizing fields inside the particle. Thus, an external field can apply a torque on a linearly magnetizable spheroidal particle that does not possess a permanent dipole moment.

In a previous analysis, Kao (1961) used an energetic approach to calculate the torque on a linearly magnetizable spheroidal particle, which is given by:

$$\vec{\tau} = \frac{dU}{d\theta}\hat{e} \times \hat{H} = \frac{2\pi ab^2 \mu_2 (\mu_1 - \mu_2)(G_x - G_y)\sin(2\theta)\vec{H}^2}{3(\mu_2 - (\mu_2 - \mu_1)G_x)(\mu_2 - (\mu_2 - \mu_1)G_y)}\hat{e} \times \hat{H} \tag{3.9}$$

where $\hat{e}$ is the principal axis of the particle, and the shape-dependent parameters are given by:

$$G_x = \frac{ab^2}{2}\int_0^{\infty} \frac{ds}{(s + a^2)^{3/2}(s + b^2)} \quad \text{and} \quad G_y = \frac{ab^2}{2}\int_0^{\infty} \frac{ds}{(s + a^2)^{1/2}(s + b^2)^2} \tag{3.10}$$

More complex geometries, on the other hand, require the solution of a magnetostatic boundary value problem. Although the above approximations will suffice for most practitioners of biological applications, for completeness a more rigorous analysis of force calculation is presented here. The Kelvin volume force density (Melcher 1981) can be calculated by integrating the M and H fields over the object's volume:

$$\vec{f} = \mu_0 (\vec{M} \cdot \nabla)\vec{H} \tag{3.11}$$

It should be recognized that the H fields in Equation 3.11 are those that would have been observed in the absence of the manipulated object. Alternatively, the force can also be calculated by integrating the total B and H fields over the object's bounding surfaces using the Maxwell stress tensor (Rosensweig 1985, Rinaldi and Brenner 2002), which is expressed in vector form as:

$$\vec{\vec{T}} = \vec{B}\vec{H} - 1/2\mu_0 \vec{H}^2 [I] \tag{3.12}$$

where $[I]$ is the identity matrix, and a vector dyadic product is assumed between the $\vec{B}$ and $\vec{H}$ fields. From classical mechanics, the divergence of a stress tensor is the body force: $\vec{f} = \nabla \cdot \vec{\vec{T}}$; thus, integration over the bounding surfaces leads to the following expression of force and demonstrates the equivalence of the two approaches:

$$\vec{F} = \int_V \vec{f}\, dV = \int_V \nabla \cdot \vec{\vec{T}}\, dV = \oiint_S \vec{\vec{T}} \cdot \hat{n} ds \tag{3.13}$$

The Maxwell stress tensor is preferred in most problems with complex geometries due to ease of integrating over only two dimensions instead of three; however, the volume force density approach and surface integral approaches both lead to equivalent solutions, though in some cases one has to be careful (Rinaldi and Brenner 2002). The torque is determined by integrating the force-couple over the material surface, using the relationship:

$$\vec{\tau} = \int_V \vec{r} \times \vec{f}\, dV = \int_V \vec{r} \times \nabla \cdot \vec{\vec{T}}\, dV = \oiint_S \vec{r} \times (\vec{\vec{T}} \cdot \hat{n})\, dS \tag{3.14}$$

where $\vec{r}$ is assumed to be taken with respect to the geometric or mass center of the object.

The surface/volume integration approaches described above are necessary for exact analysis of complex geometries; however, for many applications the material's shape is quite reasonably approximated as a sphere, rod, or disk, which permits the use of more convenient analytical solutions with negligible error (Stratton 1941). Of particular interest is the case where the material is exposed to a magnetic field that is reasonably uniform on the length scale of the body, whose well-known solution is a material with uniform magnetization. These limiting cases allow treatment of the material as a point dipole, which considerably simplifies the computation of force and torque.

3.2.2 Dynamics of Particles in Fluids

In general, fluid motion and hydrodynamic interactions between particles are quite complex, and a rigorous analysis should incorporate techniques from computational fluid dynamics to analyze the fluid velocity distributions and particle-fluid hydrodynamic coupling, especially in turbulent flows. A rigorous treatment is outside the scope of this review; therefore, the treatment of fluid interactions will be presented in a relatively simple limiting case analysis of low Reynolds number flow. The Navier–Stokes equation for an incompressible fluid without convective acceleration is given by:

$$\rho \frac{\partial \vec{v_f}(\vec{r})}{\partial t} = -\nabla p(\vec{r}) + \eta \nabla^2 \vec{v_f}(\vec{r}) + \vec{f}\,(\vec{r}) \tag{3.15}$$

where ρ is the fluid mass density, p is the pressure distribution, η is the fluid viscosity, and $\vec{f}$ is the applied force which in this case is predominantly the magnetic force. The simplest solution can be obtained by ignoring multiparticle interactions, and using the assumption of low Reynolds number flow, which allows the inertial terms to be ignored. This results in the classical equation for Stokes drag on a spherical particle (Rosensweig 1985):

$$\vec{F} = 6\pi\eta a(\vec{v_f} - \vec{v_p}) \tag{3.16}$$

which is commonly used to approximate the trajectory of particles inside slowly moving fluids. Shape correction factors are required to calculate the drag on elongated particles as a function of its orientation relative to the fluid flow. Likewise, the torsional motion of spherical particles can be determined using similar assumptions, leading to:

$$\vec{\tau} = 8\pi\eta a^3(\vec{\omega_f} - \vec{\omega_p}) \tag{3.17}$$

where the subscripts f and p refer to the fluid and particle, respectively. For simplicity, the velocity of the fluid is often calculated independently of the particle motion, by taking into account the pressure fields, geometric constraints, and boundary conditions of the fluid container. One of the earliest examples of this type of analysis for magnetic separation was performed by following the logic used to analyze electrostatic separation (Bean 1971, Watson 1973, Clarkson and Kelland 1978). These models ignored the possibility of particle buildup on the magnetic elements that could influence both the magnetic forces and hydrodynamics. Self-consistent fluid velocity calculations were first attempted in the late 1970s (Friedlaender, Takayasu, Rettig, and Kentzer, 1978) to model the particle buildup around the magnetic elements. This approach was extended in later works to calculate the separation efficiency as a function of various parameters, including the particle and wire capture radii, and fluid flow speed; however, due to the complexity of the problem it was recognized that these first attempts were mere approximations that did not model the fluid velocity in a completely self-consistent manner. The effects of particle–particle and particle–wall hydrodynamic interactions have also been for the most part ignored even though it is recognized that these effects can play an important role in particle separation systems (Yavuz et al. 2006). In order to overcome this problem, an alternative phenomenological approach has been investigated based on the ideas of statistical mechanics and probability distributions to model the complex scattering and collision phenomena that can occur within particulate fluids.

When particles are much smaller than 1-μm, in which case the magnetostatic potential energy of the particle becomes comparable to or much smaller than thermal fluctuation energy, Brownian motion will also play an important role in particle motion. In this case, the notion of a detailed particle trajectory analysis is no longer meaningful. Instead, these systems are usually modeled using particle distribution functions to describe the probability of observing a particle as a function of space-time variables. In magnetic separation, the particle concentration is the key parameter used in the distribution functions, which is interpreted as the number of particles multiplied by the probability density of observing a particle at a given location in space-time. Usually the simplest approach is to use drift-diffusion equations to model the flux of particles in space-time, in which the particle fluxes are written as:

$$\vec{J}_{diff} = -D(\vec{r})\nabla C(\vec{r}) \quad \text{and} \quad \vec{J}_{drift} = C(\vec{r})\vec{v}_p(\vec{r}) \tag{3.18}$$

where D is the diffusion coefficient that can be a function of space, and depends on the local particle concentration among other parameters. The total flux of the particles is therefore:

$$\vec{J}_{net} = \vec{J}_{diff} + \vec{J}_{drift} \tag{3.19}$$

which can be solved by applying the mass conservation law:

$$\frac{\partial C}{\partial t} + \vec{J}_{net} = 0 \tag{3.20}$$

Using Equation 3.20, the local particle concentration can be determined as a function of the appropriate boundary conditions (e.g. the particle flux at the boundary in the direction of the surface normal is zero). It is also possible to take into account the inhomogeneity in the fluid magnetization that results from locally varying particle concentration (Erb, Sebba, Lazarides, and Yellen, 2008, Erb and Yellen 2008, Hovorka, Yellen, Dan, and Friedman, 2005), which can affect the force and particle buildup near the magnetic elements. From this brief review, it is clear that the intricate coupling between mass conservation, hydrodynamics, and magnetic forces represents a rich area of theoretical study that has seen significant progress over the last several decades. These analyses represent the most well-known attempts to guide experimental work in the field of magnetic manipulation, discussed next.

3.3 MAGNETIC SEPARATION: REVIEW AND CURRENT TRENDS

The goal of magnetic bioseparation is to remove and recover specific species from a fluid stream, which may be cells, cell fragments, proteins, DNA, virus, bacteria, and other biological species. An example of commercially available separation systems is provided in Figure 3.1. Magnetic

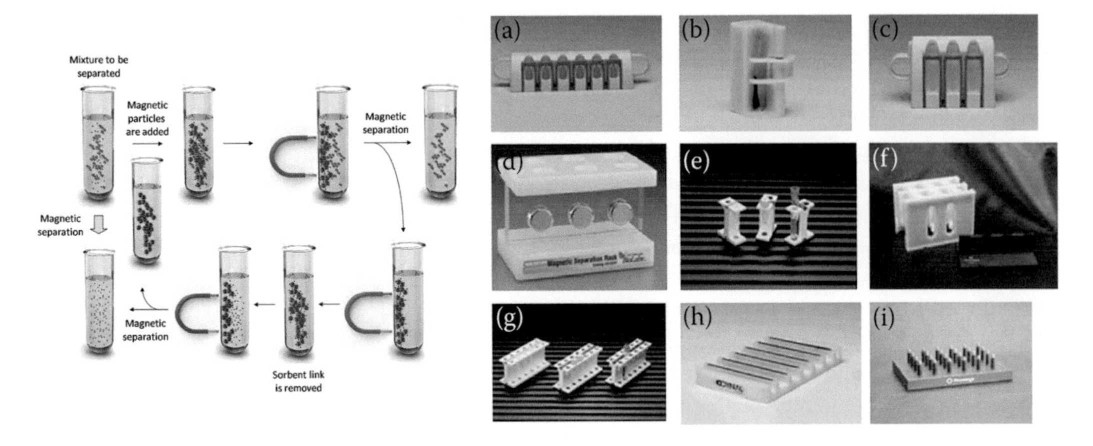

FIGURE 3.1 **(Left)** Basic schematic overview of magnetic bioseparation, **(Right)** Examples of permanent magnet array separators in existing commercial technology, images reprinted from reference (Yavuz et al. 2009).

bead polymer composites are typically used for this purpose, which are composed of a collection of paramagnetic nanograins (e.g. ~10 nm iron oxide or other magnetic nanoparticles) encapsulated inside an inert polymer matrix, such as polystyrene, PLGA, or other biocompatible materials (Lu and Schüth 2007). These polymers are utilized because they are a convenient platform for functionalizing their surfaces with specific targeting antigens and other molecular recognition elements that can strongly bind to the biological species of interest. This class of magnetic beads is typically called "superparamagnetic" because the large number of magnetic nanograins inside each bead gives the entire composite a "super large" paramagnetic response. In order to avoid irreversible flocculation and assist in the recovery of the targeted biological species post-separation, superparamagnetic beads are usually preferred over ferromagnetic beads.

3.3.1 Conventional Magnetic Separators

The two main approaches to improving the efficiency of bioseparation are to increase the force on the magnetic bead by (1) increasing its magnetic dipole moment, or (2) increasing the magnetic field gradient. However, there is an inherent trade-off in increasing the magnetic bead size (and thus its dipole moment), because larger beads have less available surface area available for biocapture and will diffuse more slowly inside the working fluid. Due to this trade-off, the optimal magnetic bead sizes used in separation are typically in the range of 0.5–5 μm, which are large enough to be separated from the fluid but small enough to efficiently diffuse in the fluid and capture the biological species in minimal time. As a result of this size constraint, the main approach to improving bioseparation has been to increase the magnetic field gradients, hence the name *high-gradient magnetic separation* (HGMS).

Magnetic field gradients scale roughly with the size of the separation apparatus, and one can use rough calculations to estimate the achievable field gradient in different separation systems. For example, a cylindrical magnetic fiber with a diameter, $d = 10$ μm, can achieve a magnetic field gradient near its surface on the order of $\mu_0 M_S/d \sim 10^4$–10^5 T/m; however, when the magnetic beads are several wire diameters away from the magnetic elements, then the magnetic field gradient is practically zero, which implies that this type of magnetic separator is efficient only when in direct contact with the fluid stream. As a consequence, one of the more popular separators to date is a steel wool filter placed directly in the fluid stream, which maximizes the volume average of field gradients throughout the fluid. This type of separation approach is a two-step process that requires a magnetic capture matrix that has high saturation magnetization, but low remanent magnetization.

In the first step, a strong external field is applied across the steel wool matrix leading to maximal field gradients throughout the fluid and efficient bead capture (i.e. capture mode). In the second step, the external field is removed and the weak remanent magnetization of the steel wool is no longer sufficient to retain the adhered particles, thus they can be removed with the fluid stream (i.e. release mode).

In some magnetic separation systems, it is desirable to achieve continuous magnetic separation. This approach requires an arrangement of external permanent magnets or electromagnets outside of a fluid container (Lacoste and Lubensky 2001, Hatch and Stelter 2001, Zborowski et al. 1999) in order to induce the magnetic beads to migrate across adjacent fluid streams toward the container walls. The flow bands nearest the container walls are separated in the outlet flow, whereas the center portion of the flow stream is typically deficient in magnetic beads and represents the waste stream. In either case (continuous or non-continuous separation systems), the goal is to selectively separate one component from a complex mixture, and then release this component post-separation for further analysis and/or operations. A variety of techniques have been developed to release the biological material attached to the collected magnetic beads by altering the salinity, pH, temperature, or other physico-chemical parameters.

3.3.2 Multiplexed Separation: Linear vs Non-linear Separation

One of the drawbacks of the magnetic separation system described above is its capability to separate only one biological species in each step. When separating several biological species from the fluid, this approach must be applied sequentially to each biological species, which not only increases the separation time but also can lead to loss of biological material. For this reason, there has been recent interest in developing approaches that can multiplex the separation process.

A multiplexed separation strategy requires the utilization of several different types of magnetic beads, wherein each bead type has a different surface reactivity that recognizes and captures a specific biological species from a common working fluid. In this case, the goal is not only to separate the magnetic beads from the working fluid, but also to separate the different magnetic bead types from one another. Some work in recent years has gone into adapting conventional separation systems to achieve multiplexed separation, in which multiple different bead types are separated by differentials in the magnetic force. Inside the same field gradient, the larger magnetic beads migrate into a different flow stream than the smaller magnetic beads. The laminar flow bands are later collected into several different fluid streams, each of which contains the species of interest. However, because all beads migrate in the same direction under the action of magnetic force, it is difficult to achieve suitable resolution between the different flow streams and achieve efficient multiplexed magnetic separation. The equations for magnetophoresis clearly elucidate this dilemma (Equations 3.6, 3.7, and 3.16). For example, the magnetic force scales with the volume of the bead; however, the friction coefficient scales with the bead radius. Thus, the differential in the magnetically induced velocity between two different bead types scales with the square of the ratio of the bead radii. Due to the linear force equations used in this technique, it can be classified as a linear separator.

Recently, an alternative approach for separation, which uses non-linear dynamics and bifurcation phenomena, has been gaining interest for separating multiple different types of biological species. Rather than increase the fields and field gradients to achieve larger differentials in the force, it has been shown that a translating periodic potential energy landscape can induce certain beads to become mobile (i.e. move across a substrate), whereas other beads are immobilized (i.e. trapped in a local region of the substrate). This type of separation modality has been demonstrated using several types of excitation, including optical (Ladavac, Kasza, and Grier, 2004), electrical (Cui, Holmes, and Morgan, 2001, Gascoyne and Vykoukal 2002, Green and Morgan 1997), and magnetic source fields (Gao, Gottron, Virgin, and Yellen, 2010, Yellen et al. 2007, Kose, Fischer, Mao, and Koser, 2009), and can potentially achieve infinite separation resolution between two different types of beads based on only minor variation in size, magnetic content, or type of material attached to the beads.

In general, a periodic potential energy landscape can be created with optical, electrical, or magnetic fields, either by creating interference patterns between two or more optical beams (Roichman, Wong, and Grier, 2007, Ladavac et al. 2004, Korda, Taylor, and Grier, 2002) or by patterning a substrate with arrays of metallic contacts or magnetic islands (Gao et al. 2010, Yellen and Virgin 2009, Yellen et al. 2007, Yellen et al. 2005). The beads are attracted to the field maxima (or minima) depending on their relative optical, electrical, or magnetic properties, as illustrated in Figure 3.2. When the potential energy landscape is time-modulated, such as by flowing beads through a static energy landscape, or by applying external fields at different frequencies to modulate the energy landscape, the beads attempt to entrain themselves in the periodic potential and remain trapped in a given potential energy minima. If the modulation is applied at a sufficiently slow rate, the beads are able to overcome viscous effects and remain trapped in a particular phase position relative to the landscape, hence the name *phase-locked motion.* On the other hand, when the modulation is too fast, the beads can no longer remain trapped in a particular phase position, and slip out of a given potential energy minima into an adjacent one, hence the name *phase-slipping motion.* The transition between phase-locked and phase-slipping motion is highly sensitive to the bead properties; thus by exploiting this non-linear dynamic behavior it is possible to implement extremely high-resolution separation.

A mathematical model based on a non-uniform harmonic oscillator can capture much of this behavior, whose equation is given by:

$$\frac{d\phi}{d\tau} = \frac{\omega}{\omega_C} - \sin(\phi) \tag{3.21}$$

where ϕ is the instantaneous phase of the bead relative to the moving landscape, ω is the driving frequency, $\tau = \omega t$ is dimensionless time, and ω_C is a characteristic frequency of the system. When the driving frequency is less than a critical frequency of the system, $\omega < \omega_C$, the bead is able to find some stable phase, ϕ, within the translating periodic potential energy landscape such that its phase velocity can equal zero, $d\phi/d\tau = 0$ (i.e. phase-locked motion). On the other hand, when the driving frequency is too fast, i.e. $\omega > \omega_C$, then no stable phase can be found and its phase-velocity can never satisfy the relation, $d\phi/d\tau = 0$, which implies that the phase is constantly moving (i.e. phase-slipping motion). The time required to travel exactly one period (i.e. move from one potential energy minima to an adjacent one) can be determined by integration Equation 3.21, leading to an expression for the time average of the phase velocity:

$$\left\langle \frac{d\phi}{d\tau} \right\rangle = \begin{Bmatrix} 0 & \omega \le \omega_c \\ -\sqrt{\omega^2 - \omega_c^2} & \omega > \omega_c \end{Bmatrix} \tag{3.22}$$

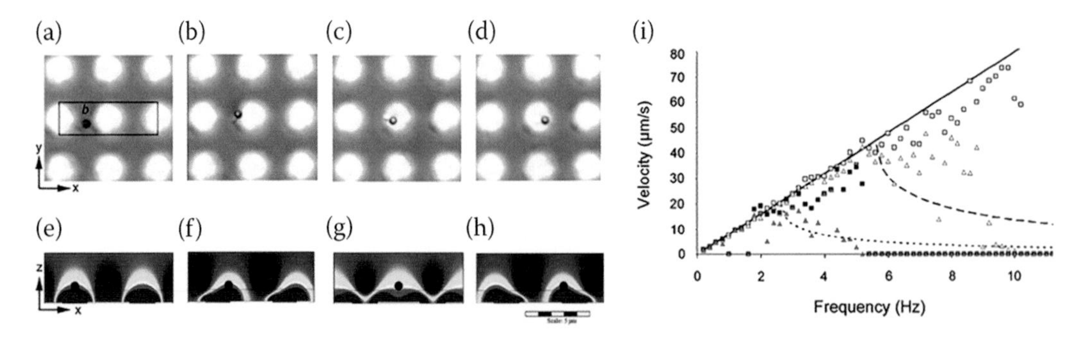

FIGURE 3.2 Example of non-linear separation system. **(Left)** A periodic potential energy landscape is used to transport a 1-μm magnetic bead across a micro-magnet array, **(Right)** The velocity of different bead types will depend on the driving frequency, allowing for some bead types to become trapped, whereas others are mobile, images adapted from reference (Yellen et al. 2007).

which clearly reveals the bifurcation of the driving frequency. Below the critical frequency, the phase velocity is zero. Above the critical frequency, the phase velocity experiences a time-average velocity, which starts off relatively small when the driving frequency is only slightly above the critical frequency, but eventually grows to equal the driving frequency in the limit of infinite frequency. In terms of separation, this result implies that the bead will have zero forward velocity only at infinite driving frequency.

In prior work, it has been shown that the critical frequency can be predicted theoretically by analyzing the properties of the bead and landscape, and it has been shown both theoretically and experimentally that highly efficient separation resolution is realizable (Yellen et al. 2007, Gao et al. 2010) (see Figure 3.2). These phenomena have been demonstrated both for the translation of magnetic beads across a micro-magnet array (Yellen et al. 2007, Gao et al. 2010) and for the translation of ferromagnetic beads across a glass surface in an external rotating field (Agayan, Smith, and Kopelman, 2008), the first of which utilizes an applied force and the second an applied torque.

While the theoretical approach presented above is certainly a good approximation in the phase-locked regime, near the phase-slipping regime it fails to adequately explain the experimental results at high driving frequencies in which the bead has been observed to reside in closed orbits (i.e. zero time average translation velocity) (Yellen et al. 2007). The underlying reason is that Equation 3.21 is an approximation that retains only one term in the force equation. However, if multiple different oscillators are simultaneously acting on the same bead, then the mutual interactions between the oscillators can induce regimes in which the bead's velocity becomes bounded even at finite driving frequency.

As one example, let's consider a traveling magnetic field wave that is produced by the superposition of the fields from an array of ferromagnetic micro-magnets and an external rotating magnetic field. For simplification, it is possible to assume the fields of the substrate behave as a periodic array of alternating line poles with magnetic pole density (Gao et al. 2010, Yellen and Virgin 2009):

$$\lambda(x) = 2\lambda_0 \sum_{n=odd} \cos\left(\frac{2\pi n}{d_m}x\right) \tag{3.23}$$

and period, d_m. Through separation of variables in two dimensions, the fields and field gradients can be determined, allowing for the force and magnetophoretic velocity to be deduced:

$$v(x,t) = v_1 \sin\left(\frac{2\pi}{d_m}x - \omega t\right) + v_3 \sin\left(\frac{6\pi}{d_m}x - \omega t\right) + v_5 \sin\left(\frac{10\pi}{d_m}x - \omega t\right) + \ldots \tag{3.24}$$

where v_i are the velocity coefficients, whose explicit derivation can be found in the references (Gao et al. 2010, Yellen and Virgin 2009). When only the first term in the equation is retained, it is possible to convert Equations 3.24 into 3.21 by using the transformation: $\phi = 2\pi x/d - \omega t$, and $\omega_C = 2\pi v_1/d_m$, for which the concept of a specific phase is clear. On the other hand, when multiple terms are retained in Equation 3.24, the notion of a specific phase is meaningless, and moreover the equation cannot be analytically integrated due to the strong non-linear coupling between the space and time terms. Numerical simulations, however, have revealed that these higher-order terms are responsible for producing a dynamics regime where closed orbits can exist at frequencies only slightly above the critical frequency. A quasi-analytical proof on the existence of closed orbits can be found in Gao et al. (2010), along with the frequency range in which closed orbits can exist. This insight is important because it can be used to experimentally design a separation apparatus that can achieve infinite separation resolution between two bead types, in which one bead type is mobile (i.e. in an open orbit) and can be transported across the array, while a different bead type is immobile (i.e. in a closed orbit) and experiences only oscillatory motion about a given micro-magnet.

The field of non-linear magnetic separation systems is still in the early stages, and there are many new ideas under exploration. Some examples include the analysis of the bead motion on a two-dimensional (2D) substrate when the magnetic field is applied at an angle relative to the substrate magnetization. In this case, the bead motion will remain entrained along the direction of the substrate magnetization below a critical angle. However, when the field is applied above a critical angle, the beads will move in a direction perpendicular to the substrate magnetization (Tahir, Gao, Virgin, and Yellen, 2011). By tuning this angle parameter, it may be possible to induce two different bead types to move in orthogonal directions. Another potential research direction consists of applying multifrequency magnetic fields to an ensemble of different beads. Depending on the strengths and frequencies of the excitation fields, it may be possible to induce two different bead types to move in opposite directions. This type of magnetic control is only possible in non-linear separation systems, and thus remains a fertile field of exploration.

3.4 BIOMECHANICS AND FORCE SPECTROSCOPY

Although the most common biomedical application of magnetic force is in material separation and drug delivery, there is a growing trend of using the convenient features of magnetic materials and remote force manipulation in other areas of biological analysis, particularly in biomechanics and single-molecule force spectroscopy (SMFS). Aided by the advent of the atomic force microscope (AFM), the field of biophysics has experienced a revolution in the last two decades through the adoption of new techniques for applying $>10^{-9}$ N forces on single molecules (Best et al. 2003, Marszalek, Li, Oberhauser, and Fernandez, 2002b, a, Clausen-Schaumann, Rief, Tolksdorf, and Gaub, 2000, Rief, Gautel, and Gaub, 2000, Carrion-Vazquez et al. 1999, Oberhauser, Marszalek, Carrion-Vazquez, and Fernandez, 1999, Rief, Clausen-Schaumann, and Gaub, 1999, Rief, Gautel, Oesterhelt, Fernandez, and Gaub, 1997), ligand-receptor complexes (MacKerell and Lee 1999, Moy, Florin, and Gaub, 1994, Lee, Kidwell, and Colton, 1994, Paci, Caflisch, Pluckthun, and Karplus, 2001, Riener et al. 2003, Izrailev, Stepaniants, Balsera, Oono, and Schulten, 1997, Zlatanova, Lindsay, and Leuba, 2000, Raab et al. 1999), lipid bilayers (Gueta, Barlam, Shneck, and Rousso, 2006, Benoit and Gaub 2002, Czajkowsky, Iwamoto, and Shao, 2000, Tamm et al. 1996, Mou, Yang, and Shao, 1995, Yang and Shao 1995, Mou, Yang, Huang, and Shao, 1994, Thormann et al. 2007), and other biologically relevant materials (Jiang et al. 2009, Grandbois, Dettmann, Benoit, and Gaub, 2000, Shao, Shi, and Somlyo, 2000, Thie et al. 1998). In the general approach, one end of the molecule (or complex) is attached to a solid support, such as a glass slide, and the other end is attached to the AFM cantilever tip. By monitoring the deflection of a cantilever and the displacement of the stage (or tip), it is possible to deduce the applied force and the state of molecular extension, respectively. This data is used to construct force extension curves, which can reveal the mechanical properties of single molecules, including the elasticity, plasticity, and other mechanical features. For example, the mechanical properties of individual DNA and RNA molecules (both single-stranded and double-stranded) and other biopolymers have been measured through force extension curves. The rupture force between various ligand-receptor complexes, such as biotin and streptavidin, has also been measured as a function of different loading rates. The elasticity of phospholipids membranes has also been characterized using SMFS techniques. A review of recent progress in the field of SMFS can be found in the references (Neuman and Nagy 2008, Ritort 2006).

3.4.1 Brief Comparison of SMFS Techniques

While it has been a boon to the field of biophysics, the AFM approach has several drawbacks that hinder its adoption, particularly in the low force and low loading rate regimes. Specifically, thermal fluctuations in the cantilever and low-frequency mechanical drift of the piezo stage hinder the ability to apply controlled forces below 10 pN and loading rates below 10^2 pN/s. In order to access these biologically relevant forces and force loading regimes, which tend to be in the picoNewton range and applied on the timescale of minutes to hours, several alternative force approaches have been developed, including the biomembrane force probes (BFP) (Erdmann, Pierrat, Nassoy, and Schwarz, 2008, Ritort 2006, Merkel, Nassoy, Leung, Ritchie, and Evans, 1999, Evans and Ritchie 1997), optical tweezers (Steven et al. 2008, Williams 2006, Coirault, Pourny, Lambert, and Lecarpentier, 2003, Curtis, Koss, and Grier, 2002, Bockelmann, Thomen, Essevaz-Roulet, Viasnoff, and Heslot, 2002, Bustamante, Smith, Liphardt, and Smith, 2000, Wang, Yin, Landick, Gelles, and Block, 1997, Felgner, Muller, and Schliwa, 1995, Block, Goldstein, and Schnapp., 1990), and magnetic tweezers (Mosconi et al. 2009b, Ribeck and Saleh 2008, de Vries, Krenn, van Driel, and Kanger, 2005, Danilowicz, Greenfield, and Prentiss, 2005, Gosse and Croquette 2002, Haber and Witz 2000, Bausch, Moller, and Sackmann, 1999). In these cases, a colloidal bead or liposome replaces the AFM tip, and force is applied to the bead by mechanical pressure, optical fields, or magnetic fields. The capabilities and limitations of each of these force application techniques can be found in several reviews (Neuman and Nagy 2008, Ritort 2006).

Of these, magnetic tweezers offer several important advantages that are becoming increasingly recognized in the field of biophysics. First, the lack of direct mechanical contact between the force actuator and the probe (i.e. colloidal bead) allows for SMFS techniques to be massively parallelized, enabling entire histograms to be obtained in a single experiment. This feature is particularly important when measuring sub-picoNewton forces, in which thermal fluctuations strongly influence the measurements. A second advantage arises from the fundamental difference between magnetic tweezers and other approaches, namely that magnetic tweezers are inherently a force-controlled actuator whereas AFM, BFP, and OT are inherently displacement-controlled actuators. In other words, AFM, BFP, and OT techniques control the displacement of the cantilever tip, pipette tip, or optical trap, respectively, and infer the applied force through calibration of the cantilever, membrane, or optical trap stiffness. Magnetic tweezers, on the other hand, can apply a direct force on a colloidal bead; whereas the displacement of the bead (and thus extension of the molecule) must be measured through other means, such as an optical measurement, e.g. Newton fringing patterns of the bead or an evanescent wave. As a result, magnetic tweezers are much less susceptible to low-frequency drift of the mechanical components, and can therefore access loading rate regimes several orders of magnitude lower ($<10^{-3}$ pN/s) than is possible by other approaches. Finally, magnetic tweezers have the unique ability to apply torque as well as force on colloidal beads, which assists the analysis of mechanical torsion in biophysical systems.

3.4.2 Magnetic Tweezers: Recent Results and Current Trends

One of the first applications of magnetic tweezers developed in the 1950s was to measure the viscoelastic properties of the cytoplasm (Crick and Hughes 1949, Crick 1950). In the 1990s, the technique was adopted to analyze the rheological properties of actin filaments in suspension, first in one dimension (Ziemann, Rädler, and Sackmann, 1994), and then later extended to 2D translational and rotational motions using custom designed magnetic micromanipulators (Amblard, Maggs, Yurke, Pargellis, and Leibler, 1996a, b). Around the same time, magnetic tweezers were also implemented to study the elastic properties of phospholipid membranes (Heinrich and Waugh 1996), and the elastic properties of single DNA molecules (Smith et al. 1992, Haber and Witz 2000). Spurred on by these successes, there has been a substantial increase in scientific activity over the last decade both in improving the magnetic manipulation instrumentation as well as in using magnetic tweezers to derive fundamental insights into biophysical processes. See Figure 3.3 for a schematic of typical magnetic tweezer instruments and their applications in biophysical experimentation. For example, near-field magnetic tweezers were developed to apply forces in excess of 200 nN (Yan, Skoko, and Marko, 2004), magnetic force microscopes were developed to apply force in three dimensions (Fisher et al. 2006, Fisher et al. 2005), and sophisticated electromagnetic coupling mechanisms have been used to integrate microscope-based particle tracking

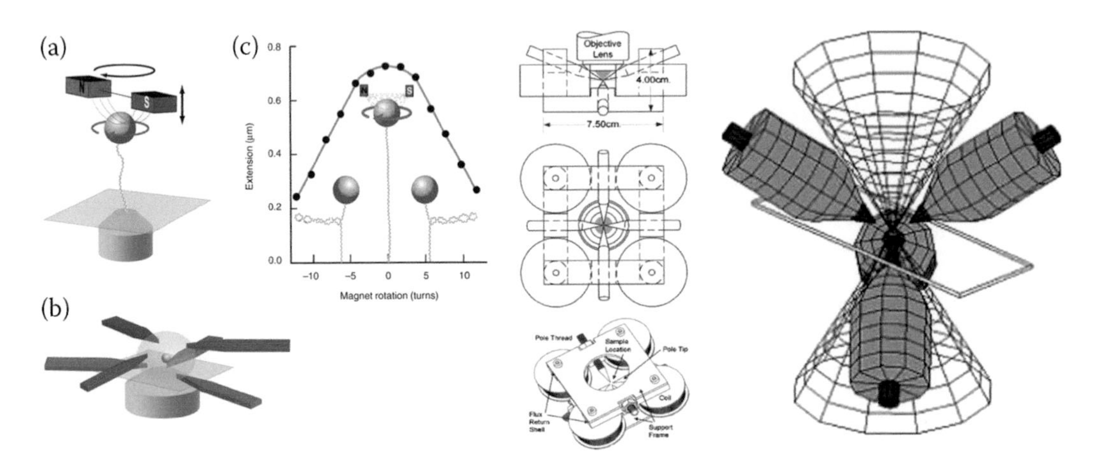

FIGURE 3.3 **(Left)** Schematic overview of force and torque application for analysis of DNA supercoiling, reprinted from reference (Neuman and Nagy 2008), **(Right)** Examples of pole piece construction and electromagnet design for multidimensional force microscopy, reprinted from reference (Fisher et al. 2005).

systems in digital feedback loops to improve the control over the multidimensional forces and torques on a magnetic bead (Gosse and Croquette 2002). The convenient ability to apply magnetic torque was also utilized to observe the mechanical properties of supercoiled double-stranded DNA (dsDNA) (Strick et al. 1998), as well as to uncover mechano-chemical transduction mechanisms in biological systems, including the effect of topoisomerases on unwinding DNA supercoils (Strick, Croquette, and Bensimon, 2000), the effect of the molecular motor DNA gyrase on introducing negative supercoils into DNA (Gore et al. 2006), and the mechano-chemical synthesis of ATP using magnetic force (Itoh et al. 2004). Magnetic forces have also been utilized to measure the rupture strength of ligand-receptor pairs, such as biotin and streptavidin, in massively parallel, enabling an entire histogram to be obtained in identical experimental conditions (Danilowicz et al. 2005).

In recent years, however, there have been relatively few transformative ideas introduced in the field of magnetic tweezers beyond just increasing the fields and the field gradients. One side effect of these design improvements is the introduction of larger spatial variation in the fields and field gradients, causing the magnetic force applied to each bead to change as a function of position. In addition, normal variations in the beads' volumes and their magnetic susceptibilities (e.g. magnetic content within each bead) have necessitated tedious calibration procedures to determine the true magnetic force. Typically, the calibration procedure consists of using the Brownian motion of the bead to infer the applied magnetic force; however, this approach requires high-resolution optical techniques and lengthy analysis times on the order of several minutes for each bead.

Recently, the present author has introduced a fundamentally different approach to magnetic tweezer technology, which uses "self-repulsion" of colloidal particles away from an interface using a technique described as *imaginary magnetic tweezing*. The approach is named such because it utilizes the repulsive image forces that are produced on a colloidal particle placed nearby a planar interface. This classical problem in magnetostatics can be solved by method of images, in which the boundary conditions of the sphere/plane geometry can be represented as the interaction between a true sphere and an imaginary sphere reflected across the planar boundary. The image force is repulsive when the magnetic permeability of the fluid, μ_f, is larger than that of the substrate, μ_s, which can be achieved by using a ferrofluid (a dispersion of ~10 nm magnetic nanoparticles suspended in an aqueous carrier fluid).

There are several advantages for adopting this approach. First, the use of ferrofluid permits the image force to be applied to non-magnetic colloidal beads, which have magnetic permeability nearly equal to vacuum, μ_0, and allows for elimination of any magnetic permeability variation between different beads. Second, the field gradient is produced by the bead's own image appearing in the substrate, which creates substantially more uniform field gradients than is possible by conventional magnetic tweezer approaches. Third, the force is controlled by an external uniform field, which allows highly controllable loading rates to be applied in massively parallel with instrumentation as simple as a solenoid coil. The ferrofluid permeability can be measured *a priori*, and it is reasonable to assume that the permeability throughout the fluid is homogeneous. Due to these advantages, the only major source of variation is the volume differences from bead to bead, which is much easier to control (<2% size variation is common in microsphere synthesis) than is controlling the amount of magnetic material within each bead.

A simple analysis reveals how the variability in the magnetic force should scale with the bead size. Recall the effective magnetic moment of a non-magnetic bead with volume V exposed to a uniform external magnetic field, $\vec{H}$, given in Equation 3.6. When the bead is nearby a non-magnetic substrate such as glass, the bead/substrate interaction can be modeled through a method of images as the interaction between a true dipole and an image dipole, whose moment is magnified by the factor $(\mu_f - \mu_0)/(\mu_f + \mu_0)$ (Panofsky and Phillips 1955). Because the image dipole is always oriented in a repulsive manner to the true dipole regardless of the field orientation, the bead/substrate force is always in the direction of the surface normal, $\hat{n}$. The bead/substrate interaction is modeled as the force on a point dipole interacting with the field gradient of its own image, leading to the following expression for magnetic force:

$$\vec{F_m} = \frac{3\pi\mu_0 d^2 \vec{H}^2}{8(1+2\varepsilon)^4}\left(\frac{\mu_f-\mu_0}{\mu_f+\mu_0}\right)\left(\frac{\mu_f-\mu_0}{\mu_f+2\mu_0}\right)^2 \hat{n} \tag{3.25}$$

where d is the bead diameter, Δ is the bead/substrate separation distance, and $\varepsilon = \Delta/d$ is the dimensionless separation distance between the surface of the bead and substrate. This derivation demonstrates that the force should scale with the square of the bead diameter.

Calibration of the imaginary magnetic tweezer can be accomplished using confocal microscopy techniques and an assumed balance between magnetic and gravitational forces. Virtually any aqueous-based ferrofluid can be used, such as the EMG series from Ferrotec (Nashua, NH), and monodisperse colloidal beads can be purchased from a variety of vendors (e.g. Thermo Fisher Scientific, Fremont, CA). In the calibration procedure, the density of the beads, ρ_p, must be independently measured by a sedimentation analysis in different ferrofluid concentrations. This technique allows the bead's density to be determined by titration of the sink/float transition point in ferrofluid, whose mass density of the ferrofluid, ρ_f, can be measured in a microbalance. The gravitational force can then be used to calibrate the magnetic force, through a magneto-gravitational force balance. Specifically, when strong magnetic fields are applied to the colloidal suspension, the image force levitates the beads sufficiently far away from the substrate surface such that the only compensating force remains the effective gravitational force, with negligible contributions from all other surface forces. Thus, by incorporating a force balance: $\vec{F_m} = -\vec{F_\rho}$, an analytical relationship can be derived for the equilibrium bead/substrate separation distance, given by:

$$\varepsilon = \frac{1}{2}\left[\sqrt[4]{\frac{18\mu_0 \vec{H}^2}{8dg|\rho_f-\rho_p|}\left(\frac{\mu_f-\mu_0}{\mu_f+\mu_0}\right)\left(\frac{\mu_0-\mu_f}{\mu_0+2\mu_f}\right)^2} - 1\right] \tag{3.26}$$

In Equation 3.26, the only free parameter is the ferrofluid permeability, μ_f. This parameter can be experimentally determined by finding the best fit in the ε vs $\vec{H}$ relationship. Quantitative analysis of bead/substrate separations are obtained directly in a confocal microscope, with one example

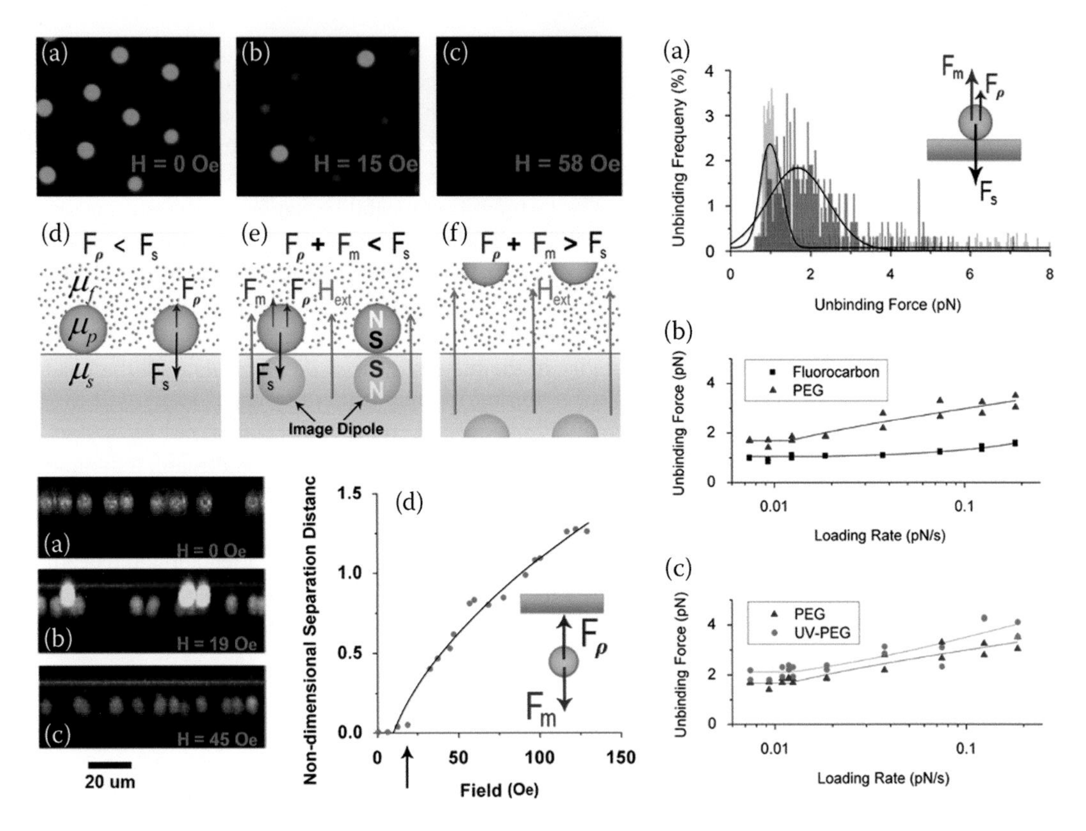

FIGURE 3.4 (Upper Left) Illustration of imaginary magnetic force tweezers and experimental images, **(Lower Left)** Depiction of confocal microscopy images of levitating beads and force calibration curve, **(Right)** Histogram and force loading rate diagrams of unbinding data for plastic beads to various surfaces, images adapted from reference (Yang et. al 2011).

calibration curve shown in Figure 3.4 (Yang, Erb, Wiley, Zauscher, and Yellen, 2011). Images of the beads show that they are repelled from the glass surface (red line) as the strength of the external field is increased. The match between Equation 3.26 and the experimental data allowed for the determination of the ferrofluid permeability, which is usually in the range of $\mu_f = 1–2$. The calibration of the ferrofluid permeability is a necessary step due to the potential for particle-particle aggregation and other aging effects that can change the fluid properties over time.

Once the calibration parameters of the ferrofluid have been established, the mean unbinding force can be measured using a magnetic force ramp that can range from 10^{-4} to 10^2 pN/s, even reaching down to the quasi-equilibrium force loading regime; i.e. one that is insensitive to the pulling rate. This approach can measure the rupture force between any receptor-ligand bond, or can determine the adhesion strength between a bead and virtually any surface. Video tracking and image software analysis are convenient tools for determining the number of beads unbinding from the surface at each time point, which is used to construct a histogram of the mean unbinding force as function of the loading rate. An unbinding event is identified as the instant when the bead's fluorescent intensity suddenly decreased, as it is pushed away from the substrate. The advantage of this approach is that each experiment can identify a large number of unbinding events (typically 300–1000) in identical experimental conditions, as is demonstrated in the histogram in Figure 3.4.

In this example, we tested the adhesion force of colloidal beads to fluorinated and pegylated surfaces, and we discovered that at very low loading rates (<0.02 pN/s), the unbinding force is independent of the loading rate, which implies that unbinding occurs under quasi-equilibrium conditions. The lowest force loading rate we tested is one to two orders of magnitude lower than other previously reported linear force ramps (Merkel et al. 1999). The equilibrium unbinding force for the fluorocarbon and PEG substrate is the sum of the magnetic and gravitational force, which was determined to be 0.938 pN and 1.752 pN, respectively. At higher loading rates, the unbinding force depended exponentially on the loading rate, which is consistent with prior studies on the loading rate dependence of intermolecular forces (Evans and Ritchie 1997).

This brief review of magnetic tweezers demonstrates one of the more recent applications of magnetic force and torque in biology, and is currently one of the hottest trends in magnetic manipulation. It is certain that the use of magnetic techniques in biophysical research will continue over the next several decades and lead to new breakthroughs in understanding the behavior of single molecules as well as more complex biological systems.

3.5 MAGNETIC ASSEMBLY: APPLICATIONS IN MICROARRAYS AND TISSUE ENGINEERING

Another more recent application of magnetic forces has been in the assembly of structures that have relevance to biological sensors and tissue engineering constructs. The biosensor line of research has relied heavily on advances in lithographic approaches for producing micro-scale field gradients, and has recently been used to construct genomic (Barbee and Huang 2008), proteomic (Yellen and Friedman 2004a, b), and cellulomic arrays (Tanase et al. 2005, Lai et al. 2010). In this approach, a ferromagnetic array is patterned on planar substrates by the photolithographic lift-off technique. Magnetic beads introduced to the substrate are attracted to the field gradients of the ferromagnetic pattern and are thereby assembled into periodic structures. Arrays of cells attached to magnetic nanowires (Tanase et al. 2005), as well as proteins and other molecules attached to magnetic beads (Barbee and Huang 2008, Yellen and Friedman 2004a, b) have been constructed in a massively parallel manner using this approach. See Figure 3.5 for some examples of magnetically driven assembly of biological materials. It has also been demonstrated that the ferromagnetic properties of magnetic thin films can be exploited to "programmably" assemble biomaterials onto 2D surfaces (Yellen and Friedman 2004a, b). This process utilizes an external bias field to alter the regions of local magnetic field maxima and minima above a magnetically textured surface, and induce either attraction or repulsion of magnetic beads to/from specific surface sites.

One of the major advantages of magnetic manipulation is that bio-patterning can take place on non-planar substrates. In a recent demonstration, a magnetolithographic approach was used to pattern periodic biological structures on the inside of cylindrical glass tubes (Bardea, Baram, Tatikonda, and Naaman, 2009, Kumar, Bardea, Shai, Yoffe, and Naaman, 2010) for the purpose of developing devices that can carry out sequential biochemical reactions on a micro-scale. Biochemical patterning of interior fluidic surfaces can be accomplished by positive or negative lithography approaches. In the positive lithography approach, magnetic nanoparticles attached to specific biochemicals are deposited directly onto the capillary walls by placing external permanent magnets around the exterior of a capillary. The negative lithography approach, on the other hand, uses a ferrofluid as a diffusion mask, which prevents biochemicals from depositing on specific sites in the interior of a capillary. This negative magnetic masking approach was originally developed for self-aligned photolithographic applications (Yellen, Fridman, and Friedman, 2004, Yellen, Friedman, and Barbee, 2004), and more recently was adopted for biochemical patterning non-planar surfaces.

Magnetic forces have also been exploited to produce cellular sheets (Krebs et al. 2009, Ino et al. 2007, Ito et al. 2005a, b, 2004, Tranquillo et al. 1996) and other tissue constructs that can later be implanted into the body. In some cases, the cells are magnetically functionalized by exposing them to magnetic nanoparticles that are internalized by endocytosis. The "magnetic cells" can then be guided by magnetic field gradients into particular geometries of relevance to angiogenesis and microvascularization (Ito et al. 2005b). The inverse approach has also recently been investigated, in which an inert ferrofluid has been used to apply force to a suspension of non-magnetic cells. The

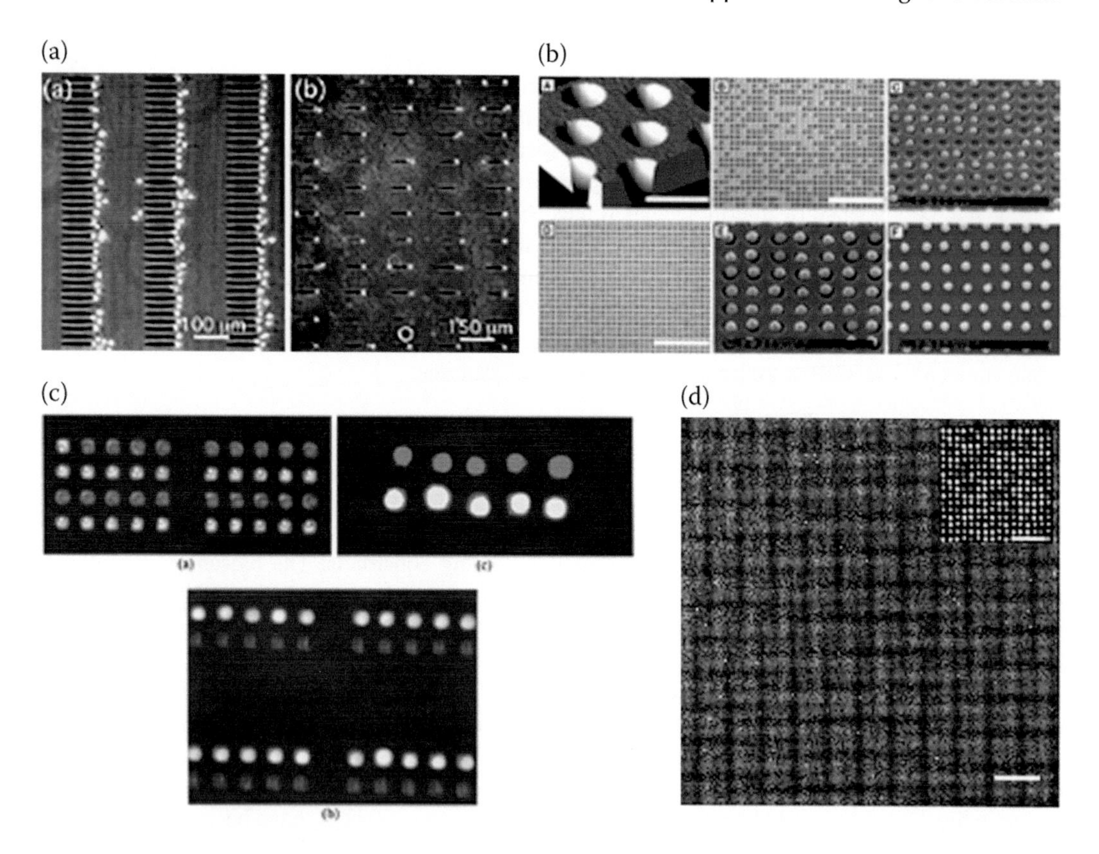

FIGURE 3.5 **(Upper Left)** Patterning cellular arrays, taken from reference (Tanase et al. 2005), **(Upper Right, Lower Right)** Programmable patterning of multi-component protein arrays, reprinted from reference (Yellen and Friedman 2004a), **(Lower Left)** Patterning DNA arrays by statistical assembly, reprinted from reference (Barbee and Huang 2008).

main idea is to use an external magnetic field to drive the cells to interact with one another via dipole–dipole interactions into cellular chain structures, which are an ideal geometry for inducing tubulogenesis (Krebs et al. 2009). There are advantages and disadvantages to both of these approaches. On the one hand, cells with internalized magnetic nanoparticles can be arranged into tissue structures inside a biocompatible matrix, such as collagen. Once the structures are formed, it is hoped that the cells can shed the excess nanoparticles by exocytosis; however, it is far from clear whether this approach will succeed because the large amount of magnetic material inside the cell can potentially damage the cell's interior organelles and interfere with normal cellular processes. On the other hand, cells that are organized into chain structures inside a ferrofluid can create similar structures in a biocompatible collagen matrix; however, the ability to remove the ferrofluid from the extracellular matrix is also difficult to accomplish.

Magnetofection (Mykhaylyk, Antequera, Vlaskou, and Plank, 2007; Mykhaylyk, Zelphati, Rosenecker, and Plank, 2008, Santori, Gonzalez, Serrano, and Isalan, 2006, Kamau et al. 2006, Isalan, Santori, Gonzalez, and Serrano, 2005) is another technique that has been gaining in popularity, in which viruses or DNA plasmids attached to magnetic nanoparticles are deposited on tissue culture substrates in desired patterns. Some works have studied the efficacy of magnetofection for eventual in vivo applications, such as using pulsed magnetic fields (Kamau et al. 2006) and analyzing the effect of magnetic forces on transfection rates. Other works have focused on generating polymerase chain reaction (PCR) products in a cell culture monolayer in a multiplexed format (Santori et al. 2006, Isalan et al. 2005). The magnetofection strategy has also been used as an analytical tool to study the effects of small interfering RNA (siRNA) (Mykhaylyk et al. 2008) and other gene-controlling therapeutics on eventual cell expression.

3.6 SUMMARY

Starting from relatively simple magnetic manipulation approaches for applications in bioseparation, the use of magnetic techniques has become increasingly sophisticated over the last several decades, making its way into diverse fields, including biophysics, drug delivery, and other bioanalytical tools. The field of magnetic separation has undergone significant changes both in the development of improved magnetic separation equipment and in the development of new separation strategies to improve the efficiency of material capture. The ability to apply controlled force on single magnetic beads has also become increasingly sophisticated as compared to the early studies in the 1950s. Nowadays, it is applied to study the basic mechanical properties of single molecules and reveal the complex interplay between mechanical force and biochemical responses of living systems. Magnetic assembly approaches are also becoming increasingly sophisticated for patterning various types of surfaces with molecules and cells, and they are being applied to various tissue-engineering applications. From recent trends, it is clear that magnetic manipulation approaches are an "attractive" area of study, and one should expect many new ideas to be introduced over the next several decades.

PROBLEMS

1. Assume a point dipole in vacuum with permanent magnetic moment, $\vec{m}$, is located at the origin with its moment aligned along the z-axis. Derive an expression for its field everywhere in Cartesian coordinates and in spherical coordinates.

Answer:

$$\vec{B}(x, y, z) = \frac{\mu_0 |\vec{m}|}{4\pi r^5}\{3xz\cdot\hat{x} + 3yz\cdot\hat{y} + (3z^2 - r^2)\hat{z}\}$$

$$\vec{B}(r, \theta, \phi) = \frac{\mu_0 |\vec{m}|}{4\pi r^3}\{2\cos\theta\cdot\vec{r} + \sin\theta\cdot\hat{\theta}\}$$

2. Derive a general expression for the magnetostatic potential energy between two point dipoles in vacuum separated by a distance r along the z-axis. Assume the orientation of the first dipole is θ_1, φ_1 and the second dipole is θ_2, φ_2.

Answer:

$$U(\theta_1, \theta_2, \phi_1, \phi_2, r) = -\mu_0 \frac{|\vec{m}_1||\vec{m}_2|}{4\pi r^3}(2\cos\theta_1\cos\theta_2 - \sin\theta_1\sin\theta_2\cos(\phi_1 + \phi_2))$$

3. Using the result from problem 2, derive a general expression for the central force between the two point dipoles.

Answer:

$$F(\theta_1, \theta_2, \phi_1, \phi_2, r) = -\frac{3\mu_0 |\vec{m}_1||\vec{m}_2|}{4\pi r^4}(2\cos\theta_1\cos\theta_2 - \sin\theta_1\sin\theta_2\cos(\phi_1 - \phi_2))$$

4. Using the result from problem 2, derive a general expression for the magnetic torque between the two point dipoles.

Answer:

$$\vec{T}(\theta_1, \theta_2, \phi_1, \phi_2, r) = \mu_0 \frac{|\vec{m}_1||\vec{m}_2|}{4\pi r^3}(2 \sin\theta_1 \cos\theta_2 + \cos\theta_1 \sin\theta_2 \cos(\phi_1 - \phi_2))\hat{m}_1 \times \hat{B}_{21}$$

5. Two identical paramagnetic beads of diameter, d, and effective magnetic susceptibility, χ, are separated from one another along the z-axis by an initial distance, r_0. Assume one is free to move and the other is fixed at the origin. If an external magnetic field, H_{ext}, is applied along the z-direction, how long will it take for the two beads to form a doublet in a fluid of viscosity, η? For this problem, assume the moment of the bead is due entirely to the external field and the fluid is static.

   Answer:

$$t = \frac{72\eta}{5\mu_0\chi^2 H_{ext}^2}\left\{\left(\frac{r_0}{d}\right)^5 - 1\right\}$$

REFERENCES

Amblard, F., A. C. Maggs, B. Yurke, A. N. Pargellis, and S. Leibler. 1996a. Subdiffusion and anomalous local viscoelasticity in actin networks. *Physical Review Letters* 77 (21), 4470.

Agayan, R. R., R. G. Smith, and R. Kopelman. 2008. Slipping friction of an optically and magnetically manipulated microsphere rolling at a glass-water interface. *Journal of Applied Physics* 104 (5), 054915-11.

Amblard, F., B. Yurke, A. Pargellis, and S. Leibler. 1996b. A magnetic manipulator for studying local rheology and micromechanical properties of biological systems. *Review of Scientific Instruments* 67 (3), 818–827.

Bahaj, A. S., P. A. B. James, and F. D. Moeschler. 1996. High gradient magnetic separation of motile and non-motile magnetotactic bacteria. *IEEE Transactions on Magnetics* 32 (5), 5106–5108.

Barbee, K. D., and X. H. Huang. 2008. Magnetic assembly of high-density DNA arrays for genomic analyses. *Analytical Chemistry* 80 (6), 2149–2154.

Bardea, A., A. Baram, A. K. Tatikonda, and R. Naaman. 2009. Magnetolithographic patterning of inner walls of a tube: A new dimension in microfluidics and sequential microreactors. *Journal of the American Chemical Society* 131 (51), 18260–18262.

Bausch, A. R., W. Moller, and E. Sackmann. 1999. Measurement of local viscoelasticity and forces in living cells by magnetic tweezers. *Biophysical Journal* 76 (1), 573–579.

Bean, C. P. 1971. Theory of magnetic filtration. *Bulletin of the American Physical Society* 16, 350.

Benoit, M., and H. E. Gaub. 2002. Measuring cell adhesion forces with the atomic force microscope at the molecular level. *Cells Tissues Organs* 172 (3), 174–189.

Best, R. B., D. J. Brockwell, J. L. Toca-Herrera, A. W. Blake, D. A. Smith, S. E. Radford, and J. Clarke. 2003. Force mode atomic force microscopy as a tool for protein folding studies. *Analytica Chimica Acta* 479 (1), 87–105.

Block, S. M., L. S. Goldstein, and B. J. Schnapp. 1990. Bead movement by single kinesin molecules studied with optical tweezers. *Nature* 348 (6299), 348–352.

Bockelmann, U., Ph. Thomen, B. Essevaz-Roulet, V. Viasnoff, and F. Heslot. 2002. Unzipping DNA with optical tweezers: High sequence sensitivity and force flips. *Biophysical Journal* 82 (3), 1537–1553.

Bucak, S., D. A. Jones, P. E. Laibinis, and T. A. Hatton. 2003. Protein separations using colloidal magnetic nanoparticles. *Biotechnology Progress* 19 (2), 477–484.

Bustamante, C., S. B. Smith, J. Liphardt, and D. Smith. 2000. Single-molecule studies of DNA mechanics. *Current Opinion in Structural Biology* 10 (3), 279–285.

Carrion-Vazquez, M., A. F. Oberhauser, S. B. Fowler, P. E. Marszalek, S. E. Broedel, J. Clarke, and J. M. Fernandez. 1999. Mechanical and chemical unfolding of a single protein: A comparison. *Proceedings of the National Academy of Sciences of the United States of America* 96 (7), 3694–3699.

Clarkson, C. J., and D. R. Kelland. 1978. Theory and experimental verification of a model for high gradient magnetic separation. *IEEE Transactions on Magnetics* 14:97–103.

Clausen-Schaumann, H., M. Rief, C. Tolksdorf, and H. E. Gaub. 2000. Mechanical stability of single DNA molecules. *Biophysical Journal* 78 (4), 1997–2007.

Coirault, C., J. C. Pourny, F. Lambert, and Y. Lecarpentier. 2003. Optical tweezers in biology and medicine. *M S-Medecine Sciences* 19 (3), 364–367.

Crick, F. H. C. 1950. The physical properties of cytoplasm: A study by means of the magnetic particle method. Part II. Theoretical treatment. *Experimental Cell Research* 1 (4), 505–533.

Crick, F. H. C., and A. F. W. Hughes. 1949. The physical properties of cytoplasm: A study by means of the magnetic particle method. *Experimental Cell Research* 1:36–80.

Cui, L., D. Holmes, and H. Morgan. 2001. The dielectrophoretic levitation and separation of latex beads in microchips. *Electrophoresis* 22 (18), 3893–3901.

Curtis, J. E., B. A. Koss, and D. G. Grier. 2002. Dynamic holographic optical tweezers. *Optics Communications* 207 (1–6), 169–175.

Czajkowsky, D. M., H. Iwamoto, and Z. F. Shao. 2000. Atomic force microscopy in structural biology: From the subcellular to the submolecular. *Journal of Electron Microscopy* 49 (3), 395–406.

Danilowicz, C., D. Greenfield, and M. Prentiss. 2005. Dissociation of ligand-receptor complexes using magnetic tweezers. *Analytical Chemistry* 77 (10), 3023–3028. doi:10.1021/ac050057+.

Dauer, R. R., and E. H. Dunlop. 1991. High gradient magnetic separation of yeast. *Biotechnology and Bioengineering* 37 (11), 1021–1028.

Erb, R. M., and B. B. Yellen. 2008. Concentration gradients in mixed magnetic and nonmagnetic colloidal suspensions. *Journal of Applied Physics* 103:07A312-3.

Erb, R. M., D. S. Sebba, A. A. Lazarides, and B. B. Yellen. 2008. Magnetic field induced concentration gradients in magnetic nanoparticle suspensions: Theory and experiment. *Journal of Applied Physics* 103 (6), 063916-5.

Erdmann, T., S. Pierrat, P. Nassoy, and U. S. Schwarz. 2008. Dynamic force spectroscopy on multiple bonds: Experiments and model. *EPL (Europhysics Letters)* 81 (4), 48001.

Evans, E., and K. Ritchie. 1997. Dynamic strength of molecular adhesion bonds. *Biophysical Journal* 72 (4), 1541–1555.

Felgner, H., O. Muller, and M. Schliwa. 1995. Calibration of light forces in optical tweezers. *Applied Optics* 34 (6), 977–982.

Fisher, J. K., J. R. Cummings, K. V. Desai, L. Vicci, B. Wilde, K. Keller, C. Weigle, G. Bishop, R. M. Taylor, C. W. Davis, R. C. Boucher, E. T. O'Brien, and R. Superfine. 2005. Three-dimensional force microscope: A nanometric optical tracking and magnetic manipulation system for the biomedical sciences. *Review of Scientific Instruments* 76 (5), pp.053711

Fisher, J. K., J. Cribb, K. V. Desai, L. Vicci, B. Wilde, K. Keller, R. M. Taylor, J. H. Ii, K. Bloom, E. Timothy O'Brien, and R. Superfine. 2006. Thin-foil magnetic force system for high-numerical-aperture microscopy. *Review of Scientific Instruments* 77 (2), 023702–023709.

Forbes, Z. G., B. B. Yellen, K. A. Barbee, and G. Friedman. 2003. An approach to targeted drug delivery based on uniform magnetic fields. *IEEE Transactions on Magnetics* 39 (5), 3372–3377.

Friedlaender, F., M. Takayasu, J. Rettig, and C. Kentzer. 1978. Particle flow and collection process in single wire HGMS studies. *IEEE Transactions on Magnetics* 14 (6), 1158–1164.

Friedman, G., and B. B. Yellen. 2005. Magnetic separation, manipulation and assembly of solid phase in fluids. *Current Opinion in Colloid & Interface Science* 10 (3–4), 158–166.

Gao, L., N. Gottron III, L. Virgin, and B. B. Yellen. 2010 The synchronization of superparamagnetic beads driven by a micro-magnetic ratchet. *Lab on a Chip* 10 (16), 2108–2114.

Gascoyne, P. R. C., and J. Vykoukal. 2002. Particle separation by dielectrophoresis. *Electrophoresis* 23 (13), 1973–1983.

Gaster, R. S., D. A. Hall, C. H. Nielsen, S. J. Osterfeld, H. Yu, K. E. Mach, R. J. Wilson, B. Murmann, J. C. Liao, S. S. Gambhir, and S. X. Wang. 2009. Matrix-insensitive protein assays push the limits of biosensors in medicine. *Nature Medicine* 15 (11), 1327-U130. doi:10.1038/nm.2032.

Gore, J., Z. Bryant, M. D. Stone, M. N. Nollmann, N. R. Cozzarelli, and C. Bustamante. 2006. Mechanochemical analysis of DNA gyrase using rotor bead tracking. *Nature* 439 (7072), 100–104.

Gosse, C., and V. Croquette. 2002. Magnetic tweezers: Micromanipulation and force measurement at the molecular level. *Biophysical Journal* 82 (6), 3314–3329.

Grandbois, M., W. Dettmann, M. Benoit, and H. E. Gaub. 2000. Affinity imaging of red blood cells using an atomic force microscope. *Journal of Histochemistry & Cytochemistry* 48 (5), 719–724.

Green, N. G., and H. Morgan. 1997. Dielectrophoretic separation of nano-particles. *Journal of Physics D: Applied Physics* 30 (11), L41–L44.

Gueta, R., D. Barlam, R. Z. Shneck, and I. Rousso. 2006. Measurement of the mechanical properties of isolated tectorial membrane using atomic force microscopy. *PNAS* 103 (40), 14790–14795. doi:10. 1073/pnas.0603429103.

Haber, C., and D. Witz. 2000. Magnetic tweezers for DNA micromanipulation. *Review of Scientific Instruments* 71:4561–4569.

Hatch, G. P., and R. E. Stelter. 2001. Magnetic design considerations for devices and particles used for biological high-gradient magnetic separation (HGMS) systems. *Journal of Magnetism and Magnetic Materials* 225 (1–2), 262–276.

Heinrich, W., and R. E. Waugh. 1996. A picoNewton force transducer and its application to the measurement of the bending stiffness of the phospholipid membranes. *Annals of Biomedical Engineering* 24:595–605.

Hovorka, O., B. Yellen, N. Dan, and G. Friedman. 2005. Self-consistent model of field gradient driven particle aggregation in magnetic fluids. *Journal of Applied Physics* 97 (10), 10Q306–13Q306.

Hultgren, A., M. Tanase, C. S. Chen, and D. H. Reich. 2004. High-yield cell separations using magnetic nanowires. *IEEE Transactions on Magnetics* 40 (4), 2988–2990.

Ino, K., A. Ito, and H. Honda. 2007. Cell patterning using magnetite nanoparticles and magnetic force. *Biotechnology and Bioengineering* 97 (5), 1309–1317.

Isalan, M., M. I. Santori, C. Gonzalez, and L. Serrano. 2005. Localized transfection on arrays of magnetic beads coated with PCR products. *Nature Methods* 2 (2), 113–118.

Ito, A., Y. Takizawa, H. Honda, K.-I. Hata, H. Kagami, M. Ueda, and T. Kobayashi. 2004. Tissue engineering using magnetite nanoparticles and magnetic force: Heterotypic layers of cocultured hepatocytes and endothelial cells. *Tissue Engineering* 10 (5–6), 833–840. doi:10.1089/1076327041348301.

Ito, A., E. Hibino, C. Kobayashi, H. Terasaki, H. Kagami, M. Ueda, T. Kobayashi, and H. Honda. 2005a. Construction and delivery of tissue-engineered human retinal pigment epithelial cell sheets, using magnetite nanoparticles and magnetic force. *Tissue Engineering* 11 (3–4), 489–496.

Ito, A., K. Ino, M. Hayashida, T. Kobayashi, H. Matsunuma, H. Kagami, M. Ueda, and H. Honda. 2005b. Novel methodology for fabrication of tissue-engineered tubular constructs using magnetite nanoparticles and magnetic force. *Tissue Engineering* 11 (9–10), 1553–1561. doi:10.1089/ten.2005.11.1553.

Itoh, H., A. Takahashi, K. Adachi, H. Noji, R. Yasuda, M. Yoshida, and K. Kinosita. 2004. Mechanically driven ATP synthesis by F1-ATPase. *Nature* 427 (6973), 465–468.

Izrailev, S., S. Stepaniants, M. Balsera, Y. Oono, and K. Schulten. 1997. Molecular dynamics study of unbinding of the avidin-biotin complex. *Biophysical Journal* 72 (4), 1568–1581.

Jackson, J. D. 1975. *Classical Electrodynamics.* 2nd ed. New York, NY: John Wiley.

Jeong, K., L. Petrakis, M. Takayasu, and F. Friedlaender. 1982. High gradient magnetic separation I: The removal of solids from shale oils. *IEEE Transactions on Magnetics* 18 (6), 1692–1694.

Jiang, Y., M. Rabbi, M. Kim, C. H. Ke, W. Lee, R. L. Clark, P. A. Mieczkowski, and P. E. Marszalek. 2009. UVA generates pyrimidine dimers in DNA directly. *Biophysical Journal* 96 (3), 1151–1158. doi:10.1016/j.bpj.2008.10.030.

Kamau, S. W., P. O. Hassa, B. Steitz, A. Petri-Fink, H. Hofmann, M. Hofmann-Amtenbrink, B. von Rechenberg, and M. O. Hottiger. 2006. Enhancement of the efficiency of non-viral gene delivery by application of pulsed magnetic field. *Nucleic Acids Research* 34 (5), p.e40

Kao, K. C. 1961. Some electromechanical effects on dielectrics. *British Journal of Applied Physics* 12 (11), 629–632.

Kelland, D., Y. Hiresaki, F. Friedlaender, and M. Takayasu. 1981. Diamagnetic particle capture and mineral separation. *IEEE Transactions on Magnetics* 17 (6), 2813–2815.

Khalil, K., A. Sagastegui, M. Ahmed, Y. Yang, B. J. Wiley, and B. B. Yellen. 2010. Phase behavior of 2-D colloidal spin lattices. *Nature Communications* 3, 794

Korda, P. T., M. B. Taylor, and D. G. Grier. 2002. Kinetically locked-in colloidal transport in an array of optical tweezers. *Physical Review Letters* 89 (12), 128301.

Kose, A. R., B. Fischer, L. Mao, and H. Koser. 2009. Label-free cellular manipulation and sorting via biocompatible ferrofluids. *PNAS* 106 (51), 21478–21483. doi:10.1073/pnas.0912138106.

Krebs, M. D., Randall M. E., B. B. Yellen, B. Samanta, A. Bajaj, V. M. Rotello, and E. Alsberg. 2009. Formation of ordered cellular structures in suspension via label-free negative magnetophoresis. *Nano Letters* 9 (5), 1812–1817. doi:10.1021/nl803757u.

Kumar, T. A., A. Bardea, Y. Shai, A. Yoffe, and R. Naaman. 2010. Patterning gradient properties from submicrometers to millimeters by magnetolithography. *Nano Letters* 10 (6), 2262–2267.

Lacoste, D., and T. C. Lubensky. 2001. Phase transitions in a ferrofluid at magnetic-field-induced microphase separation. *Physical Review E* 64 (4), 041506.

Ladavac, K., K. Kasza, and D. G. Grier. 2004. Sorting mesoscopic objects with periodic potential landscapes: Optical fractionation. *Physical Review E* 70 (1).pp. 010901

Lai, M. F., C. Y. Chen, C. P. Lee, H. T. Huang, T. R. Ger, and Z. H. Wei. 2010. Cell patterning using microstructured ferromagnetic thin films. *Applied Physics Letters* 96 (18), 183701.

Lee, G. U., D. A. Kidwell, and R. J. Colton. 1994. Sensing discrete streptavidin biotin interactions with atomic force microscopy. *Langmuir* 10 (2), 354–357.

Leuba, S. H., M. A. Karymov, M. Tomschik, R. Ramjit, P. Smith, and J. Zlatanova. 2003. Assembly of single chromatin fibers depends on the tension in the DNA molecule: Magnetic tweezers study. *Proceedings of the National Academy of Sciences of the United States of America* 100 (2), 495–500.

Lionnet, T., S. Joubaud, R. Lavery, D. Bensimon, and V. Croquette. 2006. Wringing out DNA. *Physical Review Letters* 96 (17), 4. doi:17810210.1103/PhysRevLett.96.178102.

Liu, Q., and F. J. Friedlaender. 1994. Fine particle processing by magnetic carrier methods. *Minerals Engineering* 7 (4), 449–463.

Lu, A.-H., E. L. Salabas, and F. Schüth. 2007. Magnetic nanoparticles: Synthesis, protection, functionalization, and application. *Angewandte Chemie International Edition* 46 (8), 1222–1244.

MacKerell, A. D., and G. U. Lee. 1999. Structure, force, and energy of a double-stranded DNA oligonucleotide under tensile loads. *European Biophysics Journal with Biophysics Letters* 28 (5), 415–426.

Marszalek, P. E., H. B. Li, A. F. Oberhauser, and J. M. Fernandez. 2002a. Chair-boat transitions in single polysaccharide molecules observed with force-clamp AFM. *Biophysical Journal* 82 (1), 41A–41A.

Marszalek, P. E., H. B. Li, A. F. Oberhauser, and J. M. Fernandez. 2002b. Chair-boat transitions in single polysaccharide molecules observed with force-ramp AFM. *Proceedings of the National Academy of Sciences of the United States of America* 99 (7), 4278–4283.

McNaughton, B. H., K. A. Kehbein, J. N. Anker, and R. Kopelman. 2006. Sudden breakdown in linear response of a rotationally driven magnetic microparticle and application to physical and chemical microsensing. *The Journal of Physical Chemistry B* 110 (38), 18958–18964. doi:10.1021/jp060139h.

Melcher, J. R. 1981. *Continuum Electromechanics*. Cambridge, MA: MIT Press.

Merkel, R., P. Nassoy, A. Leung, K. Ritchie, and E. Evans. 1999. Energy landscapes of receptor-ligand bonds explored with dynamic force spectroscopy. *Nature* 397 (6714), 50–53.

Mosconi, F., J. F. Allemand, D. Bensimon, and V. Croquette. 2009a. Measurement of the torque on a single stretched and twisted DNA using magnetic tweezers. *Physical Review Letters* 102 (7), 4. doi:07830110.1103/PhysRevLett.102.078301.

Mosconi, F., J. F. Allemand, D. Bensimon, and V. Croquette. 2009b. Measurement of the torque on a single stretched and twisted DNA using magnetic tweezers. *Physical Review Letters* 102 (7), 078301–078304.

Mou, J. X., J. Yang, C. Huang, and Z. F. Shao. 1994. Alcohol induces interdigitated domains in unilamellar phosphatidylcholine bilayers. *Biochemistry* 33 (33), 9981–9985.

Mou, J. X., J. Yang, and Z. F. Shao. 1995. Atomic-force microscopy of cholera-toxin B-oligomers bound to bilayers of biologically relevant lipids. *Journal of Molecular Biology* 248 (3), 507–512.

Moy, V. T., E. L. Florin, and H. E. Gaub. 1994. Adhesive forces between ligand and receptor measured by Afm. *Colloids and Surfaces a-Physicochemical and Engineering Aspects* 93:343–348.

Mykhaylyk, O., Y. S. Antequera, D. Vlaskou, and C. Plank. 2007. Generation of magnetic nonviral gene transfer agents and magnetofection in vitro. *Nature Protocols* 2 (10), 2391–2411.

Mykhaylyk, O., O. Zelphati, J. Rosenecker, and C. Plank. 2008. siRNA delivery by magnetofection. *Current Opinion in Molecular Therapeutics* 10 (5), 493–505.

Nakamura, M., K. Decker, J. Chosy, K. Comella, K. Melnik, L. Moore, L. C. Lasky, M. Zborowski, and J. J. Chalmers. 2001. Separation of a breast cancer cell line from human blood using a quadrupole magnetic flow sorter. *Biotechnology Progress* 17 (6), 1145–1155.

Neuman, K. C., and A. Nagy. 2008. Single-molecule force spectroscopy: Optical tweezers, magnetic tweezers and atomic force microscopy. *Nature Methods* 5 (6), 491–505. doi:10.1038/nmeth.1218.

Oberhauser, A. F., P. E. Marszalek, M. Carrion-Vazquez, and J. M. Fernandez. 1999. Single protein misfolding events captured by atomic force microscopy. *Nature Structural Biology* 6 (11), 1025–1028.

Oberteuffer, J. A. 1974. Magnetic separation: A review of principles, devices, and applications. *IEEE Transactions on Magnetics* 10 (2), 223–238.

Paci, E., A. Caflisch, A. Pluckthun, and M. Karplus. 2001. Forces and energetics of hapten-antibody dissociation: A biased molecular dynamics simulation study. *Journal of Molecular Biology* 314 (3), 589–605.

Panofsky, W. K. H., and M. Phillips. 1955. *Classical Electricity and Magnetism*. New York: Addison Wesley.

Raab, A., W. H. Han, D. Badt, S. J. Smith-Gill, S. M. Lindsay, H. Schindler, and P. Hinterdorfer. 1999. Antibody recognition imaging by force microscopy. *Nature Biotechnology* 17 (9), 902–905.

Ribeck, N., and O. A. Saleh. 2008. Multiplexed single-molecule measurements with magnetic tweezers. *Review of Scientific Instruments* 79 (9), 6. doi:09430110.1063/1.2981687.

Rief, M., M. Gautel, F. Oesterhelt, J. M. Fernandez, and H. E. Gaub. 1997. Reversible unfolding of individual titin immunoglobulin domains by AFM. *Science* 276 (5315), 1109–1112.

Rief, M., H. Clausen-Schaumann, and H. E. Gaub. 1999. Sequence-dependent mechanics of single DNA molecules. *Nature Structural Biology* 6 (4), 346–349.

Rief, M., M. Gautel, and H. E. Gaub. 2000. Unfolding forces of titin and fibronectin domains directly measured by AFM. *Elastic Filaments of the Cell* 481:129–141.

Riener, C. K., C. M. Stroh, A. Ebner, C. Klampfl, A. A. Gall, C. Romanin, Y. L. Lyubchenko, P. Hinterdorfer, and H. J. Gruber. 2003. Simple test system for single molecule recognition force microscopy. *Analytica Chimica Acta* 479 (1), 59–75.

Rinaldi, C., and H. Brenner. 2002. Body versus surface forces in continuum mechanics: Is the Maxwell stress tensor a physically objective Cauchy stress? *Physical Review E* 65 (3), 036615.

Ristic, D., M. Modesti, T. van der Heijden, J. van Noort, C. Dekker, R. Kanaar, and C. Wyman. 2005. Human Rad51 filaments on double- and single-stranded DNA: Correlating regular and irregular forms with recombination function. *Nucleic Acids Research* 33 (10), 3292–3302.

Ritort, F. 2006. Single-molecule experiments in biological physics: Methods and applications. *Journal of Physics: Condensed Matter* 18 (32), R531–R583. doi:10.1088/0953-8984/18/32/r01.

Roichman, Y., V. Wong, and D. G. Grier. 2007. Colloidal transport through optical tweezer arrays. *Physical Review E* 75 (1). pp.011407

Rosensweig, R. E. 1985. *Ferrohydrodynamics*. New York: Cambridge University Press.

Santori, M. I., C. Gonzalez, L. Serrano, and M. Isalan. 2006. Localized transfection with magnetic beads coated with PCR products and other nucleic acids. *Nature Protocols* 1 (2), 526–531.

Setchell, C. H. 1985. Magnetic separations in biotechnology—A review. *Journal of Chemical Technology and Biotechnology. Biotechnology* 35 (3), 175–182.

Shao, Z. F., D. Shi, and A. V. Somlyo. 2000. Cryoatomic force microscopy of filamentous actin. *Biophysical Journal* 78 (2), 950–958.

Smith, S. B., L. Finzi, and C. Bustamante. 1992. Direct mechanical measurements of the elasticity of single DNA-molecules by using magnetic beads. *Science* 258 (5085), 1122–1126.

Steven, B. S., J. R. Mofitt, Y. R. Chemla, and C. Bustamante. 2008. Recent advances in optical tweezers. *Annual Review of Biochemistry* 77 (1), 205–228

Stratton, J. A. 1941. *Electromagnetic Theory*. New York: McGraw-Hill Book Company.

Strick, T. R., J. F. Allemand, D. Bensimon, and V. Croquette. 1998. Behavior of supercoiled DNA. *Biophysical Journal* 74 (4), 2016–2028.

Strick, T. R., V. Croquette, and D. Bensimon. 2000. Single-molecule analysis of DNA uncoiling by a type II topoisomerase. *Nature* 404 (6780), 901–904.

Tahir, M. A., L. Gao, L. N. Virgin, and B. B. Yellen. 2011. Transport of superparamagnetic beads through a two-dimensional potential energy landscape. *Physical Review E* 841 (1), 011403.

Takayasu, M., N. Duske, S. Ash, and F. Friedlaender. 1982. HGMS studies of blood cell behavior in plasma. *IEEE Transactions on Magnetics* 18 (6), 1520–1522.

Tamm, L. K., C. Bohm, J. Yang, Z. F. Shao, J. Hwang, M. Edidin, and E. Betzig. 1996. Nanostructure of supported phospholipid monolayers and bilayers by scanning probe microscopy. *Thin Solid Films* 285:813–816.

Tanase, M., E. J. Felton, D. S. Gray, A. Hultgren, C. S. Chen, and D. H. Reich. 2005. Assembly of multicellular constructs and microarrays of cells using magnetic nanowires. *Lab on a Chip* 5 (6), 598–605.

Thie, M., R. Rospel, W. Dettmann, M. Benoit, M. Ludwig, H. E. Gaub, and H. W. Denker. 1998. Interactions between trophoblast and uterine epithelium: Monitoring of adhesive forces. *Human Reproduction* 13 (11), 3211–3219.

Thomas, T. E., S. J. R. Abraham, A. J. Otter, E. W. Blackmore, and P. M. Lansdorp. 1992. High gradient magnetic separation of cells on the basis of expression levels of cell surface antigens. *Journal of Immunological Methods* 154 (2), 245–252.

Thompson, J., and Michael, T. 2008. Single-molecule magnetic tweezer tests on DNA: Bounds on topoisomerase relaxation. *Proceedings of the Royal Society A: Mathematical, Physical and Engineering Science* 464 (2099), 2811–2829. doi:10.1098/rspa.2008.0132.

Thormann, E., A. C. Simonsen, L. K. Nielsen, and O. G. Mouritsen. 2007. Ligand receptor interactions and membrane structure investigated by AFM and time-resolved fluorescence microscopy. *Journal of Molecular Recognition*20 (6), 554–560.

Tranquillo, R. T., T. S. Girton, B. A. Bromberek, T. G. Triebes, and D. L. Mooradian. 1996. Magnetically orientated tissue equivalent tubes: Application to a circumferentially orientated media-equivalent. *Biomaterials* 17:349.

Wang, M. D., H. Yin, R. Landick, J. Gelles, and S. M. Block. 1997. Stretching DNA with optical tweezers. *Biophysical Journal* 72 (3), 1335–1346.

Watson, J. F. P. 1973. Magnetic filtration. *Journal of Applied Physics* 44:4209–4213.

Watson, J. H. P., and D. C. Ellwood. 1994. Biomagnetic separation and extraction process for heavy metals from solution. *Minerals Engineering* 7 (8), 1017–1028.

Williams, M. C. 2006. Optical tweezers: Measuring piconewton forces. *Biophysics Textbook Online.*

Yan, J., D. Skoko, and J. F. Marko. 2004. Near-field-magnetic-tweezer manipulation of single DNA molecules. *Physical Review E* 70 (1), 5. doi:10.1103/PhysRevE.70.011905.

Yang, J., and Z. F. Shao. 1995. Recent advances in biological atomic-force microscopy. *Micron* 26 (1), 35–49.

Yang, Ye, R. M. Erb, B. J. Wiley, S. Zauscher, and B. B. Yellen. 2011. Imaginary magnetic tweezers for massively parallel surface adhesion spectroscopy. *Nano Letters* 11 (4), 1681–1684.

Yavuz, C. T., J. T. Mayo, W. W. Yu, A. Prakash, J. C. Falkner, S. Yean, L. Cong, H. J. Shipley, A. Kan, M. Tomson, D. Natelson, and V. L. Colvin. 2006. Low-field magnetic separation of monodisperse Fe_3O_4 nanocrystals. *Science* 314 (5801), 964–967. doi:10.1126/science.1131475.

Yavuz, C. T., A. Prakash, J. T. Mayo, and V. L. Colvin. 2009. Magnetic separations: From steel plants to biotechnology. *Chemical Engineering Science* 64 (10), 2510–2521.

Yellen, B. B., and G. Friedman. 2004a. Programmable assembly of colloidal particles using magnetic microwell templates. *Langmuir* 20 (7), 2553–2559.

Yellen, B. B., and L. N. Virgin. 2009. Nonlinear dynamics of superparamagnetic beads in a traveling magnetic field wave. *Physical Review E (Statistical, Nonlinear, and Soft Matter Physics)* 80:011402–011406.

Yellen, B. B., G. Friedman, and K. A. Barbee. 2004. Programmable self-aligning ferrofluid masks for lithographic applications. *IEEE Transactions on Magnetics* 40 (4), 2994–2996.

Yellen, B. B., G. Fridman, and G. Friedman. 2004. Ferrofluid lithography. *Nanotechnology* 15 (10), S562–S565.

Yellen, B. B., O. Hovorka, and G. Friedman. 2005. Arranging matter by magnetic nanoparticle assemblers. *PNAS* 102 (25), 8860–8864.

Yellen, B. B., Z. G. Forbes, D. S. Halverson, G. Fridman, K. A. Barbee, M. Chorny, R. Levy, and G. Friedman. 2005. Targeted drug delivery to magnetic implants for therapeutic applications. *Journal of Magnetism and Magnetic Materials* 293 (1), 647–654.

Yellen, B. B., R. M. Erb, H. S. Son, R. Hewlin Jr, H. Shang, and G. U. Lee. 2007. Traveling wave magnetophoresis for high resolution chip based separations. *Lab on a Chip* 7 (12), 1681–1688.

Yellen, B. R., and G. Friedman. 2004b. Programmable assembly of heterogeneous colloidal particle arrays. *Advanced Materials* 16 (2), 111–115.

Zborowski, M., L. Sun, L. R. Moore, P. S. Williams, and J. J. Chalmers. 1999. Continuous cell separation using novel magnetic quadrupole flow sorter. *Journal of Magnetism and Magnetic Materials* 194 (1–3), 224–230.

Zborowski, M., L. R. Moore, P. S. Williams, and J. J. Chalmers. 2002. Separations based on magnetophoretic mobility. *Separation Science and Technology* 37 (16), 3611–3633.

Ziemann, F., J. Rädler, and E. Sackmann. 1994. Local measurements of viscoelastic moduli of entangled actin networks using an oscillating magnetic bead micro-rheometer. *Biophysical Journal* 66 (6): 2210–2216.

Zlatanova, J., S. M. Lindsay, and S. H. Leuba. 2000. Single molecule force spectroscopy in biology using the atomic force microscope. *Progress in Biophysics & Molecular Biology* 74 (1–2), 37–61.

de Vries, A. H. B., B. E. Krenn, R. van Driel, and J. S. Kanger. 2005. Micro magnetic tweezers for nanomanipulation inside live cells. *Biophysical Journal* 88 (3), 2137–2144. doi:10.1529/biophysj.104.052035.

4 Colloidal Interactions of Magnetic Nanoparticles

O. Thompson Mefford and Steven L. Saville

CONTENTS

4.1 COLLOIDAL PROPERTIES OF AQUEOUS SUSPENSIONS OF MAGNETITE

Colloidal stability is a central factor in determining the effectiveness of nanoparticles in biomedical applications. Nanoparticles that do not remain stable throughout the timeframe of a treatment may have significantly reduced effects. This chapter defines colloids, colloidal stability, and what mechanisms influence their stability.

A colloid is defined as a substance, usually in the sub-micron size regime, that is suspended in a solvent due to the random Brownian motion of the particles (Russel, Saville, and Schowalter 1989). This random motion often causes collisions amongst particles, and the interparticle interactions that occur between particles during a collision event define the overall stability of a colloidal system. If interparticle interactions between colloids allow for close contact of colloid cores, then adhesion may occur. Once adhesion occurs, flocculation, followed by settling of the colloids out of solution, may occur. When designing a colloidal nanoparticle complex for biomedical applications it is important to consider the colloidal stability of these systems and overall colloidal interactions. There are many potential biomedical applications for magnetic nanoparticles, including cell separation (Honda et al. 1998), biomolecular imaging (Mahmoudi, Serpooshan, and Laurent 2011, Jiang et al. 2009), drug delivery vessels (Neoh and Kang 2011, Yallapu et al. 2011, Yang et al. 2011, Sun, Lee, and Zhang 2008, Soppimath et al. 2001), magnetic hyperthermia (Jeun et al. 2012, Guardia et al. 2012, Al-Hayek 2012, Stone et al. 2011, Laurent et al. 2011, Xie et al. 2009, Tasci et al. 2009, Hiergeist et al. 1999, Viroonchatapan et al. 1995, Jordan et al. 1993), magnetic resonance imaging (MRI) contrast enhancement (Jung and Jacobs 1995, Chouly, Pouliquen, Lucet, Jeune, & Jallet, 1996, Cheng et al. 2005, Duan et al. 2008, Matsumoto and Jasanoff 2008, Carroll et al. 2009), magnetic particle imaging (MPI) (Ferguson, Minard, and Krishnan 2009, Ferguson, Khandhar, and Krishnan 2011, 2012, Ferguson et al. 2011), and many others. All of these applications significantly depend on the ability for these nanoparticle systems to have a defined colloidal arrangement over the course of their use. It is therefore crucial to understand the ability of these systems to remain evenly dispersed, specifically identifying the relationship between particle aggregation and interparticle potentials.

There are several mechanisms that cause attraction between magnetic nanoparticles suspended in solution, with the main attractive forces being magnetic and van der Waals interactions. There are three origins of van der Waals attractive forces: Keesom forces, which describe permanent dipole–permanent dipole interactions of colloids (French 2000, Rosenholm 2010); Debije forces, which describe permanent dipole–induced colloidal interactions (Liang et al. 2007); and London forces, which describe the instantaneous–induced dipole colloidal interactions (Xia, Monteiro-Riviere, and Riviere 2010, Gibbs et al. 2011). There are several types of dipoles, but for the purposes of defining colloidal stability, electrical and magnetic dipoles are to be considered. An electric dipole is when a substance has both positive and negative ends, where dipolar interactions lead to the interactions of a positive section with a negative section. Magnetic dipoles are similar with the exception that instead of electric charges, particles contain areas of north and south poles. Keesom and Debije forces are both important only at very short ranges (<2 nm); however, London forces act at a much larger length scale (5–10 nm). This is why describing the London attractive forces of colloids is often given the general name van der Waals attractive forces. Because there are constant long-range attractive forces acting between chemically similar colloids suspended in solution, it is crucial to stabilize these colloids with equal or greater long-range repulsion mechanisms. The most common ways to provide these stabilizing mechanisms are with an electrostatic double layer or with absorbed or covalent bonded molecules that provide steric stabilization.

4.2 EXAMPLES OF STABILIZED NANOPARTICLE SYSTEMS

As discussed in previous sections, the understanding of colloidal stability of magnetic nanoparticles is complicated, with various mechanisms to understand and overcome. Understanding the mechanisms that cause and prevent aggregation of these nanoparticles allows researchers to synthesize magnetic nanoparticle systems unique for their applications, specifically taking advantage of end uses in designing their stability mechanisms. While a broad overview is provided below, if the reader seeks additional information, please refer to these references on the basics of colloidal interactions: Odenbach (2009), Berg (2012), Stokes and Evans (1996), Russel, Saville, and Schowalter (1987), Evans and Wennerström (1999), and Israelachvili (2011).

4.2.1 Electrostatically Stabilized Magnetite Nanoparticles

Several researchers have taken advantage of electrostatic repulsion in the synthesis on magnetite nanoparticles, with the most common synthetic method of producing iron oxide nanoparticles being the Massart or coprecipitation method (Massart 1981, Massart et al. 1995). This technique involves the coprecipitation of iron salts (typically ferrous and ferric chlorides) in a basic solution (Cheng et al. 2005, Kim et al. 2001) yielding electrostatically charged magnetite nanoparticles in the 5–15 nm size range. In addition, this synthesis is extremely quick, as going from raw materials to an iron oxide nanoparticle aqueous suspension usually takes less than an hour. As the entire process is conducted in aqueous media, it is a very forgiving and relatively easy process for those who are new to the field.

However, there are several drawbacks to the production of iron oxide particles via coprecipitation, including uncontrollable nanoparticle core shape and very wide particle size polydispersity. What has been claimed as "monodisperse" in the early literature, is not acceptable by modern standards. Scientists have attempted to improve this method through higher temperature and pressures, but results are less than ideal. In addition, the long-term stability of the resulting particles has issues at a neutral pH. To counteract this effect, most researchers do not use coprecipitated magnetic nanoparticles in their raw form, instead modifying them with several types of surface coatings to improve their stability and usability. For example, Ma et al. (2003) synthesized coprecipitation magnetite nanoparticles and then modified the surface with 3-aminopropyl-triethoxysilane (APTS) to create a monolayer of an amino-silane. This free amine group aids in the conjuncture to proteins, biomolecules, drugs, and other biomedical applications. The modification of these particles with APTS not only increased their stability, but also further increased their usefulness in their end-use application. Other examples of small molecule coatings that rely on electrostatic repulsion are citrate (Liu et al. 2009), sodium oleate (Kim et al. 2001), and humic acid (Illes and Tombacz 2003, 2006, Tombacz et al. 2004). Each of these systems has unique end-use applications that heavily rely on surface charge to achieve stability.

Nonetheless, there are several limitations to electrostatically stabilized nanoparticles, mainly the detrimental effects of solution electrolyte concentrations on nanoparticle stability. That is to say, electrostatic repulsion is an effective method for the stabilization in low-ionic-strength media, such as deionized water, but is typically inadequate in biologically relevant media. For example, Wiogo et al. (2011) performed stability studies on magnetite nanoparticles stabilized with electrostatic repulsion generated from carboxylic acid molecules on the surface. The stability of this suspension was measured in water, phosphate-buffered solution, and cell-culture media. These studies demonstrated the negative effect of ion concentration present in solution on the electrostatic double layer used for nanoparticle stability. Their results indicated that these nanoparticles displayed no visible change in zeta or size over time in deionized water, while they displayed significant decrease in the zeta potential and increase in size in both phosphate-buffered saline and cell-culture media.

4.2.2 Sterically Stabilized Magnetite Nanoparticles

To overcome the limitations of electrostatically stabilized nanoparticles, many researchers have turned to sterically stabilized nanoparticles to compete against adhesion interactions. There are countless types of sterically based stabilizing mechanisms, the bulk of which are centered on polymer modification of nanoparticles. There are two main types of polymer-modified nanoparticles, and they are classified by how they interact with the nanoparticle core. The first type of polymer coating is usually referred to as *physisorbed,* which refers to the loose physical absorption of polymers to the surface of magnetite. This type of surface chemistry gained popularity due to its use of modification and ability to quickly modify magnetite nanoparticles for biocompatibility. The simplest types of magnetite nanoparticle systems employing this surface chemistry are dextran coated coprecipitation particles, which are commonly used as T_2 MRI contrast agents (Jung and Jacobs 1995, Grimm et al. 2000). Resovist and Feridex were the first magnetite nanoparticles to gain approval from the FDA (Kedziorek and Kraitchman 2010), but have since lost popularity due to poor characterization and biological stability, as well as poor performance in the market. There are several other types of physisorbed polymer coatings for magnetite; one of the more promising new areas is in amphiphilic polymers. For the purpose of physisorption to magnetite, most polymers of this type are triblock, containing both hydrophobic and hydrophilic sections. The hydrophobic sections of the polymer adsorb to the hydrophobic surface coating of magnetite (usually ligands used in nanoparticle synthesis), allowing for the hydrophilic sections to interact with the surrounding solution. This type of modification allows for easy transition of hydrophobic magnetite nanoparticles to hydrophilic solvents (Saleh et al. 2005, Gonzales and Krishnan 2007). This type of coating method can also be significantly more complex, involving multiple layers of physisorbed layers. For example, Yallapu et al. (2011) synthesized coprecipitation nanoparticles, followed by multiple surface coatings to allow for a multifunctional nanoparticle for imaging and cancer therapy purposes. This group was successfully able to coat magnetite nanoparticles with β-cyclodextrin, followed by a pluronic poly (ethylene glocol)-poly (propylene glycol) copolymer that allowed for water solubility. This combination of layers allowed for drug loading of curcumin in the inner layer while maintaining water solubility due to the hydrophilic outer layer.

The other main type of polymer coating commonly used to impart colloidal stability to magnetite is chemisorbed polymers. This type of polymer coating employs a strongly binding group to the surface of magnetite, often referred to as an anchor group, and utilizes a polymer "brush" to stabilize the particles. This has gained popularity over other types of surface coatings due to the variability in polymer types, the ability to tune surface coverage, and the ability to employ the free end group for specific targeting (Stone et al. 2011). In the case of magnetite, a popular polymer for the coating of magnetite nanoparticle is polyethylene glycol (PEG), due to its low cost, ease of modification, and biocompatibility (Shen et al. 1993). PEG is known as a "stealth" polymer due to the lack of response of the human body to its presence (Neoh and Kang 2011, Needham et al. 1992). There are several other types of polymer surface coating types, such as polydimethyl siloxane (PDMS) (Wilson et al. 2005, Mefford et al. 2008a, b), polyvinyl alcohol (PVA) (Kumar et al. 2000, Lee, Isobe, and Senna 1996, Mahmoudi et al. 2008, 2009), and poly(*N*-isopropylacrylamide) (PNIPAM) (Guo et al. 2005, Isojima et al. 2008, Purushotham and Ramanujan 2010), along with many others. Copolymers have also been explored heavily, taking advantage of combining the properties of multiple polymer types for their end-use application. This is especially important in stimulus-responsive drug delivery, where block copolymers are used to store drugs until a stimulus is applied, as well as stabilize the magnetite nanoparticle inside biological conditions. For example, several researchers have explored the potential uses of PNIPAM as a drug delivery system, as this poly displays lower critical solution temperature (LCST) properties (Wong et al. 2008, Zhang, Srivastava, and Misra 2007, Zhang and Misra 2007, Chen, Liu, Gong, Huang, & Chen, 2011). Essentially, PNIPAM is soluble in water until it reaches a certain temperature (usually around 35 °C) where it becomes no longer soluble in water. Drugs encapsulated in this polymer layer will then be excreted into the solution, therefore effectively creating a thermo-sensitive drug delivery device. Copolymers of this polymer are often used in order to retain stability in solution after the LCST is achieved. Specifically, Aqil et al. (2008) synthesized a block copolymer consisting of two random copolymers of PNIPAM—polyacrylic acid (PAA-PNIPAM) and poly(acrylate methoxy-poly (ethylene glycol)) (PAM-PEG). PNIPAM was copolymerized with PAA in order to raise its LCST to 38 °C, while PAM was copolymerized with PEG to improve its water solubility while retaining its stealth ability. Once this block copolymer was attached to the surface of magnetite via the carboxylic acid groups present in PAA, magnetic hyperthermia is then used to raise the temperature of the system past 38 °C, the LCST, essentially collapsing the inner polymer layer onto the surface of magnetite while the PAM-PEG polymer layer remains extended in solution. If drugs were encapsulated in this inner polymer layer, using magnetic hyperthermia to stimulate the collapsing of the inner polymer layer would trigger the release of the drug into the surrounding solution, essentially creating a triggered drug release system.

4.2.3 Stabilizing Magnetic Nanoparticles Through the Use of Liposomes

Another popular method to stabilize magnetic nanoparticles is through the use of liposomes. A liposome is a synthesized structure consisting of a lipid bilayer of amphophilic molecules. The most common type of liposome used to encapsulate magnetite consists of a hydrophilic head and a hydrophobic tail (Figure 4.1). A liposome encapsulates a region of aqueous solution or nanoparticle inside a hydrophobic outer layer. Using this mechanism, hydrophilic molecules cannot readily pass through the lipids. However, hydrophobic molecules can be trapped inside the lipid bilayer. Using this method, magnetic nanoparticles can be sterically stabilized with the drug delivery mechanism, the lipid bilayer. There are also several types of drug delivery mechanisms using liposome-encapsulated magnetite, including controlled pH delivery and magnetic hyperthermia. For pH delivery, liposomes can be designed to have that low or high pH to electrostatically encapsulate drugs that are charged in solution. As protons naturally diffuse through the outer barrier neutralizing the liposome, the drug will also be neutralized, allowing it to freely pass into surrounding solution (Bogdanov, Martin, Weissleder, and Brady, 1994, Couvreur and Vauthier 2006). For magnetic hyperthermia stimulated delivery, the heat generated from the magnetic nanoparticle stimulates the release of the encapsulated drug within the hydrophobic bilayer (Viroonchatapan et al. 1995, Kawai et al. 2005, Kong and Dewhirst 1999). For example, Zhu et al. (2009) synthesized thermo-responsive magnetic liposomes encapsulated with methotrexate (common chemotherapy drug). The liposomes used in this study were a temperature-sensitive 1,2-dipalmitoyl-sn-glycero-3-phosphocholine and cholesterol. Using magnetic hyperthermia to trigger drug release, more than 80% of methotrexate was released from the liposome within 30 minutes of a temperature increase from 37 °C to 41 °C, thus showing its potential as a stimulus-responsive drug delivery agent.

4.2.4 Polyelectrolyte Stabilized Magnetite Nanoparticles

Another extremely effective method for stabilizing magnetite nanoparticles combines the electrostatic and steric stabilization mechanism into electrosteric stabilization. This method usually involves coating the surface of magnetite with a polyelectrolyte, or a polymer containing multiple charged groups (Figure 4.2).

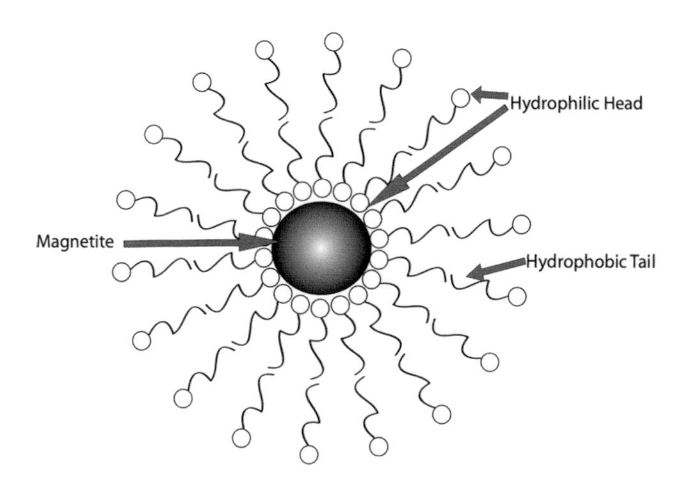

FIGURE 4.1 Typical encapsulation of a magnetite nanoparticle with a liposome.

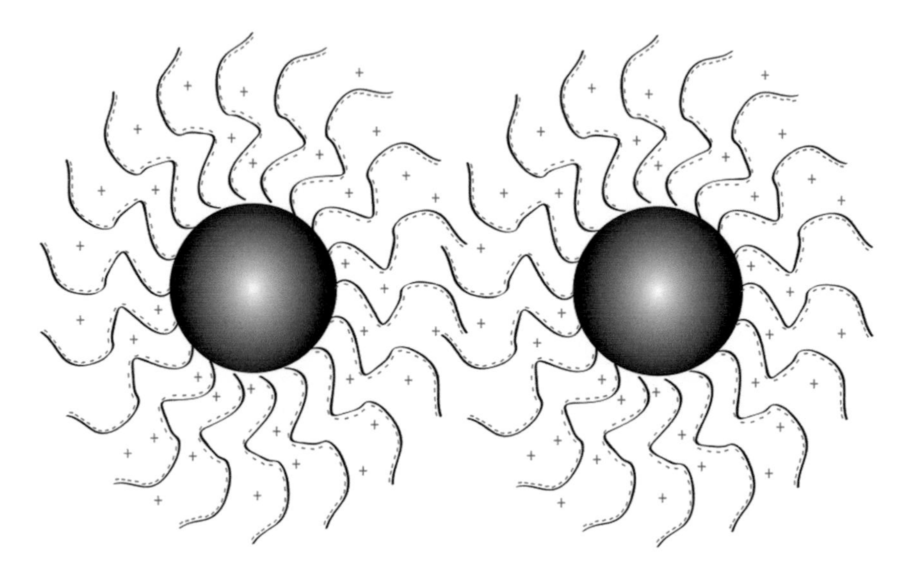

FIGURE 4.2 Typical stabilization of magnetite using a polyelectrolyte polymer brush that incorporates both electrostatic and steric repulsion.

One of the more common polymer types used for this type of nanoparticle stabilization is polyethyelimine (PEI) (Hajdu et al. 2009, Wang et al. 2009). These nanoparticle systems have displayed both high electrostatic and steric repulsion, essentially negating most detrimental aspects of each type of coating (Hajdu et al. 2009). Furthermore, the nature of these charged polymers may have increased importance in biological applications, where recent research has shown that polyelectrolyte coatings may increase the proton relaxivity for MRI contrast agents (Duan et al. 2008) and the cell uptake mechanism (Petri-Fink et al. 2008). Finally, another possible advantage of using polyelectrolyte surface coatings is the potential use of electrostatic interactions between the polymer and deliverable agents. For example, Steitz et al. (2007) explored the interactions of DNA and PEI-coated magnetite nanoparticles. They were successfully able to prove that electrostatic interaction between PEI and DNA was sufficient to remain complexed under biological conditions. They also were able to display significantly improved gene transfection of DNA into cells under the influence of an external magnetic field.

4.3 MECHANISMS OF COLLOIDAL INSTABILITY

There are several factors that influence the point at which dispersed particles tend to aggregate, known as the critical flocculation point (CFP), which refers to a set of conditions where the affinity of the stabilized particles to the dispersion medium decreases to a point where stabilizing mechanisms are no longer sufficient to overcome attractive forces. It is important to understand the CFP for a dispersed nanoparticle system if long-term stability is required for the end-use applications. The main factors that influence the CFP are nanoparticle concentration, nanoparticle size, dissolved ions in solution, polymer molecular weight, polymer surface coverage, and solvent affinity for stabilizing polymer.

4.3.1 Nanoparticle Concentration

An important factor in determining the interparticle interactions in a colloidal suspension is the concentration of nanoparticles. The greater the concentration of the solution, the greater the chance of collision between two particles. In some cases, a weakly stabilized system may be observed stable in dilute concentrations and may experience heavy aggregation at higher concentrations. In general terms, the more dilute a particle system, the more stable. This is more evident when considering that concentration is the inverse of the average particle–particle separation in a solution. Assuming spherical particles and uniform viscosity of the suspension media, one can find that the aggregation time is relative to the concentration to the 3/5 power (Valberg and Butler 1987). More over, the formation of linear aggregates in DC magnetic fields was found to follow similar dilution effects (Saville et al. 2013).

4.3.2 Nanoparticle Size

The size of the nanoparticle core also plays a very important role in the instability mechanisms present in a colloidal system. As particle size increases, the number of atoms increases in each nanoparticle, increasing the van der Waals attractive forces. Also, in the case of magnetic nanoparticles, the core size has a dramatic effect of the saturation magnetization and therefore the magnetic attraction between particles inside a field (Goya et al. 2003, Goya et al. 2008, Liu and Zhang 2001). This effect becomes extremely important when magnetite core size is above 20–25 nm, where the generally accepted size limitation occurs for superparamagnetic nanoparticles (Pankhurst et al. 2003). Because large core sizes of magnetite are essential for many applications, special consideration is needed for these systems to efficiently stabilize them over long periods of time.

4.3.3 Dissolved Ions in Solution

Dissolved ions in a colloidal suspension of magnetic nanoparticles play a key role in stability for both electrostatically and sterically stabilized systems. For electrostatically stabilized systems, as discussed above, dissolved ions in solution can screen surface charges, decreasing the thickness of the electrical double layer. This decrease will effectively limit the electrostatic repulsion of the system, causing nanoparticles to aggregate. For magnetite nanoparticles stabilized with steric repulsion, the effect of dissolved ions in solution is less direct. It is important to understand the steric polymer binding chemistry, as dissolved ions in solution may have a greater affinity for the surface of magnetite than the stabilizing polymer anchor group. In this situation, the polymer surface coating may be displaced, decreasing the effectiveness of the polymer at steric stabilization. This is especially the case in biological media, where phosphates and other biological salts can readily display several polymer anchor types (Goff et al. 2009, Saville et al. 2012b, Fang, Bhattarai, Sun, and Zhang, 2009).

4.3.4 Polymer Surface Coverage

The brush density is controlled by several factors, including the amount of ligands in solution, the polymer molecular weight, and the anchor group size (Butterworth, Illum, and Davis, 2001, Milner 1991). Any deficiencies in the polymer brush density, which can be caused by improper surface modification or ligand displacement, will result in an increase in the ease of nanoparticle core collision and therefore increase the likelihood of aggregation. Several researchers have also discovered an effect commonly referred to as *lateral migration* of surface ligands when high enough brush densities are not achieved. For example, Ionita et al. (2007) explored the possibility of ligands laterally diffusing from one area of a nanoparticle surface to another. Their results indicate that under the right conditions, bare patches on nanoparticle surfaces (areas of little polymer coating) that experience higher van der Waals attractive forces may agglomerate with nearby ligands laterally diffusing around these combination points to accommodate nanoparticle core collision. It is therefore important to achieve maximum polymer surface coverage in most cases to prevent any permanent clustering.

4.4 CHARACTERIZATION OF COLLOIDAL STABILITY

There are several different ways to characterize the stability of a nanoparticle system in suspension. These techniques range from monitoring the size of the nanoparticle in solution, measuring the surface charge, and measuring the polymer surface coverage.

4.4.1 Thermogravimetric Analysis

Thermogravimetric analysis (TGA) is a simple technique that measures weight loss as a function of temperature. Commonly employed to determine the degradation temperature of small molecules and polymers, it is also commonly used to measure ratios of inorganic or organic components of a system. In nanoparticle characterization, it is commonly used to determine the polymer-to-nanoparticle ratio. Using this technique, the amount of polymer per square nanometer of nanoparticle surface can be calculated. This value is known as the brush density and is an extremely important value when determining nanoparticle stability. As discussed in Section 1.1.2.4, surface coverage of the nanoparticle greatly determines the extent at which nanoparticle cores may collide, with any deficiencies in surface coverage significantly increasing the instability of the system. Generally, the higher the surface coverage of the system, the more stable the system. There is a limit to the value, however, determined by the steric hindrance of the system, which is a function of diffusion of the polymer to the nanoparticle surface, bulkiness of the polymer head group, and size of the nanoparticle. While this method is not a direct way to measure nanoparticle stability, it can successfully measure surface coverage changes in a nanoparticle system.

A crucial aspect of nanoparticle stability provided steric repulsion is retention of the surface coating in biological media, and TGA can measure loss in ligand surface coverage as a function of media concentration. However as this is a bulk measurement, it mainly gives the mass ratio of organic to inorganic material. Because of this, it is difficult to account for unbound ligand still in solution and residual sacrificial ligand remaining from ligand exchange. This issue was highlighted by measurements of ligand exchange via radio nucleotide techniques by Davis, Cole, Ghelardini, Powell, and Mefford, (2016; Davis et al. 2014). Nonetheless, this gives researchers an understanding of not only the stability of a ligand anchor group to displacement, but can also be conferred to overall nanoparticle stability.

4.4.2 Zeta Potential

The zeta potential, as described in Section 1.1.5.2, is the value of the electric potential at the slipping plane. The further from a value of zero the zeta potential, the more stable the particle. Particle systems with a zeta potential above +30 mV and below –30 mV are generally considered electrostatically stable. One of the greatest contributors to zeta potential is the pH of the solution, which determines the concentration and charge of free ions in solution. Therefore, it is crucial to understand both the zeta potential and pH of the solution, and both are usually reported together. The most common way to represent the electrostatic stability of a nanoparticle suspension is to run an isoelectric point analysis, where the pH of a system is varied (usually pH 2–12) while recording the zeta potential at each pH. At a certain pH, the zeta potential will reach 0 mV, generally considered the point where a colloidal system is least stable. Understanding the isoelectric point of a colloidal system can lead to a greater understanding of the overall stability of the system, and pH ranges that achieve stability.

Electrically charged particles exhibit electrokinetic effects, and can thus be influenced by the presence of an applied electric field (Berg 2010). The most important electrokinetic effect when determining the zeta potential of a system is known as *electrophoresis,* the movement of a charged particle through a suspension under the influence of an applied electric field. This effect basically means that negatively charged particles will move toward the cathode and positively charged particles will move toward the anode. Zeta potential of the system dramatically affects the electrophoresis of a nanoparticle suspension by determining the velocity at which the particles move through solution. Using this velocity, or electrophoretic mobility, the zeta potential of a nanoparticle system can be calculated using the Henry equation, given by:

$$U_E = \frac{2\varepsilon Z f(k)}{3\eta} \tag{4.1}$$

where U_E is the electrophoretic mobility, ε is the dielectric constant of the medium, Z is the zeta potential, η is the viscosity of the medium, and $f(k)$ is defined as the Henry function. The Henry function is usually defined as either 1 or 1.5, depending on the type of system. For particle systems larger than 200 nm in moderately concentrated electrolyte solutions, the Henry function used is 1.5 and is commonly referred to as the Smoluchowski approximation. For small particles in low dielectric constant media or non-aqueous media, the Henry function used is 1, usually referred to the Huckel approximation. Zeta potential therefore is not typically directly measured, rather calculated by measuring the electrophoretic mobility.

4.4.3 Scattering Techniques

The next three sections outline basic scattering experiments to probe the structure of suspensions of particles utilizing light, x-rays, and neutrons. Each scattering technique has its own advantages and disadvantages. Light scattering works best for dilute suspensions with good optical transmission. X-ray scattering can scatter in optically opaque samples, but good contrast between the particles and the media is required for good characterization. For example, polymer brushes on particles will not provide sufficient contrast compared to the suspension media, thus x-ray scattering is most appropriate for characterization of the inorganic core of these materials. Finally, neutron scattering is more versatile in contrast ability, as organic materials and the media can be produced with deuterium rather than hydrogen to alter the contrast ratio to probe various parts of the colloidal suspension including information on the organic stabilizing shell (Herrera et al. 2010).

The principles behind each these scattering techniques are essentially the same with the understanding that light and x-rays follow electromagnetic mechanics, where neutrons have a mass and wavelength following the de Broglie relationship (Chu and Liu 2000). The sections are organized in order of approachability for the average researcher, where light scattering is commonly found in most nanomaterial laboratories and x-ray scattering is found on most university campuses; specialized high-intensity sources such as synchrotrons and neutron-scattering require specialized facilities with a neutron source. With light waves in the range of hundreds of nanometers, x-rays typically are 1–2 Å, and neutrons around 5 Å (Chu and Liu 2000). One additional feature of small-angle neutron scattering (SANS) is the ability to contrast match out features using deuterated solvents. Depending on the scattering angle used, these three techniques can produce overlapping scattering information providing detail on the structure of the particles and the suspension.

4.4.3.1 Dynamic Light Scattering

One of the most common ways to characterize nanoparticle size and stability is using dynamic light scattering (DLS), also known as photon correlation spectroscopy (PCS) or quasi-elastic light scattering (QELS). While there are several techniques that DLS uses to measure nanoparticle hydrodynamic size, the most common mode is measuring the fluctuations in scattered light intensity due to particles moving about solution due to random Brownian motion. When a laser is passed through a colloidal solution, the intensity of the observed light is the vectorial sum of the light scattered from each suspended colloid in solution (Berg 2010). The intensity of this light varies, based upon whether the scattered light from the particles is constructive or destructive, and therefore depends on the position of each colloid in solution. As the colloids undergo random Brownian motion, the positions as well as the intensity of scattered light are constantly changing. Therefore, the change in light intensity is related to the rate of diffusion by each colloid in solution, which is governed by several factors including the size of the colloid and the viscosity of the medium. The positions of these particles are then correlated with respect to time to give an intensity correlation function, given by the equation:

$$G(\tau) = lim_{T \to \infty} I(t) I(t + \tau) dt \tag{4.2}$$

where T is the integration time of the experiment, and τ is the time shift between correlation events. The resulting correlation function decays over time as the colloids move throughout solution. The correlation function can then be related to the diffusion rate of a particle by the equation:

$$G(\tau) = A_0 + Ae^{-\Gamma\tau} \tag{4.3}$$

where A_0 is the background signal level, A is an instrument constant, τ is the time shift or the time between correlation events, and Γ is the decay constant in the correlation function, given by:

$$\Gamma = Q^2 D \tag{4.4}$$

where Q is the magnitude of the scattering wave vector and D is the diffusion coefficient of a particle. Once the diffusion constant for the system has been calculated, the particle hydrodynamic radius can be calculated using the Stokes–Einstein equation, given by:

$$r = \frac{kT}{6\pi\mu D} \tag{4.5}$$

where μ is the viscosity of the medium. It is important to note here that this equation is for calculating the hydrodynamic radius for monodisperse particles, and the calculation of polydisperse samples becomes significantly more complicated.

While the main use for DLS is characterization of nanoparticle hydrodynamic size, it is also very useful in determining nanoparticle stability. Because clustering of nanoparticles results in an increase in the overall hydrodynamic size, comparing results from theoretical calculations presented in Section 4.4.6.2 to experimental results can give an indication of clustering. For example, particles measured to be significantly larger than theoretically calculated may be severely clustered. Furthermore, DLS experiments can be repeated over long periods of time without sample disruption or changes. This can give an indication of the time frame in which nanoparticle instability occurs. For example, it is possible to observe hydrodynamic size as a function of time, with significant increases in nanoparticle size indicating active clustering of nanoparticle systems. This is also a useful technique when determining the stability of nanoparticle systems in biological media, where the size of nanoparticles can be monitored over long periods of time in synthetic biological conditions. For instance, the size of a nanoparticle complex can be measured in phosphate buffered saline (PBS) or fetal bovine serum (FBS) over 24 hours by simply taking measurements over the time frame (Saville et al. 2012a). A steady increase in the hydrodynamic size may indicate instability in the media as clustering occurs. It is important to note that though DLS is a common and preferred technique, it is also flawed. Due to the way it calculates size, DLS results are extremely sensitive to a small number of aggregates. Therefore, it is important to review all aspects of DLS data, including the correlation function, to determine if results are significant.

4.4.3.2 Small-Angle X-Ray Scattering

Small-angle x-ray scattering (SAXS) has been used to measure multiple aspects of magnetic nanoparticles including particle size (Nappini et al. 2010) and structure of suspension of materials (Santiago-Quinones, Acevedo, and Rinaldi 2009), including chaining (Butter, Bomans, Frederik, Vroege, & Philipse, 2003) and clusters of particles (Balasubramaniam et al. 2014). For example, the sizes of uncoated and citrate coated cobalt ferrite particles were characterized as DLS, TEM, and SAXS. As both TEM and SAXS are heavily influenced by heavier elements and not organics, the TEM and SAXS size measurements were very similar for the coated and non-coated samples. However, the measured hydrodynamic radius from the DLS showed differences in coated and uncoated samples (Nappini et al. 2010). Liquid crystals of particles have also been measured with SAXS. Santiago-Quiñones et al. (2009) demonstrated the influence of the weight fraction of cobalt ferrite particles in *m*-cresol. The structure of nanocomposites of magnetic particles has also been measured to determine the distribution of the particles within a sample (Bonini, Lenz, Falletta, & Ridi, 2008; Bonini, Lenz, Giorgi, & Baglioni, 2007). Furthermore the structure of assemblies of these materials can also be measured via SAXS (Farrell et al. 2006, Klokkenburg et al. 2007, Holm and Weis 2005). Magnetic nanoparticles that have been assembled into small clusters for use as T_2 MRI contrast agents were characterized via SAXS to determine cluster size (Balasubramaniam et al. 2014) and then calculate theoretical calculations based on SAXS-T_2 relationships established by Carroll et al. (2010).

4.4.3.3 Small-Angle Neutron Scattering

SANS is a rather new technique used to measure nanoparticle size and colloidal stability. As discussed above, with the correct scattering contrast, information can be gained regarding the structure or the organic stabilizing layer on the inorganic particle. For example Amstad et al. (2011) demonstrated the fitting of SANS data to determine the thickness of the stabilizing polymer on particles as well as interacting liposomes. Dennis et al. (2009) demonstrated the thickness of a dextran shell on magnetite nanoparticles via SANS. Changes in the solubility of the polymer brush due to thermal changes were observed for cobalt ferrite particles coated in poly(*N*-isopropylacrylamide) (pNIPAM) (Herrera et al. 2010).

SANS is also effective in measuring the structure of the magnetic particles themselves. Sara Majetich's group has led many of these efforts in showing disordered as well as face-centered cubic nanoparticle crystals via neutron scattering experiments (Farrell et al. 2006). Also they have utilized polarized SANS to reveal spin canting on the surface of 9 nm magnetite nanoparticles in a 1.2 T field. The group was able to demonstrate that the thickness of this canted shell was dependent on temperature and applied field (Krycka et al. 2010).

4.5 BACKGROUND ON INTERPARTICLE POTENTIAL THEORY USING DLVO THEORY

Understanding particle stability has a long history that many researchers have tried to fully understand. Debye and Huckel reported in 1923 the first theory for a distribution of charges in an ionic solution, which Levine and Dube extended to colloidal dispersions in 1940. This theory posited that particles dispersed in a medium should experience large short-range repulsive forces and small long-range attractive forces. This theory had several limitations, and it was not until Derjaguin, Landau, Verway, and Overbeek (DLVO) in 1941 and 1948 introduced their theory that interparticle potentials were more fully understood. These researchers modeled the stability of colloids as the sum of all attractive and repulsive forces, which they considered to be van der Waals attractive forces (V_A) and electrostatic repulsive forces (V_E). The total interparticle force is then calculated by the equation:

$$V_T = V_A + V_E \tag{4.6}$$

4.5.1 Van der Waals Attractive Forces

Van der Waals forces occur between all molecules and particles. These forces are usually attractive for similar materials and can be repulsive for unlike materials. These attractive forces in the case of colloidal dispersions, depending on their strength, can lead to combination or aggregation. Calculating the van der Waals forces between two like objects has been well-established, but can become more complicated in the case of nanoparticles due to the size of the nanoparticle and their separation length scales being very similar (Min et al. 2008). Most often, the total van der Waals attractive force between nanoparticles is calculated as the potential energy sum over all interacting nanoparticle pairs. The size of the nanoparticle, the surface-to-surface distance of the nanoparticle cores, and the Hamaker constant govern the overall van der Waals potential energy between dispersed spherical nanoparticles. The potential energy of the system from the van der Waals forces can then be calculated by:

$$V_{VdW} = -\frac{1}{6}A_{eff}\left(\frac{2a^2}{r^2 - 4a^2} + \frac{2a^2}{r^2} + ln\left(\frac{r^2 - 4a^2}{r^2}\right)\right) \tag{4.7}$$

where A_{eff} is the effective Hamaker constant, a is the core particle radius, and r is the center-to-center distance of two particle complexes. The Hamaker constant is a material-dependent interaction parameter that gives an indication of the overall attractiveness of two approaching bodies. The effective Hamaker constant accounts for any retardation effects, and can be calculated by the equation (Mefford et al. 2008b):

$$A_{eff} = \frac{3}{4}k_B T\left(\frac{\varepsilon_1 - \varepsilon_2}{\varepsilon_1 + \varepsilon_2}\right)^2 + \frac{3h\omega}{16\sqrt{2}}\left(\frac{(n_1{}^2 - n_2{}^2)^2}{(n_1{}^2 + n_2{}^2)^{3/2}}\right)F(\delta). \tag{4.8}$$

where ε_1 and ε_2 are the dielectric constants of the medium and the substrate, n_1 and n_2 are the refractive indexes of the medium and substrate, h is Planck's constant, k_B is Boltzmann's constant, and ω is the absorption frequency. $F(\delta)$ is a function that accounts for the dispersion energy decaying faster at large separation distances (>10 nm) due to retardation effects, and is calculated by the equation:

$$F(\delta) \approx \left(1 + \left(\frac{\pi\delta}{4\sqrt{2}}\right)^{3/2}\right)^{-2/3} \tag{4.9}$$

where δ is a dimensionless distance related to the nanoparticle core surface-to-surface distance Δ_{ss}, given by the equation:

$$\delta = n_2(n_1^2 + n_2^2)^{1/2}\frac{\Delta_{ss}\,\omega}{c} \tag{4.10}$$

where c is the speed of light (3×10^8 m/s).

4.5.2 Electrostatic Repulsive Force

A main consideration of DLVO theory is nanoparticle repulsive force due to electrostatic interactions. Electrostatic repulsion arises from nanoparticles having a surface charge due to the presence of ionic species either through dissociation of ionic surface groups or the adsorption of an ionic additive such as a surfactant. When a nanoparticle contains sufficient surface charge, counterions present in the medium are therefore attracted to the nanoparticle surface. The layer of counterions closely surrounding the nanoparticle and attached through electrostatic forces is known as the Stern layer. Following this immediate nanoparticle surface and counterion layer is the diffuse layer, containing both positive and negatively charged species with an abundance of counterions. These layers form the electric double layer, and reach a maximum value at the nanoparticle surface and decrease as a function of increasing distance away from the nanoparticle surface (Figure 4.3). When a nanoparticle moves through a medium, a layer of liquid remains attached to the particle, and its boundary with the medium is known as the slipping plane. The value of the electric potential at this slipping plane is known as the zeta potential, and is commonly measured to give an indication of both the surface charge of a nanoparticle system and its stability based upon electrostatic repulsion. The electrostatic potential can therefore be calculated by the equation:

$$V_e = \frac{2\pi R_c \varepsilon_2 \varepsilon_0 \psi_0^2 \ln(1 + e^{-k\Delta_{ss}})}{k_B T} \tag{4.11}$$

where ψ_0 is the surface potential, ε_0 is the permittivity of free space, and $1/k$ is the Debye length. It can be seen from this equation that the electrostatic repulsion of a nanoparticle is extremely dependent on the surface potential of the nanoparticle and the Debye length. This results in high electrostatic repulsion for systems containing nanoparticles with high surface potentials with polar coatings (i.e. citrate-coated gold) and negligible electrostatic repulsion in systems containing nanoparticles with low surface potential and nonpolar coatings (i.e. PEG-coated magnetite).

4.5.3 Extended DLVO Theory

In its original form, DLVO theory accounted for only repulsive forces due to any electrostatic repulsive potentials, but has since been modified to account for steric repulsion interactions as well as magnetic interactions (Napper 1970, 1977, Napper and Netschey 1971). Using each of these components, the total potential energy of a system is given by the equation:

$$V_{Total} = V_{vdW} + V_M + V_{ES} + V_{OSM} \tag{4.12}$$

where V_{vdW} is the van der Waals attractive forces, V_M is the magnetic attractive forces, V_{ES} is the electrostatic repulsive forces, and V_{OSM} is the osmotic repulsive forces.

4.5.4 Magnetic Attractive Force

In magnetic nanoparticles, the magnetic attractive force is an important parameter to understand. The magnetic interactions of this system can be accounted for by the equation:

$$V_M = -\frac{1}{k_B T}\frac{8\pi\mu_0 R_c^6 M_s^2}{9(\Delta_{cc} + 2R_c)^3} \tag{4.13}$$

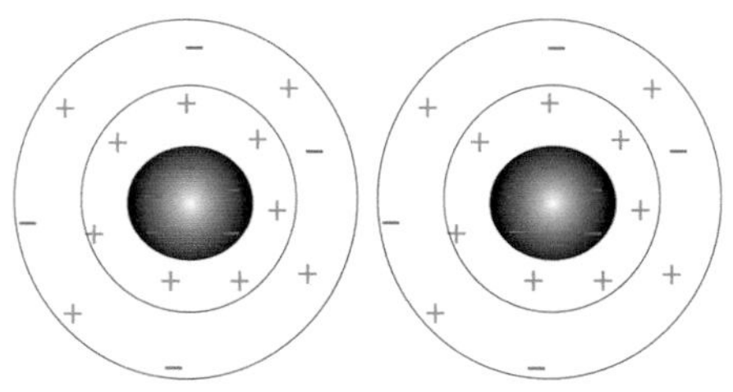

FIGURE 4.3 Electrical double layer stabilizing mechanism for electrostatically stabilized magnetite nanoparticles.

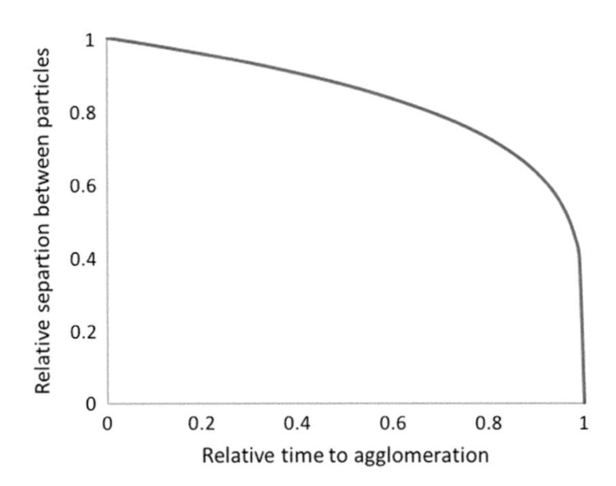

FIGURE 4.4 Interparticle separation relative to the time to agglomeration, adapted from (Valberg and Butler 1987).

where μ_0 is the permeability of free space ($1.26*10^{-6}$ m*kg*s^{-2}*A^{-2}), M_s is the saturation magnetization, and Δ_{cc} is the center-to-center separation distance. The equation is based on the assumption that a saturating magnetic field is present, the nanoparticles possess a single-domain crystalline structure, and particles are in dipole alignment. This equation effectively calculates the maximum magnetic attractive force between two magnetic nanoparticles. It can also be seen from this equation that the magnetic attraction between two particles significantly increases with particle core size and saturation magnetization. This attractive potential therefore becomes extremely important to understand with larger particle sizes (>15 nm), as the magnetic attractive force can easily overcome small-molecule steric repulsion or short-range electrostatic repulsion.

As seen in Figure 4.4, unstabilized particles will be drawn toward each other due to dipole interactions. The time required for aggregation to occur is dependent upon the interparticle spacing (the relative concentration of the suspension), viscosity of the suspending media, diameter of the particle, viscosity of the suspending media, and the magnitude of the remnant magnetic moment of the particle.

4.5.5 Steric Repulsion

Another common way to stabilize nanoparticles, especially those in consideration for biomedical applications, is steric repulsion provided by surfactants or polymers. In the case of biomedical applications, ion concentrations may be high enough to essentially screen electrostatic charges, limiting their effectiveness at stabilizing nanoparticles. This, combined with other biological effects such as protein adsorption, limits the effectiveness of electrostatically stabilized nanoparticles for biological applications (Fang et al. 2009, Petri-Fink et al. 2008). To negate this effect, nanoparticles are often coated with surfactants, which use small molecules or polymeric chains to stabilize nanoparticle systems (Figure 4.5). Surfactants limit the approach of nanoparticle cores in two ways, providing both osmotic repulsion and volume restriction. In the case of osmotic repulsion, when surfactant molecules overlap as in the case of colliding nanoparticle cores, an unfavorable mixing of the surfactant molecules occurs, increasing the osmotic pressure of the system. This increase in the osmotic pressure at the surfactant interface increases the diffusion of solvent to the area of overlap separating the nanoparticles and therefore limiting further interdiffusion of surfactant molecules. In the case of volume restriction, when surfactant molecules collide, there is a loss of configurational entropy due to the reduction in volume for free reptation of the surfactant molecules. This furthermore restricts the collision of nanoparticle cores and improves the overall stability of the nanoparticle system.

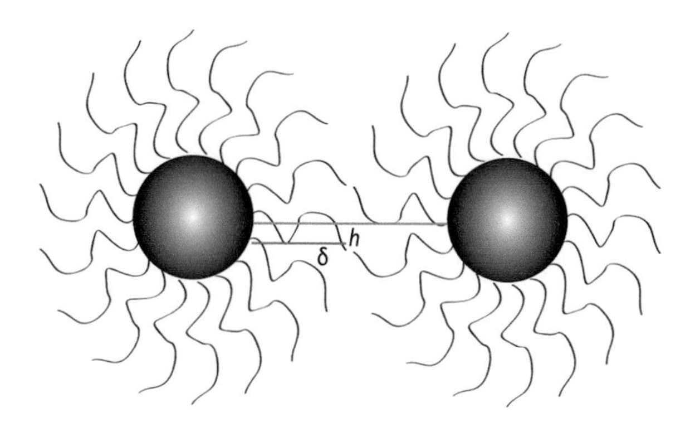

FIGURE 4.5 Typical structure for sterically stabilized magnetite nanoparticles, where δ is the length of the polymer chain and h is the surface-to-surface separation distance.

The composition of the surfactant used is also important to understand, and depending on the application, both small-molecule and polymer-based surfactants are employed. The main factors of steric stabilization are the surfactant's average length (usually governed by number of backbone units or molecular weight), number of surfactant molecules per unit area (packing density), surfactant–solvent interaction parameter (Flory–Huggins parameter), and mode of surfactant–nanoparticle surface attachment. The thickness of the surfactant layer plays a key role in the steric stabilization of nanoparticles, as longer surfactants provide larger steric repulsion due to the enhancement of solvated volume around the nanoparticle. This not only increases the distance a nanoparticle core needs to travel to collide with another core, but also increases the osmotic repulsion between the two nanoparticle cores. The surfactant packing density also is important to consider, as this parameter also determines the volume available for free reptation. Using these parameters, the particle–particle osmotic potential of repulsion can be calculated by the equation:

$$V_{OSM} = \frac{4\pi_* k_B T}{3V}\varphi^2\left(\frac{1}{2} - \chi\right)\left(\delta - \frac{h}{2}\right)\left(3R + 2\delta + \frac{h}{2}\right);\quad \delta < h < 2\delta \tag{4.14}$$

where V is the molecular volume of water, χ is the Flory–Huggins parameter, and φ is the volume fraction of the polymer coating calculated by the equation:

$$\varphi = \frac{6w_s}{\pi\rho_s N_p (D_H^3 - D_C^3)} \tag{4.15}$$

where w_s is the mass of the polymer shell, ρ_s is the density of the polymer, N_p is the number of nanoparticles in the system, D_H is the hydrodynamic diameter of a nanoparticle, and D_c is the core size of the nanoparticle. D_H is usually measured using techniques such as DLS and D_C is usually measured using transmission electron microscopy (TEM).

One final parameter is important to consider, and that is the elastic repulsion generated by interpenetrating polymer brushes at very small interparticle separation distances ($h < \delta$). This effect is due to the interpenetrating polymer brushes at such close distances compressing each other, preventing extension into the surrounding solution. This loss of configurational entropy can be described by the equations:

$$V_{ELAS} = \frac{2\pi k_B T \varphi^2}{V}\left(\frac{1}{2} - \chi\right)\left(\frac{2V_a^2}{V_c} - V_a\right) 0 < h < \delta \tag{4.16}$$

$$V_a = \frac{\pi\delta(-24R^2h + 36\delta R^2 - 12Rh^2 + 28\delta^2 R - 6h^2\delta + 3\delta^3 + 8h\delta^2)}{12(2R + h)} \tag{4.17}$$

$$V_c = \frac{\pi(24R^2\delta^2 - 12R^2h^2 + 24\delta^3 R - 8Rh^3 + 6\delta^4 - h^4)}{12(2R + h)} \tag{4.18}$$

where V_a is the volume of overlap in the osmotic region ($h > \delta$) and V_c is the volume of overlap in the elastic region ($h < \delta$). While this effect is not negligible, in most cases of polymer-coated nanoparticles the osmotic repulsion is significant enough to prevent distances close enough for this effect to occur.

4.5.6 Calculating Colloidal Stability Using Star-Like Polymers

Nanoparticle colloids with polymer brushes may also be considered as star-like polymers. This method of calculating the steric repulsion provided from a polymeric coating has advantages over the osmotic approach due to allowance of further calculation of the size of the complexes. First developed by Likos (2006), this method for calculating the steric repulsion provided by a polymer coating by considering star-like polymers was based on the star-like polymer model developed by Daoud and Cotton (1982). This model effectively calculates particle pair interactions, with the steric repulsion as a function of inter-particle distance given by the equation:

$$V_s = \begin{cases} \frac{5}{18} f(R_c)^{3/2}\left[-ln\left(\frac{\Delta_{cc}}{\psi}\right) + \frac{1}{1+\frac{\sqrt{f(R_C)}}{2}}\right] & \text{for} \quad \Delta_{cc} < \psi \\ \frac{5}{18} f(R_c)^{3/2}\frac{1}{1+\frac{\sqrt{f(R_C)}}{2}}\frac{\psi}{\Delta_{cc}}e^{\sqrt{f(R_c)}(\Delta_{cc}-\psi)/2\psi} & \text{for} \quad \Delta_{cc} \geq \psi \end{cases} \tag{4.19}$$

where $f(R_c)$ is the number of surfactant chains per nanoparticle, Δ_{cc} is the nanoparticle center-to-center separation distance, and ψ is a scaling factor defined as 1.3 times the radius of gyration (R_g). Using this equation combined with other equations for particle attraction can give the overall potential energy diagram, depicted in Figure 4.6.

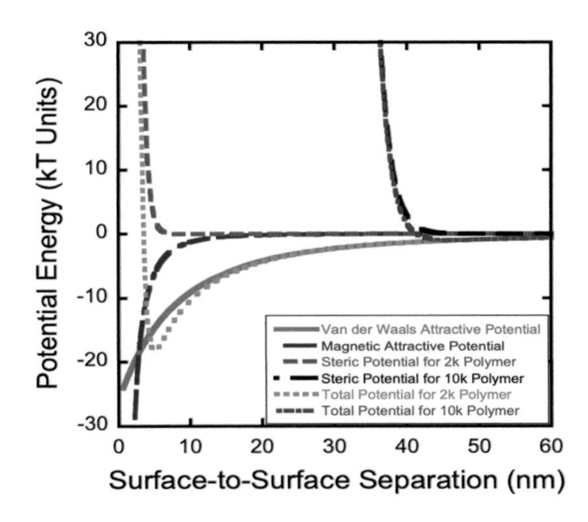

FIGURE 4.6 The total potential energy of a 24 nm in diameter magnetite nanoparticle system coated with PEG of 2000 and 10,000 g/mol molecular weight.

4.5.6.1 Radius of Gyration

The radius of gyration of a colloid is important because it describes the size of a nanoparticle suspension without describing the shape, and characterizes the spatial extension of a colloid (Hiemenz and Rajagopalan 1997). The squared radius of gyration for a polymer is defined as the average squared distance of any point in the polymer chain from its center of mass. The radius of gyration can be calculated using the equation:

$$R_g = \sqrt{\frac{I}{M}} \tag{4.20}$$

where I is the moment of inertia for the polymer-coated nanoparticle, and M is the mass of the complex. The moment of inertia for the polymer-coated nanoparticle can be calculated using the equation:

$$R_g = \sqrt{\frac{I}{M}} = \sqrt{\frac{\frac{4\pi R_c^5 \rho_m}{5} + \frac{4\pi A m_s}{a_s^{1/v}\left(2+\frac{1}{v}\right)\left(R_H^{(2+1/v)} - R_c^{(2+1/v)}\right)}}{\rho_m \frac{4}{3}\pi R_c^3 + f(R_C)\frac{M_n}{N_a}}} \tag{4.21}$$

where ρ_m is the density of the nanoparticle, m_s is the mass of a segment of polymer, a_s is the size of a monomer in a polymer chain, v is the Flory exponent, M_n is the number average molecular weight of the polymer, N_a is Avogadro's number, and A is a constant of proportionality calculated by the equation:

$$A = \frac{3\, x\, 4^{1/v} f(R_c)^{(3v-1/2v)}}{32} \tag{4.22}$$

The hydrodynamic radius (R_H) is an important term here, and is usually measured by such techniques as DLS. It can also be approximated, which is detailed in the next section.

4.5.6.2 Calculating the Hydrodynamic Radius

The theoretical hydrodynamic diameter of polymer-coated particles can be calculated using the blob model (Mefford et al. 2008a), which is based on a model for star polymers by Daoud and Cotton (1982) and assumes concentric shells with a constant number of blobs in each shell. The blob diameter is a continuous function of distance from the surface. The hydrodynamic radius can then be described as:

$$R_H(r) = \left(\frac{8\, x\, N_k\, x\, f(r)^{(1-v)/2v}}{3\, x\, 4^{1/v}\, x\, v} L_k^{1/v} + r^{1/v}\right)^v \tag{4.23}$$

where ν is the Flory exponent, r is the radius of the nanoparticle, and N_k is the number of Kuhn segments in one of the corona chains, L_k is the Kuhn segment length, and $f(r)$ is the number of polymer chains per particle. The number of Kuhn segments is defined by:

$$N_k = n / c_\infty \tag{4.24}$$

where n is the number of backbone bonds in a chain, and c_∞ is the characteristic ratio of the polymer. The Kuhn segment length, L_k, is defined as:

$$L_k = c_\infty * l_0 \tag{4.25}$$

where l_0 is the average length of a backbone bond and $f(r)$ is the number of corona chains per particle, and can be calculated using the equation:

$$f(r) = 4\pi r^2 \sigma \tag{4.26}$$

where σ is the surface density of chains on the particle and a function of the particle size distribution, given by the equation:

$$\sigma = \frac{W_P N_{av} \rho_m}{M_n W_m} \int_0^\infty \left(\frac{3}{r}\right) P(r) dr \tag{4.27}$$

where W_P is the weight fraction of polymer, N_{av} is Avagadro's number, ρ_m is the density of the nanoparticle, M_n is the number average molecular weight of the polymer, W_m is the weight fraction of the nanoparticle, and $P(r)$ is the distribution function of the core particle size.

4.6 SUMMARY

It is hoped that readers of this guide will benefit from the increased understanding of the importance of colloidal stability in the biomedical applications of magnetic nanoparticles. Failing to recognize the necessary material engineering required to produce robust systems stabile in biological media could result in severe consequences. Moreover, these challenges are compounded by the use of magnetic materials as magnetic attractive forces typically have a much longer range than van der Waals attractive forces. In addition, electrostatic stabilized systems are challenging due to interaction with the immune system as well as electrostatic screening from ions in the media. For best results, those working in this field are encouraged to take a comprehensive approach while carefully characterizing their systems through each step of the development process.

REFERENCES

Al-Hayek, S. S. 2012. Experimental and Theoretical Study of Magnetic Hyperthermia. 1–198.

Amstad, E., J. Kohlbrecher, E. Müller, T. Schweizer, M. Textor, and E. Reimhult. 2011. Triggered release from liposomes through magnetic actuation of iron oxide nanoparticle containing membranes. *Nano Letters* 11 (4), 1664–1670.

Aqil, A., S. Vasseur, E. Duguet, C. Passirani, J. P. Benoit, R. Jerome, and C. Jerome. 2008. Magnetic nanoparticles coated by temperature responsive copolymers for hyperthermia. *Journal of Materials Chemistry* 18 (28), 3352–3360.

Balasubramaniam, S., S. Kayandan, Y.-N. Lin, D. F. Kelly, M. J. House, R. C. Woodward, T. G. St Pierre, J. S. Riffle, and R. M. Davis. 2014. Toward design of magnetic nanoparticle clusters stabilized by biocompatible diblock copolymers for T_2-weighted MRI contrast. *Langmuir* 30 (6), 1580–1587.

Berg, J. C. 2010. *An Introduction to Interfaces and Colloids: The Bridge to Nanoscience*. Toh Tuck Link, Singapore: World Scientific Publishing Co. Pte. Ltd.

Berg, J. C. 2012. *An Introduction to Interfaces and Colloids*. Toh Tuck Link, Singapore:World Scientific Publishing Co. Pte. Ltd.

Bogdanov, A. A., C. Martin, R. Weissleder, and T. J. Brady. 1994. Trapping of dextran-coated colloids in liposomes by transient binding to aminophospholipid—Preparation of ferrosomes. *Biochimica Et Biophysica Acta-Biomembranes* 1193 (1), 212–218. doi:10.1016/0005-2736(94)90350-6.

Bonini, M., S. Lenz, R. Giorgi, and P. Baglioni. 2007. Nanomagnetic sponges for the cleaning of works of art. *Langmuir* 23 (17), 8681–8685. doi:10.1021/la701292d.

Bonini, M., S. Lenz, E. Falletta, and F. Ridi. 2008. Acrylamide-based magnetic nanosponges: A new smart nanocomposite material. *Langmuir* 24 (21), 12644–12650.

Butter, K., P. H. H. Bomans, P. M. Frederik, G. J. Vroege, and A. P. Philipse. 2003. Direct observatioof dipolar chains in iron ferrofluids by cryogenic electron microscopy. *Nature Materials* 2 (2), 88–91.

Butterworth, M. D., L. Illum, and S. S. Davis. 2001. Preparation of ultrafine silica- and PEG-coated magnetite particles. *Colloids and Surfaces A-Physicochemical and Engineering Aspects* 179 (1), 93–102. doi:10.1016/s0927-7757(00)00633-6.

Carroll, M. R. J., R. C. Woodward, M. J. House, W. Y. Teoh, R. Amal, T. L. Hanley, and T. G. St Pierre. 2009. Experimental validation of proton transverse relaxivity models for superparamagnetic nanoparticle MRI contrast agents. *Nanotechnology* 21 (3), 035103. doi:10.1088/0957-4484/21/3/035103.

Carroll, M. R. J., R. C. Woodward, M. J. House, W. Y. Teoh, R. Amal, T. L. Hanley, and T. G. St Pierre. 2010. Experimental validation of proton transverse relaxivity models for superparamagnetic nanoparticle MRI contrast agents. *Nanotechnology* 21, 035103.

Chen, J. C., M. Z. Liu, H. H. Gong, Y. J. Huang, and C. Chen. 2011. Synthesis and self-assembly of thermoresponsive PEG-b-PNIPAM-b-PCL ABC triblock copolymer through the combination of atom transfer radical polymerization, ring-opening polymerization, and click chemistry. *Journal of Physical Chemistry B* 115 (50), 14947–14955. doi:10.1021/jp208494w.

Cheng, F. Y., C. H. Su, Y. S. Yang, C. S. Yeh, C. Y. Tsai, C. L. Wu, M. T. Wu, and D. B. Shieh. 2005. Characterization of aqueous dispersions of Fe_3O_4 nanoparticles and their biomedical applications. *Biomaterials* 26 (7), 729–738. doi:10.1016/j.biomaterials.2004.03.016.

Chouly, C., D. Pouliquen, I. Lucet, J. J. Jeune, and P. Jallet. 1996. Development of superparamagnetic nanoparticles for MRI: Effect of particle size, charge and surface nature on biodistribution. *Journal of Microencapsulation* 13 (3), 245–255.

Chu, B., and T. Liu. 2000. Characterization of nanoparticles by scattering techniques. *Journal of Nanoparticle Research* 2 (1), 29–41.

Couvreur, P., and C. Vauthier. 2006. Nanotechnology: Intelligent design to treat complex disease. *Pharmaceutical Research* 23 (7), 1417–1450. doi:10.1007/s11095-006-0284-8.

Daoud, M., and J. P. Cotton. 1982. Star shaped polymers: A model for the conformation and its concentration dependence. *Journal de Physique* 43, 531–538.

Davis, K., B. Qi, M. Witmer, C. L. Kitchens, B. A. Powell, and O. T. Mefford. 2014. Quantitative measurement of ligand exchange on iron oxides via radiolabeled oleic acid. *Langmuir* 30 (36), 10918–10925. doi:10.1021/la502204g.

Davis, K., B. Cole, M. Ghelardini, B. A. Powell, and O. T. Mefford. 2016. Quantitative measurement of ligand exchange with small-molecule ligands on iron oxide nanoparticles via radioanalytical techniques. *Langmuir* 32 (51), 13716–13727. doi:10.1021/acs.langmuir.6b03644.

Dennis, C. L., A. J. Jackson, J. A. Borchers, P. J. Hoopes, R. Strawbridge, A. R. Foreman, J. van Lierop, C. Grüttner, and R. Ivkov. 2009. Nearly complete regression of tumors via collective behavior of magnetic nanoparticles in hyperthermia. *Nanotechnology* 20, 395103.

Duan, H. W., M. Kuang, X. X. Wang, Y. A. Wang, H. Mao, and S. M. Nie. 2008. Reexamining the effects of particle size and surface chemistry on the magnetic properties of iron oxide nanocrystals: New insights into spin disorder and proton relaxivity. *Journal of Physical Chemistry C* 112 (22), 8127–8131. doi:10.1021/jp8029083.

Evans, D. F., and H. Wennerström. 1999. *The Colloidal Domain: Where Physics, Chemistry, Biology, and Technology Meet*. New York: Wiley-VCH.

Fang, C., N. Bhattarai, C. Sun, and M. Q. Zhang. 2009. Functionalized nanoparticles with long-term stability in biological media. *Small* 5 (14), 1637–1641. doi:10.1002/smll.200801647.

Farrell, D., Y. Ijiri, C. Kelly, J. Borchers, J. Rhyne, Y. Ding, and S. Majetich. 2006. Small angle neutron scattering study of disordered and crystalline iron nanoparticle assemblies. *Journal of Magnetism and Magnetic Materials* 303 (2), 318–322. doi:10.1016/j.jmmm.2006.01.219.

Ferguson, R. M., K. R. Minard, and K. M. Krishnan. 2009. Optimization of nanoparticle core size for magnetic particle imaging. *Journal of Magnetism and Magnetic Materials* 321 (10), 1548–1551. doi:10.1016/j.jmmm.2009.02.083.

Ferguson, R. M., K. R. Minard, A. P. Khandhar, and K. M. Krishnan. 2011. Optimizing magnetite nanoparticles for mass sensitivity in magnetic particle imaging. *Medical Physics* 38 (3), 1619–1626. doi:10.1118/1.3554646.

Ferguson, R. M., A. P. Khandhar, and K. M. Krishnan. 2011. Magnetization spectroscopy of biocompatible magnetite (Fe_3O_4) nanoparticles for MPI. In *Medical Imaging 2011: Biomedical Applications in Molecular, Structural, and Functional Imaging*, edited by J. B. Weaver and R. C. Molthen.Bellingham, WA:SPIE Publications.

Ferguson, R. M., A. P. Khandhar, and K. M. Krishnan. 2012. Tracer design for magnetic particle imaging (invited). *Journal of Applied Physics* 111 (7). doi:07b31810.1063/1.3676053.

French, R. H. 2000. Origins and applications of London dispersion forces and Hamaker constants in ceramics. *Journal of the American Ceramic Society* 83 (9), 2117–2146.

Gibbs, G. V., T. D. Crawford, A. F. Wallace, D. F. Cox, R. M. Parrish, E. G. Hohenstein, and C. D. Sherrill. 2011. Role of long-range intermolecular forces in the formation of inorganic nanoparticle clusters. *Journal of Physical Chemistry A* 115 (45), 12933–12940. doi:10.1021/jp204044k.

Goff, J. D., P. P. Huffstetler, W. C. Miles, N. Pothayee, C. M. Reinholz, S. Ball, R. M. Davis, and J. S. Riffle. 2009. Novel phosphonate-functional poly (ethylene oxide)-magnetite nanoparticles form stable colloidal dispersions in phosphate-buffered saline. *Chemistry of Materials* 21 (20), 4784–4795. doi:10.1021/cm901006g.

Gonzales, M., and K. M. Krishnan. 2007. Phase transfer of highly monodisperse iron oxide nanocrystals with Pluronic F127 for biomedical applications. *Journal of Magnetism and Magnetic Materials* 311 (1), 59–62. doi:10.1016/j.jmmm.2006.10.1150.

Goya, G. F., T. S. Berquo, F. C. Fonseca, and M. P. Morales. 2003. Static and dynamic magnetic properties of spherical magnetite nanoparticles. *Journal of Applied Physics* 94, 3520.

Goya, G. F., A. Arelaro, E. Lima, and T. E. Torres. 2008. Magnetic hyperthermia with Fe_3O_4 nanoparticles: The influence of particle size on energy absorption. *IEEE Transactions on Magnetics* 44 (11), 4444–4447.

Grimm, J., N. Karger, S. Lusse, S. Winoto-Morbach, B. Krisch, S. Muller-Hulsbeck, and M. Heller. 2000. Characterization of ultrasmall, paramagnetic magnetite particles as superparamagnetic contrast agents in MRI. *Investigative Radiology* 35 (9), 553–556. doi:10.1097/00004424-200009000-00006.

Guardia, P., R. Di Corato, L. Lartigue, C. Wilhelm, A. Espinosa, M. Garcia-Hernandez, F. Gazeau, L. Manna, and T. Pellegrino. 2012. Water-soluble iron oxide nanocubes with high values of specific absorption rate for cancer cell hyperthermia treatment. *ACS Nano* 6 (4), 3080–3091. doi:10.1021/nn2048137.

Guo, J., W. L. Yang, Y. H. Deng, C. C. Wang, and S. K. Fu. 2005. Organic-dye-coupted magnetic nanoparticles encaged inside thermoresponsive PNIPAM microcapsutes. *Small* 1 (7), 737–743. doi:10.1002/smll.200400145.

Hajdu, A., E. Illes, E. Tombacz, and I. Borbath. 2009. Surface charging, polyanionic coating and colloid stability of magnetite nanoparticles. *Colloids and Surfaces A-Physicochemical and Engineering Aspects* 347 (1–3), 104–108. doi:10.1016/j.colsurfa.2008.12.039.

Herrera, A. P., C. Barrera, Y. Zayas, and C. Rinaldi. 2010. Monitoring colloidal stability of polymer-coated magnetic nanoparticles using AC susceptibility measurements. *Journal of Colloid and Interface Science* 342 (2), 540–549.

Hiemenz, P. C., and R. Rajagopalan. 1997. *Principles of Colloid and Surface Chemistry*. 3rd ed. New York: Marcel Dekker Inc.

Hiergeist, R., W. Andra, N. Buske, R. Hergt, I. Hilger, U. Richter, and W. Kaiser. 1999. Application of magnetite ferrofluids for hyperthermia. *Journal of Magnetism and Magnetic Materials* 201 (1–3), 420–422. doi:http://dx.doi.org/10.1016/S0304-8853(99)00145-6.

Holm, C., and J. J. Weis. 2005. The structure of ferrofluids: A status report. *Current Opinion in Colloid and Interface Science* 10 (3–4), 133–140.

Honda, H., A. Kawabe, M. Shinkai, and T. Kobayashi. 1998. Development of chitosan-conjugated magnetite for magnetic cell separation. *Journal of Fermentation and Bioengineering* 86 (2), 191–196. doi:10.1016/s0922-338x(98)80060-3.

Illes, E., and E. Tombacz. 2003. The role of variable surface charge and surface complexation in the adsorption of humic acid on magnetite. *Colloids and Surfaces A-Physicochemical and Engineering Aspects* 230 (1–3), 99–109. doi:10.1016/j.colsurfa.2003.09.017.

Illes, E., and E. Tombacz. 2006. The effect of humic acid adsorption on pH-dependent surface charging and aggregation of magnetite nanoparticles. *Journal of Colloid and Interface Science* 295 (1), 115–123. doi:10.1016/j.jcis.2005.08.003.

Ionita, P., A. Volkov, G. Jeschke, and V. Chechik. 2007. Lateral diffusion of thiol ligands on the surface of au nanoparticles: An electron paramagnetic resonance study. *Analytical Chemistry* 80 (1), 95–106. doi:10.1021/ac071266s.

Isojima, T., M. Lattuada, J. B. Vander Sande, and T. A. Hatton. 2008. Reversible clustering of pH- and temperature-responsive Janus magnetic nanoparticles. *ACS Nano* 2 (9), 1799–1806. doi:10.1021/nn800089z.

Israelachvili, J. N. 2011. *Intermolecular and Surface Forces*. 3rd ed. London: Academic Press.

Jeun, M., S. Lee, J. K. Kang, A. Tomitaka, K. W. Kang, Y. I. Kim, Y. Takemura, K.-W. Chung, J. Kwak, and

S. Bae. 2012. Physical limits of pure superparamagnetic Fe_3O_4 nanoparticles for a local hyperthermia agent in nanomedicine. *Applied Physics Letters* 100 (9), 092406. doi:10.1063/1.3689751.

Jiang, J. S., Z. F. Gan, Y. Yang, B. Du, M. Qian, and P. Zhang. 2009. A novel magnetic fluid based on starch-coated magnetite nanoparticles functionalized with homing peptide. *Journal of Nanoparticle Research* 11 (6), 1321–1330. doi:10.1007/s11051-008-9534-5.

Jordan, A., P. Wust, H. Fahling, W. John, A. Hinz, and R. Felix. 1993. Inductive heating of ferrimagnetic particles and magnetic fluids—Physical evaluation of their potential for hyperthermia. *International Journal of Hyperthermia* 9 (1), 51–68. doi:10.3109/02656739309061478.

Jung, C. W., and P. Jacobs. 1995. Physical and chemical properties of superparamagnetic iron-oxide MR contrast agents—Ferumoixes, ferumoxtran, and ferumoxil. *Magnetic Resonance Imaging* 13 (5), 661–674. doi:10.1016/0730-725x(95)00024-b.

Kawai, N., A. Ito, Y. Nakahara, M. Futakuchi, T. Shirai, H. Honda, T. Kobayashi, and K. Kohri. 2005. Anticancer effect of hyperthermia on prostate cancer mediated by magnetite cationic liposomes and immune-response induction in transplanted syngeneic rats. *Prostate* 64 (4), 373–381. doi:10.1002/pros.20253.

Kedziorek, D. A., and D. L. Kraitchman. 2010. Superparamagnetic iron oxide labeling of stem cells for MRI tracking and delivery in cardiovascular disease. *Methods of Molecular Biology* 660, 171–183.

Kim, D. K., Y. Zhang, W. Voit, K. V. Rao, and M. Muhammed. 2001. Synthesis and characterization of surfactant-coated superparamagnetic monodispersed iron oxide nanoparticles. *Journal of Magnetism and Magnetic Materials* 225 (1–2), 30–36. doi:10.1016/s0304-8853(00)01224-5.

Klokkenburg, M., B. H. Erné, A. Wiedenmann, A. V. Petukhov, and A. P. Philipse. 2007. Dipolar structures in magnetite ferrofluids studied with small-angle neutron scattering with and without applied magnetic field. *Physical Review E* 75 (5), 51408.

Kong, G., and M. W. Dewhirst. 1999. Hyperthermia and liposomes. *International Journal of Hyperthermia* 15 (5), 345–370.

Krycka, K. L., R. A. Booth, C. R. Hogg, Y. Ijiri, J. A. Borchers, W. C. Chen, S. M. Watson, M. Laver, T. R. Gentile, L. R. Dedon, S. Harris, J. J. Rhyne, and S. A. Majetich. 2010. Core-shell magnetic morphology of structurally uniform magnetite nanoparticles. *Physical Review Letters* 104 (20), 207203.

Kumar, R. V., Y. Koltypin, Y. S. Cohen, Y. Cohen, D. Aurbach, O. Palchik, I. Felner, and A. Gedanken. 2000. Preparation of amorphous magnetite nanoparticles embedded in polyvinyl alcohol using ultrasound radiation. *Journal of Materials Chemistry* 10 (5), 1125–1129. doi:10.1039/b000440p.

Laurent, S., S. Dutz, U. O. Häfeli, and M. Mahmoudi. 2011. Magnetic fluid hyperthermia: Focus on superparamagnetic iron oxide nanoparticles. *Advances in Colloid and Interface Science* 166 (1–2), 8–23. doi:http://dx.doi.org/10.1016/j.cis.2011.04.003.

Lee, J., T. Isobe, and M. Senna. 1996. Preparation of ultrafine Fe_3O_4 particles by precipitation in the presence of PVA at high pH. *Journal of Colloid and Interface Science* 177 (2), 490–494. doi:10.1006/jcis.1996.0062.

Liang, Y., N. Hilal, P. Langston, and V. Starov. 2007. Interaction forces between colloidal particles in liquid: Theory and experiment. *Advances in Colloid and Interface Science* 134–135, 151–166. doi:http://dx.doi.org/10.1016/j.cis.2007.04.003.

Likos, C. N. 2006. Soft matter with soft particles. *Soft Matter* 2 (6), 478–498. doi:10.1039/b601916c.

Liu, C., and Z. J. Zhang. 2001. Size-dependent superparamagnetic properties of Mn spinel ferrite nanoparticles synthesized from reverse micelles. *Chemistry of Materials* 13 (6), 2092–2096.

Liu, J., Z. K. Sun, Y. H. Deng, Y. Zou, C. Y. Li, X. H. Guo, L. Q. Xiong, Y. Gao, F. Y. Li, and D. Y. Zhao. 2009. Highly water-dispersible biocompatible magnetite particles with low cytotoxicity stabilized by citrate groups. *Angewandte Chemie-International Edition* 48 (32), 5875–5879. doi:10.1002/anie.200901566.

Ma, M., Y. Zhang, W. Yu, H.-Y. Shen, H.-Q. Zhang, and N. Gu. 2003. Preparation and characterization of magnetite nanoparticles coated by amino silane. *Colloids and Surfaces A: Physicochemical and Engineering Aspects* 212 (2–3), 219–226. doi:10.1016/s0927-7757(02)00305-9.

Mahmoudi, M., A. Simchi, M. Imani, A. S. Milani, and P. Stroeve. 2008. Optimal design and characterization of superparamagnetic iron oxide nanoparticles coated with polyvinyl alcohol for targeted delivery and imaging. *Journal of Physical Chemistry B* 112 (46), 14470–14481. doi:10.1021/jp803016n.

Mahmoudi, M., A. Simchi, A. S. Milani, and P. Stroeve. 2009. Cell toxicity of superparamagnetic iron oxide nanoparticles. *Journal of Colloid and Interface Science* 336 (2), 510–518. doi:10.1016/j.jcis.2009.04.046.

Mahmoudi, M., V. Serpooshan, and S. Laurent. 2011. Engineered nanoparticles for biomolecular imaging. *Nanoscale* 3 (8), 3007–3026. doi:10.1039/c1nr10326a.

Massart, R. 1981. Preparation of aqueous magnetic liquids in alkaline and acidic media. *IEEE Transactions on Magnetics* 17 (2), 1247–1248.
Massart, R., E. Dubois, V. Cabuil, and E. Hasmonay. 1995. Preparation and properties of monodisperse magnetic fluids. *Journal of Magnetism and Magnetic Materials* 149, 1–5.
Matsumoto, Y., and A. Jasanoff. 2008. T_2 relaxation induced by clusters of superparamagnetic nanoparticles: Monte Carlo simulations. *Magnetic Resonance Imaging* 26 (7), 994–998. doi:10.1016/j.mri.2008.01.039.
Mefford, O. T., M. R. J. Carroll, M. L. Vadala, J. D. Goff, R. Mejia-Ariza, M. Saunders, R. C. Woodward, T. G. St Pierre, R. M. Davis, and J. S. Riffle. 2008a. Size analysis of PDMS-magnetite nanoparticle complexes: Experiment and theory. *Chemistry of Materials* 20 (6), 2184–2191. doi:10.1021/cm702730p.
Mefford, O. T., M. L. Vadala, J. D. Goff, M. R. J. Carroll, R. Mejia-Ariza, B. L. Caba, T. G. St Pierre, R. C. Woodward, R. M. Davis, and J. S. Riffle. 2008b. Stability of polydimethylsiloxane-magnetite nanoparticle dispersions against flocculation: Interparticle interactions of polydisperse materials. *Langmuir* 24 (9), 5060–5069. doi:10.1021/la703146y.
Milner, S. T. 1991. Polymer brushses. *Science* 251 (4996), 905–914. doi:10.1126/science.251.4996.905.
Min, Y., M. Akbulut, K. Kristiansen, Y. Golan, and J. Israelachvili. 2008. The role of interparticle and external forces in nanoparticle assembly. *Nature Materials* 7 (7), 527–538.
Napper, D. H. 1970. Flocculation studies of sterically stabilized dispersions. *Journal of Colloid and Interface Science* 32 (1), 106–114.
Napper, D. H. 1977. Steric stabilization. *Journal of Colloid and Interface Science* 58 (2), 390–407.
Napper, D. H., and A. Netschey. 1971. Studies of the steric stabilization of colloidal particles. *Journal of Colloid and Interface Science* 37 (3), 528–535.
Nappini, S., M. Bonini, F. Bombelli, and F. Pineider. 2010. Controlled drug release under a low frequency magnetic field: Effect of the citrate coating on magnetoliposomes stability. *Soft Matter* 7, 1025–1037.
Needham, D., K. Hristova, T. J. McIntosh, M. Dewhirst, N. Wu, and D. D. Lasic. 1992. Polymer-grafted liposomes: Physical basis for the "stealth" property. *Journal of Liposome Research* 2 (3), 411–430. doi:10.3109/08982109209010218.
Neoh, K. G., and E. T. Kang. 2011. Functionalization of inorganic nanoparticles with polymers for stealth biomedical applications. *Polymer Chemistry* 2 (4), 747–759.
Odenbach, S. 2009. *Colloidal Magnetic Fluids: Basics, Development and Application of Ferrofluids*. Vol. 763, *Lecture Notes in Physics*. Berlin Heidelberg: Springer.
Pankhurst, Q. A., J. Connolly, S. K. Jones, and J. Dobson. 2003. Applications of magnetic nanoparticles in biomedicine. *Journal of Physics D: Applied Physics* 36, R167.
Petri-Fink, A., B. Steitz, A. Finka, J. Salaklang, and H. Hofmann. 2008. Effect of cell media on polymer coated superparamagnetic iron oxide nanoparticles (SPIONs): Colloidal stability, cytotoxicity, and cellular uptake studies. *European Journal of Pharmaceutics and Biopharmaceutics* 68 (1), 129–137. doi:10.1016/j.ejpb.2007.02.024.
Purushotham, S., and R. V. Ramanujan. 2010. Thermoresponsive magnetic composite nanomaterials for multimodal cancer therapy. *Acta Biomaterialia* 6 (2), 502–510. doi:10.1016/j.actbio.2009.07.004.
Rosenholm, J. B. 2010. Critical comparison of molecular mixing and interaction models for liquids, solutions and mixtures. *Advances in Colloid and Interface Science* 156 (1–2), 14–34. doi:10.1016/j.cis.2010.02.005.
Russel, W. B., D. A. Saville, and W. R. Schowalter. 1987. *Colloid Dispersions*. Cambridge: Cambridge University.
Russel, W. B., D. A. Saville, and W. R. Schowalter. 1989. *Colloidal Dispersions*, edited by G.K. Batchelor. Cambridge:The Press Syndicate of the University of Cambridge.
Saleh, N., T. Phenrat, K. Sirk, B. Dufour, J. Ok, T. Sarbu, K. Matyiaszewski, R. D. Tilton, and G. V. Lowry. 2005. Adsorbed triblock copolymers deliver reactive iron nanoparticles to the oil/water interface. *Nano Letters* 5 (12), 2489–2494. doi:10.1021/nl0518268.
Santiago-Quinones, D. I., A. Acevedo, and C. Rinaldi. 2009. Magnetic and magnetorheological characterization of a polymer liquid crystal ferronematic. *Journal of Applied Physics* 105 (7), 07B512.
Saville, S. L., R. C. Stone, B. Qi, and O. T. Mefford. 2012a. Investigation of the stability of magnetite nanoparticles functionalized with catechol based ligands in biological media. *Journal of Materials Chemistry* 22 (47), 24909.
Saville, S. L., R. C. Stone, B. Qi, and O. T. Mefford. 2012b. Investigation of the stability of magnetite nanoparticles functionalized with catechol based ligands in biological media. *Journal of Materials Chemistry* 47 (22), 24909–24917.

Saville, S. L., R. C. Woodward, M. J. House, A. Tokarev, J. Hammers, B. Qi, J. Shaw, M. Saunders, R. R. Varsani, T. G. St Pierre, and O. T. Mefford. 2013. The effect of magnetically induced linear aggregates on proton transverse relaxation rates of aqueous suspensions of polymer coated magnetic nanoparticles. *Nanoscale* 5 (5), 2152–2163.

Shen, T., R. Weissleder, M. Papisov, A. Bogdanov, and T. J. Brady. 1993. Monocrystalline iron-oxide nanocompounds (MION)—Physicohemical properties. *Magnetic Resonance in Medicine* 29 (5), 599–604.

Soppimath, K. S., T. M. Aminabhavi, A. R. Kulkarni, and W. E. Rudzinski. 2001. Biodegradable polymeric nanoparticles as drug delivery devices. *Journal of Controlled Release* 70(1–2), 1–20.

Steitz, B., H. Hofmann, S. W. Kamau, P. O. Hassa, M. O. Hottiger, B. von Rechenberg, M. Hofmann-Amtenbrink, and A. Petri-Fink. 2007. Characterization of PEI-coated superparamagnetic iron oxide nanoparticles for transfection: Size distribution, colloidal properties and DNA interaction. *Journal of Magnetism and Magnetic Materials* 311 (1), 300–305. doi:10.1016/j.jmmm.2006.10.1194.

Stokes, R. J., and D. F. Evans. 1996. *Fundamentals of Interfacial Engineering*. New York: Wiley-VCH.

Stone, R., T. Willi, Y. Rosen, O. T. Mefford, and F. Alexis. 2011. Targeted magnetic hyperthermia. *Therapeutic Delivery* 2 (6), 815–838. doi:doi:10.4155/tde.11.48.

Sun, C., J. Lee, and M. Zhang. 2008. Magnetic nanoparticles in MR imaging and drug delivery. *Advanced Drug Delivery Reviews* 60 (11), 1252–1265. doi:10.1016/j.addr.2008.03.018.

Tasci, T. O., I. Vargel, A. Arat, E. Guzel, P. Korkusuz, and E. Atalar. 2009. Focused RF hyperthermia using magnetic fluids. *Medical Physics* 36 (5), 1906. doi:10.1118/1.3106343.

Tombacz, E., Z. Libor, E. Illes, A. Majzik, and E. Klumpp. 2004. The role of reactive surface sites and complexation by humic acids in the interaction of clay mineral and iron oxide particles. *Organic Geochemistry* 35 (3), 257–267. doi:10.1016/j.orggeochem.2003.11.002.

Valberg, P. A., and J. P. Butler. 1987. Magnetic particle motions within living cells. Physical theory and techniques. *Biophysical Journal* 52 (4), 537–550.

Viroonchatapan, E., M. Ueno, H. Sato, I. Adachi, H. Nagai, K. Tazawa, and I. Horikoshi. 1995. Preparation and characterization of dextran magnetite-incorporated thermosensitive liposome—An online flow system for quantifying magnetic responsiveness. *Pharmaceutical Research* 12 (8), 1176–1183. doi:10.1023/a:1016216011016.

Wang, Z. Y., G. Liu, J. Sun, B. Y. Wu, Q. Y. Gong, B. Song, H. Ai, and Z. W. Gu. 2009. Self-assembly of magnetite nanocrystals with amphiphilic polyethylenimine: Structures and applications in magnetic resonance imaging. *Journal of Nanoscience and Nanotechnology* 9 (1), 378–385. doi:10.1166/jnn.2009.J033.

Wilson, K. S., J. D. Goff, J. S. Riffle, L. A. Harris, and T. G. St Pierre. 2005. Polydimethylsiloxane-magnetite nanoparticle complexes and dispersions in polysiloxane carrier fluids. *Polymers for Advanced Technologies* 16 (2–3), 200–211. doi:10.1002/pat.572.

Wiogo, H. T. R., M. Lim, V. Bulmus, J. Yun, and R. Amal. 2011. Stabilization of magnetic iron oxide nanoparticles in biological media by fetal bovine serum (FBS). *Langmuir* 27 (2), 843–850. doi:10.1021/la104278m.

Wong, J. E., A. K. Gaharwar, D. Muller-Schulte, D. Bahadur, and W. Richtering. 2008. Dual-stimuli responsive PNiPAM microgel achieved via layer-by-layer assembly: Magnetic and thermo-responsive. *Journal of Colloid and Interface Science* 324 (1–2), 47–54. doi:10.1016/j.jcis.2008.05.024.

Xia, X. R., N. A. Monteiro-Riviere, and J. E. Riviere. 2010. An index for characterization of nanomaterials in biological systems. *Nature Nanotechnology* 5 (9), 671–675. doi:10.1038/nnano.2010.164.

Xie, J., J. Huang, X. Li, S. Sun, and X. Chen. 2009. Iron oxide nanoparticle platform for biomedical applications. *Current Medicinal Chemistry* 16 (10), 1278–1294.

Yallapu, M. M., S. F. Othman, E. T. Curtis, B. K. Gupta, M. Jaggi, and S. C. Chauhan. 2011. Multi-functional magnetic nanoparticles for magnetic resonance imaging and cancer therapy. *Biomaterials* 32 (7), 1890–1905. doi:10.1016/j.biomaterials.2010.11.028.

Yang, X., H. Hong, J. J. Grailer, I. J. Rowland, A. Javadi, S. A. Hurley, Y. Xiao, Y. Yang, Y. Zhang, R. J. Nickles, W. Cai, D. A. Steeber, and S. Gong. 2011. cRGD-functionalized, DOX-conjugated, and 64Cu-labeled superparamagnetic iron oxide nanoparticles for targeted anticancer drug delivery and PET/MR imaging. *Biomaterials* 32 (17), 4151–4160. doi:10.1016/j.biomaterials.2011.02.006.

Zhang, J., and R. D. K. Misra. 2007. Magnetic drug-targeting carrier encapsulated with thermosensitive smart polymer: Core-shell nanoparticle carrier and drug release response. *Acta Biomaterialia* 3 (6), 838–850. doi:10.1016/j.actbio.2007.05.011.

Zhang, J. L., R. S. Srivastava, and R. D. K. Misra. 2007. Core shell magnetite nanoparticles surface encapsulated with smart stimuli-responsive polymer: Synthesis, characterization, and LCST of viable drug-targeting delivery system. *Langmuir* 23 (11), 6342–6351. doi:10.1021/la0636199.

Zhu, L., Z. Huo, L. Wang, X. Tong, Y. Xiao, and K. Ni. 2009. Targeted delivery of methotrexate to skeletal muscular tissue by thermosensitive magnetoliposomes. *International Journal of Pharmaceutics* 370 (1–2), 136–143. doi:10.1016/j.ijpharm.2008.12.003.

5 Magnetic Characterization: Instruments and Methods

Cindi L. Dennis

CONTENTS

5.1 INTRODUCTION

This chapter will review magnetic instrumentation and magnetic measurements, including the techniques and the types of information that can be learned from them. Unfortunately, due to space constraints, this chapter cannot be exhaustive for all techniques and example applications. Instead, a subset has been selected that is useful for the characterization of magnetic nanoparticles. In addition, some non-magnetic techniques will be highlighted that are critical for understanding the magnetic behavior of magnetic nanoparticles, as well as highlighting some more specialized magnetic techniques. Finally, this chapter does assume a familiarity with the basic concepts of magnetism, as discussed in Chapter 1.

5.2 INSTRUMENTATION

As with any experimental measurement method, it is necessary to understand the basics of how the instrument works in order to be able to properly prepare and mount samples, as well as to identify common artifacts.

5.2.1 DC Magnetometer

The DC magnetometer is one of the most common magnetic characterization instruments in use. It measures the static magnetic properties of the sample. There are two main variants of this, plus a hybrid that combines aspects of both. The first is a vibrating sample magnetometer (VSM) and the second is the superconducting quantum interference device (SQUID) magnetometer. The hybrid instrument is the SQUID VSM.

These instruments fundamentally operate on Faraday's law, which states that a changing magnetic field induces an electric field, as follows:

$$\nabla \times \vec{E} = -\frac{\partial \vec{B}}{\partial t} \tag{5.1}$$

where E is the electric field vector, B is the magnetic flux vector, and t is the time. Another special rule defines the sign of this induced electric field, Lenz's law, which states that the changing magnetic field through a coil induces a current in that same coil in the direction that opposes the magnetic field change. In a well-designed DC magnetometer, the magnetic field is uniform and constant through the range of motion of the sample, so the sample's magnetic moment is not affected by the movement. Therefore, the moment detection is done indirectly by measuring the voltage change in a coil induced as the sample's non-uniform (dipole) magnetic field moves through it. In a VSM, the coil, or pickup coil, is typically made of copper, and the voltage generated in the loop is directly measured. In a SQUID (or SQUID VSM), the pickup coil is made out of a superconductor (e.g. NbTi), with a very small insulating break junction through which the current must quantum mechanically tunnel to complete the loop. The voltage change is measured across this tunnel junction, more formally known as a Josephson junction. Since the applied magnetic field is a static DC field, the sample must be moved to generate the changing magnetic field. An AC field would also induce a voltage change in the pickup coil that may dominate the sample signal. In a VSM (or SQUID VSM), the changing magnetic field is generated by vibrating the sample at a fixed frequency (e.g. 75 Hz or 14 Hz) and averaging the voltage amplitude measured for a specified length of time, typically a few seconds. For practical reasons, this frequency is never 50 Hz or 60 Hz or a multiple thereof, to avoid interference from the electrical mains. In the SQUID, the changing magnetic field is generated by stepping the sample through the coil, and measuring the voltage at the different steps. The voltage versus position data is then fit to a specific functional form that assumes that the sample is a point dipole moment, and the moment is extracted (Boekelheide and Dennis 2016). The measurement time here is typically about a minute. The voltage measured in both instruments is then calibrated against a known standard reference material (e.g. Ni sphere, Ni disk, Pt cylinder) to convert it into a magnetic moment.

In both a VSM and a SQUID, the pickup coils are never found singly, but rather in pairs. The primary pickup coil pair always has the plane of the coil perpendicular to the static magnetic field, because the sample magnetization generally tracks the applied field direction. Furthermore, the coils are generally wrapped in the same direction, so that the voltage measured in each is additive for better signal-to-noise ratios. However, secondary coils are also used, such as in a vector magnetometer, to measure the magnetization along a direction perpendicular to the applied magnetic field (Figure 5.1). A secondary coil can also be in conjunction with the primary coil pair, such as in the SQUID magnetometer, to form a second-derivative configuration (Figure 5.2). In this case, a second pickup coil pair, which is wrapped in the opposite direction from the first pickup coil pair, is placed on either side of the first coil pair. The second coil acts to remove any contributions from a linear gradient in the static magnetic field because it is wrapped in the opposite direction. (A constant field is not measured at all, due to Lenz's law.) Hence, this configuration is called a second-derivative coil because anything constant or linear in the magnetic field is removed.

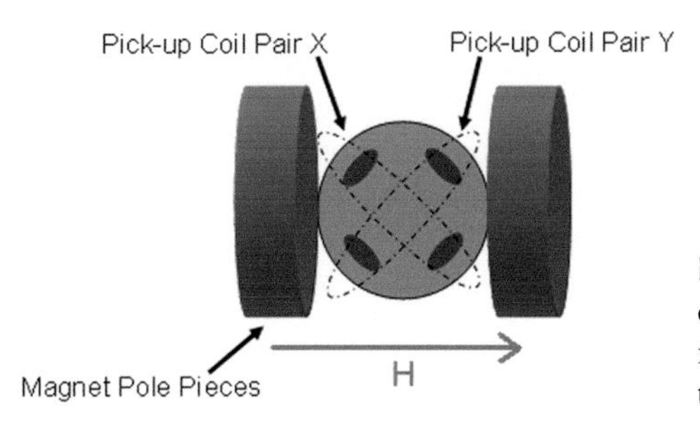

FIGURE 5.1 Schematic of the vector magnetometer coil configuration, with pickup coil pairs measuring the magnetic moment both parallel and perpendicular to the field.

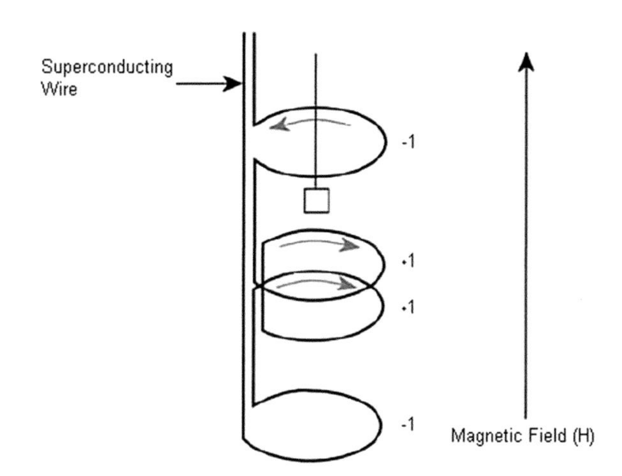

FIGURE 5.2 Schematic of the second-derivative coil configuration, as is used in a SQUID magnetometer. The box indicates a sample and the line is the direction it moves through the coils, courtesy of Quantum Design, Inc. (c) 2020.

Both the single pair and the second-derivative configuration have issues. First, if the sample is not perfectly centered in the coil pair, i.e. if the sample rod is not perfectly straight but kinks slightly to one side toward a single coil, then the measurement will be off. This is especially important when trying to determine the vector magnetization within the plane. Second, in the second-derivative configuration, if your sample holder does not continue well beyond the end of your sample, like a substrate mounted on a piece of tape or thread inside a straw, but rather ends abruptly, like in a liquid capsule, then you will measure a false ferromagnetic signal simply due to a background that is not constant nor linear in magnetic field. When measuring very dilute colloids containing magnetic nanoparticles, it can be very easy to center on this false ferromagnetic signal from the end of the liquid capsule rather than center on the sample itself. Third, the very sensitivity of the SQUID detectors themselves can be an issue, if care is not taken. For example, the dust coming out of many industrial HVAC systems can contain iron oxide particles. Also, stainless steel tweezers can transfer magnetic material onto your sample via contact (Abraham, Frank, and Guha 2005). The first can be fought by either very high-quality HEPA filters, or by simply starting with a very clean workspace prior to mounting your sample. The second can be dealt with only by using non–stainless steel tweezers, such as plastic or gold coated. However, these can, with time, pick up contaminants from the samples themselves, which can then be transferred. An alternative is to coat stainless steel tweezers with clear nail polish, which can be removed with acetone after use, and then recoated. (Many of the colored nail polishes contain magnetic material, and must be avoided.) This prevents cross-contamination without requiring the purchase of a large number of disposable tweezers. Finally, both superconducting magnets and electromagnets have artifacts (Atzmony,

Bennett, and Swartzendruber 1995). Once a current is applied to a superconducting magnet, the magnetic flux can be trapped within it, generally at pinning sites such as dislocations. The amount of flux trapped depends on the magnetic field generated, but can be as large as 4.8 kA/m (60 Oe). This can be a significant offset if the behavior of interest is near zero field. In contrast, moments measured in large fields in an electromagnet can display an image effect due to saturation of the iron in the pole pieces (Higgins et al. 2008). This results in an apparent decrease in the magnetization.

Finally, there are some general rules that indicate when to use one type of DC magnetometer over another. First, SQUIDs are typically used for very low moment samples due to their high sensitivity. Also, because SQUIDs have superconducting magnets, they are used for measurements that need magnetic fields greater than 2 T, which is the practical limit for electromagnets. In contrast, electromagnet-based VSMs are used when total measurement time matters, because they are significantly faster. (This is the case for measuring first-order reversal curves—see section on DC Hysteresis Loops.) VSMs are also typically used for large moment (>1 emu or $>1 \times 10^{-3}$ A-m^2) samples. The hybrid SQUID VSM is typically used for low moment or large fields, and also has significant speed increases over a conventional SQUID.

5.2.2 Alternating Gradient Force Magnetometer

The alternating gradient force magnetometer (AGFM), also commonly referred to as the alternating gradient magnetometer, operates substantially differently in terms of the moment detection. Simply, a static DC field is applied to the sample, in conjunction with a small alternating field gradient. The field gradient must be small, or it will perturb the magnetic moment so that it cannot be assumed to be constant during the measurement. This gradient applies a force to the sample, as given below.

$$\vec{F} = m \bullet \nabla \vec{B} \tag{5.2}$$

where F is the force vector, m is the magnetic moment, and B is the magnetic flux vector. The alternating current generating the field causes this force to oscillate at the same frequency. The resulting displacement is then directly measured, usually through a mechanical cantilever system operating at resonance. This is then calibrated against a known standard reference material to convert the displacement into a moment. Most modern AGFMs use a piezoelectric for the cantilever, as it generates an electrical signal when deformed. This avoids the need for laser interferometry or microscopes for determining amplitude. However, care must be taken in the choice of material so that the cantilever itself is not affected by the applied magnetic field, the gradient, or temperature.

As with the DC magnetometers, sample positioning and mounting are critical for accuracy. When the sample is off-center, it is subject to additional alternating field gradients in directions other than the primary one. (It is not possible to ever eliminate all components of the gradient except one; it is only possible to minimize them in the center of the coil so that they are negligible.) Furthermore, if the sample is not securely mounted, the displacement measured will be smaller than the actual displacement as the sample will also rotate about the most secure attachment point, or it may eventually come loose altogether. Finally, the mass of the sample plays an important role, as it changes the resonant frequency of the system. Too large of a mass can shift the AGFM outside of its optimum range, resulting in significant errors to the reported moment.

Finally, an AGFM bridges the gap between a VSM and a SQUID. The total measurement time of an AGFM is fast, like a VSM, but its sensitivity typically lies between that of a VSM and a SQUID. Like the VSM, it is typically limited in magnetic field range due to the use of an electromagnet. However, its main limitation is that it cannot handle large mass samples. So, its usage falls between that of a SQUID and a VSM.

5.2.3 AC Susceptibility

As defined in Chapter 1, the DC susceptibility (χ_{dc}) is given by:

$$\chi_{dc} = \frac{M}{H} \tag{5.3}$$

where M is the scalar magnetization and H is the magnetic field magnitude. This is the equation used when the field is static and known, and the magnetic moment is measured at that fixed field. In the limit of small oscillating (AC) fields about a fixed DC external field (H_{ext}), the susceptibility becomes a dynamic AC susceptibility:

$$\chi_{ac} = \left.\frac{dM}{dH}\right|_{H\_} \tag{5.4}$$

In this limit, χ_{ac} is the slope of the hysteresis loop at a particular DC field, which may be non-zero, and is, therefore, also known as the differential susceptibility.

AC susceptibility instruments are similar to the DC magnetometers (VSM and SQUID) in that they also use pickup coils to measure the change in voltage to determine the magnetization of the sample. The difference comes in how the changing magnetic field is generated. In DC magnetometers, the changing magnetic field is generated by moving the sample in the pickup coil. In an AC susceptometer, the magnetic field itself is oscillated while the sample is left centered within the coils. Therefore, an AC susceptometer (Figure 5.3) starts with a primary excitation field coil to generate this time-varying magnetic field. The time-dependent (dynamic) properties of the sample are determined by measuring the voltage change, which includes both the amplitude as well as the phase angle, in the primary pickup coil with a lock-in amplifier. However, because the magnetic field and the sample are now varying with time, the applied field is also measured in the primary pickup coils. For this reason, an empty coil pair called the secondary compensation coil is also present, which measures the signal induced by the applied field so that this background can be removed from the signal. (This secondary compensation coil is "empty" because it never sees the sample.)

Within AC susceptibility instruments, there are two variants: longitudinal and transverse (Figure 5.4). The longitudinal AC susceptometer is the most common AC susceptibility instrument in use. In this instrument, the AC field is applied parallel to the DC bias field, so that the AC properties are measured parallel (or longitudinal) to the bias field. In contrast, in the transverse AC susceptometer, the AC field is applied perpendicular to the DC bias field, so that the AC

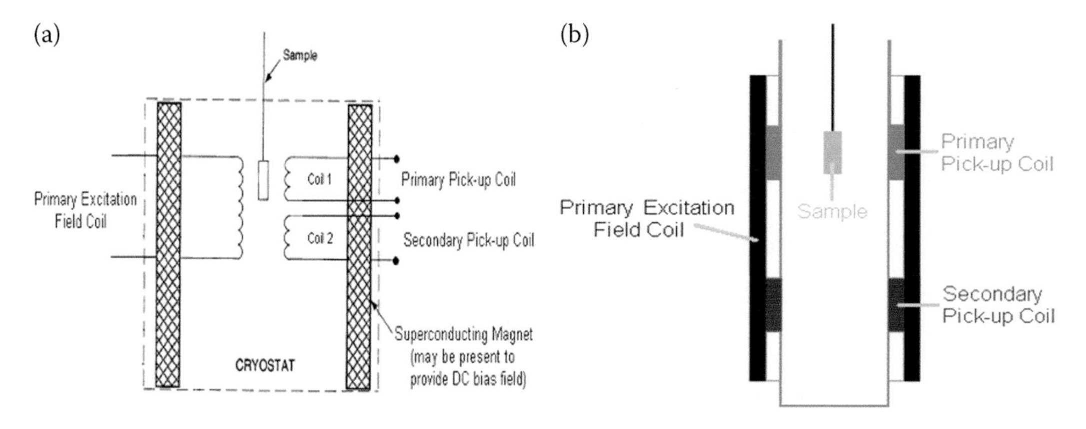

FIGURE 5.3 **(a)** Schematic of and **(b)** Cross-section view of an example AC susceptibility coil configuration, courtesy of LakeShore Cyrotronics, Inc.

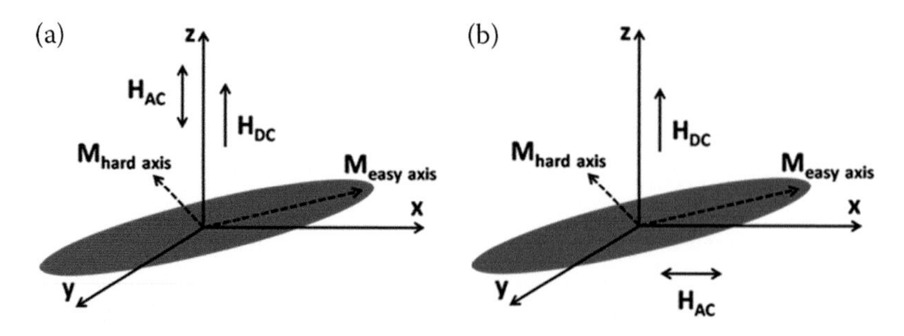

FIGURE 5.4 Schematic of the **(a)** Longitudinal and **(b)** Transverse AC susceptibility configuration, courtesy of N.F. Huls.

properties are measured normal (or transverse) to the bias field. However, both measure the real (χ') and imaginary (χ'') components of the susceptibility, as a function of frequency, static magnetic bias field, and AC field magnitude. (The temperature of the sample may be controlled separately.) The real component χ' is in-phase with the applied AC field, while the imaginary component χ'' is out of phase. The relative angle or lag between the two is indicative of the ability of the sample's magnetization to track the field at a given frequency. χ'' is also related to the energy absorbed by the sample from the AC field.

The primary error in AC susceptibility measurements is in the determination of the phase angle ϕ, as this defines the relative intensity of the real and imaginary components.

$$\begin{aligned} \chi' &= \chi \cos\phi \qquad & \chi'' &= \chi \sin\phi \\ \chi &= \sqrt{\chi'^2 + \chi''^2} \qquad & \phi &= \tan^{-1}\left(\frac{\chi''}{\chi'}\right) \end{aligned} \tag{5.5}$$

This phase angle is also critical to measuring any frequency shifts in the susceptibility. One method is to set the system up in a zero-loss state, and work away from that state. For example, the zero-loss state in a ferromagnetic system might be well above the Curie temperature, where the sample is paramagnetic, and then measure the susceptibility as a function of decreasing temperature. Alternatively, a zero-loss state might be in a DC magnetic field ($f = 0$ Hz), where the magnetization is static and therefore "in phase" with the field, and then measure the susceptibility as a function of increasing frequency. In addition, because the voltage induced in the pickup coil is proportional to the frequency of oscillation, lower frequencies typically have lower signal-to-noise ratios than higher frequencies. However, elimination of other sources of noise near the frequency of interest (e.g. electrical mains operate at 50 Hz or 60 Hz) can help to increase signal-to-noise ratios at a given frequency.

Finally, the AC susceptometer is used for studying the dynamic properties of the sample, such as relaxation processes, irreversible domain wall movement, or energy losses. This is distinct from the DC properties measured by the DC magnetometers or the AGFM, so it fills a particular niche.

5.2.4 Mössbauer Spectroscopy

Mössbauer spectroscopy, while a common technique for composition determination, can also be used for magnetic characterization. Specifically, Mössbauer spectroscopy is based on the Mössbauer effect, which measures resonant and recoil-free emission or absorption of gamma rays by atomic nuclei. The most common form of Mössbauer spectroscopy measures the absorption of gamma rays, and it is this which is used for iron oxides.

In its simplest form (Figure 5.5), a Mössbauer spectrometer has a radioactive source that emits gamma rays toward a solid sample. (The sample must be solid to eliminate recoil so that you have

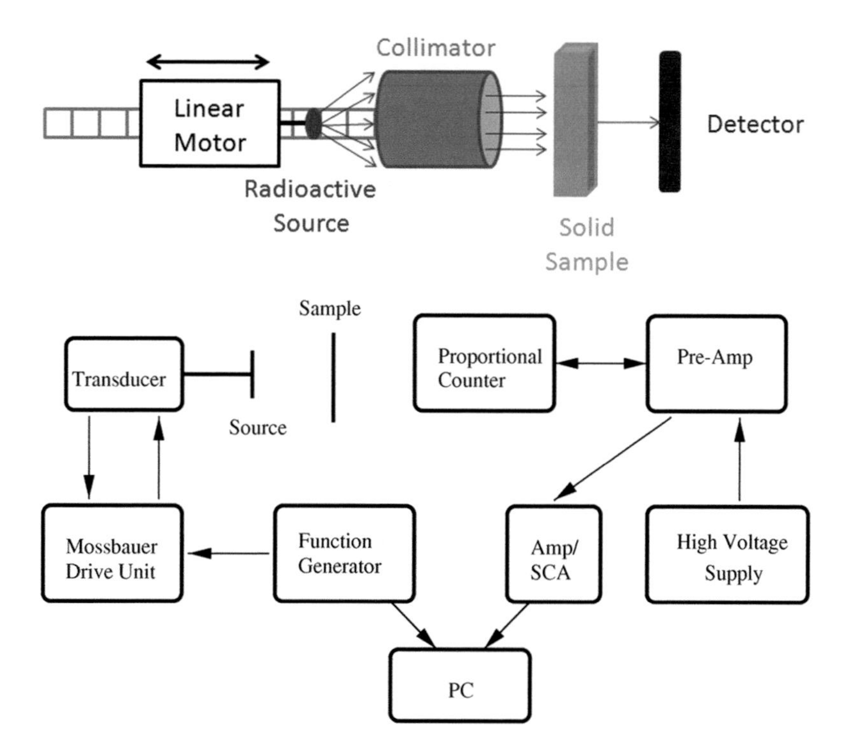

FIGURE 5.5 Schematic of **(Top)** The gamma ray path and **(Bottom)** Control and signal detection configuration in a Mössbauer spectrometer, courtesy of J. van Lierop.

resonant absorption.) Behind the sample is a detector, which measures the intensity of the gamma rays transmitted through the sample. When the energy of the gamma ray is resonant with an atomic energy level, some of the gamma rays are absorbed, resulting in a drop in the measured intensity. Therefore, a typical Mössbauer spectrum (Figure 5.6) shows dips in the data, rather than peaks. However, just to confuse non-experts, in the literature, these dips are typically referred to as peaks, or more accurately, "peaks in absorbance."

The atoms that can be studied with Mössbauer spectroscopy are limited by the choice of radioactive source. As the gamma ray energy needs to be identical to a nuclear energy level for absorption, it is most common to choose a source that decays into the atom to be studied. For example, Co^{57} is used to study Fe^{57} atoms, because Co^{57} decays through electron capture to an excited state of Fe^{57}, which then decays to the ground state by emitting a gamma ray (14.4 keV). Fe^{57} is a naturally occurring isotope (~2% of all iron isotopes), so Co^{57} is the source of choice for studying iron oxides. In addition, the half-life of Co^{57} is 272 days, which is long enough to be practical. However, because the local environment around any individual Fe^{57} atom may not be just more Fe^{57} atoms, it is necessary to change the energy of the gamma ray to detect any shifts in the energy levels. This is done by using the Doppler effect, quantified by:

$$f = \left(\frac{c}{c + v_s}\right) f_0 \tag{5.6}$$

where f is the observed frequency at the sample, f_0 is the emitted frequency by the source, c is the speed of light, and v_s is the velocity of the source. (This equation applies only to a system where the sample is fixed.) The Co^{57} source is placed on a linear motor and accelerated through a range of velocities (typically ±10 mm/s) to change the energy of the gamma ray and effectively scan the energy through a range centered about the initial energy. (Recall that $E = hf$, where E is the gamma

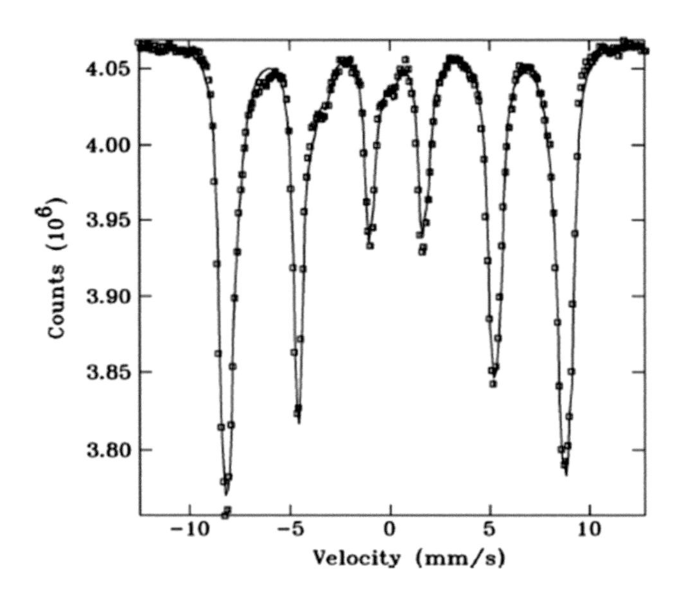

FIGURE 5.6 Example Mössbauer spectrum of 98% Fe_3O_4 and 2% $Fe(OH)_2$ magnetic nanoparticles, measured at 10 K.

ray energy, h is Planck's constant, and f is the frequency.) Therefore, a typical Mössbauer spectrum (Figure 5.6) shows the energy scale in terms of the velocity.

While Mossbauer spectroscopy is one of the most sensitive techniques available, its resolution is limited by three factors:

1. Beam collimation: It is not possible to create a perfectly collimated beam of gamma rays from a radioactive source. Instead, there is a finite divergence angle after the beam leaves the collimator, which induces a cosine shift. If the beam geometry is the same for the calibration sample (typically, an α-Fe foil) and the sample, this can be accounted for. (For high count rates, the beam half-angle can be as large as 14° and cause the velocity calibration [mm/s per channel] to differ by 2% or so from the expected value.)
2. Non-linearities in the linear motor: The addition of a laser interferometer to the linear motor can be used to calibrate the velocity scale as well as to document any non-linearities in the linear motor.
3. α-Fe foil calibration: No two α-Fe foils are exactly identical. Therefore, although the foil calibration helps account for the beam collimation errors, which are larger, the expected accuracy on any system calibrated by this method is approximately 1%.

Finally, Mössbauer spectroscopy, as the above description makes clear, is used to measure small changes in the atomic energy levels that originate from the atomic environment. It is the number, position, and intensity[1] of these peaks that provide information about the chemical environment (such as oxidation state changes or the effect of different ligands on a particular atom) and the magnetic environment of the absorbing nuclei. The three most common measurements are the isomer or chemical shift, quadrupole splitting, and hyperfine (or magnetic) splitting. The last is also sometimes referred to as the Zeeman shift.

The isomer shift is a relative measurement describing the shift in energy of the whole spectrum (Figure 5.7), which can be either positive or negative.[2] This change in the nuclear excited state depends on the nucleus itself, as well as on the size of the nucleus and the density of s-electrons at the nucleus. These changes originate from the addition of one or more neutrons to an atom, or by changing the oxidation state. (Changes in temperature and pressure also affect the size of the nucleus.)

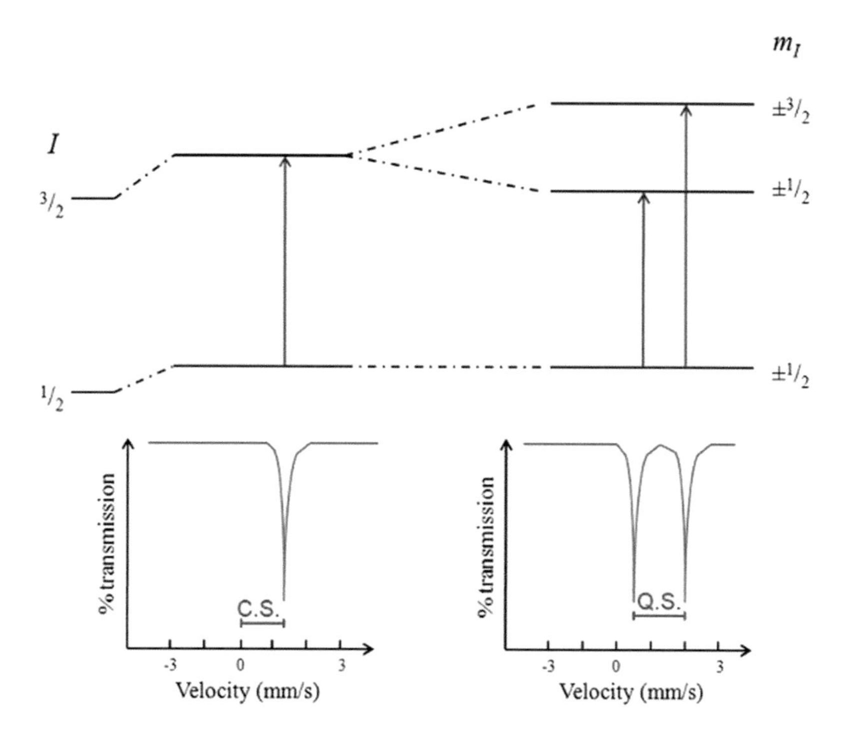

FIGURE 5.7 Chemical shift (C.S.) and quadrupole splitting (Q.S.) of the nuclear energy levels and corresponding Mössbauer spectra, courtesy of J. van Lierop.

Technically, only electrons in s-orbitals matter because, with the spherical shape centered at the nucleus, there is a non-zero density of electrons in the nucleus. However, the p, d, and other shells may influence the s-electron density through screening. For example, ferric iron (Fe^{+3}) has a smaller positive isomer shift than ferrous iron (Fe^{+2}) due to greater screening by d-electrons in ferrous iron.

The electric quadrupole splitting depends upon the non-radially symmetric electrons (i.e. non s-shell electrons) producing an electric field gradient. For example, assume that the Fe^{57} atom is located at the center of a regular octahedron (a typical configuration). If the six nearest neighbor atoms are identical, the electric field gradient is zero. If the lattice is distorted either through pressure or temperature changes, or if different atoms with different charge configurations occupy the nearest neighbor positions, then we have changed the electric field gradient seen by Fe^{57} atom. This electric field gradient splits a single state into two, producing a doublet in the Mossbauer spectrum (Figure 5.7). Specifically, Fe^{57} has an excited state with an angular quantum number I of 3/2. Therefore, changes in the electric field gradient split the 3/2 to 1/2 transition into two spin substates: $m_I = \pm 1/2$ and $m_I = \pm 3/2$. This separation between the states can be used to measure the sign and magnitude of this electric field gradient.

The hyperfine splitting results from the presence of an "effective" magnetic field at the Fe^{57} nucleus. This field most commonly originates from the electrons on the Fe^{57} atom itself, although it is possible to apply an external field large enough to see this effect. Since the neutron also has a spin, the magnetic field at the nucleus favors certain spin states over others, as described by the Zeeman effect. The Zeeman effect is simple magnetostatics, where

$$\begin{aligned} E &= -\vec{\mu} \cdot \vec{B} \\ E &= -g\mu_N \vec{I} \cdot \vec{B} = -g\mu_N B m_I \end{aligned} \tag{5.7}$$

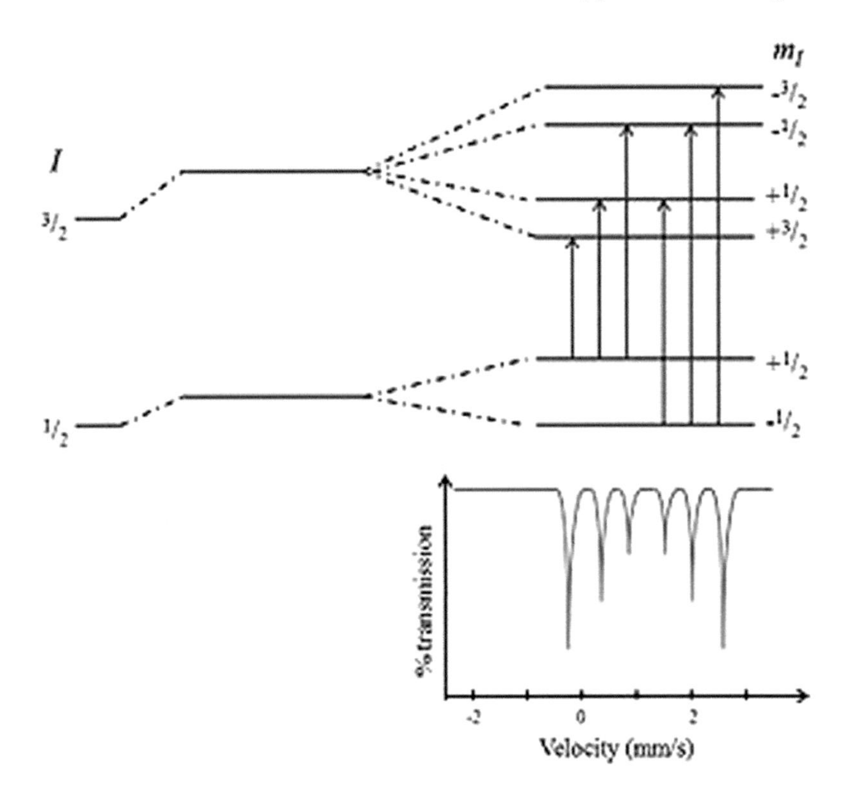

FIGURE 5.8 Magnetic splitting of the nuclear energy levels and the corresponding Mössbauer spectrum, courtesy of J. van Lierop.

where μ is the dipole moment, μ_N is the nuclear magneton, g is the gyromagnetic ratio, B is the magnetic flux,[3] I is the angular momentum quantum number, and m_I is the spin. This energy splitting causes a nucleus with angular momentum I to split into $(2I + 1)$ sub-levels. For example (Figure 5.8), Fe^{57} with an angular momentum state $I = 3/2$ will split into four non-degenerate sub-states with m_I equal to +3/2, +1/2, –1/2, and –3/2. However, quantum transition rules dictate that a transition can occur between the excited state and ground state only when m_I changes by 0 or 1. This gives six possible transitions for a 3/2 to 1/2 transition. In practice, the "effective" magnetic field very closely follows that of the total magnetization in ferromagnetic materials, or the sub-lattice magnetization in antiferromagnetic materials. This means that Mössbauer spectroscopy can be used to directly determine the dependence of the magnetization on temperature, pressure, phase and/or doping (Bødker et al. 1999), crystallite size or grain structure (Bahl et al. 2006), and crystal lattice orientation. In particular, for magnetic nanoparticles, Mössbauer spectroscopy is very useful for temperature-dependent effects, such as identifying blocking temperatures (Desautels, Cadogan, and van Lierop 2009), or magnetic relaxation phenomena (Mørup 1983, Van Lierop and Ryan 2001).

5.2.5 Specialized Techniques

Other more specialized techniques are also useful for the analysis of magnetic nanoparticle systems. There are three categories that have not yet been discussed: resonance techniques, imaging techniques, and those requiring specialized facilities, such as synchrotrons or neutron sources. The resonant techniques include ferromagnetic resonance (FMR) (Kittel 1967) and electron paramagnetic resonance (EPR), also known as electron spin resonance (ESR) (Atherton 1993), and they measure damping constants and attempt frequencies of the system. The imaging methods include special techniques with magnetic fields in transmission electron microscopy (aka Lorentz TEM) (Williams and Carter 1996), such as Fresnel or Foucault imaging, which can help determine magnetic structures and properties on single magnetic nanoparticles or very small collections of them. Magneto-optical Kerr effect (MOKE) (Park et al. 2003, Acremann et al. 2000), magneto-optical Faraday effect (Gornakov et al. 2004), and magnetic force microscopy (MFM) (Sarid 1994) are three additional imaging techniques that can also highlight large-scale magnetic structure within fixed systems. Finally, high-intensity x-ray (Glatter and Kratky 1982) and neutron (Chatterji 2005) scattering, both polarized (Fischer 2011, Krycka et al. 2012a, b) and unpolarized, can provide a wealth of information about ensemble magnetic structures. Interested readers are encouraged to examine the references for more information.

5.2.6 Physical Characterization

Finally, no discussion on the characterization of magnetic nanoparticles would be complete without the physical characterization necessary for interpreting the results of the magnetic characterization. These include (but are not limited to) determination of:

1. Composition through x-ray diffraction (XRD) (Cullity 1978) for non-iron oxide system or Mössbauer spectroscopy[4] (May 1971, Long 1993) for iron oxide–based systems or Raman spectroscopy (McCreery 2000). This allows for comparison with similar systems or with bulk values for the same system.
2. Size and size distribution through TEM for core size, Scherrer analysis (Cullity 1978) in XRD for crystallite size, dynamic light scattering (DLS)/photon correlation spectroscopy (PCS) (Pecora 1985) or nanoparticle tracking analysis using Stokes' equations (Malloy and Carr 2006, Saveyn et al. 2010) for hydrodynamic radius determination. This allows for estimation of blocking temperatures (Néel 1949, Bean and Livingston 1959) and anisotropies as well as for analyzing magnetic property distributions.
3. Surface charge through ς-potential measurements (Hunter 2001, Tantra, Schulze, and Quincey 2010) within DLS. This allows for analysis of interactions and stability.
4. Sample mass through densitometry (Boyes 2003) and freeze-drying to determine total mass of material in solution, thermogravitimetric analysis (TGA) (Brown 2001) for separating surfactant mass from magnetic material mass, and inductively coupled plasma (ICP) (Hill 2006), especially with mass spectrometry (ICP-MS) (Thomas 2013) for determination of elemental content, like Fe, Ni, or Co. This allows for normalization[5] of results.

Again, this is not an exhaustive list, but is a good starting point for understanding the results of the magnetic characterization.

5.3 MEASUREMENT METHODS

Now, with an understanding of how the instrumentation works and common artifacts, the measurement methods themselves can be described. There are two aspects that are critical to any measurement method: understanding the assumptions behind the theoretical analysis and knowing how to identify when your sample meets these conditions. These will be discussed for each method.

5.3.1 DC Hysteresis Loops

A hysteresis loop is simply a plot of the magnetization versus applied magnetic field. These are typically DC measurements, performed on a DC magnetometer (VSM, SQUID) or AGFM. The

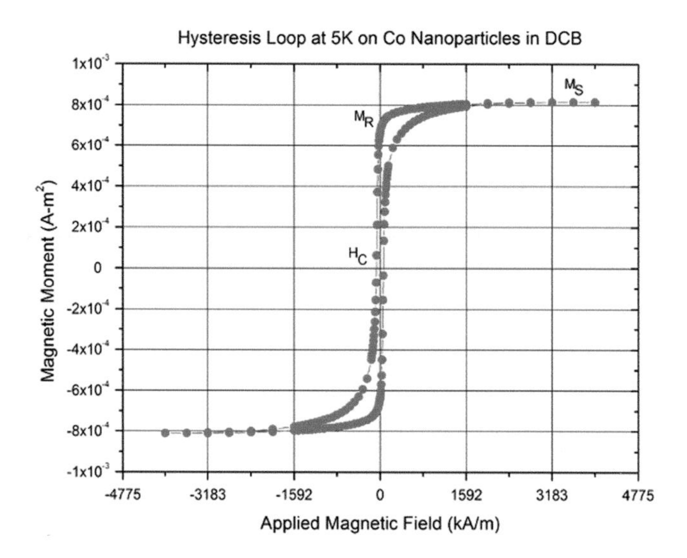

FIGURE 5.9 Example hysteresis loop of a magnetic nanoparticle system, with the major parameters indicated.

most common type of hysteresis loop is also known as a major loop. As a measurement, the major loop is very simple: Starting from a positive saturating magnetic field, the induced moment is measured while decreasing the field to negative saturation, and then back to positive saturation.

The interpretation of this measurement, however, is anything but simple. As discussed in Chapter 1, the most basic parameters, however, are readily evident from the plot (Figure 5.9). These include the saturation magnetization (M_S), the remanent magnetization (M_R), and the coercivity (H_C). (Note: The coercivity is not an intrinsic property of the material, while the switching field is. Furthermore, although the coercivity and the switching field are often equated, they are not necessarily the same (Kahler, Bennett, and Della Torre 2006).) In some cases, an exchange bias may be present shifting the loop along the magnetic field axis by the bias field (H_B).

Ultimately, however, all of the information about the magnetic nanoparticles is contained within this loop. The difficulty is in separating it into its components. In a few simple cases, the hysteresis loops can be understood quite well, or a reasonable estimate of some parameters can be achieved. For example, if we have a true paramagnet (or superparamagnet) where the atoms (or nanoparticles) are not interacting, we can fit the entire curve with the quantum mechanical prediction for paramagnetism where g (free electron g-factor) is assumed to be 2 and J (quantum mechanical spin) is assumed to be ½. (The difference between paramagnetism and super-paramagnetism is in the size of the unit magnetic moment.) This is given by:

$$M_T = Nm \tanh\left(\frac{mB}{k_B T}\right) \tag{5.8}$$

where M_T is the total magnetic moment, N is the number of particles, m is the moment per particle, B is the magnetic flux, [7]k_B is the Boltzmann constant, and T is the temperature. The semi-classical version of this equation, also known as the Langevin theory of paramagnetism, is given by:

$$M_T = Nm\left[\coth\left(\frac{mB}{k_B T}\right) - \left(\frac{k_B T}{mB}\right)\right] \tag{5.9}$$

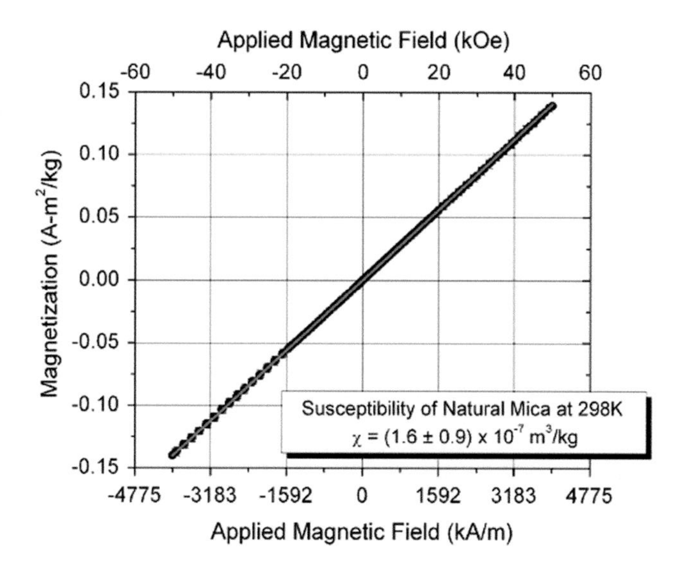

FIGURE 5.10 Example hysteresis loop of a paramagnetic system of natural mica. The line indicates the fit Equation 5.3 for the susceptibility.

For most paramagnetic samples, the moment will not saturate in reasonably achievable laboratory fields, so only the susceptibility (Equation 5.3) can be calculated (Figure 5.10). In contrast, the moment will generally saturate in laboratory magnetic fields for a superparamagnetic sample (which must, by definition, be non-interacting). For example, the moment per nanoparticle for a well-dispersed system of 4 nm Fe_3O_4 nanoparticles is 963 μ_B (Figure 5.11).

At the other end of the spectrum, we can estimate the anisotropy from (typically) a strongly interacting and narrow polydispersity system. This is easiest with a clear uniaxial anisotropy, which results in an easy and hard axis (Figure 5.12). For a strongly interacting system, this strong uniaxial anisotropy will dominate any thermal randomization, and result in almost a straight line

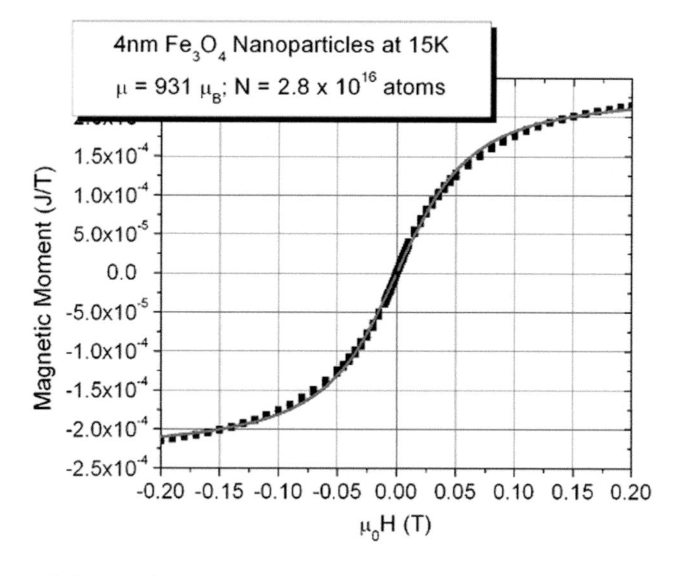

FIGURE 5.11 Example hysteresis loop of a superparamagnetic system of well-dispersed iron oxide nanoparticles. The line indicates the fit to Equation 5.9.

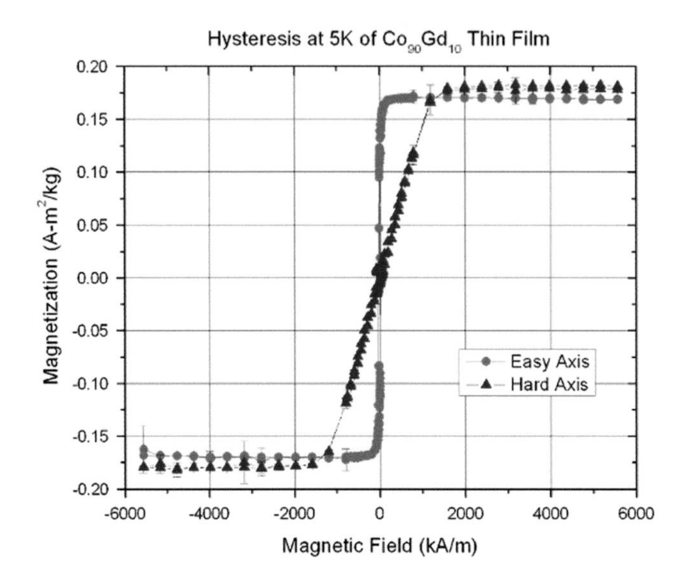

FIGURE 5.12 Example hysteresis loop of a magnetic thin film of $Co_{90}Gd_{10}$, showing easy and hard axes. Note: The saturation magnetization values along the easy and hard axes are within error of each other.

for the hysteresis until near saturation. (If the sample is fixed so that the nanoparticles cannot physically rotate, then the anisotropy of both the easy and hard axes can be determined.) The anisotropy field (H_K) can then be estimated by the point at which this straight line hysteresis would sharply turn over into saturation. (If the system is not strongly interacting or has too large of a size distribution, then the transition from linear slope into saturation will be very broad, and the resulting error renders this estimate worthless.) The anisotropy field is related (Chikazumi 1964) to the anisotropy constant by:

$$\mu_0 H_K = \frac{2K_{eff}}{M_S} \tag{5.10}$$

where μ_0 is the permeability of free space, K_{eff} is the effective anisotropy constant, and M_S is the saturation magnetization.

Both of these extremes are based on the presence or absence of interactions between the magnetic nanoparticles. So, it is critical to estimate when they can be neglected. Again, this information can be determined from the hysteresis loop itself, by adding one additional piece of information: the virgin curve. This is an additional quadrant, which can be measured on a sample that has never seen a magnetic field before: Starting from zero applied magnetic field, the induced moment is measured while increasing the field to positive saturation. This is the virgin curve. The major loop then follows.

Before examining the virgin curve in context of the major loop, it is important to know what type of magnet is in the instrument. As previously mentioned, a superconducting magnet will have trapped flux that offsets the actual magnetic field. This offset must be accounted for as it can be considerable and is dependent on the magnetic field history of the magnet since its last "reset." In addition, what is defined as "negligible interactions" is determined by the resolution of the instrument you are using. Now, for a non-interacting magnetic nanoparticle system, after the first zero-field data point (which may be positive or negative depending on if the sample has ever seen any stray fields during sample preparation or loading), the virgin curve should be identical to the major loop (Taketomi and Shull 2002). (If trapped flux is not taken care of, then the curve will be parallel but offset from the major loop.) Non-negligible interactions mean that

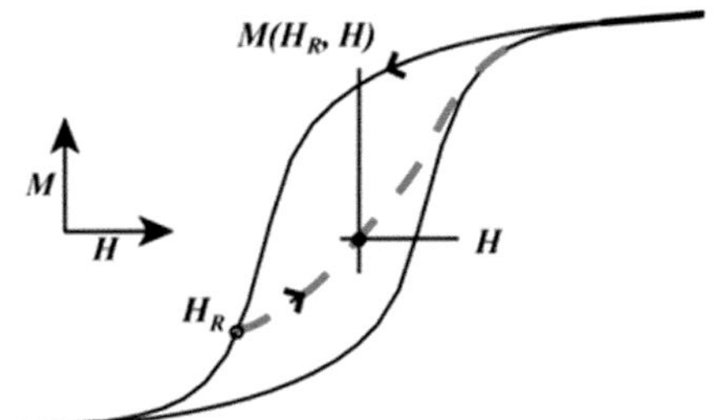

FIGURE 5.13 Example minor loop, specifically, a first-order reversal curve (dashed line), courtesy of J.E. Davies.

the virgin curve will diverge from the major loop. The larger the deviation is, the stronger the interactions are.

Finally, beyond these simple cases, it is very difficult to separate out the different contributions to a hysteresis major loop. Since this is the majority of samples, modeling can be used to help shed light on the different contributions. For more information, interested readers should consult the following references: Stoner and Wohlfarth (1948), Schaller et al. (2009a, b, 2010), Aranda et al. (2009), Dorfbauer et al. (2007), and Chantrell et al. (1996).

Additional information can be gained from hysteresis loops beyond the major loop. The next step is to include minor loops to fill in the center of the major loop. One type of minor loop, also known as a first-order reversal curve (FORC), is measured by stopping the major loop at a particular field and then increasing the applied field while measuring the induced magnetic moment (Figure 5.13). These loops are only interesting when the magnetization is not completely reversible, but rather has an irreversible component. (If the magnetization were completely reversible, like in a pure paramagnet or superparamagnet, the minor loop would just re-trace the major loop.) By filling the entire major loop from positive saturation to negative saturation with these FORCs, enough information is gathered to quantify (Davies et al. 2004, 2005a, b, 2008, 2009, Dumas et al. 2007, Katzgraber et al. 2002) the amount (the FORC distribution ρ) of irreversible switching in a given sample. (Typically, the major loop delineates the outer boundary of the FORCs, but not always (Davies et al. 2008).) For this reason, the experimental determination of ρ does not depend on any models, but can be used to measure any sample that exhibits hysteresis. To fully analyze FORC measurements, ρ is interpreted in a framework of hysteresis models (Pike, Roberts, and Verosub 1999), particularly the Preisach model (Preisach 1935, Mayergoyz 1991, Torre 1999). Ultimately, the utility of FORC lies in its ability to take detailed *macroscopic* measurements and turn them into information on *microscopic* reversal mechanisms. More importantly for many magnetic nanoparticle systems, the interpretation of the ρ may also quantify (Ciureanu et al. 2008, Dobrotă and Stancu 2013) the distributions in magnetic properties. As applications of magnetic nanoparticles become more developed, it becomes even more critical to detect if there are variations in the magnetic properties of materials produced by a given synthesis procedure, and then to quantify that variation. Therefore, FORC is particularly useful for quality control, in that it identifies the property distributions and how it varies within a single batch as well as from batch to batch. This is also useful for comparing the nanoparticles from different synthesis methods (Willard et al. 2004) (e.g. co-precipitation, thermodecomposition, polyol, electrochemical), since they are used to control the parameters, most especially size, morphology, and crystallinity.

To truly understand FORC, it is necessary to have some understanding of its origins and assumptions, as well as errors inherent in the analysis. The basis of FORC analysis, the Preisach model (Preisach 1935), assumes that the major hysteresis loop of a ferromagnetic sample is comprised of an infinite set of building blocks called hysterons. Each hysteron ($\hat{\gamma}_{H_{down},H_{up}}$) has a square hysteresis loop, with two parameters associated with it (Figure 5.14): H_{down}, the reversal field where the magnetization goes from positive to negative saturation and H_{up}, the reversal field

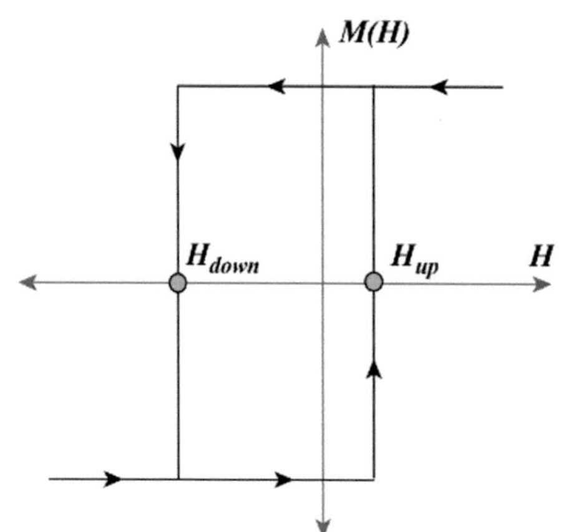

FIGURE 5.14 Example hysteron indicating up and down reversal fields, courtesy of J.E. Davies.

where the magnetization goes from negative to positive saturation. (Every hysteron is assumed to have the same saturation magnetization.) In the context of magnetic nanoparticles, a single hysteron may (Dobrotă and Stancu 2013) represent a single nanoparticle (assuming that the nanoparticles themselves are single-domain). Otherwise, a hysteron may represent a single magnetic domain. Furthermore, there is a normalized distribution function ρ that describes how many hysterons with that specific reversal field combination (H_{up}, H_{down}) are in the major loop. From this, the magnetization M at a given applied field $H_{applied}$ can be written as a summation of hysterons:

$$M = \iint_{H_{up} \geq H_{down}} \rho\left(H_{down}, H_{up}\right) \hat{\gamma}_{H_{down}, H_{up}} \, dH_{down} \, dH_{up} \tag{5.11}$$

The question remains: how is the distribution function ρ (Pike 2003) determined?

For ease of understanding, this calculation of ρ will be first represented graphically. Because each hysteron has a unique H_{up} and H_{down}, they can be plotted on a grid where the axes are H_{down} and H_{up} (Figure 5.15). For any given major loop, there are limits on the magnitude of the characteristic reversal fields for the hysterons, which are represented by the rightmost vertical and bottommost horizontal line. The 45° boundary is where $H_{up} = H_{down}$. For simplicity, we define $H_{up} \geq H_{down}$.

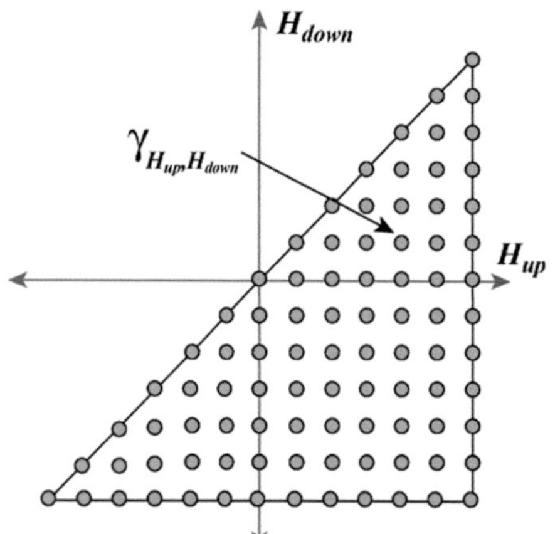

FIGURE 5.15 Grid where each point (H_{up}, H_{down}) corresponding to a hysteron, courtesy of J.E. Davies.

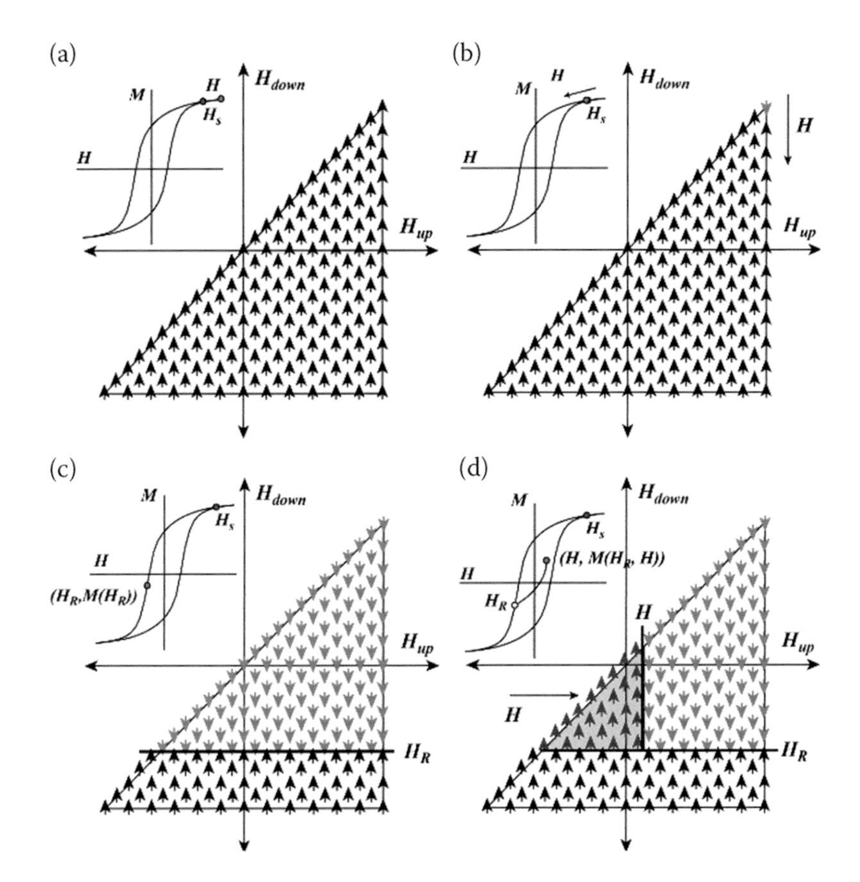

FIGURE 5.16 Graphical representation of FORC distribution for **(a)** $H > H_s$, **(b)** $H = H_s$ where the hysteron with the maximum H_{down} begins to reverse, **(c)** H at some reversal field H_R, and **(d)** H traversing back up to positive saturation along a first-order reversal curve, courtesy of J.E. Davies.

Pictorially, the FORC distribution can now be calculated for any minor loop within a major loop. First, the applied field is set higher than the saturation field (H_S) of the major loop (Figure 5.16a), so that the hysterons are all "up." Once the field is lowered below H_S, hysterons with the largest H_{down} values reverse first (Figure 5.16b). As the field is reduced further to H_R, with the magnetization tracing the major loop, more hysterons with progressively smaller values of H_{down} reverse (Figure 5.16c). The magnetization at this point is denoted as $M(H_R)$. Next, the applied field is increased from H_R, with the magnetization tracing out the minor loop, and hysterons with progressively larger values of H_{up} reverse (Figure 5.16d). The magnetization at this point is denoted as $M(H_R, H)$. The difference in these two magnetizations, $M(H_R)–M(H_R, H)$, given by the shaded triangle in Figure 5.16d, represents the number of hysterons with $H_{down} \geq H_R$ and $H_{up} \leq H$ (since all hysterons have the same saturation magnetization).

Mathematically, this is summarized as follows. The magnetization at any given H is the sum of the "up" and the "down" hysterons:

$$M(H) = \iint_{S^+} \rho(H_{up}, H_{down})\hat{\gamma}_{H_{up},H_{down}} dH_{up} dH_{down} + \iint_{S^-} \rho(H_{up}, H_{down})\hat{\gamma}_{H_{up},H_{down}} dH_{up} dH_{down} \tag{5.12}$$

where S^+ and S^- denote the regions in Figure 5.16 where the hysterons are "up" or "down," respectively. By assigning the hysteron operator $\hat{\gamma}_{H_{down},H_{up}}$ a value of 1 for "up" or –1 for "down," this simplifies to:

$$M(H) = \iint_{S^+} \rho(H_{up}, H_{down})dH_{up}dH_{down} - \iint_{S^-} \rho(H_{up}, H_{down})dH_{up}dH_{down} \tag{5.13}$$

The difference $M(H_R)$–$M(H_R, H)$ can now be expressed as:

$$M(H_R) - M(H_R, H) = -2\iint_{T(H_R,H)} \rho(H_{up}, H_{down})dH_{up}dH_{down} \tag{5.14}$$

where the integration is over the shaded triangle [$T(H_R, H)$] shown in Figure 5.16d. Solving for $\rho(H_R, H)$ yields:

$$\rho(H_R, H) = -\frac{1}{2}\frac{\partial[M(H_R, H) - M(H_R)]}{\partial H_R \partial H} \tag{5.15}$$

Since $M(H_R)$ is not a function of H, its contribution is zero. This leaves:

$$\rho(H_R, H) = -\frac{1}{2}\frac{\partial M(H_R, H)}{\partial H_R \partial H} \tag{5.16}$$

By knowing the magnetization at a given (H_R, H) the percentage of hysterons with (H_R, H) can be experimentally determined. In addition, the FORC distribution is also a measure of the change in the local susceptibility (Martínez-García et al. 2014) as H_R is changed. This means that if the slope becomes more positive with decreasing H_R then $\rho(H_R, H) > 0$. However, if the slope becomes less positive as H_R is decreased then $\rho(H_R, H) < 0$. The sign of the FORC distribution in a given region is indicative of the physical reversal processes actually occurring.

Equation 5.16 defines the FORC distribution ρ utilized in the subsequent examples. The only assumption made in this calculation is that the sample is composed of a set of hysterons, each with a square hysteresis loop with the same saturation magnetization and a characteristic pair of reversal fields. This assumption is independent of the hysteresis model. In fact, only simple systems like non-interacting single-domain particles (superparamagnets below their blocking temperature) yield the same distributions in both the FORC and the basic Preisach model (Preisach 1935). (Certain features, like regions where $\rho(H_R, H) < 0$, can occur in FORC distributions (Davies 2009, Stancu, Andrei, and Stoleriu 2006), but are not seen in the basic Preisach model.)

In practice, a large number of FORCs with different values of H_R from positive saturation to negative saturation must be measured to calculate the FORC distribution. For ease in numerically calculating the mixed partial derivative, the H_R values are normally evenly spaced and this step size is equal to the step size in the applied field along a minor loop. Other numerical issues (Heslop and Muxworthy 2005), including noise in the data, which can cause significant local changes in the FORC distribution, are typically handled by smoothing the data prior to differentiation. Also, the distribution at the boundary where $H = H_R$ is problematic. It is typically evaluated purely in the limit as $H \rightarrow H_R$, although some extensions of the dataset to $H < H_R$ have been suggested (Acton et al. 2007) for deriving information on the reversal behavior at $M(H_R)$ (i.e. the reversible component of the magnetization).

Furthermore, the large number of FORCs that must be measured means that, experimentally, the instrument used for these measurements must have a rapid field change as well as rapid moment measurement. For example, on a VSM, AGFM, or SQUID VSM, a set of FORCs can take easily two days to measure. In contrast, the relative field change rates and measurement times on a SQUID mean that the same set of FORCs can take two weeks or more. In practice, the set of FORCs is generally performed on the descending branch of the major loop, but it can also be performed on the ascending branch. Normally, the two branches would be expected to be mirror images, but if the sample switches by different methods (Nikitenko et al. 1998) between ascending

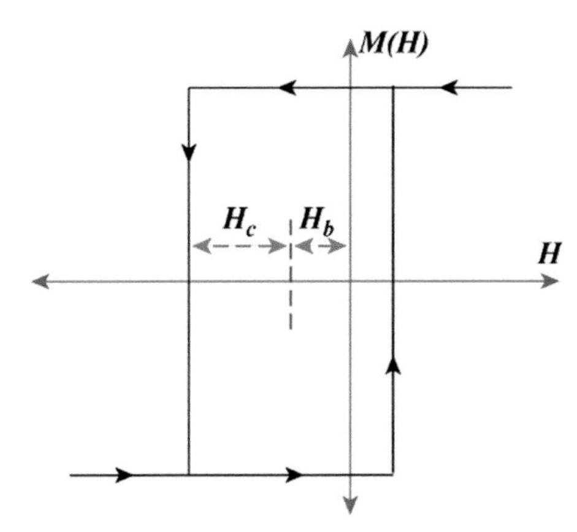

FIGURE 5.17 Hysteron represented in terms of its bias (H_b) and coercivity (H_c), courtesy of J.E. Davies.

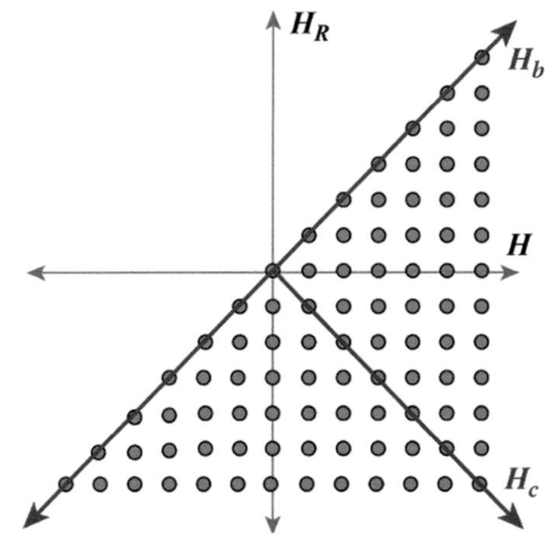

FIGURE 5.18 Field plot showing the relation between H–H_R and H_b–H_c coordinates, courtesy of J.E. Davies.

and descending (e.g. domain nucleation and growth versus rotation), then different minor loops would result.

Finally, the descriptive H_{up} and H_{down} are not directly relevant in discussions on magnetic nanoparticle properties. Instead, it is easier to compare (Figure 5.17) their coercivity (H_C, loop half width) and bias field (H_b, loop shift from origin). Fortunately, H_{up} and H_{down} can be rotated into H_C and H_b through a coordinate transformation (Figure 5.18):

$$H_c = \frac{H - H_R}{2} \quad H_b = \frac{H + H_R}{2} \tag{5.17}$$

This rotation of the coordinate system also means that the system can be analyzed in terms of the distributions of local coercivity and bias. These distributions also imply other property distributions, like anisotropy (Oliva, Bertorello, and Bercoff 2004) and interactions. For example, the coercivity of a nanoparticle is defined (O'Handley 2000) by its intrinsic anisotropy as well as by the size and shape of the particle. The biasing of a nanoparticle can be accounted for by the interactions of that nanoparticle with its neighbors (Pike, Roberts, and Verosub 1999, Muxworthy and Williams 2005) through, for example, dipolar fields or exchange interactions. For ease in visual comparison, the full FORC distribution (Figure 5.19a) is often projected onto just one axis, for a coercivity (Figure 5.19b) or a bias field distribution (Figure 5.19c).

For example, consider two magnetic nanoparticle systems. The first (Cheng et al. 2007b) is composed of narrow (10%) polydispersity ε-Co nanoparticles 10 nm in diameter coated with oleic

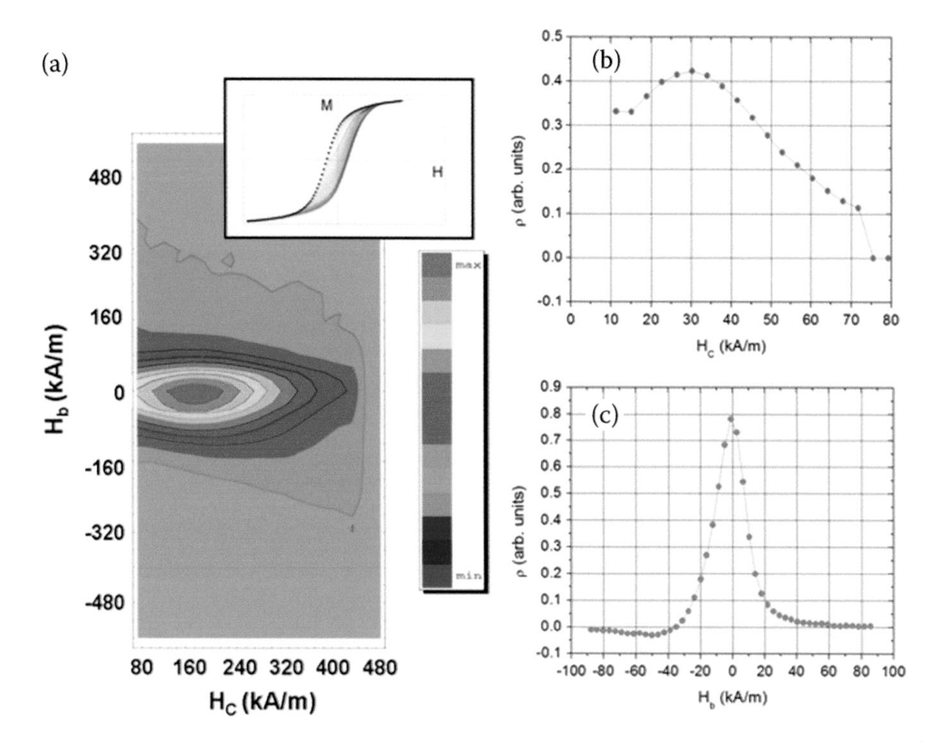

FIGURE 5.19 **(a)** Full, **(b)** H_C-only, and **(c)** H_b-only FORC distribution for ε-Co nanoparticles at 150 K. The inset to **(a)** shows the original first-order reversal curves.

acid while the second (Grüttner et al. 2007) is composed of broad (30%) polydispersity Fe_3O_4 nanoparticles 44 nm in diameter coated with dextran and cross-linked with maleimide groups (Figure 5.20). While the minor loops are definitely different between the two samples, the FORC distributions show glaring differences. First, the location on the FORC diagram of the largest intensity (or highest concentration of hysterons): For the Co nanoparticles, this is a strip centered about $H_b = 0$ kA/m and $H_c = 0$ kA/m, and is much narrower along the H_b axis than along the H_c axis; for the Fe_3O_4 nanoparticles, this area is off-centered around $H_b = 8$ kA/m and $H_c = 48$ kA/m, but is only slightly elongated along the H_c axis. This difference can be traced back to their equilibrium spacing. We know that the Co nanoparticles have a very short surfactant chain (~3 nm) on the surface (Cheng et al. 2009). Therefore, the Co particles can physically be quite close together, resulting in a strong dipolar coupling. The Fe_3O_4 nanoparticles, however, have an order of magnitude thicker coating (~24 nm) (Dennis et al. 2008), thereby preventing the particles from getting as close together. This greatly reduces the dipolar coupling between the Fe_3O_4 nanoparticles. Therefore, the interaction coupling is much more uniform for the Co nanoparticles than for the Fe_3O_4 nanoparticles. However, the bias (interaction) field is still non-zero in the Fe_3O_4 nanoparticles, in agreement with small-angle neutron scattering measurements (Dennis et al. 2008, 2009), which establish the existence of strong interactions and their spacing. Second, only the strongest outliers in coercivity for the Co nanoparticles are visible in the FORC distribution. In contrast, there is significantly more variation in the coercivities in the Fe_3O_4 nanoparticles, due in part to the 3 × larger polydispersity.

Other variations on FORCs exist (Béron et al. 2007, Béron, Ménard, and Yelon 2008, Béron, Pirota, and Knobel 2011, Andrei, Caltun, and Stancu 2006), including using the ascending rather than the descending major loop. In addition, hysteresis measurements can be extended another step beyond FORCs, to second-order reversal curves (SORCs) (Bodale, Stoleriu, and Stancu 2010, Stancu, Andrei, and Stoleriu 2006). In these measurements, the field is reversed part-way through a

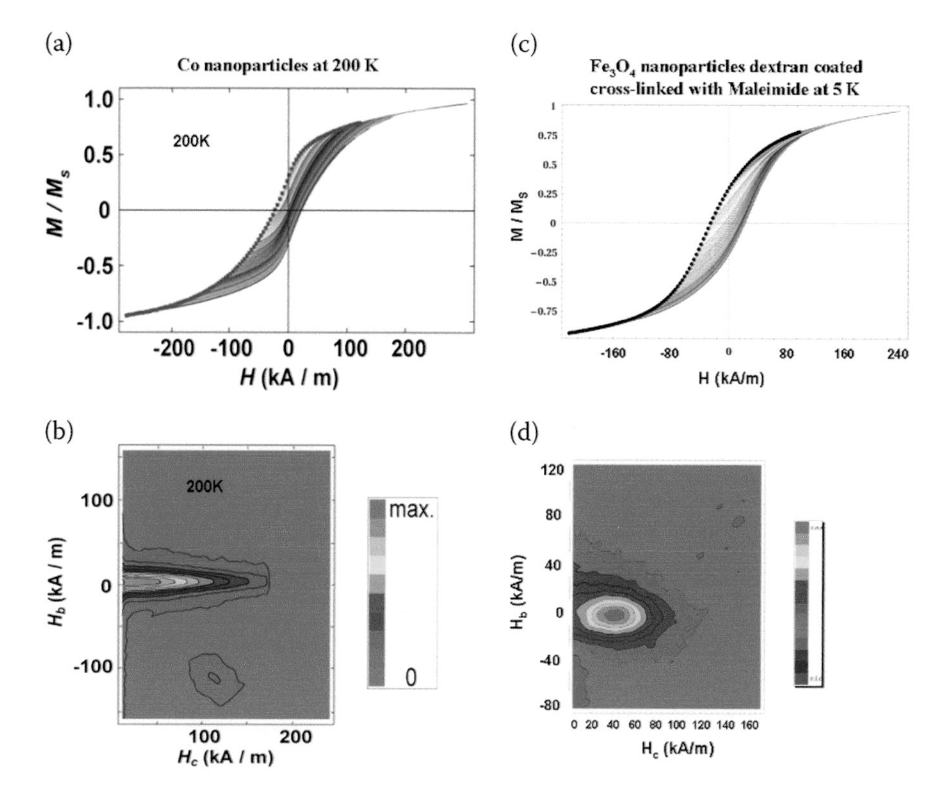

FIGURE 5.20 FORC data on ε-Co nanoparticles at 200 K showing the **(a)** Minor loops and **(b)** Distribution in H_b–H_c. FORC data on Fe_3O_4 nanoparticles coated with dextran and cross-linked with maleimide groups at 25 K showing the **(c)** Minor loops and **(d)** Distribution in H_b–H_c. For the Co, the measurements were performed with an H_R step size of 4 kA/m (50 Oe). For the Fe_3O_4, the measurements were performed with an H_R step size of 2 kA/m (25 Oe).

FORC. Interested readers are directed to the references for more information on the analysis of SORC datasets.

Finally, FORC is not the only way to estimate magnetic distributions within a nanoparticle system. For a superparamagnetic (well-dispersed and non-interacting) system, it is also possible to estimate the distribution of blocking temperatures[6] within a nanoparticle system (Bean and Livingston 1959). The blocking temperature for a single monodomain nanoparticle is defined by the measurement time scale τ_{meas}, as well as the ratio of the thermal energy available ($k_B T_B$) to overcome the anisotropy energy barrier ($K_{eff} V_M$). It can be calculated by:

$$\tau_{meas} = \tau_0 e^{\left(k_{eff} V_M / k_B T_B\right)} \tag{5.18}$$

where τ_0 is the frequency with which the moment attempts to cross its energy barrier (and is 1 ns for a truly non-interacting system), K_{eff} is the effective anisotropy, V_M is the magnetic volume of the particle, k_B is Boltzmann's constant, and T is the temperature. Given the exponential dependence upon size combined with the existence of a size distribution for every nanoparticle system, even those which self-assemble and are commonly mislabeled as "monodisperse" rather than the correct "narrowly polydisperse," the blocking temperature for a nanoparticle system is never an exact value. Measurement of the blocking temperature distribution can be done by normalizing the major loops for temperature, using a horizontal axis of *H*/*T*. When all of the nanoparticles are superparamagnetic, the magnetization versus *H*/*T* curves will overlap, regardless of the temperature. As the temperature is lowered from a completely superparamagnetic state to a completely blocked state, it is possible to identify at each temperature the total magnetization of all thermally blocked nanoparticles. Once normalized to the total magnetization in the completely blocked state, a distribution in blocking temperatures can be generated. This is, of course, also related to distributions in size and anisotropy. While a time-consuming measurement method, it is the most direct measurement of the blocking temperature distribution.

5.3.2 Magnetization versus Angle (Torque)

Magnetization versus angle (M versus θ) plots are instrumental in calculating the torque applied in the magnetic system. Torque can also be measured directly using a torquemeter, but given that magnetic torques are often quite small, especially for magnetic nanoparticles, it is more common to measure, using a vector magnetometer (see Figure 5.1), the magnetization as a function of the angle θ between the applied magnetic field and the magnetization vector, and then calculate the torque. The torque (τ) is

$$\vec{\tau} = \vec{r} \times \vec{F} \tag{5.19}$$

where r is the moment arm and F is the force. In a magnetic system, this means that

$$\begin{gathered}\vec{\tau} = \vec{m} \times \vec{B} \\ \tau = mB \sin \theta\end{gathered} \tag{5.20}$$

where m is the moment and B is the magnetic flux.[3] More importantly, however, the torque is also the derivative of the energy. Specifically, in magnetism, the torque is the derivative of the anisotropy energy (E_A) with respect to the angle between the magnetization and the applied magnetic field.

$$\tau = \frac{\partial E_A}{\partial \theta} \tag{5.21}$$

To measure the vector magnetization experimentally, the instrument must have at least two coils oriented at 90° to each other within the plane in which the sample or magnet rotates (Figure 5.1). If rotating the sample rather than the magnet itself, a perfectly straight rod must be used; otherwise, the background must account for the anisotropy artifact originating from the sample always being closer to one pickup coil in a pair at any given angle. When rotating the magnet, this is not a problem because the sample stays fixed relative to the pickup coils and the applied magnetic field is uniform. The two pickup coils are necessary to measure both components of the magnetization as a function of the angle of the magnetic field (relative to a starting point). The magnetization can be measured at any static applied field; however, only if the applied magnetic field is also a saturating field will the calculated anisotropy be a real anisotropy. Otherwise, the calculated anisotropy is simply an effective anisotropy.

From the vector components of the magnetization data as a function of the applied field angle, the magnetization angle can be calculated with trigonometry, using the same point of origin as the field angle. The difference between the applied field angle and the angle of the magnetization in the plane is the angle θ between the magnetization vector and the field. This combined with the saturation magnetization magnitude and the applied field magnitude means the torque can be calculated from Equation 5.20.

The final step is to calculate the derivative of the anisotropy energy with respect to the angle. This function can then be used to fit the resulting data to determine the anisotropy constant (effective or real) and its associated error. Unfortunately, there is no absolute determination of the magnetocrystalline anisotropy energy, as it depends on the crystal structure and the size of the particle (Skomski 2008). However, phenomenological models do exist for a number of simple systems (Skomski 2008). We'll consider the most common case in magnetic nanoparticles: a particle with a uniaxial anisotropy. (There are also higher-order terms that may play a role; however, these are typically negligible compared to the first term.) The uniaxial anisotropy energy is:

$$E_A = K_u V_m \sin^2(\theta - \phi_u) \tag{5.22}$$

where K_u is the uniaxial anisotropy, V_m is the volume of the magnetic component of the particle, θ is the angle between the magnetic field and the magnetization, and φ_u is a constant. The torque then becomes:

$$\tau = K_u V \sin\left[2(\theta - \phi_u)\right] \tag{5.23}$$

Phenomenologically, with magnetic nanoparticle colloids, a second term is often observed, which originates from the chains that form in the presence of a magnetic field (Dennis et al. 2018). This is a unidirectional anisotropy, which has one preferred direction—along the length of the chains in the direction of the initial DC magnetic field. This anisotropy appears as the chains start to cluster, and they can no longer freely rotate. Therefore, the original field direction is the preferred orientation. This unidirectional anisotropy takes the form:

$$E_A = K_d V_m \cos\left(\theta - \phi_d\right) \tag{5.24}$$

where K_d is the unidirectional anisotropy and φ_d is a constant. The torque then becomes:

$$\tau = K_d V \sin(\theta - \phi_d) \tag{5.25}$$

The two terms individually look like Figure 5.21a, and combined they give a linear appearance, as shown in Figure 5.21b. Typically, the volume is divided out to normalize the torque for comparison across different systems.

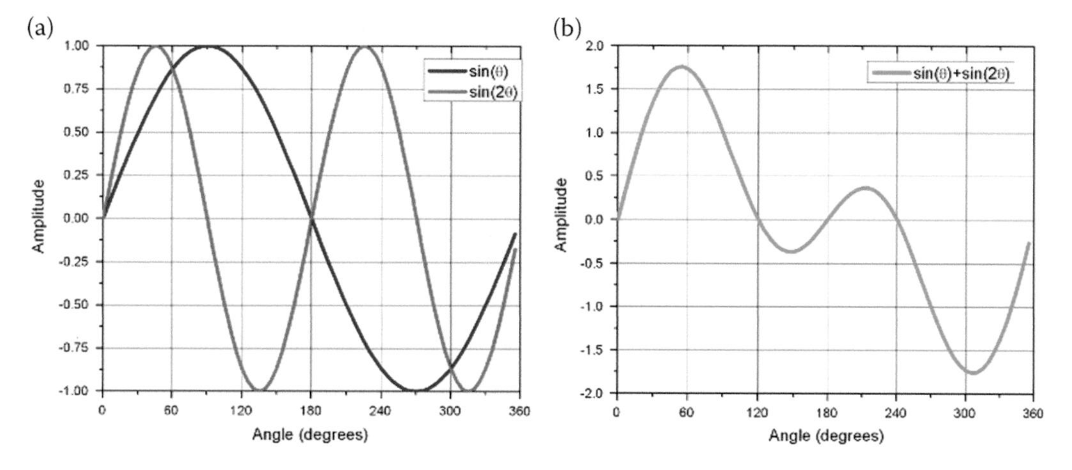

FIGURE 5.21 **(a)** Uniaxial and unidirectional components of the torque as a function of angle, **(b)** Linear combination of uniaxial and unidirectional components of the torque, in equal amounts.

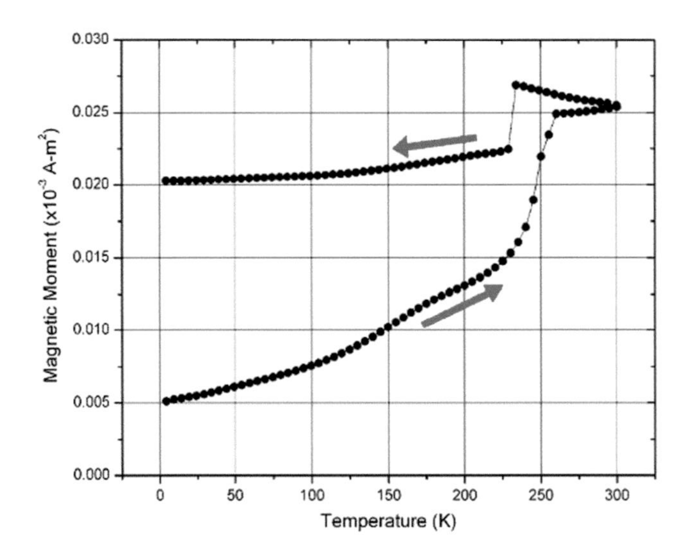

FIGURE 5.22 Magnetization versus temperature characterization in H = 15.9 kA/m (200 Oe) after cooling to 4.2 K in zero field of 10 nm diameter Co nanoparticles. The arrows indicate the direction of the measurement. (M versus T of DCB alone shows a purely diamagnetic signal.)

For example, consider a magnetic nanoparticle system composed of 10 nm ε-Co nanoparticles dispersed in 1,2-dichlorobenzene (DCB). This is an interacting system, which displays (Figure 5.22) in low fields of 15.9 kA/m (200 Oe) a continuous rise in the magnetization at ~250 K during sample warming, and a discontinuous drop in the magnetization at ~230 K during sample cooling. These temperatures correspond well to the differential scanning calorimetry analysis (Cheng et al. 2007a) for the melting and supercooling points of the DCB. The continuous rise is associated with a transition between two different spin rotation mechanisms: Néel rotation of the magnetic moment with respect to the nanoparticles and/or chains at lower temperatures when the solvent is frozen and physical Brownian rotation of the entire nanoparticle and/or chain at higher temperatures when the solvent melts.

The discontinuous drop is anomalous, so vector magnetometry is used to determine (Dennis et al. 2007) if the magnetization vector is changing as a function of temperature. Three separate measurements were made. First, the magnetization as a function of angle was measured with a saturating field (H = 1194 kA/m [15,000 Oe]), which quantifies the magnitude and type of anisotropy in the sample. Second, the measurements were repeated above (240 K) and below (220 K) the transition at the relevant field below saturation (H = 15.9 kA/m [200 Oe]), which mimics the actual conditions of the M versus T measurements, and tells how much of the sample is rotating with the applied field.

To quantify the type and magnitude of the anisotropies in the sample, the torque data shown in Figures 5.23 and 5.24 are fit to the sum of the uniaxial and unidirectional anisotropies, given by:

$$\tau = K_u \sin\left[2(\theta - \phi_u)\right] + K_d \sin(\theta - \phi_d) + C \tag{5.26}$$

where τ is the torque on the system, K_u is the uniaxial anisotropy, K_d is the unidirectional anisotropy, θ is the angle, φ_u and φ_d are the offset angles for the uniaxial and unidirectional anisotropy respectively, and C is a constant offset. (For nanoparticles, the offset angles and offset constant are only fitting parameters; they have no physical meaning.) In this system, the volume is defined to be 1 because the same sample is measured.

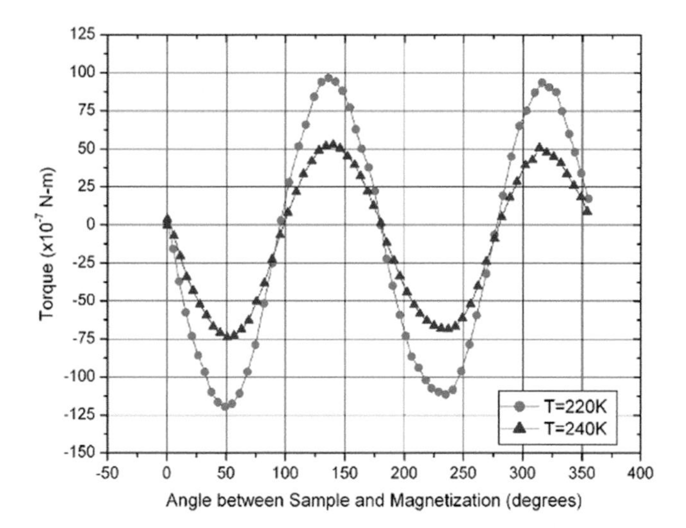

FIGURE 5.23 Vector magnetometry measurements during cooling of the torque versus angle between the magnetization and a sample of 10 nm diameter ε-Co nanoparticles as synthesized in 1,2-dichlorobenzene. The applied field was 1194 kA/m (15,000 Oe)—a saturated condition. (The sample angle is defined with respect to the initial magnetization direction.)

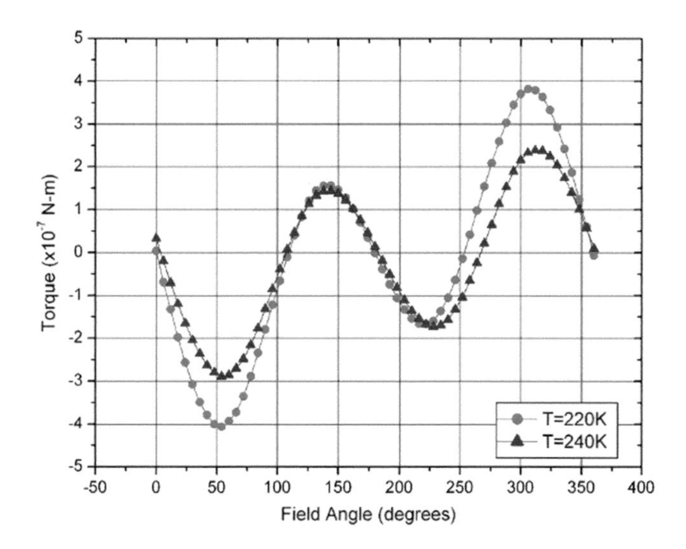

FIGURE 5.24 Vector magnetometry measurements during cooling of the torque versus field angle of 10 nm diameter Co nanoparticles as synthesized in 1,2-dichlorobenzene. The applied field was 15.9 kA/m (200 Oe)—a non-saturated condition.

Under saturation (H = 1194 kA/m [15,000 Oe]), only a uniaxial anisotropy is present (Figure 5.23). The unidirectional anisotropy is negligible at constant value of 1.6×10^{-7} J, regardless of temperature. At 240 K (after warming from 150 K to 300 K and then cooling to just above the supercooling point of the DCB [~230 K]), the uniaxial anisotropy is $61.1(7) \times 10^{-7}$ J. After cooling through the solvent supercooling point from 240 K to 220 K, the uniaxial anisotropy increases to $104.2(9) \times 10^{-7}$ J. This significant change in the anisotropy when crossing the liquid-solid phase transition is the result of turning off the Brownian motion. This locks the nanoparticles and their chains (Cheng et al. 2005) into position, raising the anisotropy energy of any future adjustments.

Under non-saturated conditions (H = 15.9 kA/m [200 Oe]), only effective anisotropies can be determined. More importantly, the two harmonics in the data (Figure 5.24) can be described as a sin(θ) piece, which represents the unreversed component of the magnetization due to chain formation and a sin(2θ) piece, which represents an effective anisotropy of the saturated nanoparticles. At 240 K (after warming from 150 K to 300 K and then cooling to just above the supercooling point of the solvent), the first harmonic has a magnitude of $(0.70 \pm 0.03) \times 10^{-7}$ J while the second harmonic is $(2.33 \pm 0.03) \times 10^{-7}$ J. After cooling through the solvent supercooling point from 240 K to 220 K, the first harmonic has a magnitude of $(1.65 \pm 0.04) \times 10^{-7}$ J while the second harmonic is $(3.04 \pm 0.04) \times 10^{-7}$ J. This change in the second harmonic indicates that there is also a slight increase in the effective anisotropy under the M versus T measurement conditions as the liquid-to-solid transition is traversed. These values are lower than those for the saturated case simply because not all of the nanoparticles are contributing. As for the first harmonic, this changes by a factor of 2.4 as the temperature cools through the supercooling point of the DCB, corresponding to a 2.4 × increase in the magnitude of the unreversed component of the magnetization during traversal of the liquid-solid transition. This significant change in the unreversed component across the discontinuous drop in magnetization originates from the competition between magnetic dipole alignment and lattice crystallization stresses. It is also present in high-concentration samples that are strongly interacting (Dennis et al. 2018). In the latter, clustering prevents the chains from rotating with the magnetic field, while strong interactions and magnetocrystalline anisotropies keep the moment point along the easy axis, increasing the unidirectional or unreversed component.

5.3.3 DC Magnetization versus Temperature

The magnetization versus temperature (M versus T) plots are often definitive in identifying the particular types of magnetism present, as well as their characteristic temperatures. For example, identifying a ferromagnet from an antiferromagnet is most readily done this way. An antiferromagnet's transition to a paramagnet occurs by approaching infinity at the Néel temperature (T_N). In practice, due to finite temperature steps in any instrument, the approach is more of a cusp, as shown for the antiferromagnet MnF_2 in Figure 5.25. Above T_N, the magnetization drops off like $1/T$. A ferromagnet transitions into a paramagnet at the Curie temperature (T_C), which is a sudden decrease in the magnetization, as shown for a NiFe sample in Figure 5.26.

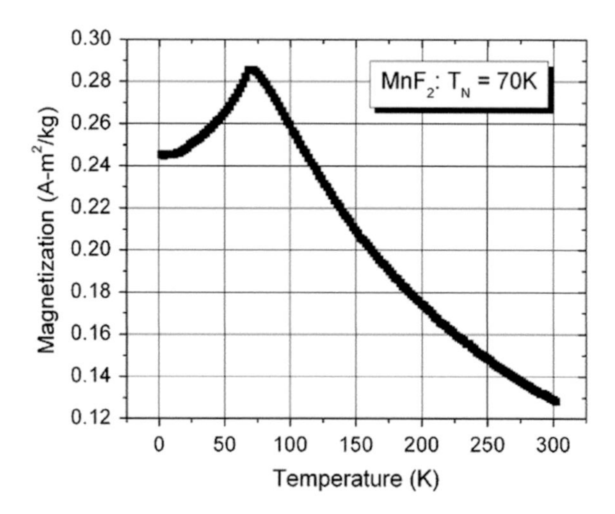

FIGURE 5.25 Example M versus T plot of the antiferromagnet MnF_2 showing its Néel temperature at 70 K.

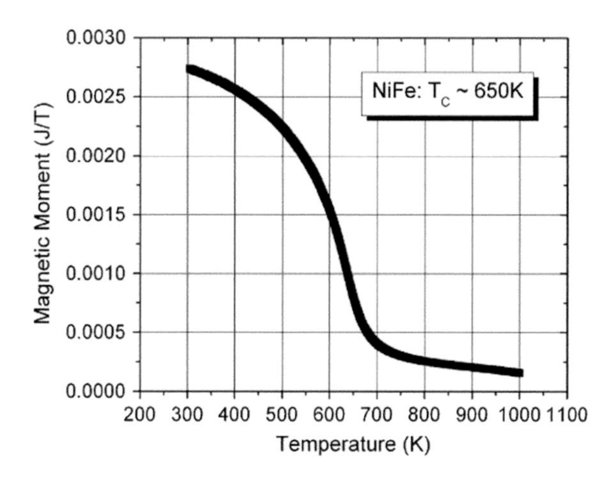

FIGURE 5.26 Example M versus T plot of the ferromagnet NiFe, showing its Curie temperature around 650 K.

For a pure paramagnetic sample, if the magnetic energy of a nanoparticle (mB) is significantly smaller than the thermal energy (k_BT), then Equation 5.9 can be simplified to:[3]

$$M_T = \frac{Nm^2B}{3k_BT} \tag{5.27}$$

where M_T is the total magnetization, m is the moment per particle, N is the number of particles per volume, B is the magnetic flux, k_B is Boltzmann's constant, and T is the temperature. However, above both the Néel and Curie temperatures, where the sample is also paramagnetic, this equation is invalid because of the offset from 0 K of the paramagnetic behavior as well as due to spatial quantization from the crystal structure. The Weiss theory of ferromagnetism (Weiss 1948, 1907) modifies this equation to be:

$$M_T = \frac{Nm^2B}{3k_B(T - T^*)} \tag{5.28}$$

where T^* is the characteristic temperature, either T_C or T_N. For ferromagnets, T^* is always positive. However, for antiferromagnets, T^* can be negative. Finally, right around the characteristic temperature for any system (0 K for a paramagnet, T_C for a ferromagnet, and T_N for an antiferromagnet), very weak interactions between moments (or nanoparticles) may become visible. Therefore, phenomenologically, an exponential γ is added to the characteristic temperature:

$$M_T = \frac{Nm^2B}{3k_B(T - T^*)^{\gamma}} \tag{5.29}$$

For pure non-interacting systems, $\gamma = 1$. However, for interacting systems, $\gamma \neq 1$. This interaction can be ferromagnetic, in which case T^* is positive, or antiferromagnetic, in which case T^* is negative. Since Curie discovered this effect experimentally, this equation is often referred to as the Curie–Weiss law.

Beyond this, the magnetization versus temperature plots can be broken into four different measurement categories (El-Hilo and O'Grady, 1990):

1. Zero-field cooled magnetization (ZFCM): The sample is cooled in zero field from room temperature (or another temperature that is higher than the transition temperature of the

system) to the base temperature of the instrument, and then a small DC magnetic field is applied. The magnetization is then measured at this field while warming the sample to above the transition temperature.
2. Field-cooled magnetization (FCM): This typically follows immediately after a ZFCM curve. The sample is cooled in a non-zero magnetic field from room temperature (or another temperature that is higher than the transition temperature of the system) to the base temperature of the instrument, while measuring the magnetization.
3. Isothermal remanent magnetization (IRM): The sample is cooled in zero field from room temperature (or another temperature that is higher than the transition temperature of the system) to the base temperature of the instrument, and then a small DC magnetic field is applied. This field is left on for a specific period of time[7] (Swartzendruber et al. 1997, Bennett et al. 1997), and then the field is reset to zero. After another specified period of time, the magnetization is measured. (It's a remanent magnetization since it is measured in zero field after the application of a non-zero field.)
4. Thermoremanent magnetization (TRM): The sample is cooled in a magnetic field from room temperature (or another temperature that is higher than the transition temperature of the system) to the base temperature of the instrument (or another temperature below the transition temperature). After a specified period of time, the magnetic field is reset to zero. After another specified period of time, the magnetization is measured. (It's also a remanent magnetization since it is measured in zero field after the application of a non-zero field.)

As described in the FORC measurement method, the magnetization can be broken into two components, the reversible and irreversible component. In nanoparticles, these two components are determined by these four measurements. The difference between the ZFCM and the IRM is the reversible component.

$$M_{rev} = ZFCM - IRM \tag{5.30}$$

In a superparamagnet, the reversible component is the unblocked nanoparticles. The IRM is the portion of nanoparticles for which the field is large enough that the energy barrier between "up" and "down" can be surmounted. Meanwhile, the TRM is the irreversible component of the magnetization. (In a superparamagnet, the irreversible component is the blocked nanoparticles.) Therefore, the FCM can be represented as:

$$FCM = ZFCM - IRM + TRM \tag{5.31}$$

By separating out these different components as a function of temperature, it is also possible to determine the amount of blocked versus unblocked particles at any given temperature. However, a great deal of care must be taken to ensure that the temperature is stable prior to every measurement, and that the field is really reset to zero (especially for a superconducting magnet). It is not valid to estimate the blocking temperature from the peak in the ZFCM at low fields, as shown by Equation 5.31, because this peak is dependent upon not only the measurement time scale but also the applied field.

5.3.4 Susceptibility versus Field

The susceptibility as a function of field comes in two flavors: longitudinal and transverse. When the AC field is applied in the same direction as the DC bias field, the longitudinal susceptibility is measured. For most magnetic nanoparticle systems, especially at room temperature where the sample has no coercivity and is superparamagnetic when highly diluted, the measured susceptibility results in a peak at zero field, which is also the switching field (Figure 5.27).

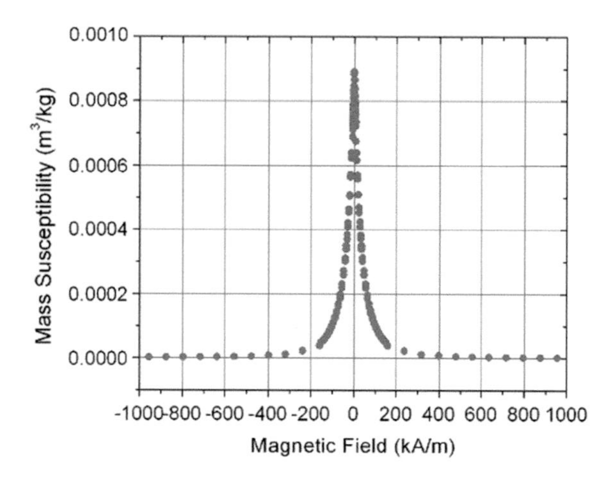

FIGURE 5.27 Longitudinal susceptibility of unblocked iron oxide nanoparticles at 15 K.

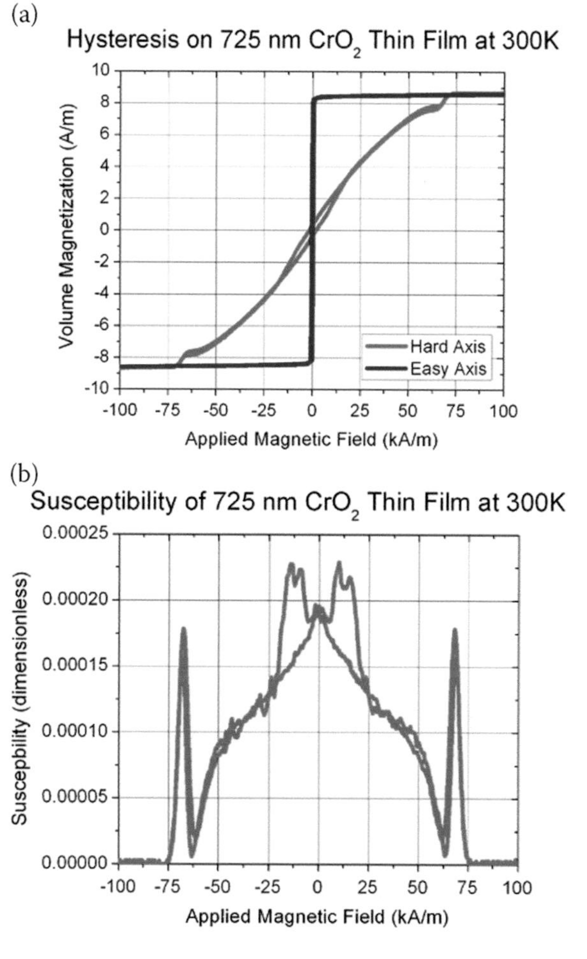

FIGURE 5.28 **(a)** Hysteresis loop at 300 K along the hard and easy axis and **(b)** The longitudinal susceptibility along the hard axis of CrO_2 thin film at 300 K, courtesy of N.F. Huls.

However, for a ferromagnetic system (or a blocked system if the nanoparticles are dilute enough to be superparamagnetic), the longitudinal susceptibility is more complicated, with multiple peaks. For clarity, this is shown in Figure 5.28 with a thin film of CrO_2 aligned along the hard axis (Frey et al. 2006). As in the paramagnetic system, the peaks occur at the switching fields. These peaks are typically symmetric in most magnetic nanoparticle systems, but could be asymmetric if exchange bias is present, for example. Therefore, longitudinal susceptibility measurements at fixed frequency as a function of field are useful for identifying the switching field(s) present in the system. In addition, information about the switching field distribution is contained in the width of the peak; a narrow peak has a narrower distribution than a broad peak. The limits in each case are a δ-function for the ideal case with single switching field and a peak so broad it cannot be discerned because every nanoparticle has a different switching field.

In contrast, when the AC field is applied perpendicular to the direction of the DC bias field, the transverse susceptibility is measured. The transverse susceptibility as a function of field (Figure 5.29) will also have peaks (Aharoni et al. 1957), and additional peaks may appear beyond those which correspond to a switching field. These extra peaks are a measure of the anisotropy of the system, as each peak represents an effective anisotropy field (Equation 5.10). In addition, it is also common in magnetic nanoparticle systems to observe only two anisotropy peaks, but with a marked asymmetry in peak height. This has been attributed (Poddar et al. 2008, Frey et al. 2007) to merging of one of the anisotropy peaks with a switching peak.

Furthermore, these anisotropy peaks are not necessarily symmetric in field position either, even if the switching peaks are. This can be due (Pierce et al. 2004) to symmetry breaking on different surfaces due to crystallographic faces or surfactant attachments, interparticle interactions, or anisotropy distributions within the system.

For example, two interesting systems are composed (Yu et al. 2005) of dimers or clusters of iron oxide-Au nanoparticles. Figure 5.30a shows the dimer configuration, with a 9 nm iron oxide nanoparticle in contact with an 8 nm Au nanoparticle, forming a structure that looks like a dumbbell. In the cluster configuration, shown in Figure 5.30b, the 8 nm Au nanoparticle forms the "center" of the cluster or "flower," and is surrounded by multiple 9 nm iron oxide nanoparticles, which are "petals."

There are three immediately noticeable differences between the dimers and clusters in the transverse susceptibility data as a function of DC bias field (Frey et al. 2009, 2007). The first is that the anisotropy peak heights are symmetric in the dimer system, but highly asymmetric in the

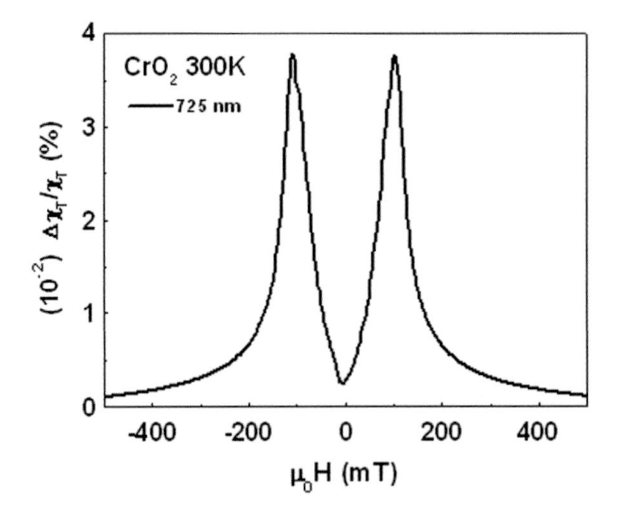

FIGURE 5.29 Transverse susceptibility of CrO_2 thin film at 300 K, courtesy of N.F. Huls.

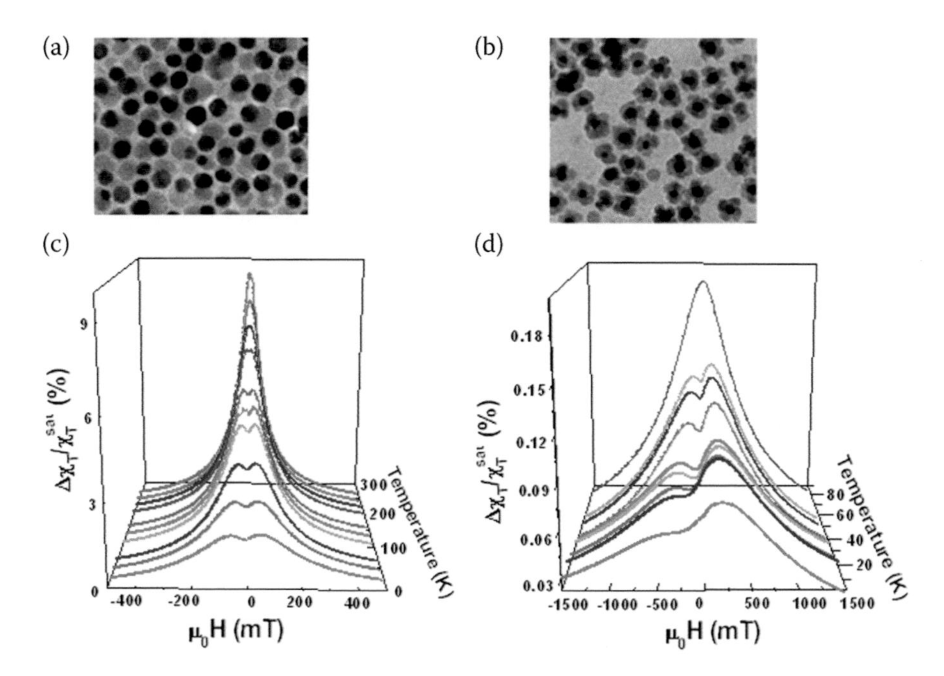

FIGURE 5.30 Transverse susceptibility as a function of DC bias field and temperature of iron oxide–Au dimers (dumbbells) and clusters (flowers with Au nanoparticle in center), courtesy of N.F. Huls.Transmission electrom micrographs are shown in (a) for the dumbbells and (b) for the flowers. The transverse susceptibility data is shown in (c) for the dumbbells and (d) for the flowers.

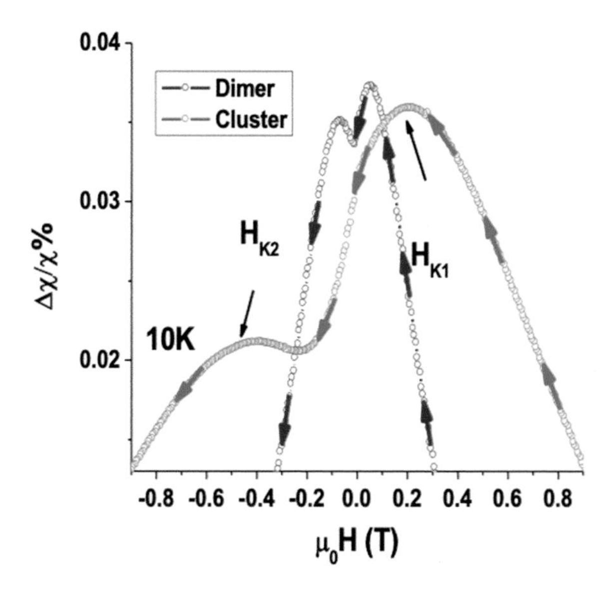

FIGURE 5.31 Detailed transverse susceptibility as a function of DC bias field at 30 K of iron oxide–Au dimers (dumbbells) and clusters (flowers with Au nanoparticle in center), courtesy of N.F. Huls.

cluster system. Second, the anisotropy field (peak location) is larger in the cluster system than the dimer system (see Figure 5.31 for more detail), but the dimer system peaks merge into a single switching peak at a lower temperature than in the cluster system. This is consistent with a lower effective anisotropy in the dimer system, which only has one Au–iron oxide interface. This allows the dimer to generally maintain its spherical symmetry. The symmetry between peaks in the dimer is also indicative of weak dipolar interactions between different dimers. In contrast, the interactions between neighboring iron oxide nanoparticle "petals" in a cluster particle are expected to be much stronger, potentially with direct exchange present. This will strongly enhance the effective anisotropy, shifting the peaks to larger fields. In addition, the symmetry breaking occurring at each of the multiple Au–iron oxide interfaces, resulting in enhanced anisotropy as well as potential spin frustration at the surfaces, may contribute to the asymmetry between different peaks.

5.3.5 Susceptibility versus Frequency

The susceptibility as a function of frequency also has a longitudinal and transverse component. However, as a function of frequency, almost all of the work to date has been on longitudinal susceptibility. Within this, there are two cases to consider. The first is relaxation processes within a single nanoparticle (Dormann, Bessais, and Fiorani 1988). When a field is applied to a single monodomain nanoparticle, and the field is not aligned with the direction of magnetization (which is assumed to be along the easy axis), then the magnetization of the nanoparticle must rotate to align with the field. This can be done by either rotating the entire particle while the magnetization remains fixed relative to the crystal structure (called Brownian rotation), or by rotating the magnetization alone while the crystal structure does not move (called Néel rotation). Both of these have characteristic frequencies, which are quite different.

Brownian rotation has a characteristic relaxation time (τ_B) of:

$$\tau_B = \frac{3V_h\eta_0}{k_BT} \tag{5.32}$$

where V_h is the hydrodynamic radius of the particle and η_0 is the viscosity of the solution. Néel rotation has a characteristic relaxation time (τ_N) of (Néel 1955):

$$\tau_N = \tau_0 e^{KV_M/k_BT} \tag{5.33}$$

where τ_0 is an attempt time, K is the anisotropy, and V_M is the magnetic volume of the particle (Schaller et al. 2010, Eberbeck et al. 2011). It is critical to note, however, that both of these equations are only valid in the limit where there are no interactions between magnetic nanoparticles and where any applied magnetic fields are very small. In the case where interactions are present, the Néel relaxation time is modified (Shtrikman and Wohlfarth 1981) according to the Vogel–Fulcher law (τ_{VF}) and becomes:

$$\tau_{VF} = \tau_0 e^{KV_M/k_B(T-T_0)} \tag{5.34}$$

where T_0 is interaction temperature. The Vogel–Fulcher law is only valid in the limit of weak interactions between nanoparticles. Chain formation in a magnetic nanoparticle system would violate this assumption. In most magnetic nanoparticle systems, both Brownian and Néel rotation are present due to property distributions. However, one always dominates, resulting in an effective relaxation time. So, when examining longitudinal susceptibility data as a function of frequency at fixed temperature, peaks in the data are indicative of this effective relaxation time. Because fitting exponentials can be tricky, a simpler method is to plot the location of this peak in frequency as a function of temperature. Specifically, a plot of $\ln(\tau_{VF})$ versus $1/T$ should result in a straight line whose slope is equal to KV_M.

$$\ln \tau_{VF} = \ln \tau_0 + \frac{KV_M}{k_B(T-T_0)} \tag{5.35}$$

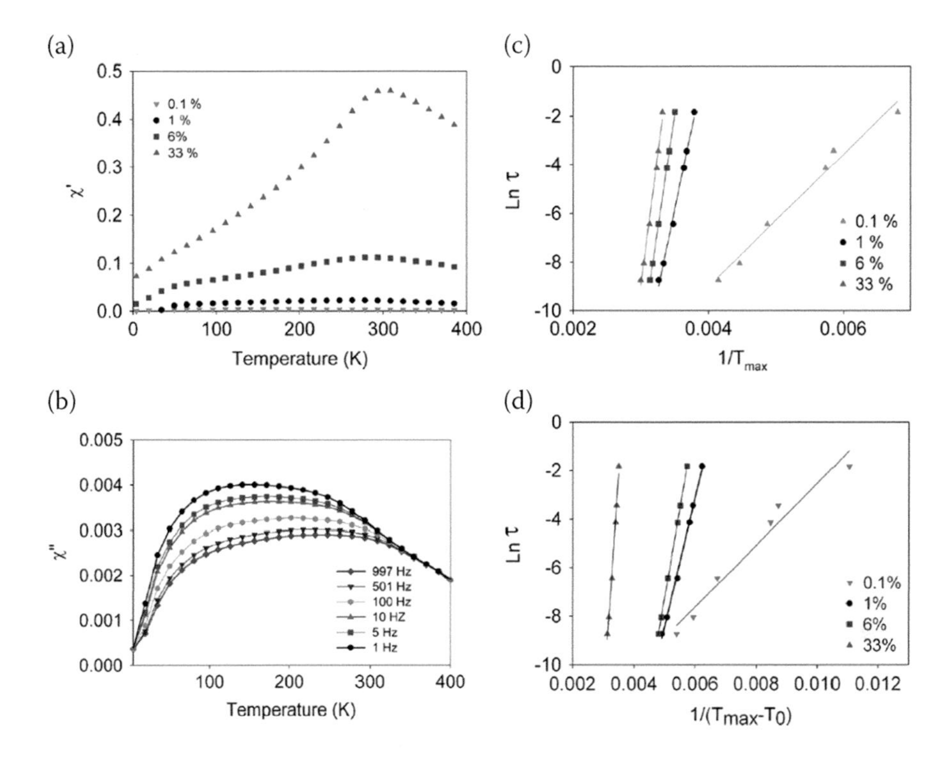

FIGURE 5.32 **(a)** Real and **(b)** Imaginary components of the longitudinal susceptibility as a function of temperature, **(c)** Néel-Arrhenius fit and **(d)** Vogel-Fulcher fit of logarithm of the peak frequency as a function of inverse temperature, courtesy of C. Rinaldi.

(If $T_0 = 0$, this reduces back to the non-interacting case, which in natural logarithm form, is sometime referred to as the Néel–Arrhenius law.) However, the validity of these equations on any given magnetic nanoparticle system can be determined most easily from the value of τ_0 in the fit. For non-interacting systems, τ_0 should be 10^{-9} seconds (which corresponds to an attempt frequency of 1 GHz) (Brown 1963, Kalmykov et al. 2010). However, if interactions are present, this attempt time will change dramatically, by multiple orders of magnitude, away from this non-interacting attempt frequency. The latter can clearly be seen by plotting the susceptibility as a function of concentration (del Castillo and Rinaldi 2009), as shown in Figure 5.32.

The τ_0 from the Néel–Arrhenius fits are 2×10^{-9} seconds, 2×10^{-23} seconds, 8×10^{-30} seconds, and 2×10^{-32} seconds for the 0.1–33% concentrations, respectively. The τ_0 from the Vogel–Fulcher fits are 2×10^{-7} seconds, 4×10^{-16} seconds, 5×10^{-20} seconds, and 6×10^{-31} seconds for the 0.1–33% concentrations, respectively. Only the 0.1% sample of iron oxide nanoparticles embedded in a rigid polymer matrix has a reasonable τ_0, confirming the absence of interactions in that sample, as well as the rapid transition from weak to strong interactions in the other samples.

5.4 SUMMARY

In conclusion, some of the most common magnetic characterization instruments, both in terms of how they work and their most common sources of errors, have been described. Furthermore, selected measurement methods useful in magnetic nanoparticle analysis have been described, including the information that can be gained and common models for fitting the data with their underlying assumptions. Interested readers are encouraged to examine the references for more information.

5.6 DISCLAIMER

The use of specific trade names does not imply endorsement of products or companies by NIST. The trade names are used to fully describe the experimental procedures.

5.5 ACKNOWLEDGEMENTS

The author thanks Julie Borchers, Larry Bennett, and Robert Shull for careful reading of the manuscript; James O'Brien, Ron Manus, Joseph Cappella, Erik Samwell, and Thomas Kent for discussions over the years on instrumentation; Johan van Lierop for permission to reprint his figure on Mössbauer spectroscopy; Natalie Huls for discussions on transverse susceptibility and permission to reprint her figures; Carlos Rinaldi for permission to reprint his figures on longitudinal susceptibility; and Joseph Davies for discussions on FORC and permission to reprint his figures.

NOTES

1 The relative intensities of the various peaks reflect the relative concentrations of compounds in a sample and can be used for semi-quantitative analysis.
2 When all three effects are observed simultaneously:

i. i. the isomer shift is given by the average of all lines;
ii. ii. the quadrupole splitting:
   a. If all four excited substates are equally shifted [two substates are lifted and other two are lowered) is given by the shift of the outer two lines relative to the inner two lines, with the innermost lines ignored. If the shifting of four substates is not equal, then the quadrupole splitting is often extracted using fitting software where all six lines are taken into account.

3 B (magnetic flux in T) and H (magnetic field in A/m), both the symbols and the names, are often used interchangeably in the literature and textbooks; however, they are not the same. Instead, they are related by the equation: $B = \mu_0(H+M)$, where M is the volume magnetization and μ_0 is the permeability of free space. However, since it is often true that $M \ll H$ (especially with nanoparticles), M can be ignored so $B = \mu_0 H$. Unfortunately, when interchanging B and H, the μ_0 is not always explicitly included. So, it is common to see, for example, $E = mB\cos\theta$ or $E = mH\cos\theta$. However, when B and H are interchanged, the μ_0 is required for units. For this reason, it is common to see data presented with "$\mu_0 H$" in Tesla on the x-axis, instead of "magnetic field" in Amperes per meter, even though magnetic field in A/m was actually measured.
4 The three Mössbauer parameters (isomer shift, quadrupole splitting, and hyperfine splitting) can be used to identify a particular compound by comparing it to known spectra.
5 Normalization of magnetic nanoparticles is non-trivial, because of the different possible normalization methods and what they mean. (1) Normalization to total mass of the sample, including any solvent yields a saturation magnetization that is not directly comparable to anything that does not have the same concentration. However, the concentration by volume, mass, or atomic percentage must also be specified. (2) Normalization to total mass of the particles, including surfactant, yields a saturation magnetization which again depends upon concentration unless the solvent background is removed from the magnetic signal. This is most useful in applications, because an application only considers the particle with everything attached to it. (3) Normalization to only the total magnetic mass of the particles yields a saturation magnetization which is directly comparable to the bulk magnetic material only if the solvent background as well as any background from the surfactant and any other non-magnetic material is removed from the magnetic signal. This is the normalization method of choice for comparing synthesis methods to bulk values.
6 When determining the blocking temperature, it is assumed that the particles are fixed. Therefore, Brownian rotation will not play a role. This is a reasonable assumption in almost every case except when the particles are dried. If they are dried in powder form, they are interacting, and therefore, the concept of a blocking temperature does not apply! Otherwise, the nanoparticles must be dispersed in something else (e.g. water, wax, epoxy) which is either already a solid (e.g. wax or epoxy), or is solid (e.g. ice) at the relevant low temperatures.
7 The time before measurement must be specified, as there are two additional time-dependent effects that are also relevant: (1) magnetic viscosity or after-effect, which is due to a thermally assisted process of crossing an energy barrier and (2) accommodation or reputation, which occurs when the magnetic field is repeatedly cycled in a minor loop between two values of the magnetic field, causing magnetic drift toward a stable loop as the cycling continues. After-effect changes are proportional to log time (see Chapter 1 on superparamagnets), while

accommodation changes are proportional to the log number of cycles; therefore, if the cycles are regular in time, accommodation can also have a log time behavior that is hard to differentiate experimentally from after-effect.

REFERENCES

Abraham, D. W., M. M. Frank, and S. Guha. 2005. Absence of magnetism in hafnium oxide films. *Applied Physics Letters* 87 (25), 252502.

Acremann, Y., C. H. Back, M. Buess, O. Portmann, A. Vaterlaus, D. Pescia, and H. Melchior. 2000. Imaging precessional motion of the magnetization vector. *Science* 290 (5491), 492–495.

Acton, G., Q.-Z. Yin, K. L. Verosub, L. Jovane, A. Roth, B. Jacobsen, and D. S. Ebel. 2007. Micromagnetic coercivity distributions and interactions in chondrules with implications for paleointensities of the early solar system. *Journal of Geophysical Research: Solid Earth* 112 (B3). doi:10.1029/2006jb004655.

Aharoni, A., E. H. Frei, S. Shtrikman, and D. Treves. 1957. The reversible susceptibility tensor of the Stoner-Wohlfarth model. *Bulletin of the Research Council of Israel* 6A, 215.

Andrei, P., O. Caltun, and A. Stancu. 2006. Rate dependence of first-order reversal curves by using a dynamic Preisach model of hysteresis. *Physica B: Condensed Matter* 372 (1), 265–268. doi:https://doi.org/10.1016/j.physb.2005.10.063.

Aranda, G. R., O. Chubykalo-Fesenko, R. Yanes, J. Gonzalez, J. J. del Val, R. W. Chantrell, Y. K. Takahashi, and K. Hono. 2009. Coercive field and energy barriers in partially disordered FePt nanoparticles. *Journal of Applied Physics* 105 (7), 07B514.

Atherton, N. M. 1993. *Principles of Electron Spin Resonance*. New York: Ellis Horwood Limited.

Atzmony, U., L. H. Bennett, and L. J. Swartzendruber. 1995. Is there a paramagnetic Meissner-Ochsenfeld effect? *IEEE Transactions on Magnetics* 31 (6), 4118–4120.

Bahl, C. R. H., M. F. Hansen, T. Pedersen, S. Saadi, K. H. Nielsen, B. Lebech, and S. Mørup. 2006. The magnetic moment of NiO nanoparticles determined by Mössbauer spectroscopy. *Journal of Physics: Condensed Matter* 18 (17), 4161.

Bean, C. P., and J. D. Livingston. 1959. Superparamagnetism. *Journal of Applied Physics* 30 (4), S120–S129.

Bennett, L. H., L. J. Swartzendruber, F. Vajda, E. Della Torre, and J. H. Judy. 1997. Aftereffect and accommodation anisotropy in metal-particle and metal-evaporated recording media. *IEEE Transactions on Magnetics* 33 (5), 4173–4175.

Bodale, I., L. Stoleriu, and A. Stancu. 2010. Reversible and irreversible components evaluation in hysteretic processes using first and second-order magnetization curves. *IEEE Transactions on Magnetics* 47 (1), 192–197.

Boekelheide, Z., and C. L. Dennis. 2016. Artifacts in magnetic measurements of fluid samples. *AIP Advances* 6 (8), 085201.

Boyes, W. 2003. *Instrumentation Reference Book*. 3rd ed. Burlington, MA: Elsevier Science.

Brown, M. E. 2001. *Introduction to Thermal Analysis: Techniques and Applications*. New York: Springer Science & Business Media.

Brown, Jr, W. F. 1963. Thermal fluctuations of a single-domain particle. *Physical Review* 130 (5), 1677.

Béron, F., L. Clime, M. Ciureanu, D. Ménard, R. W. Cochrane, and A. Yelon. 2007. Reversible and quasireversible information in first-order reversal curve diagrams. *Journal of Applied Physics* 101 (9), 09J107.

Béron, F., D. Ménard, and A. Yelon. 2008. First-order reversal curve diagrams of magnetic entities with mean interaction field: A physical analysis perspective. *Journal of Applied Physics* 103 (7), 07D908.

Béron, F., K. R. Pirota, and M. Knobel. 2011. Probing the interdependence between irreversible magnetization reversal processes by first-order reversal curves. *Journal of Applied Physics* 109 (7), 07E308.

Bødker, F., S. Mørup, S. W. Charles, and S. Linderoth. 1999. Surface oxidation of cobalt nanoparticles studied by Mössbauer spectroscopy. *Journal of Magnetism and Magnetic Materials* 196, 18–19.

Chantrell, R. W., G. N. Coverdale, M. El Hilo, and K. O'Grady. 1996. Modelling of interaction effects in fine particle systems. *Journal of Magnetism and Magnetic Materials* 157, 250–255.

Chatterji, T. 2005. *Neutron Scattering from Magnetic Materials*. Amsterdam, The Netherlands: Elsevier.

Cheng, G., D. Romero, G. T. Fraser, and A. R. H. Walker. 2005. Magnetic-field-induced assemblies of cobalt nanoparticles. *Langmuir* 21 (26), 12055–12059.

Cheng, G., C. L. Dennis, R. D. Shull, and A. R. Hight Walker. 2007a. Influence of the colloidal environment on the magnetic behavior of cobalt nanoparticles. *Langmuir* 23 (23), 11740–11746.

Cheng, G., C. L. Dennis, R. D. Shull, and A. R. Hight Walker. 2009. Probing the growth and aging of colloidal cobalt nanocrystals: A combined study by transmission electron microscopy and magnetic measurements. *Crystal Growth and Design* 9 (8), 3714–3720.

Cheng, G. G., Cindi L. D., R. D. Shull, A. R. Hight Walker. 2007b. Influence of the colloidal environment on the magnetic behavior of cobalt nanoparticles. *Langmuir* 23 (23), 11740–11746.

Chikazumi, S. 1964. *Physics of Ferromagnetism*. New York: John Wiley & Sons.

Ciureanu, M., F. Béron, P. Ciureanu, R. W. Cochrane, D. Ménard, A. Sklyuyev, and A. Yelon. 2008. First order reversal curves (FORC) diagrams of Co nanowire arrays. *Journal of Nanoscience and Nanotechnology* 8 (11), 5725–5732.

Cullity, B. D. 1978. *Elements of X-ray Diffraction*. 2nd ed. New York: Addison-Wesley Publishing Company.

Davies, J. E., O. Hellwig, E. E. Fullerton, G. Denbeaux, J. B. Kortright, and K. Liu. 2004. Magnetization reversal of Co/Pt multilayers: Microscopic origin of high-field magnetic irreversibility. *Physical Review B* 70 (22), 224434.

Davies, J. E., O. Hellwig, E. E. Fullerton, J. S. Jiang, S. D. Bader, G. T. Zimanyi, and K. Liu. 2005. Anisotropy dependence of irreversible switching in Fe/Sm Co and Fe Ni/Fe Pt exchange spring magnet films. *Applied Physics Letters* 86 (26), 262503.

Davies, J. E., J. Wu, C. Leighton, and K. Liu. 2005b. Magnetization reversal and nanoscopic magnetic-phase separation in $La_{1-x}Sr_xCoO_3$. *Physical Review B* 72 (13), 134419.

Davies, J. E., O. Hellwig, E. E. Fullerton, and K. Liu. 2008. Temperature-dependent magnetization reversal in (Co/Pt)/Ru multilayers. *Physical Review B* 77 (1), 014421.

Davies, J. E., O. Hellwig, E. E. Fullerton, M. Winklhofer, R. D. Shull, and K. Liu. 2009a. Frustration driven stripe domain formation in Co/Pt multilayer films. *Applied Physics Letters* 95 (2), 022505.

Dennis, C. L., G. Cheng, K. A. Baler, B. B. Maranville, A. R. H. Walker, and R. D. Shull. 2007. The influence of temperature on the magnetic behavior of colloidal cobalt nanoparticles. *IEEE Transactions on Magnetics* 43 (6), 2448–2450.

Dennis, C. L., A. J. Jackson, J. A. Borchers, R. Ivkov, A. R. Foreman, J. W. Lau, E. Goernitz, and C. Cruettner. 2008. The influences of collective behavior on the magnetic and heating properties of iron oxide nanoparticles. *Journal of Applied Physics* 103, 07A319.

Dennis, C. L., A. J. Jackson, J. A. Borchers, P. J. Hoopes, R. Strawbridge, A. R. Foreman, J. van Lierop, C. Grüttner, and R. Ivkov. 2009. Nearly complete regression of tumors via collective behavior of magnetic nanoparticles in hyperthermia. *Nanotechnology* 20 (39), 395103.

Dennis, C. L., A. J. Jackson, J. A. Borchers, C. Gruettner, and R. Ivkov. 2018. Correlation between physical structure and magnetic anisotropy of a magnetic nanoparticle colloid. *Nanotechnology* 29 (21), 215705. doi:10.1088/1361-6528/aab31d.

Desautels, R. D., J. M. Cadogan, and J. van Lierop. 2009. Spin dynamics in $CoFe_2O_4$ nanoparticles. *Journal of Applied Physics* 105 (7), 07B506.

Dobrotă, C.-I., and A. Stancu. 2013. What does a first-order reversal curve diagram really mean? A study case: Array of ferromagnetic nanowires. *Journal of Applied Physics* 113 (4), 043928.

Dorfbauer, F., R. Evans, M. Kirschner, O. Chubykalo-Fesenko, R. Chantrell, and T. Schrefl. 2007. Effects of surface anisotropy on the energy barrier in cobalt–silver core–shell nanoparticles. *Journal of Magnetism and Magnetic Materials* 316 (2), e791–e794.

Dormann, J. L., L. Bessais, and D. Fiorani. 1988. A dynamic study of small interacting particles: Superparamagnetic model and spin-glass laws. *Journal of Physics C: Solid State Physics* 21 (10), 2015.

Dumas, R. K., C.-P. Li, I. V. Roshchin, I. K. Schuller, and K. Liu. 2007. Magnetic fingerprints of sub-100 nm Fe dots. *Physical Review B* 75 (13), 134405.

Eberbeck, D., F. Wiekhorst, S. Wagner, and L. Trahms. 2011. How the size distribution of magnetic nanoparticles determines their magnetic particle imaging performance. *Applied Physics Letters* 98 (18), 182502.

El-Hilo, M., and K. O'Grady. 1990. Components of magnetisation of a fine particle system. *IEEE Transactions on Magnetics* 26 (5), 1807–1809.

Fischer, P. 2011. Exploring nanoscale magnetism in advanced materials with polarized X-rays. *Materials Science and Engineering: R: Reports* 72 (5), 81–95.

Frey, N. A., S. Srinath, H. Srikanth, M. Varela, S. Pennycook, G. X. Miao, and A. Gupta. 2006. Magnetic anisotropy in epitaxial CrO_2 and CrO_2/Cr_2O_3 bilayer thin films. *Physical Review B* 74 (2), 024420.

Frey, N. A., S. Srinath, H. Srikanth, C. Wang, and S. Sun. 2007. Static and dynamic magnetic properties of composite Au-Fe_3O_4 nanoparticles. *IEEE Transactions on Magnetics* 43 (6), 3094–3096.

Frey, N. A., M. H. Phan, H. Srikanth, S. Srinath, C. Wang, and S. Sun. 2009. Interparticle interactions in coupled Au–Fe_3O_4 nanoparticles. *Journal of Applied Physics* 105 (7), 07B502.

Glatter, O., and O. Kratky. 1982. *Small Angle X-ray Scattering*. New York: Academic Press.

Gornakov, V. S., Y. P. Kabanov, V. I. Nikitenko, O. A. Tikhomirov, A. J. Shapiro, and R. D. Shull. 2004. Chirality of a forming spin spring and remagnetization features of a bilayer ferromagnetic system. *Journal of Experimental and Theoretical Physics* 99 (3), 602–612.

Grüttner, C., K. Müller, J. Teller, F. Westphal, A. Foreman, and R. Ivkov. 2007. Synthesis and antibody conjugation of magnetic nanoparticles with improved specific power absorption rates for alternating magnetic field cancer therapy. *Journal of Magnetism and Magnetic Materials* 311 (1), 181–186.

Heslop, D., and A. R. Muxworthy. 2005. Aspects of calculating first-order reversal curve distributions. *Journal of Magnetism and Magnetic Materials* 288, 155–167.

Higgins, A. K., C. D. Graham, R. M. Strnat, and C. H. Chen. 2008. Apparent image effect in closed-circuit magnetic measurements. *IEEE Transactions on Magnetics* 44 (11), 3269–3272.

Hill, S. J. 2006. *Inductively Coupled Plasma Spectrometry and Its Applications*. Oxford, UK: Wiley-Blackwell.

Hunter, R. 2001. *Foundations of Colloid Science*. 2nd ed. Oxford, UK: Oxford University Press.

Kahler, G. R., L. H. Bennett, and E. Della Torre. 2006. Coercivity and the critical switching field. *Physica B: Condensed Matter* 372 (1-2), 1–4.

Kalmykov, Y. P., W. T. Coffey, U. Atxitia, O. Chubykalo-Fesenko, P.-M. Déjardin, and R. W. Chantrell. 2010. Damping dependence of the reversal time of the magnetization of single-domain ferromagnetic particles for the Néel-Brown model: Langevin dynamics simulations versus analytic results. *Physical Review B* 82 (2), 024412.

Katzgraber, H. G., F. Pazmandi, C. R. Pike, K. Liu, R. T. Scalettar, K. L. Verosub, and G. T. Zimanyi. 2002. Reversal-field memory in the hysteresis of spin glasses. *Physical Review Letters* 89 (25), 257202.

Kittel, C. 1967. *Introduction to Solid State Physics*. 3rd ed. New York: Wiley.

Krycka, K., J. Borchers, Y. Ijiri, R. Booth, and S. Majetich. 2012a. Polarization-analyzed small-angle neutron scattering. II. Mathematical angular analysis. *Journal of Applied Crystallography* 45 (3), 554–565.

Krycka, K., W. Chen, J. Borchers, B. Maranville, and S. Watson. 2012b. Polarization-analyzed small-angle neutron scattering. I. Polarized data reduction using Pol-Corr. *Journal of Applied Crystallography* 45 (3), 546–553.

Long, G. J. 1993. *Mössbauer Spectroscopy Applied to Magnetism and Materials Science*. New York: Plenum Press.

Malloy, A., and B. Carr. 2006. NanoParticle tracking analysis–The Halo™ system. *Particle & Particle Systems Characterization* 23 (2), 197–204.

Martínez-García, J. C., M. Rivas, D. Lago-Cachón, and J. A. García. 2014. FORC differential dissection of soft biphase magnetic ribbons. *Journal of Alloys and Compounds* 615, S276–S279.

May, L. 1971. *An Introduction to Mössbauer Spectroscopy*. New York: Plenum Press.

Mayergoyz, I. D. 1991. The classical Preisach model of hysteresis. In *Mathematical Models of Hysteresis*, 1–63. New York: Springer-Verlag.

McCreery, R. L. 2000. *Raman Spectroscopy for Chemical Analysis*. New York: John Wiley & Sons.

Muxworthy, A., and W. Williams. 2005. Magnetostatic interaction fields in first-order-reversal-curve diagrams. *Journal of Applied Physics* 97 (6), 063905.

Mørup, S. 1983. Magnetic hyperfine splitting in Mössbauer spectra of microcrystals. *Journal of Magnetism and Magnetic Materials* 37 (1), 39–50.

Nikitenko, V. I., V. S. Gornakov, L. M. Dedukh, Y. P. Kabanov, A. F. Khapikov, A. J. Shapiro, R. D. Shull, A. Chaiken, and R. P. Michel. 1998. Asymmetry of domain nucleation and enhanced coercivity in exchange-biased epitaxial NiO/NiFe bilayers. *Physical Review B* 57 (14), R8111.

Néel, L. 1949. Theory of magnetic viscosity of fine grained ferromagnetics with application to baked clays. *Annals of Geophysics* 5 (99-136), 41.

Néel, L. 1955. Some theoretical aspects of rock-magnetism. *Advances in Physics* 4 (14), 191–243.

Oliva, M. I., H. R. Bertorello, and P. G. Bercoff. 2004. Switching field of partially exchange-coupled particles. *Physica B: Condensed Matter* 354 (1), 203–208. doi:https://doi.org/10.1016/j.physb.2004.09.067.

O'Handley, R. C. 2000. *Modern Magnetic Materials: Principles and Applications*. New York: Wiley.

Park, J. P., P. Eames, D. M. Engebretson, J. Berezovsky, and P. A. Crowell. 2003. Imaging of spin dynamics in closure domain and vortex structures. *Physical Review B* 67 (2), 020403.

Pecora, R. 1985. *Dynamic Light Scattering: Applications of Photon Correlation Spectroscopy*. New York: Plenum Press.

Pierce, J. P., M. A. Torija, Z. Gai, J. Shi, T. C. Schulthess, G. A. Farnan, J. F. Wendelken, E. W. Plummer, and J. Shen. 2004. Ferromagnetic stability in Fe nanodot assemblies on Cu (111) induced by indirect coupling through the substrate. *Physical Review Letters* 92 (23), 237201.
Pike, C. R. 2003. First-order reversal-curve diagrams and reversible magnetization. *Physical Review B* 68 (10), 104424.
Pike, C. R., A. P. Roberts, and K. L. Verosub. 1999. Characterizing interactions in fine magnetic particle systems using first order reversal curves. *Journal of Applied Physics* 85 (9), 6660–6667.
Poddar, P., M. B. Morales, N. A. Frey, S. A. Morrison, E. E. Carpenter, and H. Srikanth. 2008. Transverse susceptibility study of the effect of varying dipolar interactions on anisotropy peaks in a three-dimensional assembly of soft ferrite nanoparticles. *Journal of Applied Physics* 104 (6), 063901.
Preisach, F. 1935. Über die magnetische Nachwirkung. *Zeitschrift für physik* 94 (5-6), 277–302.
Sarid, D. 1994. *Scanning Force Microscopy: With Applications to Electric, Magnetic, and Atomic Forces (Oxford Series on Optical and Imaging Sciences)*. Oxford, UK: Oxford University Press.
Saveyn, H., B. De Baets, O. Thas, P. Hole, J. Smith, and P. Van Der Meeren. 2010. Accurate particle size distribution determination by nanoparticle tracking analysis based on 2-D Brownian dynamics simulation. *Journal of Colloid and Interface Science* 352 (2), 593–600.
Schaller, V., G. Wahnström, A. Sanz-Velasco, P. Enoksson, and C. Johansson. 2009a. Monte Carlo simulation of magnetic multi-core nanoparticles. *Journal of Magnetism and Magnetic Materials* 321 (10), 1400–1403.
Schaller, V., G. Wahnström, A. Sanz-Velasco, S. Gustafsson, E. Olsson, P. Enoksson, and C. Johansson. 2009b. Effective magnetic moment of magnetic multicore nanoparticles. *Physical Review B* 80 (9), 092406. doi:10.1103/PhysRevB.80.092406.
Schaller, V., G. Wahnström, A. Sanz-Velasco, P. Enoksson, and C. Johansson. 2010Determination of nanocrystal size distribution in magnetic multicore particles including dipole-dipole interactions and magnetic anisotropy: A Monte Carlo Study. *AIP Conference Proceedings*1311(1): 42
Shtrikman, S., and E. P. Wohlfarth. 1981. The theory of the Vogel-Fulcher law of spin glasses. *Physics Letters A* 85 (8-9), 467–470.
Skomski, R. 2008. *Simple Models of Magnetism*. New York: Oxford University Press.
Stancu, A., P. Andrei, and L. Stoleriu. 2006. Magnetic characterization of samples using first-and second-order reversal curve diagrams. *Journal of Applied Physics* 99 (8), 08D702.
Stoner, E. C., and E. P. Wohlfarth. 1948. A mechanism of magnetic hysteresis in heterogeneous alloys. *Philosophical Transactions of the Royal Society of London. Series A, Mathematical and Physical Sciences* 240 (826), 599–642.
Swartzendruber, L. J., L. H. Bennett, F. Vajda, and E. Della Torre. 1997. Relationship between the measurement of accommodation and after-effect. *Physica B: Condensed Matter* 233 (4), 324–329.
Taketomi, S., and R. D. Shull. 2002. Experimental study of magnetic interactions between colloidal particles in magnetic fluids. *Journal of Applied Physics* 91 (10), 8546–8548.
Tantra, R., P. Schulze, and P. Quincey. 2010. Effect of nanoparticle concentration on zeta-potential measurement results and reproducibility. *Particuology* 8 (3), 279–285.
Thomas, R. 2013. *Practical Guide to ICP-MS: A Tutorial for Beginners*. CRC Press.
Torre, E. D. 1999. *Magnetic Hysteresis*. Piscataway, NJ: Wiley-IEEE Press.
Van Lierop, J., and D. H. Ryan. 2001. Mössbauer spectra of single-domain fine particle systems described using a multiple-level relaxation model for superparamagnets. *Physical Review B* 63 (6), 064406.
Weiss, P. 1907. L'hypothèse du champ moléculaire et la propriété ferromagnétique. *Journal of Physics: Theories and Applications*. 6 (1), 661.
Weiss, P. R. 1948. The application of the Bethe-Peierls method to ferromagnetism. *Physical Review* 74 (10), 1493.
Willard, M. A., L. K. Kurihara, E. E. Carpenter, S. Calvin, and V. G. Harris. 2004. Chemically prepared magnetic nanoparticles. *International Materials Reviews* 49 (3-4), 125–170.
Williams, D. B., and C. B. Carter. 1996. *Transmission Electron Microscopy*. New York: Plenum Press.
Yu, H., M. Chen, P. M. Rice, S. X. Wang, R. L. White, and S. Sun. 2005. Dumbbell-like bifunctional Au–Fe_3O_4 nanoparticles. *Nano Letters* 5 (2), 379–382.
del Castillo, V. L. C.-D., and C. Rinaldi. 2009. Effect of sample concentration on the determination of the anisotropy constant of magnetic nanoparticles. *IEEE Transactions on Magnetics* 46 (3), 852–859.

Section II

Magnetic Particle Applications

6 Synthesis and Functionalization of Magnetic Particles

Erika C. Vreeland and Dale L. Huber

CONTENTS

6.1 INTRODUCTION

Nanoparticles are never the thermodynamic ground state of a system but contain a great deal of energy in the form of surface energy (Huber 2005). The system's overall energy would always be lowered by allowing nanoparticles to form macroscopic structures. They are then appropriately viewed as inherently unstable structures that are able to remain only because the pathway to lower their energy is blocked or slowed by the presence of a surfactant coating on the surface. Figure 6.1 shows a general sketch of the structure of a magnetic nanoparticle and the species that are used to assemble it. Because nanoparticles are not the ground state, the method of their formation has a dramatic impact on the final product including such fundamental properties as size, size dispersity, crystallinity, and crystal phase. Reproducibility in preparation is a particular issue for nanoparticles, because they reside on a steeply sloped energy regime, where even small increases in size can lead to a dramatic decrease in surface energy. This is unlike most familiar preparation methods where the desired product is in a stable or at least meta-stable region of the energy landscape.

The preparation of nanoparticles is an exercise in the kinetic trapping of growing species in the desired size range, and while the details of the methods used to affect this trapping range widely, the general approaches have some fundamental similarities. There are two general approaches; one is to build up particles to the desired size and the other is to break them down to the desired size. In chemical approaches, we build particles up during the course of some chemical reaction, whereas, in what are termed physical approaches, we begin with bulk materials and, through the addition of tremendous amounts of energy, tear them down to the nanoscale.

While there are some physical methods of creating nanoparticles that will be addressed here, the majority of approaches are chemical in nature. Physical approaches are those that involve grinding, evaporations, or other processes where the material's size changes, but identity does not. In contrast, chemical methods involve chemical transformations that generate nanoparticles that consist of a chemical species that did not exist at the outset. There are a number of reasons why physical approaches are used less often than chemical, but they generally come down to quality and quantity of materials. Approaches such as grinding are plagued by the inefficiency of grinding at the nanoscale, meaning that size distributions tend to be broad and grinding times quite long (Kumar, Tiwary, and Biswas 2018). Evaporation techniques can quickly produce particles of narrow size distribution but are limited by the area of the surface onto which the particles are being deposited (Gangopadhyay et al. 1992). While there are exceptions, these difficulties have severely limited the usefulness of physical preparation of magnetic nanoparticles for biomedical applications.

In contrast, a large number of chemical approaches are successfully used to synthesize and functionalize magnetic nanoparticles (Yu et al. 2004, Park et al. 2004, Kataby et al. 1999, Watt et al. 2017, McGrath et al. 2017, Smith and Wychick 1980). These reactions may be formal reductions, oxidations, decompositions, metatheses (i.e. a reaction where components are exchanged between compounds, but without a change in formal charge of any of the species), or combinations of these (Huber 2005). In any case, the basic components are similar, as is the approach to size control. The reactions contain the same basic components, follow similar mechanisms, and yield similar, though far from identical, results. The real art and science is in the details of synthesis.

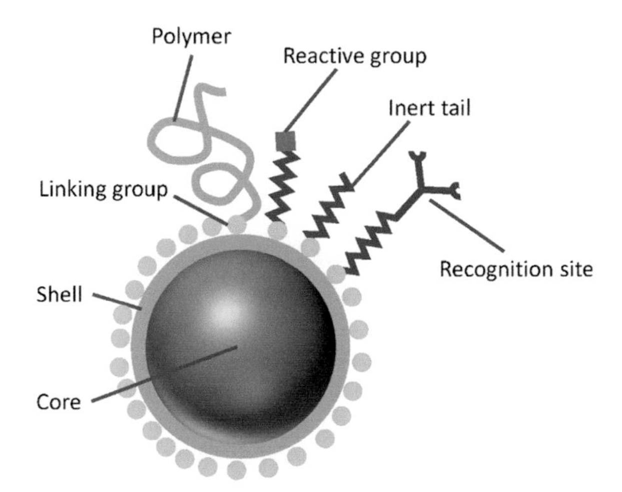

FIGURE 6.1 Schematic representation of the various potential parts of a nanoparticle synthesized and functionalized specifically for a biotechnology application.

6.2 MATERIALS CHOICE

The magnetic particles commonly used in biotechnology contain at least one of the three traditional magnetic elements: iron, nickel, and cobalt. While measurable magnetism has been demonstrated in a number of other materials ranging from noble metals to organic molecules, the effects tend to be too weak for practical use (Nealon et al. 2012, Miller 2014). Because the number of elements is so small, the question then becomes whether to use the metal, an alloy, or an oxide. Metals tend to have the strongest magnetism, though alloys come close and in some cases do exceed the magnetic properties of their parent metals, and oxides tend to have magnetism that is suppressed relative to the parent metal. Finely divided metals, however, tend to be very reactive, sometimes even pyrophoric (Huber 2005). As a practical matter, this sensitivity to oxygen and water precludes the use of pure magnetic metals in most biotechnology applications. Coatings to prevent or slow oxidation have been employed and include organic coatings (McGrath et al. 2017) and inorganic coatings (Herrmann et al. 2009), including a native oxide layer (Watt et al. 2018b). This is still not an extremely popular approach as a coated metal nanoparticle often has magnetic properties comparable to or even weaker than an oxide particle. Alloying, however, can lead to real improvements. One common example is to use the more stable iron platinum in place of iron (Sun et al. 2000), though there are a number of other combinations of magnetic-noble metal alloys (Wu et al. 2016). These oxidatively stable alloys tend to have magnetism that is somewhat suppressed relative to the pure metal but are greater than the oxides.

By far the most common materials in biotechnology applications are metal oxides, particularly iron oxides. The oxides are readily synthesized, stable for long periods of time in physiological conditions, and are still strongly magnetic. Magnetite (Fe_3O_4) is the most commonly formed iron oxide at or near ambient temperatures, but maghemite (gamma-Fe_2O_3) is also reasonably common. Magnetite has the advantage of slightly higher saturation magnetization, while maghemite, being fully oxidized to iron (III), is more oxidatively stable. In the end, the choice between the two often is a question of which has a more convenient synthesis for a specific application. In contrast, cobalt and nickel oxides tend to be anti-ferromagnetic or paramagnetic and are not generally useful magnetic materials, though they find use as dopants in iron oxides (Fellows et al. 2018).

Iron oxides dominate the literature of magnetic particles in biological and medical applications, because of their ease of synthesis, their low cost, and especially, their biocompatibility (Jain et al. 2008). Iron and its oxides are generally very well-tolerated even in relatively large doses

(Wang 2011, De Haro et al. 2015), while the other magnetic elements are of concern. A number of metal alloys are used in their bulk form in implants; the concern is that they are more bioavailable in fine particle form. For this reason, few researchers pursue these materials for in vivo use. In contrast, several iron oxide nanoparticle-based products have been granted regulatory approval for injection into humans (Wang 2011).

6.3 PHYSICAL METHODS OF NANOPARTICLE FORMATION

6.3.1 Mechanical Grinding

The standard approach for mechanically grinding materials into fine powders is ball milling. In this method, the material to be ground is placed in a sealed vessel with spheres of a grinding medium. The vessel is then shaken vigorously for some prescribed length of time to reduce the particle size to the desired level. Ball milling can quickly and easily reduce materials to powders of about a micron, but to reduce the size further, in particular to the tens of nanometers of interest for most biotechnology applications, requires the input of tremendous energy. The plot of surface area versus size in Figure 6.2 pictorially demonstrates the increase in energy in the form of surface energy as the total surface energy in a system is proportional to the total surface area. Inputting this tremendous amount of energy mechanically undoubtedly changes the underlying material, so that crystalline order often suffers when high-energy ball milling is pursued. As a practical matter, inputting this energy mechanically can also require more vigorous shaking than is customary and is referred to as high-energy milling. Addition of surfactant is sometimes used to reduce the amount of surface energy in the system, making it easier to reduce particle size in the nanoscale regime. This approach is termed surfactant-assisted ball milling (Chakka et al. 2006). While milling approaches can be tedious and time-consuming, they are fairly straightforward for brittle oxides (Goya 2004), but can require more complicated methods for ductile metals. Metals don't generally grind well, as the particles tend to stick and recombine under the applied pressure of the grinding medium. To overcome this issue, metals can be ground when mixed with an added brittle material such as alumina, to create a nanocomposite of the two. Metals can also be formed in situ by reacting an oxide with a more reactive metal (e.g. iron oxide milled with aluminum) (Pardavi-Horvath and Takacs 1995). Unfortunately, these methods for producing metals create a composite material that consists of large quantities of non-magnetic material. In the end, grinding is a difficult way to produce high-quality magnetic nanoparticles for biotechnology applications and is therefore relatively rarely used.

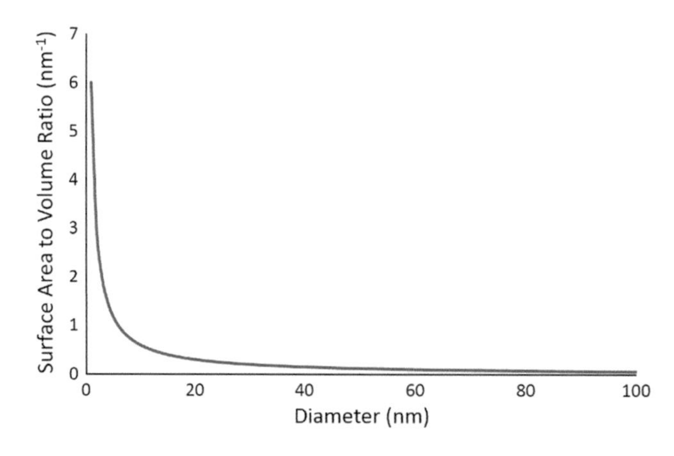

FIGURE 6.2 Graph of the surface area to volume ratio for nanoparticles of varying diameter assuming spherical shape.

6.3.2 Metal Evaporation

Evaporating metal onto a surface is a popular method of forming supported thin films of metals, but if the coating is extremely thin, the metal generally dewets the surface and forms discrete nanoparticles. This method has frequently been used to make supported nanoparticles. The difficulty is that if one wants to make appreciable quantities of particles that are solution, then evaporation onto a solid substrate is not useful. There are clever approaches to depositing metal directly into a solvent by either rotating a drum of viscous solvent to constantly refresh the surface and therefore allowing continuous deposition (Nakatani et al. 1987), or by co-depositing frozen solvent with the metal to grow a frozen matrix of solution-borne nanoparticles (Klabunde et al. 1974). Very good quality particles can be made in this way, including stoichiometrically controlled alloys or core-shell particles, but very specialized equipment is required, and the quantities produced are still fairly limited. To date these approaches have been used to produce very specialized structures for fundamental scientific studies and have not been broadly applied due to the difficulty and expense of producing particles in this manner.

6.3.3 Sonication

A newly reported physical approach to the formation of nanoparticles is the application of ultrasound to bulk metals. Ultrasound can be used to drive bubble formation and collapse in a solvent, which leads to strong mechanical forces and temperatures that can be as high as thousands of degrees Celsius. These forces are known to be capable of tearing particles into pieces and can easily generate micron-scale particles. However, it has been shown that particles cannot be continuously broken into smaller and smaller pieces (Suslick and Price 1999). When particles reach a critical size, in the sub-micron range, the forces across the particles can no longer break them efficiently. Additionally, the same forces can smash particles together, leading to the generation of an equilibrium size (Prozorov, Prozorov, and Suslick 2004).

It is also well-known that when a bubble forms near a surface, the surface can break the spherical symmetry of the bubble, leading to a flattened side near the surface. When this aspherical bubble collapses, it generates a microjet that impinges on the surface (Suslick and Price 1999). This is the mechanism by which ultrasonic cleaners work. It has recently been demonstrated that when the surface of a bulk metal is coated with an appropriate surfactant, that metal can be ejected from the surface and directly captured as nanoparticles (Watt et al. 2018a). While this hasn't yet appeared in the literature as a method for synthesizing magnetic nanoparticles, given the simplicity of the approach, it can be expected to be seen very soon.

6.4 CHEMICAL METHODS OF NANOPARTICLE FORMATION

Chemical synthesis has a wide range of applicability because there is such a wide choice of potential components. The basic mechanism of the vast majority of these approaches is similar and is described in some detail below. Similarly, there are a handful of basic components that make up a chemical synthesis of magnetic nanoparticles, including solvent, surfactant, and metal precursor. This section will introduce the fundamental problems with nanoparticle synthesis, describe the popular approaches to mechanistically understand the reactions, discuss the various components that make up most reactions, and then discuss specific approaches to nanoparticle synthesis.

As mentioned briefly before, nanoparticles are not the ground state of any reaction and instead are commonly referred to as a kinetic product. This means that the final product is determined by the kinetics of the reaction, not the thermodynamics. This contrasts with the way more traditional chemical reactions are described by high-energy reagents that react to form a lower energy product, usually after the addition of some activation energy. The situation with nanoparticle generating reactions, shown graphically in Figure 6.3, is that the desired product has an energy that lies well above the thermodynamically favored product, a bulk material. In this generalized scheme, we see the fundamental problem, that larger nanoparticles are lower energy than smaller particles. To produce nanoparticles of a desired size, one must prevent them from continuing to form a lower energy product. To do this, the kinetics of the reaction must be tuned to stop the reaction while the desired product is present. The details of how this is done fill the rest of this chapter, but there are some common themes. First is the use of a surfactant. Looking at the inset of Figure 6.3, we see that every atomic addition to a growing nanoparticle can be viewed as having its own small activation energy. Surfactants can be used to increase the energy barrier to add an atom to a nanoparticle, slowing the march down the curve toward larger sizes. While this doesn't directly exert kinetic control, it helps considerably. The other aspect to kinetic control is careful tuning of concentrations, temperatures, and times. The details of how these parameters are tuned to achieve the desired effects are dependent upon the preferred outcome and the growth mechanism of the nanoparticle forming reaction.

6.4.1 The LaMer Mechanism

LaMer and Dinegar published a paper discussing a proposed mechanism for a low size dispersity synthesis of an aqueous sulfur sol in 1950, which has become the basis for how we understand most nanoparticle syntheses (Lamer and Dinegar 1950). The explanation is simple and elegant, and while it neglects certain effects, it qualitatively describes many common syntheses of narrow size dispersity nanoparticles. Figure 6.4 shows a modified version of LaMer's diagram that conceptually describes the three stages of a reaction. In the first stage, a reaction is occurring to generate a monomeric species that will eventually coalesce to form nanoparticles. In LaMer's paper, this is referred to as a dissolved sulfur species, but in our discussion could represent an iron oxide. The concentration is zero at time zero, but quickly grows and exceeds the solubility in the solvent, becoming a supersaturated solution. Most systems, however, require some significant supersaturation to begin forming particles, so even upon crossing the saturation threshold, no stable particles are formed (transient particles below the minimum size for a stable nucleus may form but quickly dissolve).

The second stage of the reaction begins when the concentration crosses a critical nucleation concentration. This is the concentration above which the formation of stable nuclei is thermodynamically favored. This is the beginning of particle formation, and in most syntheses is coincident

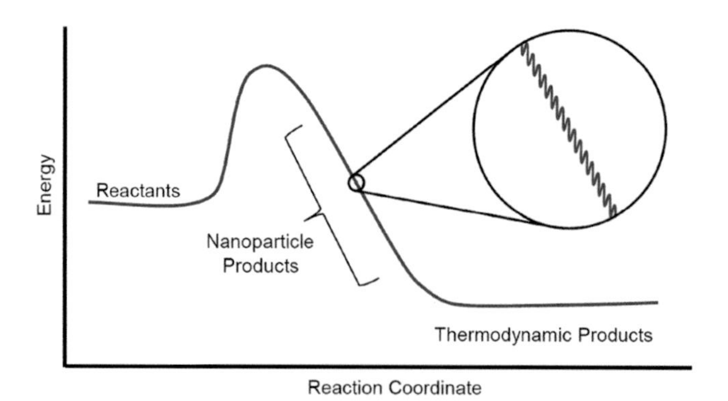

FIGURE 6.3 Energy versus reaction coordinate for a generic synthesis of a nanoparticle. Reactants begin at relatively high energy and increase in energy as they are given activation energy, then begin to decrease in energy as they grow. Successful size control requires that the nanoparticles stop growing in an energy landscape where growing leads to a lower overall energy. The inset shows that small activation energies are required for each subsequent addition of an atom, which does provide some thermodynamic support for stopping a reaction before the thermodynamic, bulk product is reached.

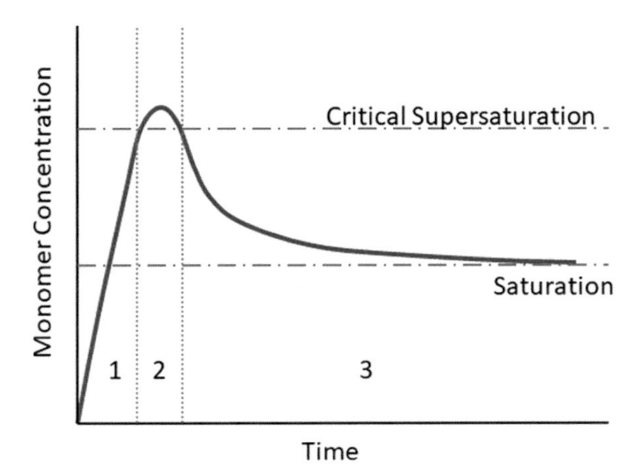

FIGURE 6.4 A diagram of the LaMer mechanism with the three phases of the reaction labeled. In phase 1 monomer concentration increases but no particles are formed. In phase 2 there is a burst of nucleation, which is followed by phase 3 where nanoparticles grow without further nucleation occurring.

with a dramatic color change. During the second stage, the concentration may initially continue to increase but will eventually decrease as there is finally a mechanism, the formation of particles, to partially relieve the supersaturation. The concentration eventually crosses below the critical concentration for nucleation, and the second stage of the reaction ends.

During the third stage, the concentration is below the concentration where nucleation is preferred so no new nuclei are formed, and the number of nanoparticles is constant. Still, the solution is supersaturated, so growth occurs on the existing particles, which continues to relieve the degree of supersaturation. This final stage of the reaction then continues, with the particles continuing to grow without the formation of new particles until the reagents are consumed or the reaction is stopped.

The argument that LaMer made based upon this framework is that the key to narrow size dispersity is a short burst of nucleation followed by a relatively long period of growth without nucleation. The reason for this is that if the time between when the first and last nuclei are formed is relatively short, then they will both have had approximately the same amount of time to grow and should be approximately the same size. On the other hand, if there is a long period of nucleation, then the earliest nuclei will have had much longer to grow and would be expected to be much larger than those particles that were nucleated later, resulting in a broad dispersity in sizes. While the reaction that LaMer reported had this kind of behavior without altering the conditions during the synthesis, ensuring a period of growth without nucleation is sometimes done by lowering the temperature after nucleation (Dabbousi et al. 1997). This lowers the rate of formation of the monomeric species and allows the concentration to more quickly drop below the threshold for nucleation and remain there. Successful application of this approach is sometimes referred to as "separation of nucleation and growth," and while this is descriptive of the approach it should not be taken too literally. The two steps are never fully separated as growth cannot be avoided when nucleation is occurring, though growth can occur without nucleation.

6.4.2 Ostwald Ripening

While the LaMer mechanism established the requirement for temporal separation of the elementary steps of nucleation and growth to ensure low size dispersity, it did not explain the inhomogeneity that evolves in the system following the consumption of the monomeric species by the growing nanoparticles. In 1896, Wilhelm Ostwald first described the spontaneous process in a liquid suspension whereby smaller particles dissolve and redeposit onto larger particles (Ostwald 1897). This phenomenon, termed Ostwald ripening, can be explained by considering the surface effects that play a critical role in many kinetic processes at the nanoscale. The atoms on the surface of a particle are energetically less stable than those that are well-ordered in the particle's interior. This means that smaller particles, with their greater surface area to volume ratio, are energetically less stable than larger particles. A simplified version of the Young-Laplace equation describes the relationship between the pressure that a particle in solution experiences as a function of its surface energy (J/m^2) or surface tension (N/m):

$$\Delta P = 2\gamma / r \tag{6.1}$$

where γ is the surface energy per unit area of the surface for a particle with radius r (Baldan 2002). From Equation 6.1, it becomes apparent that smaller nanoparticles experience a higher internal pressure than larger nanoparticles resulting from their higher surface energy. In order to reduce the overall energy of the system, molecules on the surface of a small particle will tend to detach and diffuse through solution and attach to the surface of a larger particle.

The rate of Ostwald ripening for nanoparticles with an appreciable solubility in a solvent is determined by the concentration gradient at the particle–solution interface. For small particles, the concentration of molecules at this interface is larger than the concentration in bulk solution. This results in the flux of the molecular species from the particle to the solution, leading to shrinking of small particles. Conversely, for larger particles, the concentration of molecules at the interface is less than that in the bulk solution, driving the diffusion of the molecular species from solution to the surface of the larger particles. The result is the complete dissolution of smaller particles in favor of the growth of larger ones.

Though it is typically associated with size-defocusing of nanoparticles in solution, ripening processes have been reported for increasing the average size of particles in a sample (Chen, Johnson, and Peng 2007, Zhang et al. 2015). However, this approach is typically less desirable than a process by which nanoparticle growth occurs from a continuous flux of the monomeric species in solution, as will be explored in the following section.

Other means to reduce the interfacial energy of a system include sintering and agglomeration. Sintering results from the merging of individual nanoparticles into larger, polycrystalline structures at high temperatures. While sintering of nanoparticles should generally be avoided in nanoparticle preparation, it becomes a concern only at temperatures greater than 70% of the melting point of a given material. Agglomeration of particles can also reduce the overall surface energy of a system without altering the crystalline structure of individual particles. Attractive forces between particles at the interface result in interactions that tend to be irreversible, though reversible agglomeration of nanoparticles in solution has been demonstrated as a method of size control in the synthesis of magnetic nanoparticles (Bleier et al. 2018). In this approach, magnetic nanoparticles nucleate and grow until a critical susceptibility is reached, after which, magnetic attraction between neighboring particles overcomes dispersive forces. This results in agglomeration and precipitation of the nanoparticles from solution, arresting nanoparticle growth. The magnetic agglomeration of the particles can be reversed by post-synthesis processing methods such as strong mixing, heating, sonication, and ligand exchange.

6.4.3 Extended LaMer

LaMer and Dinegar described a method by which nanoparticles with low size dispersity could be synthesized when a burst nucleation event is followed by a long growth period. However, achieving systematic size control of the nanoparticles with low size dispersity using this system presents a significant challenge. In this approach, the final size of the nanoparticles at the end of the reaction is ultimately determined by the number of nuclei formed during the nucleation event. For a given quantity of starting material, halving the number of nuclei would double the average volume of those particles when the reagent is fully consumed. Nucleation, however, is a chaotic, non-linear event that is extremely difficult to control systematically. Typical approaches to achieving reproducible control of nanoparticle size have focused on parameters reported to be influential for nucleation, such as the reaction temperature ramp rate (Guardia et al. 2010a, Park et al. 2004). However, performance properties of commercially available temperature controllers make temperature ramp rate a difficult parameter to maintain reproducibly between reactions. Additionally, environmental conditions such as altitude and humidity may affect the onset of nucleation for reactions performed under atmosphere. To realize reproducible, systematic size control of nanoparticles while accepting that the nucleation event cannot be conveniently controlled, a different approach is required.

The LaMer method described a reaction in a closed system, though it is easy to conceive of an analogous process in an open system. This process, termed the extended LaMer mechanism, uses a continuous addition of precursor to maintain a steady-state concentration of the monomer species in solution while maintaining all other parameters constant. The result is a steady growth of particles with a predictable growth trajectory that can be altered by changing details such as addition rate and ligand concentration (Vreeland et al. 2015, Fellows et al. 2018).

The extended LaMer mechanism (Figure 6.5) follows the classic LaMer mechanism in stages I and II but deviates after the nucleation event. In stage III of the extended LaMer mechanism, the concentration of monomer decreases as growth of the nanoparticles occurs. However, the continuous addition of precursor means that production of the monomer species can be maintained indefinitely. The monomer reaches steady-state concentration when the consumption of monomer by the growing nanoparticles becomes equal to the rate of production of monomer. This marks the beginning of stage IV, where steady-state growth of nanoparticles occurs. The advantage of this approach is that this fourth stage can be extended for an arbitrarily long time, allowing the growth of a wide range of particle sizes while maintaining low size dispersity. The steady concentration of monomer species suppresses undesirable ripening processes. Further, maintaining the monomer level below the critical supersaturation limit ensures that no additional nucleation events will occur. This approach is applicable to a range of nanoparticle species, though the data here was derived during the synthesis of high-quality iron oxide nanoparticles.

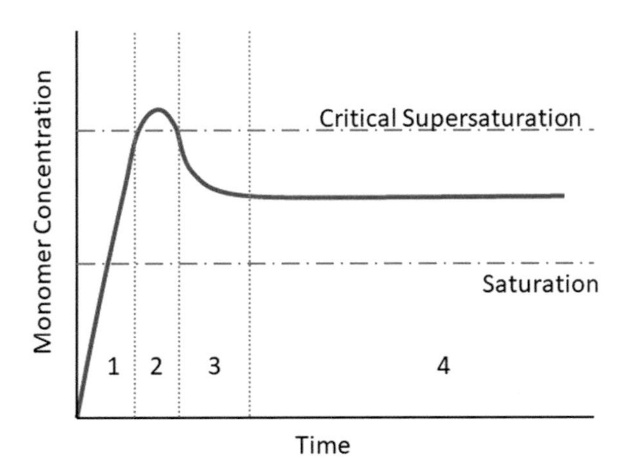

FIGURE 6.5 Diagram of the extended LaMer mechanism with the four phases of the reaction labeled. The mechanism begins in a similar way to the traditional LaMer mechanism, but with important differences in the latter times due to the continuous addition of precursor. Phase 1 has an increase in monomer concentration without the formation of nanoparticles, while phase 2 is a burst of nucleation. Phase 3 is a transitional phase where the monomer concentration decreases until the consumption of monomer equals the formation of monomer from the continuously added precursor. Phase 4 sees steady-state growth of the nanoparticles as monomer is continuously formed and consumed in the growth of the nanoparticles. In this phase, the average volume of the nanoparticles increases linearly with time.

A stoichiometric quantity of iron precursor is slowly added to a reaction flask containing a heated mixture of aliphatic hydrocarbon solvent and a long-chain fatty acid. Real time, or near real time monitoring of the reaction provides a means by which a growth rate can be calculated and a desired reaction endpoint can be predicted. Though a number of optical techniques can be applied to monitor reaction progress, small-angle x-ray scattering (SAXS) is a technique that proves ideal for quantitative measurement of nanoparticle growth. Small aliquots withdrawn from the reaction at regular intervals can be measured in 15 minutes or less and an average nanoparticle diameter calculated. In the case of iron oxide nanoparticle growth, plotting the SAXS diameter as a function of time reveals that two growth regimes are present (Figure 6.6a). In the early stages of the reaction following nucleation, rapid, catalytic growth of nanoparticles occurs. The size dispersity also decreases sharply during this time, indicative of nanoparticle size focusing. This period of catalytic growth transitions to the steady-state growth regime where the average particle volume grows linear with time. This is equivalent to diameter growing at time to the one-third power. Throughout this steady-state growth, we can see that the low size dispersity is maintained.

We have seen that nanoparticles are a kinetically determined product, which makes reproducibility in nanoparticle synthesis a significant challenge. However, the simplified kinetics of the steady-state growth stage in the extended LaMer mechanism allows for improved batch-to-batch reproducibility. Examining three identical reactions, similar trajectories for nanoparticles are observed, with diameter growing as $t^{1/3}$ for all reactions (Figure 6.6b). Additionally, a low coefficient of variation in the size of the growing particles is observed between the three reactions, which is comparable to the observed size dispersity of a single reaction. In fact, the data in the steady-state growth regime can be so well-fitted by the power law that this relationship can be used to accurately predict the final nanoparticle size with a high degree of confidence. As we discussed previously, the nucleation event is unpredictable, even with identical rates of precursor addition, and can yield differences in the number of particles produced between reactions. By taking measurements in the early stage of the steady-state growth regime, any variation in the number of nuclei formed is accounted for and the growth trajectory of an individual reaction can be reliably predicted. This allows for nanoparticles of a desired size to be synthesized reproducibly with minimal error.

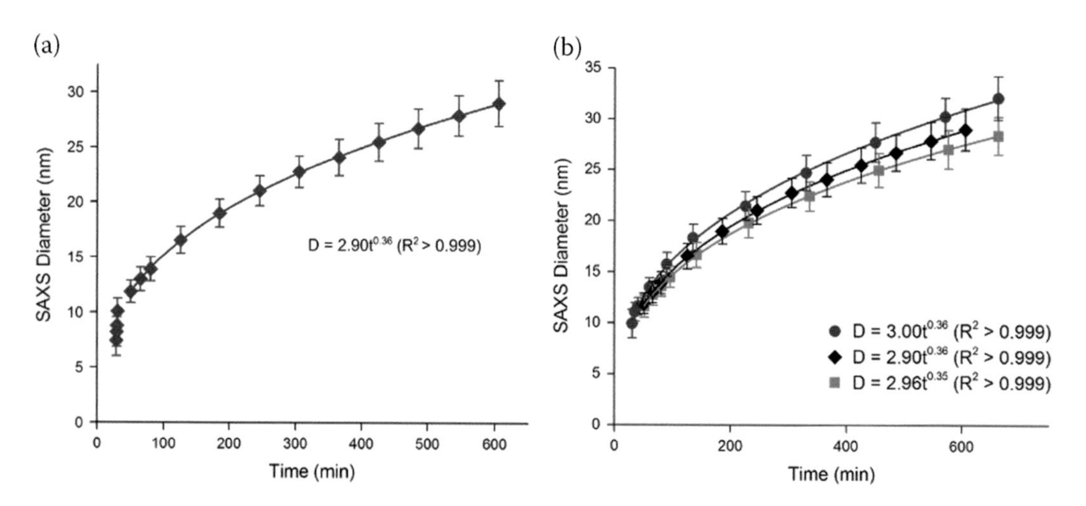

FIGURE 6.6 **(a)** A plot of nanoparticle diameter with time showing two distinct growth regimes, with a fast catalytic growth at early time and steady-state growth at later times, **(b)** A plot of the steady-state growth regimes of three different reactions demonstrating the reproducibility of the reaction, reproduced with permission from (Vreeland et al. 2015).

6.5 REACTION COMPONENTS

6.5.1 Solvent

The majority component in nearly all chemical approaches to nanoparticle synthesis is the solvent. While solvent is often viewed as an inert ingredient in a reaction, this is far from true in nanoparticle synthesis. The solvent choice can be a critical one and is an important factor in the resultant properties such as size dispersity and shape of the nanoparticles formed. The solvent can also limit the reagents that can be used, the temperature of reaction, and the kind of stabilization that can be used to kinetically trap the particles. The initial and most fundamental choice is between water and an organic solvent.

Water is an outstanding solvent for many reactions and is an obvious choice for biomedical applications, where the end application is nearly always in water. There are many metal sources for making magnetic particles that are freely soluble in water, including common salts such as halides, nitrates, acetates, and others. There are, however, a number of important drawbacks in the use of water. One is that the temperature of reaction is limited to 100 °C unless hydrothermal conditions are used, where a high-pressure reactor allows the reaction to far exceed the boiling point of water (see below for details). In many cases, a low-temperature reaction leads to less crystalline materials, whose magnetic response is lowered by the lesser crystallinity.

Magnetic materials, being metals and oxides, tend to be hydrophilic, which means that water will have strong, favorable interactions with the surfaces of growing particles. These strong interactions can differ based upon the crystal face, which results in preferred growth directions during the synthesis. The end result is often non-spherical particles that can be anything from rods, to flakes to irregularly shaped, faceted particles. While there are certainly times when these shapes are acceptable or even desirable, spheres are often preferred. These interactions can also lead to increased size dispersity. One last concern with having strong favorable interactions between water and the particles is the issue of passivating the particles. Chemists generally use some appropriate surfactant to kinetically trap the particles so that once they are formed, they remain stable. When the particles interact strongly with the solvent, any effective surfactant must interact much more strongly than the solvent, particularly given that it is generally present in orders of magnitude lower concentration. There are three common ways of dealing with this issue. One is to allow water to act as both solvent and surfactant. This is really only acceptable for a short time, as water does not present a strong barrier to agglomeration. Additionally, water passivated particles obviously cannot be dried or wholesale agglomeration can be expected. A second approach is to use a large or powerful surfactant to set up a barrier to agglomeration. A common example of this approach is the use of polar, water-soluble polymers that can have multiple favorable interactions with a particle's surface to allow it to remain attached. The final approach is to forgo attachment of a surfactant and take advantage of charge stabilization. In this approach, the aqueous solution is maintained at a pH where the surface retains a strong charge that effectively repels any particle that approaches due to their like charges. In some cases, a bulky counterion (e.g. citrate) may be associated with the surface to provide additional stabilization through steric stabilization. Charge stabilization is very common in particle synthesis in water and can be highly effective, even for long periods of time. However, the pH and ionic strength must be maintained within a window of stability, or the particles' surfaces can be neutralized leading to a loss of stability. For this reason, charge stabilized particles are often coated to provide steric stabilization before being introduced into the complex and varied aqueous environments used in typical biotechnology applications.

Syntheses in non-polar organic solvents generally do not use charge stabilization because dissociation of ions is rare in these solvents. Instead, steric stabilization is very common and works quite well in these systems. Because the particles' surfaces are hydrophilic and the solvent hydrophobic, traditional surfactants with a polar end-group and a long, non-polar tail function extremely well. Oleic acid is a prototypical surfactant for synthesis of magnetic particles in non-polar solvents.

The highly polar carboxylic end group can have a strong, chelating attachment to a metal atom on the surface of a particle, while the long alkyl chain shields the particle from the solvent and neighboring particles. The structure is reminiscent of an inverse micelle with its hydrophilic core and hydrophobic exterior. Extending this analogy can also explain the tendency of particles synthesized in non-polar solvents to be rounder that those synthesized in water. Micelles are round as this shape minimizes the contact area between the hydrophilic and hydrophobic entities, minimizing the energy of the system. Similarly, hydrophilic particles synthesized in a hydrophobic solvent tend to be rounder than their counterparts made in water. This tendency, however, has its limits and at large enough size, preferred, lower-energy facets do emerge, destroying the roundness of the particles.

Because roundness and steric stabilization are generally beneficial attributes, it may appear that synthesis in organic solvents is the better route. While for some applications where roundness is crucial that is undoubtedly true, aqueous reactions have a number of benefits that can't be ignored. The lack of a hazardous solvent is a big advantage and means that aqueous chemistry is generally cheaper and more scalable. The chemistry can often be done at room temperature, sometimes on a benchtop without concern for fume removal. Reactions in organic solvents are often done at elevated temperatures, sometimes under blankets of inert gas, and generally in a chemical fume hood to manage hazardous gasses. In the end, both water and organic solvents are very common, with aqueous approaches more common in large-scale and industrial applications.

6.5.2 Metal Precursor

The two primary concerns for choice of metal precursor are its solubility in the chosen solvent system, and whether there is a convenient reaction pathway under reasonable conditions. In practice, this means that ionic salts (in particular metal halides) are common in aqueous chemistry, while organometallic complexes are more often seen in non-polar solvents. It is also beneficial for reproducibility if the starting material is stable under ambient storage conditions and has a consistent stoichiometry. While this may not seem much to ask of a starting material, there are a number of popular metal sources that present difficulties. For example, iron pentacarbonyl decomposes at all liquid temperatures (including standard freezers) losing carbon monoxide to form iron cluster compounds, and iron oleate tends to form non-stoichiometric compounds with varying numbers of oleate ligands, and even a stable salt such as iron (III) chloride is hygroscopic and can absorb significant amounts of water that, unaccounted for, can render mole calculations inaccurate. These deficiencies can be handled by appropriate means such as freshly distilling iron pentacarbonyl, carefully washing iron oleate, and storing iron chloride in a desiccator, but one must be aware of the potential problems they may cause.

6.5.3 Surfactant

The surfactant choice is crucial for the control of the reaction kinetics, the short- and long-term stability of the particles, and the solution properties of the particles in the final application. In general, the control of reaction kinetics and short-term colloidal stability are the overriding concerns, because ligands can be and often are exchanged post-reaction.

So, if control of the reaction is the primary goal, then what are the criteria for a surfactant or mixture of surfactants? While the details depend upon the exact reaction envisioned, there are some general design principles. A stronger surfactant, one that partitions more strongly to the particles' surfaces, can be present in lesser amounts than weaker surfactants. Sometimes a coordinating solvent acts as both the solvent and a weak surfactant in a particle synthesis. Determining the optimum surfactant concentration is currently an inexact science, and with concentrations used that vary from millimolar to multimolar, it can be difficult to decide where to start. Optimizing surfactant concentration for a specific reaction is unfortunately still an

Edisonian process, as our predictive ability in reactions with such complex kinetics is still poor. A thorough reading of the literature, however, can give us approximate values to begin our trial and error. Very strong surfactants are often present in concentrations below 5%, and occasionally much lower than this. Moderately strong surfactants tend to be present in the range of 5–30%, and a weak surfactant may require concentrations from 30% to greater than 90%. While these classes may sound, and really are, fairly arbitrary, it is not difficult to determine what class a surfactant belongs in. Strong surfactants partition very strongly to the surface of the particles, generally due to the formation of a strong, covalent linkage to the surface. A familiar example would be alkanethiols serving as a surfactant for gold nanoparticle synthesis. This powerful surfactant can serve its function at low concentrations due to its very high affinity for the surface. Alkanethiols are also strong surfactants for the synthesis of magnetic metals, but not the oxides. Strong surfactants are not often used for the synthesis of magnetic metal particles, because their very strong interactions can cause a large loss in magnetism in the particles (Huber 2005). Thiols (Katabi et al. 1996) and phosphines (Yee et al. 1999) are two of the more common strong surfactants for the synthesis of magnetic metal particles. The oxides, being much less reactive, do not have any commonly used strong surfactants. Moderate-strength surfactants are more common for the oxides and are those that form more ionic linkages with the particles' surfaces. These include common acids and bases such as carboxylic acids, phosphonic acids, alkyl amines, and a number of polymers that contain these groups. Additionally, functional groups that have weaker interactions can form moderate surfactants if they are multiply present on a surfactant molecule, such as the case of polymers. These polymers commonly include polyvinylpyrilidone, polyethers, polyamines, polyacrylates, polyacrylamides, and nearly any other polymer containing significant quantities of polar groups. The weak surfactants are generally small molecules containing one or a few polar groups that form no strong bonds with the particles' surfaces. These can include a number of common functional groups like alcohol, ethers, or aprotic amines, and include a number of common solvents such as tetrahydrafuran (THF), dioxane, ethylene glycol dimethyl ether (glyme), and pyridine.

6.6 METHODS TO SYNTHESIZE METAL OXIDE NANOPARTICLES

6.6.1 Aqueous Precipitation

Aqueous precipitation of oxide particles is one of the most popular methods of forming magnetic particles. In particular, the coprecipitation of iron (III) and iron (II) salts in equal amounts is a very common approach to making magnetite (Fe_3O_4) particles, which is an oxide with equal amounts of Fe (II) and Fe (III). This reaction takes advantage of the varying solubility of iron ions as a function of pH. Iron is very soluble at low pH values and can be easily dissolved by relatively dilute (0.1 M) hydrochloric acid. This dissolution forms iron chloride, which is the species used in most coprecipitation reactions. Likewise, dissolution of iron chlorides in water forms an acidic solution upon dissociation of the iron and chloride ions. This low pH is critical to its solubility and rapid addition of base leads to an immediate precipitation of the iron as oxide particles. Size of the precipitate can be varied by addition of surfactants, initial and final pH, and concentration of reagents, including altering the ratio of iron (III) to iron (II).

Hematite (gamma-Fe_2O_3) is a magnetic iron oxide where the iron is all present as Fe (III) but forming it by aqueous precipitation is not as simple as neutralizing a solution of Fe (III) ions. This reaction typically forms a complex mix of phases with some hematite and some magnetite as well. Hematite is not the thermodynamically preferred phase of iron oxide at ambient temperature and pressure, magnetite is, and this makes it difficult to form phase pure hematite using room temperature aqueous chemistry. Thermolysis of iron salts in water has been shown to form hematite more effectively, as have hydrothermal reaction conditions (discussed below).

Aqueous precipitation is certainly the easiest, cheapest, safest, and most environmentally benign approach to the synthesis of magnetic particles, which has made it one of the most popular methods of forming magnetic particles industrially. In many applications these particles work extremely well and no other approach need be considered; however, there are some applications where the odd shapes and poor size control are an issue. In general, those applications with more stringent requirements must simply use another synthesis approach, as the kinetics of precipitation are not easy to control. The reason for this is that establishing a short nucleation event followed by a slow growth as prescribed by LaMer is extremely difficult in this type of precipitation reaction. When performing a neutralization, when nearing the equivalence point, a very small addition of base leads to a very large change in pH. This of course means a very sudden loss of solubility occurs for the iron salts, so the entire reaction occurs over a very short amount of time and a short, chaotic reaction occurs to yield particles with substantial size dispersity. The rapidity of the reaction at relatively low temperature can also lead to particles that are irregularly formed and that are not fully crystalline, which can have detrimental effects on the magnetic properties.

6.6.2 Hydrothermal Synthesis

Hydrothermal synthesis is a method of aqueous synthesis where the reaction is heated above the boiling point of water by keeping the reaction in a sealed chamber. Specialized reaction equipment is designed for this kind of reaction where many atmospheres of pressure can be generated and safely contained for an extended period of time. While similar precursors may be used in this type of reaction, there are a number of important distinctions. Because they occur in a sealed reactor, hydrothermal reactions must occur without the addition of other species, so the precipitation reactions generally occur as precursors slowly react at elevated temperatures. The high temperatures and pressures also favor improved crystallinity, and the slow reaction times can allow subtle differences in the energy of different facets to result in highly non-spherical particles. For example, rods are a frequent result of hydrothermal synthesis.

While all of this sounds very promising, hydrothermal syntheses are not without their difficulties. Obviously with sealed reactions reagents cannot be added nor can aliquots be withdrawn to monitor the reaction. Additionally, the stainless steel vessels prevent any visual clues as to the progress of the reactions.

So, although other chemical methods may be more versatile in some respects, hydrothermal syntheses are important for the ability to yield highly crystalline material in an aqueous environment. This is particularly true for the synthesis of maghemite, as this is a material that is very challenging to produce as a single-phase material below 100 °C.

It is worth noting that there is an analogous non-aqueous approach referred to as solvothermal synthesis. While the synthesis differs only in the choice of solvent, it has not found common use in the synthesis of magnetic particles. This may be because high boiling point non-polar solvents exist so that reactions can be conveniently run as high as 340 °C at atmospheric pressure. Any reaction that is run higher than this will likely produce gaseous byproducts from thermal decomposition that could potentially lead to unsafe levels of autogenic pressure to be generated. In fact, it bears mentioning that any hydrothermal or solvothermal reaction should be performed only after careful considerations of the safety of the temperature/pressure state that will be achieved in the reaction. The specialized vessels used in these reactions are colloquially referred to as "bombs" for very good reason.

6.6.3 Thermolysis in Organic Solvents

In recent years thermolysis, or thermal decomposition, of organometallic compounds in organic solvents, has become a very popular method of synthesizing magnetic metal oxides. One of the most popular methods is the thermolysis of iron oleate in high boiling point solvents such as dioctyl ether or octadecene. These reactions can produce large quantities of narrow size dispersity particles that are round and often single crystalline. The quality of the particles produced in this method is truly outstanding, so it is worth discussing the reaction in some detail. While other iron carboxylates can produce nanoparticles of similar quality, it is the iron oleate reaction that has been most studied. The first step in this approach is to synthesize the iron oleate, as it is not currently commercially available. Perhaps one reason that it is not available is the difficulty in producing it in the proper stoichiometry (Bronstein et al. 2007). The reaction itself is very easy. Oleic acid is neutralized with sodium hydroxide, and then the resultant sodium oleate is mixed with iron (III) chloride and stirred to produce iron (III) oleate and sodium chloride. Iron oleate does not crystallize, instead forming a viscous oil and, upon cooling below ambient temperatures, a glass. Crystallizing is one of the simplest and most versatile methods of purifying a crude material, but without this option the most common approach is to wash repeatedly to remove unreacted oleic acid and iron chloride. The difficulty here is knowing when to stop washing, as excessive washing can continue to remove oleic acid producing a material that is deficient in the acid.

Once the iron oleate is synthesized, it is mixed with additional oleic acid and a high boiling point solvent and heated at a controlled rate and allowed to react at temperatures near 300 °C. Generally, a temperature in excess of 280 °C is required for the reaction to occur, and temperatures as high as 360 °C are not uncommon. Typical reaction times are from 30 to 90 minutes, and they are generally conducted under an inert atmosphere. Longer reaction times tend to allow faceting and wider size dispersities through ripening processes, while relatively short reactions yield uniform spheres of magnetite. While this approach can produce large quantities of particles that are very uniform, round, and highly magnetic such as those in Figure 6.7, there are difficulties in controlling and reproducing the size in these reactions. Obviously, variations in the stoichiometry of the iron oleate can alter the final product, as can variations in the heating rate that a reaction sees. Both of these probably have their primary effects by altering the reaction in the early stages, which alters the amount of time that nucleation occurs. This in turn changes the number of nuclei, which, for a system with a fixed amount of precursor, will naturally alter the size of the finished particles.

A number of other precursors can be used to produce magnetic oxide particles, including various iron carboxylates (Bronstein et al. 2011), iron pentacarbonyl, and iron acetylacetonate (Masthoff et al. 2014, Guardia et al. 2010b). The iron carboxylates are generally used in similar ways as the detailed description above of the most common member of this class, iron oleate. Iron pentacarbonyl can be used to make both metal and oxide particles depending upon the conditions chosen (Huber et al. 2004, Hufschmid et al. 2015). For the synthesis of iron oxide, generally magnetite, iron pentacarbonyl is heated in the presence of an oxidizing agent such as a carboxylic acid. (The reaction is an oxidation, because the iron in iron pentacarbonyl is formally neutral.) The decomposition reaction is generally conducted above 120 °C, often as high as 200 °C, to obtain

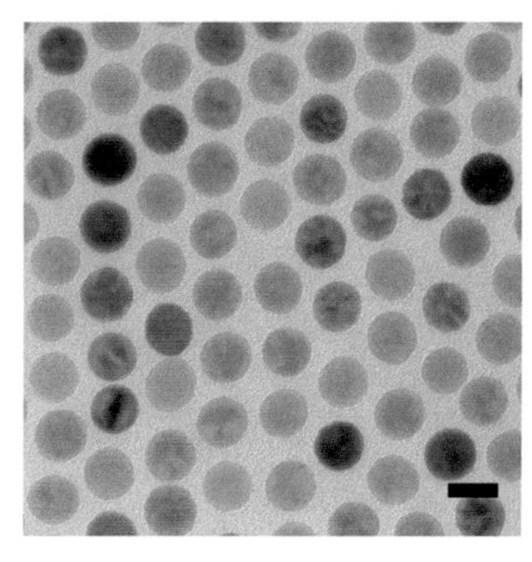

FIGURE 6.7 Round magnetite nanoparticles synthesized through the thermal decomposition of iron oleate. Scale bar is 25 nm.

rapid but controllable kinetics. The carboxylic acid is both the oxidant, and surfactant in this reaction. While this reaction produces high-quality particles, it is not as popular as other methods because iron pentacarbonyl is a reactive, toxic, and volatile liquid. Iron acetylacetonate would then appear to be an ideal reagent as it is a stable solid that is not particularly hazardous. Its primary drawback is its lack of solubility in most common solvents at ambient temperature. This can be overcome by simply adding it at room temperature and heating the solvent until the material dissolves and decomposes. When heated in oleic acid, iron acetylacetonate decomposes to form iron oleate in situ, which further decomposes to form nanoparticles with low size dispersity and excellent magnetic properties (Vreeland et al. 2015). This approach provides stoichiometric control of the iron precursor while yielding a high-quality nanocrystalline product from the iron oleate intermediate. In situ formation of iron oleate has the additional advantage of eliminating the processing and washing steps of the conventional method for preparing iron oleate that make reproducibility between reactions challenging. There also appears to be some debate in the literature as to whether the main product of this decomposition is magnetite or wüstite (FeO). It is likely that the product depends upon the details of the specific reaction in question, but wüstite is generally not desirable as it is not strongly magnetic. As it is metastable in ambient conditions, wüstite can easily be converted to Fe_3O_4 with moderate heating under atmospheric oxygen.

6.6.4 Polyol Synthesis

The polyol synthesis is a traditional method of making inert metal nanoparticles using molecules with multiple alcohol functionalities as a weak reducing agent at elevated temperatures. A typical reaction would be to heat a gold salt in the presence of a diol (e.g. ethylene glycol) in water and a surfactant to produce gold nanoparticles. When applying this reaction to iron salts, however, the reducing environment is not generally strong enough to produce pure iron nanoparticles. Instead, polyol reactions with iron salts tend to yield oxide particles (Jungk and Feldmann 2000) or iron particles with oxidized surfaces (Joseyphus et al. 2007).

6.7 METHODS TO SYNTHESIZE METAL OR ALLOY NANOPARTICLES

6.7.1 Reduction of Salts

Metal salts can be reduced to form metallic nanoparticles using a wide array of reducing agents in either water or organic solvents. The solvent is an important consideration when choosing a reducing agent, as the stronger reducing agents are not compatible with some solvents. Water in particular is incompatible with most of the well-known hydride reducing agents. One important exception is sodium borohydride, which, though it slowly decomposes in water, can be used in aqueous solutions if they are freshly prepared and used immediately. This is probably the strongest reducing agent that can be used in an aqueous solution and is therefore one of the more popular. Aqueous syntheses of metal particles have some of the same shape and size control problems seen in the oxide systems, so a compromise method is sometimes seen: the microemulsion route. In a microemulsion synthesis, an aqueous salt solution is reduced while encapsulated in water-swollen surfactant micelles. This couples two advantages: the high solubility of metal salts in water and the tendency to form round particles in organic solvents. Still, the presence of water can be problematic. Water is often avoided in the synthesis of magnetic metal nanoparticles because none of the magnetic metals is stable in the presence of water, and they generally form an oxide relatively quickly (Sun et al. 2006). The oxides on the surface of metal nanoparticles don't contribute to the magnetic properties of the particle as they tend to be disordered and have poor magnetic responses. So, while they can be tolerated in larger particles, in small particles, even a thin oxide causes a large decrease in the magnetization. This is such a significant problem, that one often sees metal particles with lower saturation magnetizations than oxide particles.

Given the difficulties with water, the reduction of metal salts in anhydrous environments has been explored. In this approach, anhydrous salt, surfactant, and solvent are mixed, and a strong reducing agent is added. One critical concern in this synthetic approach is the poor solubility of metal salts in non-polar solvents. This must be overcome by either using a coordinating solvent such as tetrahydrofuran or glyme that can provide sufficient solubility, or using a surfactant that will solubilize the salts (alkylated polyethers work well [Martino et al. 1997]). While this approach can yield very good-quality metal particles, the reaction tends to be very dilute so that the production of large quantities of nanoparticles is impractical.

6.7.2 Thermolysis/Sonochemical Decomposition

Thermal decomposition to make magnetic metal nanoparticles proceeds very much like the methods to make oxides; the primary difference is in the choice of reagents and surfactants. Because the magnetic metals all form oxides quite easily, metal precursors employed generally avoid structures that have oxygen–metal bonds. During decomposition of compounds with these metals bound to oxygen, some of the oxygen generally remains with the metal to produce oxides as in the cases of iron oleate, or iron acetylacetonate. There is one popular class of precursors that lacks a metal–oxygen bond, and also decomposes at convenient temperatures: the metal carbonyls. Carbonyls of all three magnetic metals exist and have convenient decomposition rates in the 100–200 °C range. While nickel carbonyl, $Ni(CO)_4$, is well-known, its high volatility and extreme toxicity have made it difficult to procure commercially, and its use is not recommended unless one is extremely skilled and well-versed in the handling of such acutely toxic compounds. Cobalt carbonyl, $Co_2(CO)_8$, and iron carbonyl, $Fe(CO)_5$, are both toxic, but are not of such extreme toxicity that they can't be handled using appropriate care in a well-equipped chemical laboratory. Iron and cobalt carbonyl have both been used to make metal nanoparticles individually, and as alloy particles. The general approach is to heat the materials in an appropriate high boiling point solvent such as dioctyl ether, and a surfactant (Huber et al. 2004). Oleic acid is commonly used, and while it coats the particles well, it also tends to form a thin oxide layer (Farrell et al. 2005).

While the thin oxide provided by oleic acid or some other appropriately mild oxidizer can provide some stability to the metal particles, their tendency to oxidize is so strong that additional alloying has been performed to stabilize them. Platinum and palladium alloying have been shown to stabilize both iron and cobalt toward oxidation. Iron platinum nanoparticles have been particularly well-studied and have been used in a number of biotechnology applications. As formed, iron platinum occurs in a chemically disordered face-centered cubic phase, which has very good magnetic properties. While its saturation magnetization is below that of pure iron nanoparticles, it is higher than that of any iron oxide, and is higher than most oxide-coated iron nanoparticles as well. The most common approach to make this material is to follow the approach of Sun et al., and simultaneously decompose iron carbonyl and platinum acetylacetonate in a high boiling point solvent in the presence of oleic acid and oleyl amine (Sun et al. 2003). The amount of platinum can be systematically varied in this system to alter the magnetic and oxidative properties of the resultant nanoparticles.

Another closely related synthesis is the use of ultrasound irradiation to drive the decomposition of, for example, iron carbonyl. The reaction can be viewed as similar to the thermolysis of iron carbonyl, but the heating is decidedly non-uniform, as locally very high temperatures are created where cavitation and collapse occur in the liquid (Suslick et al. 1999). The resultant high cooling rate can actually create amorphous iron nanoparticles, a material that is generally very difficult to produce (Suslick et al. 1991).

Though very magnetic metal particles have been produced, precious few have found use in biotechnology. The primary exception is iron platinum alloy particles. These have good stability in physiological conditions and have excellent magnetic properties. The pure metals are rarely used as they tend to oxidize readily and lose their magnetic properties in water or buffers. Attempts have been made to coat the particles to prevent this from occurring without tremendous success. Overcoating with noble metal has generally been unsuccessful, though some particles with carbide coatings have shown greatly enhanced stability in water (Meffre et al. 2012).

6.8 PURIFICATION

At each step of nanoparticle synthesis and functionalization, purification steps are required to isolate particles from excess reagents in solution, as well as concentrate particles and perform solvent changes. The method chosen is dictated largely by the scale of separation desired and the solvent system required.

6.8.1 Precipitation and Separation

A straightforward method for separating nanoparticles from a complex solution is precipitation of the nanoparticles, which generally requires the addition of a poor solvent for the functionalized nanoparticles and rapid precipitation from suspension by centrifugation or by placing the suspension in close proximity to a magnetic field gradient. For instance, oleic acid functionalized magnetite nanoparticles suspended in a non-polar hydrocarbon solvent such as hexane can be precipitated from solution by the addition of acetone or ethanol followed by centrifugation or magnetic separation (Vreeland et al. 2015). In this manner, small volumes of particles can be rapidly purified without complex experimental apparatus. However, practical use of this technique for production-scale purification is limited by the difficulty of scaling this process to large volumes.

6.8.2 Liquid Chromatography Separation

Size exclusion chromatography (SEC), also known as gel permeation chromatography (GPC), is a liquid chromatographic technique typically used to separate macromolecules in solution. SEC columns are packed with porous media (either polymer or silica based) that separate molecules according to their hydrodynamic volume in solution. Large molecules explore less of the pore volume of the stationary phase, allowing them to elute from the column faster than smaller molecules. In a similar manner, a suspension of nanoparticles can be purified from excess reagents by passing the mixture through an SEC column and collecting the pure nanoparticle eluent (Davis et al. 2014). Depending on the surface functionalization of the nanoparticles, a broad selection of aqueous and organic mobile phases can be used with either polymer- or silica-based stationary phases, making this a flexible purification technique.

6.8.3 Hollow Fiber Diafiltration

In the hollow fiber diafiltration process, a nanoparticle suspension is circulated through a tubular porous membrane, where the pore size determines the retention or passage of the components in solution (Sweeney, Woehrle, and Hutchison 2006). Repeated circulation of the solution with controlled replacement of the permeate results in a pure, concentrated nanoparticle solution. Commercially available membranes are best suited for use with aqueous-based suspensions. Additionally, the ability to parallelize the setup makes it an attractive option for production-scale purification processes, whereby large volumes of nanoparticle solutions can be processed simultaneously with identical process parameters.

6.9 FUNCTIONALIZATION STRATEGIES

6.9.1 Introduction

While the details of a particle functionalization strategy will depend upon the ultimate application, functionalization of nanoparticles for biomedical applications generally has two broad tasks. These are to provide colloidal stability to the particles and to impart specific functionality. Colloidal stability is covered in much more detail in Chapter 4 of this text but will be discussed in a more concise manner here as necessary. Because most biotechnology is done in near-physiological conditions, stability in complex aqueous environments is usually critical. Ideally this would mean the particles remain well-dispersed in pure water or buffers of a wide range of pH and ionic strengths, and with or without the presence of proteins. For in vivo applications, this is particularly important as the range of potential conditions that could be encountered is enormous, and significant agglomeration could have life-threatening results. Shelf-life is also a concern, so maintaining this stability for the long term is also highly desirable. The first goal of almost any functionalization strategy then is to provide water solubility to the particles in a way that is robust enough to endure whatever pH, ionic strength, and temperature excursions the material can reasonably be expected to experience.

It is also generally important to have a particle system that is biocompatible. This is critical in both in vitro and in vivo systems. The main concerns for in vitro systems is that the particles will lose their colloidal stability and agglomerate in the presence of proteins or other biomolecules or that they will participate in non-specific adsorption of proteins or other biomolecules that is inappropriate for the application. For in vivo systems, the concerns are more complex and include possibilities of provoking immune responses, agglomerating leading to an embolism, or having undesirable toxicity. In general, organic chemicals that are well hydrated, and have few or no charged species are the ones that are best tolerated in biological systems. These materials are often referred to as biocompatible or as non-fouling surfaces due to their tendency to avoid rapid, non-specific protein adsorption. The canonical example of an anti-fouling material is the water-soluble polymer polyethylene glycol, often abbreviated as PEG. This polymer may also be called polyethylene oxide, as polymers are generally named for the monomer from which they are synthesized and the polymerizations of ethylene glycol and ethylene oxide yield the same polymer. There is a trend in usage in that end-functionalized polymers and block copolymers are often named as polyethylene oxide as the anionic polymerization of this monomer is more conducive to yielding these products.

Many applications will also have targeting species bound to particles, so in these cases the functionalization strategy must provide a method for attachment of these moieties in appropriate numbers. This adds to the complexity significantly as it demands multiple functionalities in the stabilization approach and care must be taken to use compatible chemistries in the attachment to the surface and subsequent binding to targeting species.

At the most basic level, the goal of the stabilization is to prevent the particles from agglomerating and thereby providing colloidal stability. This can be accomplished in two fundamentally different ways: charge stabilization and steric stabilization. Charge stabilization is a matter of maintaining a strongly charged particle surface, so that the like charges on the particles repel neighboring particles. In steric stabilization, the particles maintain separation by having attached molecules whose steric bulk prevents the particles from touching. Both can be effective, but steric stabilization is most often used in biotechnology approaches as it is not as sensitive to pH and affords more obvious methods of attachment of targeting species.

It is worth noting that maintaining separation and colloidal stability is especially challenging in magnetic particles. There is a common misconception that superparamagnetic nanoparticles do not agglomerate magnetically because the constant realignment of their moments gives a time-averaged zero attraction. The situation is not so simple, as the moments in superparamagnets can align and interact, yielding a strong net attraction (Martin, Venturini, and Huber 2008). Of added concern is the fact that when agglomeration occurs, it can be particularly difficult to disrupt in magnetic particles as the interactions between particles can be very powerful. Agglomeration of particles is best avoided, and it is desirable to maintain colloidal stability at all times during the nanoparticle functionalization.

Following is a general discussion of the most common stabilization approaches and their relative merits. While only a subset of the approaches may be appropriate for any individual application, this should function as a general introduction to the most successful approaches.

6.9.2 Charge Stabilization

Charge stabilization is perhaps the simplest possible method of maintaining colloidal stability as there is generally no functionalization of the particles required. Generally, all that must be done is to maintain the pH of the solution in a range where the inorganic surface of the particle is highly charged. In appropriate conditions, the surface charge is stable and provides stability indefinitely. While this may sound ideal, there are significant drawbacks. If the pH is moved toward the isoelectric point of the particles, where they bear no net charge, the charge stabilization obviously ceases to function. Additionally, there are numerous species that can act as flocculants in systems like these by screening the surface charge or binding to multiple particles. Polyions in particular can be problematic in this respect, and proteins and many other biological species are polyions. Finally, because charge stabilization usually depends upon the strong charge that the inorganic surface bears, it generally precludes surface functionalization with biological targeting species.

The result of these caveats is that charge stabilization only works well in a well-controlled aqueous environment with no significant potential for pH excursions or the presence of complex biological fluids. The potential for the use of particles with colloidal stability that is purely charge-stabilized is extremely limited in biotechnology. That being said, the phenomenon is critical in many aqueous synthetic approaches, and can contribute to the stability of particles that are functionalized with charged molecules. Decrease in charge stabilization should especially be suspected when particles unexpectedly lose their colloidal stability when pH is changed.

6.9.3 Steric Stabilization

Steric stabilization is a more common approach in biological applications, as it has considerably more versatility. The approach is to coat the inorganic surface with an organic coating that masks the attraction of the cores and has no attraction itself. The goal is to provide interactions between particles that have a net repulsion at all distances. Since the magnetic cores will have a strong attraction if they are in close proximity, there is a minimum thickness of the organic coating to prevent a potentially irreversible attractive interaction from occurring. The exact distance where the magnetic attraction is significant will depend upon the details of the system such as the magnetic material and the particles' sizes but it can be calculated if the system is well-understood (Bleier et al. 2018). There is however a useful rule of thumb that can often be applied. In general, if magnetic nanoparticles are separated by a distance equal to or greater than their radius, then their magnetic interactions are negligible. So, a coating that is half the radius of the particles will generally ensure that magnetic interactions are so weak that no magnetic agglomeration will occur in the absence of an external field. In practice, the coating may function even if the thickness is significantly lower than this, depending upon the details of the system. For a 5 nm particle, a small-molecule self-assembled monolayer could easily provide the separation necessary for colloidal stability, but for a 25 nm particle, a polymer layer is often required (Saville et al. 2014). Polymers are generally very popular for particle functionalization as they are cheap and easy to make, are commercially available with numerous functionalities, and are reasonably easy to attach to surfaces.

Note that the concept of a layer thickness is not simple when one is referring to a water-soluble polymer layer. For a small-molecule self-assembled monolayer, we often assume that the layer is as thick as the molecule is long in its all-trans conformation (Bleier et al. 2018). While this may not be precisely correct, it is generally very close. In the case of a polymer, this would be woefully inaccurate. In polymer terminology, this all-trans length is referred to as the contour length and is generally much longer than the actual molecular dimension. Other length scales that are more appropriate are the radius of gyration (R_g) and the hydrodynamic radius (R_H), though even these do not generally precisely match the thickness of a polymer layer. Often, it is necessary to infer the thickness using things like changes in the size measured by dynamic light scattering (DLS).

6.9.4 Small-Molecule Surface-Bound Monolayers

As discussed above, nanoparticles are not the ground state of any system, and so a kinetic barrier to agglomeration is necessary at all times. Most systems will include some strongly bound small-molecule surfactant on the surface of the nanoparticle. These are generally present when the particles are synthesized and often remain on the particle as it is further functionalized. The continuous presence of a surface passivation layer prevents cores of the particles from making direct contact, which can lead to particle fusion and irreversible agglomeration, a process that destroys the nanostructure of the material. A common surfactant on the surface of magnetic nanoparticles, such as magnetite, is oleic acid. This molecule is inexpensive, has a high boiling point, and has a well-documented affinity for iron oxides, making it an ideal surfactant for high-temperature syntheses. It also forms a chelate with the surface, providing a reasonably strong linkage with the surface, as well as having a lengthy hydrophobic tail, which provides a significant barrier that prevents the hydrophilic cores from coming into direct contact.

Still, oleic acid has its drawbacks. It is a monofunctional molecule, and with its carboxylic acid bonded to the surface, there is no reactive species to allow conjugation of biocompatible coatings, targeting species, or other biologically relevant molecules. This means that any functionalization that occurs on top of an oleic acid monolayer is by necessity performed using non-covalent means. Additionally, though oleic acid provides good stability in many environments, the oleic acid is somewhat labile and establishes an equilibrium with some molecules free in solution while others remain on the surface. This equilibrium behavior is characteristic of surfactant systems except those with the strongest covalent surface attachments. The result of this equilibrium behavior is that nanoparticles stored in a good solvent for oleic acid, that has had excess oleic acid removed, can show decreased stability and eventual agglomeration. If excess oleic acid is repeatedly removed, oleic acid is shed from the particles to re-establish an equilibrium leading to a more diffuse coating and the potential for core-to-core contacts.

The lability of the oleic acid bond is not an insurmountable problem. First, the loss of oleic acid only occurs when the particles are dissolved in a good solvent for oleic acid. In a final biotechnology application, the solvent is generally water, in which oleic acid is insoluble. There is little need to be concerned about loss of oleic acid once the particle has been transferred to an aqueous solution. Second, the ease with which oleic acid can be removed provides an opportunity to perform a ligand exchange, where oleic acid is removed and replaced with a more desirable surfactant. The new surfactant could provide a stronger bond to the surface or have additional functionality for covalent attachment of subsequent functionality.

Ligand exchanges must be done carefully to maintain stability of the nanoparticles at all times, as any agglomeration that occurs while ligands are being removed is likely to be difficult or impossible to reverse. The general approach for a ligand exchange is to remove an excess of the initial surfactant from solution, then add a large excess of the desired new surfactant. Heat or sonication can speed the ligand exchange, but merely agitating for several days can achieve the exchange as well. There is some evidence that a slow, mild exchange over several days can lead to less agglomeration. Ligand exchanges work particularly well when the new surfactant has a stronger bond than the initial, so very successful exchanges with phosphonates (Sahoo et al. 2001, Davis et al. 2014), catechols (Stone et al. 2013, Davis et al. 2014), and silanes (Ma et al. 2003) have been reported, though other carboxylic acids can also be exchanged onto the particles' surfaces, with more difficulty (Davis et al. 2016).

Regardless of the specific species chosen as the initial surfactant, there are several roles that it must fulfill. It is there to arrest agglomeration during synthesis of the nanoparticles and prevent their irreversible agglomeration during any subsequent purification or other treatment. It often provides either a covalent linkage for further functionalization or acts as an anchored base for non-covalent interactions that allow the adsorption of functional species.

6.9.5 Small-Molecule Bilayer Coatings

Small-molecule coatings can be placed onto the initial coating of the nanoparticles using either covalent or non-covalent approaches. In practice, the non-covalent coatings work so well, that covalent attachments are rare. The result of a non-covalent attachment is a structure that is essentially a bilayer structure. A typical structure then would have a hydrophilic layer in contact with the particle's surface, with a long hydrocarbon tail (typically in the range of 10–20 carbons) extending away from the surface. These hydrocarbon tails are then in contact with another layer of hydrocarbon tails of similar length that have their associated hydrophilic headgroups extending into solution providing water solubility. Almost any surfactant with the classical structure of an alkane tail and a hydrophilic headgroup can work in this type of system and will, under appropriate conditions, form a bilayer structure. The tail can be either a single tail or a double tail as in lipids, and the head can take a variety of forms that include ionic or non-ionic hydrophilic headgroups.

One important consideration though is that the surfactant should not have high solubility as single molecules in water, meaning they should have a low critical micelle concentration (CMC). Below the CMC, every molecule in solution exists as a single isolated molecule, and at the CMC the solubility limit is reached and every additional molecule of surfactant enters into a micelle. In surfactant systems, coated particles act very much like micelles, and if the surfactant has a high CMC, then dilution of the coated nanoparticle solution can serve to strip the particles of their associated surfactants causing agglomeration. Fortunately, many commercial surfactants have low CMC values, as do naturally occurring lipids. Small-molecule surfactants that are appropriate for this type of coating approach include lipids, alkylated oligoethers, alkylated sulfates, and many others.

An important limitation on this approach is the size of the surfactant layer that is produced. Small-molecule bilayers are typically under 5 nm in thickness, meaning that for particles larger than about 10 nm, these kinds of structures may or may not provide protection from magnetic agglomeration for highly magnetic particles. For larger particles, it is often necessary to go to polymeric systems that may have similar bilayer structures but are considerably thicker.

Forming small-molecule bilayer coatings to provide water solubility can be a very simple process. Commercial surfactants are typically very good micelle formers, and the desired structure is similar in shape and size to an aqueous micelle. The process can be as simple as adding the surfactant to the hydrophobic solvent that contains the particles to be coated, adding water, and shaking or stirring the solution until the nanoparticles transfer to the aqueous phase. The transfer is generally obvious as magnetic nanoparticles are generally darkly colored, and their transfer can be seen by eye. Any emulsion that forms can usually be broken by centrifuging or adding salt to the water to drive the phase separation, or both if needed.

6.9.6 Polymer Coatings by "Grafting to" Approaches

Attachment of preformed polymers to a surface has a number of advantages for biotechnology applications. A wide range of functional and biocompatible polymers are commercially available, so they can be bought and attached to particles to yield exactly the desired properties. This can greatly simplify the functionalization chemistry. The drawback is that the density of the polymer layer that can be achieved by attaching preformed polymers to surfaces is somewhat limited. The reason for this limitation is that as a layer of polymer is forming on a surface, a concentration gradient begins to develop with a higher concentration of polymer near the surface. For further attachment to occur, each large polymer molecule must diffuse against this gradient before attaching to the surface (Huber et al. 1997). At some point, the energy penalty for moving against this gradient becomes too high and further adsorption essentially ceases. Still, the graft densities achievable are often high enough to lend stability, as a density gradient that is high enough to prevent polymers from approaching the surface is likely to be dense enough to prevent other particles from approaching the surface as well.

There are two basic approaches to the attachment of the polymers to a surface: through a single attachment point or multiple attachments. Single attachments are perhaps the canonical example but are not without problems. First, if a single attachment point is going to anchor a large-polymer molecule, it must be a very strong bond, and covalent linkages are ideal. Second, the size of the polymer limits the kinetics of attachment, as a single reactive group on a large-polymer molecule must find the surface for binding to occur. Very large polymers are rarely used for these single-point attachments, but polymers up to about 100 kDa can be used with good results.

Polymers may be attached directly to the particle surface or may be attached through a reaction with a species already present on the surface. Attaching a polymer to the surface requires a displacement of the existing surfactant in a manner that is similar to small molecules. This leads to a familiar hierarchy in attachment efficiency where amines and carboxylic acids are relatively weakly binding, but phosphonates and catechols bind more strongly. The stronger binding molecules displace more of the initial surfactant and attach more of the new polymeric surfactant (Davis et al. 2014).

When attaching a polymer to an existing functionality on a coating, a reaction with a low activation energy is highly desired, as excessive heating can change the underlying nanoparticles' morphology. The most common end terminations in these systems are probably amines and carboxylic acids. By having a surface coating that has one of these functional groups and the other on the polymer, very stable amides can be formed using well-known conjugation chemistry. A number of other chemistries have been used successfully, including reactions between isocyanates and amines and/or alcohols, epoxides and amines or alcohols, and isothiocyanates with amines (Hermanson 2013). Click chemistry approaches have also been applied including reacting thiols with alkenes, or the copper catalyzed reaction of azides with alkynes. These same reactions can be used to attach small molecules as well but are more commonly used with polymers and proteins.

Polymers that are bound through multiple attachment points do not require such strong attachments and can often be held by weaker van der Waals interactions between hydrophobic species. A good example is a diblock copolymer that has one block that is hydrophobic to provide water solubility and a second hydrophobic block that has favorable interactions with a hydrophobic surfactant on the particle surface. These molecules and the structures they form are very similar to what was discussed previously for small molecules but in this case just use much larger molecules. There is no need for the attachment points to be all in one region of the molecule, so random copolymers can be used in this situation as well. For example: monomers with hydrophobic sidechains can be incorporated into a polymer to cause it to associate with an existing hydrophobic coating (Di Corato et al. 2008), or monomers containing carboxylic acids copolymerized into the polymer can provide multiple attachments directly to the particle core. Using carboxylic acids also has the added advantage that acids that don't succeed in attaching to the surface can provide reactive sites for further functionalization.

Pre-formed polymers can even be used as initial surfactants for nanoparticle syntheses, and polar polymers such as acid-containing polymers or polyvinyl pyrrolidone have been used in this capacity. In some reactions, the polymer has even been shown to be catalytic for the particle formation reaction (Smith and Wychick 1980). While synthesizing nanoparticles in the presence of polymer is a convenient approach, most reactions that have good size and property control are done in the presence of only small-molecule surfactants.

6.9.7 Polymer Coatings by "Grafting from" Approaches

There are several advantages to growing a polymer in situ on the surface of a particle, but chief among these is the high grafting density that can be achieved by these methods. Grafting density has been shown to impact surface properties such as biocompatibility and biofouling properties, with coatings with high graft densities generally performing better (Huber et al. 2003). The drawbacks are that this approach is generally much more synthetically challenging than grafting a preformed polymer. If we take into account that this reaction is being performed on a set of particles that have been bought at great expense or synthesized with great effort, these reactions have the potential to be expensive failures if they don't perform well initially. Another problem is that it can be difficult to provide reactive groups for further functionalization reactions, though this can be achieved in well-designed reactions.

In principle, any polymerization reaction can be adapted to a surface-bound reaction, but in practice, only chain growth polymerizations, where some reactive species can cause chains to grow bound to a surface, are common. Step-growth polymerizations, which are typically condensation reactions, have a growth mechanism where monomers form dimers, dimers form trimers, etc. Because any molecule is equally likely to react, the majority of polymers form in solution and it is difficult to adapt to a surface except in unique cases where a reaction can be made to prefer the surface (Whitesell and Chang 1993) or in a reaction that is done by alternately saturating the surface with species that react with each other but not themselves (Orendorff, Huber, and Bunker 2009).

Of the chain-growth polymerizations, coordination polymerizations, which use a catalyst to drive the reaction, are rarely used for biotechnology. Catalysts tend to be poisoned by the polar, hydrophilic molecules of interest in the production of water-soluble particles. Ionic polymerizations are also fairly rare for these applications. They can also be poorly tolerant of the highly functional polymers of interest and also require extreme purity in the reagents that can be difficult to achieve in complex multicomponent systems such as these. The most common reactions by far are those that utilize radical chain growth polymerization schemes.

Standard free radical polymerizations were once the dominant method of forming a polymer layer in situ. While they remain common, controlled radical approaches have grown tremendously in popularity in recent years. Both approaches have significant advantages and the best choice depends on the details of the system. In a surface-bound free radical polymerization, a free radical initiator must be bound to a surface, and that initiator is typically heated to form the propagating free radical (though free radicals can also be initiated by illumination with UV light), which then reacts with a series of double-bond containing monomers to form a polymer. A number of approaches to binding the radical initiator to the surface can be used, and largely depend upon the functionality of the commercially available radical initiators, unless one wants to take on the fairly arduous task of custom initiator synthesis. Free radical initiators are commercially available with a number of convenient coupling chemistries including amines, carboxylic acids, alcohols, and isocyanate. Any of these can be coupled to a surface using appropriate room-temperature chemistries (radical initiators are obviously temperature-sensitive). Once attached to the surface, these radical initiators can be heated in the presence of monomer to initiate the polymerization reaction. Solutions are generally degassed as oxygen can behave as a radical scavenger, and the solvent is largely a matter of choice, though dioxane and toluene are popular choices. It is also worth noting that when a surface-bound initiator thermally decomposes, it forms two free radicals, only one of which remains on the surface. Free polymer always occurs in these reactions, and it should generally be separated from the particles at the end of the reaction. Centrifugation, filtration, and dialysis are all common methods of separating free polymers from surface-bound polymers. This free polymer is generally of about the same molecular weight as the surface-bound polymer, so analyzing its molecular weight by GPC or DLS is useful to understand the properties of the bound polymer. The polydispersity index of these polymers is generally 2 or higher, and very high graft densities can be formed.

The molecular weight of the polymer formed is controlled by the details of the kinetics, so a brief discussion of polymerization kinetics is in order. There are three basic steps in a free radical polymerization: initiation, propagation, and termination, as well as a fourth potential step: chain transfer. Initiation is defined as the formation of a free radical and addition of the first monomer. From this point on, an assumption is made that the rate of addition does not depend upon molecular weight (e.g. adding the second monomer occurs with the same rate constant as adding the 100th monomer), which makes the kinetic equations much simpler. Of course, there is some dependence of reactivity on molecular size, but the effect is generally small enough that ignoring it does not cause significant error. This simple assumption allows us to use a single-rate constant for an addition step, which is in fact a series of additions of monomers that forms the polymer. The addition continues until the growing chain is deactivated in one of two ways: termination or chain transfer. All free radical polymerizations experience termination and it is this step that destroys the radicals. There are two common ways that the chains undergo termination: recombination and disproportionation. Recombination is when two radical-bearing molecules form a new bond, thereby quenching both radicals, whereas disproportionation is the process of transferring a radical from one molecule to another radical-bearing molecule where they generally quench through the formation of a double bond. An important difference here is that recombination causes an increase in molecular weight, while disproportionation does not increase molecular weight but leads to a terminal double bond.

Finally, there is chain transfer, which in contrast to the other steps described, may or may not be a significant feature of a given polymerization reaction. Chain transfer occurs when a radical is transferred to another molecule and that molecule is able to initiate a polymerization. The mechanism for this to occur is generally that the radical on the growing chain abstracts a hydrogen atom from another molecule, leaving an unpaired electron on the second molecule. While chain transfer can be useful in lowering molecular weight, it is generally a nuisance in surface-initiated polymerization. This is because it generally transfers radicals from the surface, where they are useful, to the solution where they serve no useful purpose, but just create more unbound polymer. One exception to this is when chain transfer from solution to the surface is used to synthesize polymer layers. This approach is not in common usage but can be a convenient method of polymer growth in appropriate circumstances (Price and Huber 2013, Huber et al. 2003).

Having defined the various steps in a free radical polymerization, we can now see how the molecular weight is determined by the kinetics of the reaction. Free radicals are a highly reactive, but short-lived species, with lifetimes that are typically less than a second (McIntosh, Eager, and Spinks 1960). When a free radical polymerization is conducted, high molecular weight polymer begins to form immediately, with the number of polymer molecules slowly growing as the reaction progresses. The molecular weight of an individual chain is then determined by how many propagation reactions occur before one of the termination processes described above ends the chain's growth. From the kinetic equations in Figure 6.8 we can then determine what effect the reaction conditions have on the overall molecular weight. In general, increasing free radical concentration will yield a lower molecular weight, while increasing monomer concentration tends to lead to higher molecular weights. In addition to molecular weight, molecular weight dispersity is the other primary measure of the size of an ensemble of polymer molecules. Due to the random nature of the chain termination events, the lifetime of any particular growing chain varies significantly, which ensures a substantial dispersity in final sizes, even if the propagation rate is unchanged throughout the reaction. For a standard free radical polymerization, we expect a dispersity of at least 1.5, though it can be significantly larger. For example, when the reaction depletes monomer during the course of the reaction, the propagation rate can decrease, leading to smaller polymers being grown in the latter stages of the reaction. This variation of average size during the course of the reaction leads to an increase in size dispersity and can yield dispersities of 10 or larger if a reaction is run until monomer is completely consumed.

There is another approach to radical polymerization termed "controlled radical polymerization," or more formally "reversible-deactivation radical polymerization" (Jenkins, Jones, and Moad 2010), which refers to a collection of methods that modifies the kinetics of a radical polymerization to yield a more controlled product. These reactions include the well-known techniques of atom-transfer radical polymerization (ATRP) and reversible-addition-fragmentation chain-transfer polymerization (RAFT), as well as a number of related techniques. These reactions are often referred to as "living radical polymerizations" in the literature, although that term is no longer preferred (Jenkins, Jones, and Moad 2010). These methods share a common approach to controlling the polymerization reaction, and all have a reversible reaction that deactivates the free radical. The precise details of the various approaches are described elsewhere (Matyjaszewski and Spanswick 2005) but can be envisioned simplistically as a free radical reacting with some capping species to form a metastable dormant species that periodically dissociates reforming the active radical. In a well-controlled reaction, the vast majority of the radicals are in the dormant, deactivated state at any given time, while a small proportion of the radicals are in an active state and are adding monomer (if it is available). Individual radicals freely convert between the dormant and active state, while maintaining a low, equilibrium concentration of the active state throughout the reaction.

The growth of a polymer molecule in this approach is very different from a free radical polymerization. Here, the growth is slowed by the periods of inactivity, making initiation a relatively fast event followed by a much slower growth. If we envision an ensemble of growing chains, they would each experience infrequent bursts of growth when they enter their active radical state, interspersed with lengthy periods of dormancy where no growth occurs. So, instead of a chain initiating, growing, and terminating in less than a second, all of the chains initiate together near the outset of the reaction and grow slowly (through bursts of activity) over the course of hours, then are intentionally terminated at some point to conclude the reaction.

Controlled radical polymerizations offer a number of important advantages over traditional free radical polymerizations. First, the low concentration of active radicals at any time leads to a negligible amount of termination by recombination. Because this is the primary mode of termination in many reactions, in well-behaved systems, there is an extremely small amount of chain termination. This is what led to the use of the term "living" for these reactions, though its use is now regarded as inappropriate because termination still occurs to some small degree. Still, the termination is at such a low level that some of the advantageous features of living systems can still be present, such as low polydispersity and the ability to form diblock copolymers by addition of a second monomer after consumption of the first.

These advantages in control have led to an explosion in the use of these controlled radical polymerization techniques to functionalize surfaces. There are, however, some serious drawbacks.

1. Initiation

$$I \xrightarrow{k_d} 2(R\cdot)$$

$$R\cdot + M \xrightarrow{k_i} R\text{-}M\cdot = M_1\cdot$$

2. Propagation

$$M_n\cdot + M \xrightarrow{k_d} M_{n+1}^{\cdot}$$

3. Termination

$$M_n^{\cdot} + M_m^{\cdot} \xrightarrow{k_t} M_{n+m}$$

FIGURE 6.8 Fundamental steps in a typical radical polymerization reaction. Initiation generally proceeds by the unimolecular decomposition (with rate constant k_d), which produces two radicals that are denoted here as a dot representing the unpaired electron residing on the organic group *R*. Initiation is not complete until the first monomer addition occurs yielding an *R* group terminated molecule with a radical on the other end prepared to propagate the reaction. Propagation proceeds by adding subsequent monomers in a chain reaction forming the polymer until a chain termination event occurs. While there are several mechanisms of chain termination, recombination, the joining of two radicals into a stable molecule, is often dominant.

While free radical reactions are very versatile, and a standard initiator will work for a great many monomers, controlled radical polymerizations are generally tailored to an individual monomer or monomer class. Because there are so many unique approaches to controlled radical polymerizations, with more being developed all the time, many of the initiators used are not commercially available and require custom synthesis. Additionally, monomers that are highly functional or polar are particularly problematic, and unfortunately, these are the types of polymers used in biotechnology applications to impart water solubility and biocompatibility. These monomers tend to lead to undesirable side reactions including chain transfer and termination events. Techniques are being developed and refined all the time to improve on the controlled radical polymerization of these important materials, but the control is generally worse than in more traditional non-polar monomers.

6.9.8 Latex Particle Formation

One method of functionalization that has been popular for many years is the formation of latex particles containing magnetic nanoparticles. Though latex particles are very commonly made of polystyrene, the term *latex* does not refer to a specific material, but to a material made by an emulsion polymerization. (The exception here is the term *natural latex* that refers to the polymerizable sap from rubber plants that contains emulsified rubber that is used to make a variety of consumer products and is also a well-known allergen.) An emulsion polymerization is one where a hydrophobic monomer and a hydrophobic initiator are dispersed in an aqueous matrix with the aid of surfactants. The emulsion is generally generated and maintained by constant, strong stirring, and the emulsion is heated (or occasionally exposed to UV light) to initiate the polymerization. Latex paints therefore contain water-borne hydrophobic polymers, latex films (e.g. gloves) are formed from these types of emulsions, and latex particles are formed by solidifying the droplets in an emulsion. Because the distribution of the sizes of droplets in an emulsion can be very small, latex spheres can be manufactured that are extraordinarily uniform and reproducible. For this reason, latex spheres have a long history of use as size calibration standards in, for example, electron microscopy.

Forming magnetic latex particles can be as simple as adding hydrophobic magnetic particles to a standard latex-forming reaction. However, as the number of magnetic particles increases, several issues can arise. The addition of magnetic nanoparticles affects both the viscosity and the surface tension of the hydrophobic phase and can lead to difficulty in forming and maintaining high-quality emulsions. Additionally, highly magnetic latex particles may have poor colloidal stability due to their strong magnetic attraction. Finally, magnetic interactions between the individual nanoparticles within a latex particle can lead to undesirable changes in magnetic properties such as a change from superparamagnetism to ferromagnetism. A combination of all of these effects causes a functional limit on the amount of magnetic material that can be loaded into a latex particle that is highly dependent on the specifics of the system. Still, loadings as high as 50% by mass are not unusual for systems like this.

This approach is different from previously described functionalization strategies in that it does not lead to individual nanoparticles with a thin coating but imbeds a large number of particles into a mass of polymer. This leads to a particle in the hundreds of nanometer to micron scale that has the enhanced susceptibility of a magnetic nanoparticle, but a moment that could never be achieved with a single superparamagnetic nanoparticle. This can be highly advantageous for applications that require the application of a force using a modest magnetic field (e.g. magnetic separation of cells or biomolecules).

Another advantage of using a larger particle loaded with magnetic nanoparticles is that the surface area per volume and per moment is much lower. This means that if the goal is to magnetize a cell for collection using an antibody or other targeting molecule, less of the targeting molecule, which is typically the most expensive part of the formulation, can be used to provide a certain magnetization. An obvious disadvantage of this approach is the inability to consider in vivo applications of what are essentially small plastic beads.

As made, latex spheres are dispersed in water, so there is no awkward transfer from an organic phase to contend with, but long-term colloidal stability is not assured. Many latex dispersions are not thermodynamically stable dispersions and require vigorous agitation before use. Additional stability can be provided by functionalization with, for example, the addition of surface-bound water-soluble polymers. Biofunctionalization of latex particles is also commonplace to provide specific functionality to these particles. Many of the previously discussed approaches are used, with the obvious difference that the reactions are purely organic reactions, without the necessity of making an initial linkage to the inorganic magnetic material. The reactive functionality used for the further functionalization can be residual double bonds from a cross-linker such as divinyl benzene, or a reactive species specifically added for later functionalization.

Another method to forming polymeric microspheres is through the use of microfluidic droplet chips. In a typical approach, a flow of polymer solution in an organic solvent meets aqueous flows on both sides then is forced through a narrow orifice. This "flow focusing" leads the organic fraction to break into small droplets, which can then dry to form polymer spheres in the micron range that have very narrow size distributions. Addition of magnetic nanoparticles to the polymer solution leads to magnetic polymer spheres (Bokharaei et al. 2016). While the results of this method would not typically be called latex particles, they are very similar to the structures formed in a latex reaction.

6.9.9 Silica Coating

Until now, we've focused on organic functionality of inorganic magnetic nanoparticles, but there is one inorganic coating that is extremely useful: silica. Silica's many advantages include its chemical stability, lack of toxicity, low cost, and ease of further functionalization. The basic approach varies very little, but the details are critical to the final product. In general, the magnetic nanoparticles are dispersed in water or a water and alcohol mixture, and a silica precursor such as trimethoxysilane (TMOS) or triethoxysilane (TEOS) is added. The condensation reaction to form silica can be catalyzed by acid or base, though bases are more commonly used. The concentrations of all of these species determine the speed of the reaction, the thickness of the coating, and the likelihood that nanoparticles will be irreversibly attached to each other by silica. Silica coatings can then be broadly divided into two classes: Those that encapsulate individual nanoparticles each in their own silica shell, and those that encapsulate multiple nanoparticles into a single silica particle. Keeping nanoparticles separate requires low concentrations of nanoparticles and, generally, of the other species as well, while forming large agglomerates of particles occurs when particle and reagent concentrations are high. Silica coating individual particles retains the advantages of individual nanoparticles, while silica coating agglomerates provides many of the advantages discussed above for latex particles.

Often, silica coating is only a first step in functionalization and is followed by functionalization with organic materials. The ease with which silica is functionalized with organics is a big reason why it is so commonly used. A great many silane coupling agents, capable of forming a strong bond with silica while providing a conveniently reactive second functionality, are commercially available. The reaction that binds silane coupling agents to silica is catalyzed similarly to the silica formation reaction, generally with added base. One exception is the use of aminopropyl silane, which is known to be autocatalytic and requires no added base (White and Tripp 2000). It is worth noting that to maximize the strength of this bond, a high-temperature treatment in anhydrous conditions is used to complete the condensation reaction. Without this treatment, the coating may be susceptible to degradation upon long standing in water or exposure to non-neutral pHs.

6.10 BIOFUNCTIONALIZATION

Some in vivo applications do not require specific interactions. For example, nanoparticles that are used to image blood must only remain in circulation to be effective. Other nanoparticles can take advantage of the enhanced permeability and retention (EPR) effect in cancer tumors, though this effect and its use remain controversial (Wilhelm et al. 2016, Torrice 2016). For many systems, however, a key challenge to effective use is achieving selective delivery of systemically administered nanoparticles to a tissue of interest. This approach is attractive as it allows nanoparticles to be detected, imaged, or triggered to achieve a therapeutic effect, including localized heating or delivery of a medicine to a specific tissue. An ideal nanomedicine would target diseased tissues with high specificity, while evading uptake by healthy tissues or rapid clearance by the reticuloendothelial system (RES) (Arami et al. 2015). Modification of the nanoparticle surface with a targeting ligand that enables selective binding is then critical, as is a comprehensive understanding of how the ligand will alter the relevant properties of the nanoparticle (hydrodynamic diameter, surface charge, etc.) and the resulting pharmacokinetics and biodistribution.

An array of different organic molecules can be employed for biofunctionalization of a nanoparticle surface. These molecules can be grouped into a few general categories, including small molecules, polypeptides, nucleic acid-based aptamers, and proteins (Liu et al. 2011). Each of these classes of molecules has its associated advantages, disadvantages, conjugation chemistries, and will uniquely impact nanoparticle biodistribution. Small biomolecules and polypeptides can be used to target nanoparticles following the attachment procedures described in previous sections. Aptamers are short synthetic DNA molecules that are developed to bind to specific species and can also be attached using standard chemistries. Proteins, however, require more careful handling as their intricate structures are susceptible to damage by high temperatures, organic solvents, and pH changes. Here, we focus on protein attachment, with emphasis on monoclonal antibodies whose high selectivity and specificity have contributed to their growing use in diagnostics and bio-therapeutics (Reichert 2014). The majority of antibodies used are immunoglobulin G antibodies (IgGs), which consist of two high molecular weight polypeptides (heavy chains) and two lower molecular weight polypeptides (light chains) assembled into what is typically depicted as a Y-shaped molecule (Harris et al. 1997). As shown pictorially in Figure 6.9, the two heavy chains are linked by disulfide bonds to form the Fc (shorthand for fragment, crystallizable) domain. Each heavy chain is also disulfide bonded to a light chain in the Fab (Fragment, antigen-binding) domain. The Fab domain contains two antigen-binding sites, found at the tips of the forked portion of the Y. The Fc domain, made up of heavy chains in the base of the Y, does not take part in antigen binding.

Perhaps the simplest approach to coat a nanoparticle with protein is nonspecific adsorption onto the nanoparticle surface. For example, nonspecific adsorption of albumin is a common approach to enhance biocompatibility in nanoparticles (Zhang et al. 2012). This approach takes advantage of the fact that albumin is the most common protein in blood and is therefore largely ignored by the immune system. Other proteins though have been used to drive delivery of the nanoparticles to a specific tissue (Rodrigues et al. 2018). Whether the protein is adsorbed in its native conformation or undergoes changes in conformation, including denaturation, upon adsorption can determine the difference

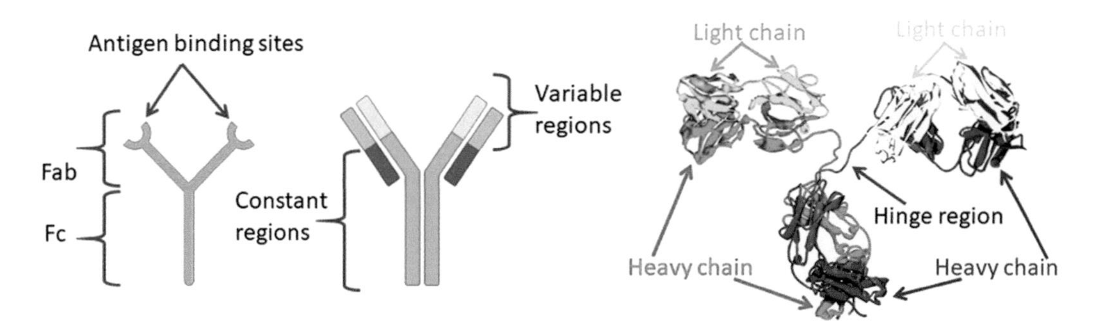

FIGURE 6.9 Three common diagrammatic representations of IgG varying from stick figure representation to one based upon the actual structure as determined by x-ray diffraction.

between effective targeting agent and a highly immunogenic nanoparticle. Conformational changes are protein-dependent and are typically driven by hydrophobic interactions between the protein and the nanoparticle surface following adsorption. Additionally, as the nanoparticle size becomes comparable to that of the protein, the radius of curvature of the nanoparticle and the size-dependent electrical double layer become increasingly important factors in driving conformational changes (Satzer et al. 2016). Moreover, protein adsorption can be reversible as a function of temperature, pH, and ionic strength of the environment. To improve the stability of the protein/nanoparticle interaction, covalent binding of the protein to the nanoparticle surface offers an attractive alternative.

The most popular approach for covalent linkage takes advantage of reactive side chains of amino acid residues found within proteins and conjugates them to appropriately functionalized nanoparticles. The most common of these include primary amines found in the lysine residue and the N-terminus of the protein, carboxylic acid in aspartic and glutamic residues and the C-terminus of the protein, and sulfhydryl in the cysteine residue. Amines react with activated carboxylate moieties to form amide bonds, where sulfhydryl moieties react with haloacetyl, alkyl halide derivatives, maleimide, pyridyl disulfide, among others, to form thioester or disulfide linkages (Hermanson 2013). Crosslinker reagents are chosen for their reactivity to the functional moieties on proteins and nanoparticle surfaces. Homobifunctional crosslinkers have the same reactivity at both ends and have the potential to link neighboring molecules, resulting in protein polymerization or nanoparticle agglomeration. In this case, heterobifunctional crosslinkers may be chosen to reduce undesired reactions. The most common "zero-length" crosslinker of this type, EDC (1-ethyl-3-(3-dimethylaminopropyl)carbodiimide), can be used to couple primary amines to carboxylic acid functionalized nanoparticles in an aqueous environment. In a simple one-step reaction, EDC is used to activate carboxylate moieties on particles, forming a reactive ester intermediate. This ester is reactive with primary amines, promoting the formation of a stable amide bond between the protein and the nanoparticle surface. NHS (*N*-hydroxysuccinimide) or sulfo-NHS is often used in conjunction with EDC to form a more stable NHS ester or sulfo-NHS ester intermediate that can improve reaction yield (Hermanson 2013). This approach is straightforward yet results in proteins bound to the surface in a variety of orientations. For conjugated antibodies, this means that a portion will have its active binding sites in the Fab region oriented in a way that will inhibit binding with the target. Directional conjugation strategies have been developed to better control the orientation of immobilized antibodies by linking antibodies through the non-binding Fc region (Vijayendran and Leckband 2001). Antibodies typically contain glycosylated residues in the Fc region that can be oxidized under mild conditions to create a hydrazide-reactive aldehyde moiety. A hydrazide-terminated heterobifunctional linker can then be used to link the antibody to the nanoparticle surface (Kumar, Aaron, and Sokolov 2008).

Simply having the ability to attach proteins is not the complete story. For all of the attachment methods described above, except the directional attachment of antibodies, there are multiple attachment points on the protein. This allows for the possibility of several non-ideal protein attachment motifs. Proteins may attach at multiple points leading to a distorted structure that could impair performance. A single protein could bridge between particles, which is essentially a cross-link between particles, which will increase the hydrodynamic radius of the particles and if common enough could lead to wholesale agglomeration. In the case of homobifunctional attachment approaches, proteins can attach to other proteins leading to similar cross-linking and agglomeration issues. Generally, these concerns lead to conjugation reactions to be performed at the lowest concentration that is practical to limit the likelihood of these kinds of cross-links.

Attachment density is also an important consideration as active binding sites have a requisite amount of space around them to allow for interactions with the target species. The goal is generally to have a small number of these binding sites on every particle's surface. Exactly how to achieve this is a difficult problem that requires careful consideration of reaction conditions in general, but the relative concentrations of the species in particular. As is often the case in nanoparticle chemistry, the general concepts are reasonably straightforward, but the parameter space is enormous, making control over the reaction extremely challenging.

6.11 SUMMARY

This chapter has attempted to summarize the synthesis and functionalization of magnetic nanoparticles for applications in biotechnology. This is a topic of enormous breadth that encompasses a tremendous number of approaches. Still, it is clear where the research is currently focused. Most magnetic nanoparticles are synthesized through wet chemical approaches, with thermolysis in organic solvents perhaps the most popular with precipitation in aqueous environments the other common approach.

Functionalization of nanoparticle surfaces is still widely varied with both covalent and non-covalent attachment approaches being common. Perhaps the strongest trend is in what is being attached. There is a continuous movement toward more complex functionality. While studies on unfunctionalized magnetic nanoparticles were once common, currently we see a great deal of intentional treatment of the surface to drive specific functionality. While they are not yet common, we are now beginning to see multi-component coatings with each species having a specific function. For example, a coating may have a stealth polymer, like PEG, but also contain a targeting IgG as well as a drug payload. There is no reason to believe that this trend will not continue with ever more complex coatings with enhanced functionalities being created.

REFERENCES

Arami, H., A. Khandhar, D. Liggitt, and K. M. Krishnan. 2015. In vivo delivery, pharmacokinetics, biodistribution and toxicity of iron oxide nanoparticles. *Chemical Society Reviews* 44 (23), 8576–8607. doi:10.1039/c5cs00541h.

Baldan, A. 2002. Review progress in Ostwald ripening theories and their applications to nickel-base superalloys Part I: Ostwald ripening theories. *Journal of Materials Science* 37 (11), 2171–2202. doi:10.1023/A:1015388912729.

Bleier, G. C., J. Watt, C. K. Simocko, J. M. Lavin, and D. L. Huber. 2018. Reversible magnetic agglomeration: A mechanism for thermodynamic control over nanoparticle size. *Angewandte Chemie* 57 (26), 7678–7681. doi:10.1002/anie.201800959.

Bokharaei, M., T. Schneider, S. Dutz, R. C. Stone, O. T. Mefford, and U. O. Häfeli. 2016. Production of monodispersed magnetic polymeric microspheres in a microfluidic chip and 3D simulation. *Microfluidics and Nanofluidics* 20 (1), 1–14. doi:10.1007/s10404-015-1693-y.

Bronstein, L. M., J. E. Atkinson, A. G. Malyutin, F. Kidwai, B. D. Stein, D. G. Morgan, J. M. Perry, and J. A. Karty. 2011. Nanoparticles by decomposition of long chain iron carboxylates: From spheres to stars and cubes. *Langmuir*. doi:10.1021/la104686d.

Bronstein, L. M., X. L. Huang, J. Retrum, A. Schmucker, M. Pink, B. D. Stein, and B. Dragnea. 2007. Influence of iron oleate complex structure on iron oxide nanoparticle formation. *Chemistry of Materials* 19 (15), 3624–3632. doi:10.1021/cm062948j.

Chakka, V. M., B. Altuncevahir, Z. Q. Jin, Y. Li, and J. P. Liu. 2006. Magnetic nanoparticles produced by surfactant-assisted ball milling. *Journal of Applied Physics* 99 (8). doi:10.1063/1.2170593.

Chen, Y., E. Johnson, and X. Peng. 2007. Formation of monodisperse and shape-controlled MnO nanocrystals in non-injection synthesis: Self-focusing via ripening. *Journal of the American Chemical Society* 129 (35), 10937–10947. doi:10.1021/ja073023n.

Dabbousi, B. O., J. Rodriguez-Viejo, F. V. Mikulec, J. R. Heine, H. Mattoussi, R. Ober, K. F. Jensen, and M. G. Bawendi. 1997. (CdSe)ZnS core–shell quantum dots: synthesis and characterization of a size series of highly luminescent nanocrystallites. *The Journal of Physical Chemistry B* 101 (46), 9463–9475. doi:10.1021/jp971091y.

Davis, K., B. Cole, M. Ghelardini, B. A. Powell, and O. T. Mefford. 2016. Quantitative measurement of ligand exchange with small-molecule ligands on iron oxide nanoparticles via radioanalytical techniques. *Langmuir* 32 (51), 13716–13727. doi:10.1021/acs.langmuir.6b03644.

Davis, K., B. Qi, M. Witmer, C. L. Kitchens, B. A. Powell, and O. T. Mefford. 2014. Quantitative measurement of ligand exchange on iron oxides via radiolabeled oleic acid. *Langmuir* 30 (36), 10918–10925. doi:10.1021/la502204g.

De Haro, L. P., T. Karaulanov, E. C. Vreeland, B. Anderson, H. J. Hathaway, D. L. Huber, A. N. Matlashov, C. P. Nettles, A. D. Price, T. C. Monson, and E. R. Flynn. 2015. Magnetic relaxometry as applied to sensitive cancer detection and localization. *Biomed Tech (Berl)* 60 (5), 445–455. doi:10.1515/bmt-2015-0053.

Di Corato, R., A. Quarta, P. Piacenza, A. Ragusa, A. Figuerola, R. Buonsanti, R. Cingolani, L. Manna, and T. Pellegrino. 2008. Water solubilization of hydrophobic nanocrystals by means of poly (maleic anhydride-alt-1-octadecene). *Journal of Materials Chemistry* 18 (17), 1991–1996. doi:10.1039/B717801H.

Farrell, D., Y. Cheng, S. Kan, M. Sachan, Y. Ding, S. A. Majetich, and L. Yang. 2005. Iron nanoparticle assemblies: Structures and magnetic behavior. In *Fifth International Conference on Fine Particle Magnetism*, edited by Q. Pankhurst, 185–195. Bristol: Iop Publishing Ltd.

Fellows, B. D., S. Sandler, J. Livingston, K. Fuller, L. Nwandu, S. Timmins, K. A. Lantz, M. Stefik, and O. T. Mefford. 2018. Extended LaMer synthesis of cobalt-doped ferrite. *IEEE Magnetics Letters* 9, 1–5. doi:10.1109/lmag.2017.2787683.

Gangopadhyay, S., G. C. Hadjipanayis, B. Dale, C. M. Sorensen, K. J. Klabunde, V. Papaefthymiou, and A. Kostikas. 1992. Magnetic properties of ultrafine iron particles. *Physical Review B (Condensed Matter)* 45 (17), 9778–9787.

Goya, G. F. 2004. Magnetic interactions in ball-milled spinel ferrites. *Journal of Materials Science* 39 (16–17), 5045–5049. doi:10.1023/B:JMSC.0000039183.99797.8d.

Guardia, P., J. Perez-Juste, A. Labarta, X. Batlle, and L. M. Liz-Marzan. 2010a. Heating rate influence on the synthesis of iron oxide nanoparticles: The case of decanoic acid. *Chemical Communications (Camb)* 46 (33), 6108–6110. doi:10.1039/c0cc01179g.

Guardia, P., N. Pérez, A. Labarta, and X. Batlle. 2010b. Controlled synthesis of iron oxide nanoparticles over a wide size range. *Langmuir* 26 (8), 5843–5847. doi:10.1021/la903767e.

Harris, L. J., S. B. Larson, K. W. Hasel, and A. McPherson. 1997. Refined structure of an intact IgG2a monoclonal antibody. *Biochemistry* 36 (7):1581–1597. doi:10.1021/bi962514+.

Hermanson, G. 2013. *Bioconjugate Techniques*. 3rd ed. Amsterdam:Academic Press.

Herrmann, I. K., R. N. Grass, D. Mazunin, and W. J. Stark. 2009. Synthesis and covalent surface functionalization of nonoxidic iron core–shell nanomagnets. *Chemistry of Materials* 21 (14), 3275–3281. doi:10.1021/cm900785u.

Huber, D. L. 2005. Synthesis, properties, and applications of iron nanoparticles. *Small* 1 (5), 482–501. doi:10.1002/smll.200500006.

Huber, D. L., K. E. Gonsalves, G. Carlson, and T. A. P. Seery. 1997. The formation of polymer monolayers: From adsorption to surface initiated polymerizations. In *Interfacial Aspects of Multicomponent Polymer Materials*, edited by D. J. Lohse, T. P. Russell and L. H. Sperling, 107–122. Boston, MA: Springer.

Huber, D. L., R. P. Manginell, M. A. Samara, B. I. Kim, and B. C. Bunker. 2003. Programmed adsorption and release of proteins in a microfluidic device. *Science* 301 (5631), 352–354. doi:10.1126/science.1080759.

Huber, D. L., E. L. Venturini, J. E. Martin, P. P. Provencio, and R. J. Patel. 2004. Synthesis of highly magnetic iron nanoparticles suitable for field structuring using a beta-diketone surfactant. *Journal of Magnetism and Magnetic Materials* 278 (3), 311–316. doi:10.1016/j.jmmm.2003.12.1317.

Hufschmid, R., H. Arami, R. M. Ferguson, M. Gonzales, E. Teeman, L. N. Brush, N. D. Browning, and K. M. Krishnan. 2015. Synthesis of phase-pure and monodisperse iron oxide nanoparticles by thermal decomposition. *Nanoscale* 7 (25), 11142–11154. doi:10.1039/c5nr01651g.

Jain, T. K., M. K. Reddy, M. A. Morales, D. L. Leslie-Pelecky, and V. Labhasetwar. 2008. Biodistribution, clearance, and biocompatibility of iron oxide magnetic nanoparticles in rats. *Molecular Pharmaceutics* 5 (2), 316–327. doi:10.1021/mp7001285.

Jenkins, A. D., R. G. Jones, and G. Moad. 2010. Terminology for reversible-deactivation radical polymerization previously called "controlled" radical or "living" radical polymerization (IUPAC Recommendations 2010). *Pure and Applied Chemistry* 82 (2), 483–491. doi:10.1351/pac-rep-08-04-03.

Joseyphus, R. J., D. Kodama, T. Matsumoto, Y. Sato, B. Jeyadevan, and K. Tohji. 2007. Role of polyol in the synthesis of Fe particles. *Journal of Magnetism and Magnetic Materials* 310 (2, Part 3), 2393–2395. doi:10.1016/j.jmmm.2006.10.1132.

Jungk, H. O., and C. Feldmann. 2000. Nonagglomerated, submicron α–Fe_2O_3 particles: Preparation and application. *Journal of Materials Research* 15 (10), 2244–2248. doi:10.1557/jmr.2000.0322.

Katabi, G., Y. Koltypin, X. Cao, and A. Gedanken. 1996. Self-assembled monolayer coatings of iron nanoparticles with thiol derivatives. *Journal of Crystal Growth* 166 (1-4), 760–762. doi:10.1016/0022-0248 (96)00091-7.

Kataby, G., M. Cojocaru, R. Prozorov, and A. Gedanken. 1999. Coating carboxylic acids on amorphous iron nanoparticles. *Langmuir* 15 (5), 1703–1708. doi:10.1021/la981001w.

Klabunde, K. J., H. F. Efner, L. Satek, and W. Donley. 1974. Preparation of an extremely active magnesium slurry for grignard-reagent preparations by metal atom-solvent cocondensations. *Journal of Organometallic Chemistry* 71 (3), 309–313. doi:10.1016/S0022-328x (00)95163-5.

Kumar, N., C. S. Tiwary, and K. Biswas. 2018. Preparation of nanocrystalline high-entropy alloys via cryomilling of cast ingots. *Journal of Materials Science*53 (19), 13411–13423. doi:10.1007/s10853-018-2485-z.

Kumar, S., J. Aaron, and K. Sokolov. 2008. Directional conjugation of antibodies to nanoparticles for synthesis of multiplexed optical contrast agents with both delivery and targeting moieties. *Nature Protocols* 3 (2), 314–320. doi:10.1038/nprot.2008.1.

Lamer, V. K., and R. H. Dinegar. 1950. Theory, production and mechanism of formation of monodispersed hydrosols. *Journal of the American Chemical Society* 72 (11), 4847–4854. doi:10.1021/ja01167a001.

Liu, R., B. K. Kay, S. Jiang, and S. Chen. 2011. Nanoparticle delivery: Targeting and nonspecific binding. *MRS Bulletin* 34 (06), 432–440. doi:10.1557/mrs2009.119.

Ma, M., Y. Zhang, W. Yu, H.-Y. Shen, H.-Q. Zhang, and N. Gu. 2003. Preparation and characterization of magnetite nanoparticles coated by amino silane. *Colloids and Surfaces A: Physicochemical and Engineering Aspects* 212 (2), 219–226. doi:https://doi.org/10.1016/S0927-7757 (02)00305-9.

Martin, J. E., E. L. Venturini, and D. L. Huber. 2008. Giant magnetic susceptibility enhancement in field-structured nanocomposites. *Journal of Magnetism and Magnetic Materials* 320 (18), 2221–2227. doi:10.1016/j.jmmm.2008.04.111.

Martino, A., M. Stoker, M. Hicks, C. H. Bartholomew, A. G. Sault, and J. S. Kawola. 1997. The synthesis and characterization of iron colloid catalysts in inverse micelle solutions. *Applied Catalysis A-General* 161 (1–2), 235–248. doi:10.1016/S0926-860x (97)00074-4.

Masthoff, I.-C., M. Kraken, D. Mauch, D. Menzel, J. A. Munevar, E. B. Saitovitch, F. J. Litterst, and G. Garnweitner. 2014. Study of the growth process of magnetic nanoparticles obtained via the non-aqueous sol–gel method. *Journal of Materials Science* 49 (14), 4705–4714. doi:10.1007/s10853-014-8160-0.

Matyjaszewski, K., and J. Spanswick. 2005. Controlled/living radical polymerization. *Materials Today* 8 (3), 26–33. doi:https://doi.org/10.1016/S1369-7021 (05)00745-5.

McGrath, A. J., C. Dolan, S. Cheong, D. A. J. Herman, B. Naysmith, F. Zong, P. Galvosas, K. J. Farrand, I. F. Hermans, M. Brimble, D. E. Williams, J. Jin, and R. D. Tilley. 2017. Stability of polyelectrolyte-coated iron nanoparticles for T2-weighted magnetic resonance imaging. *Journal of Magnetism and Magnetic Materials* 439, 251–258. doi:10.1016/j.jmmm.2017.04.026.

McIntosh, R. G., R. L. Eager, and J. W. T. Spinks. 1960. Mean lifetime of free radical chains determined by a flow technique. *Science* 131 (3405), 992–992. doi:10.1126/science.131.3405.992.

Meffre, A., B. Mehdaoui, V. Kelsen, P. F. Fazzini, J. Carrey, S. Lachaize, M. Respaud, and B. Chaudret. 2012. A simple chemical route toward monodisperse iron carbide nanoparticles displaying tunable magnetic and unprecedented hyperthermia properties. *Nano Letters* 12 (9), 4722–4728. doi:10.1021/nl302160d.

Miller, J. S. 2014. Organic- and molecule-based magnets. *Materials Today* 17 (5), 224–235. doi:10.1016/j.mattod.2014.04.023.

Nakatani, I., T. Furubayashi, T. Takahashi, and H. Hanaoka. 1987. Preparation and magnetic properties of colloidal ferromagnetic metals. *Journal of Magnetism and Magnetic Materials* 65 (2/3), 261–264.

Nealon, G. L., B. Donnio, R. Greget, J.-P. Kappler, E. Terazzi, and J.-L. Gallani. 2012. Magnetism in gold nanoparticles. *Nanoscale* 4 (17), 5244–5258. doi:10.1039/C2NR30640A.

Orendorff, C. J., D. L. Huber, and B. C. Bunker. 2009. Effects of water and temperature on conformational order in model nylon thin films. *Journal of Physical Chemistry C* 113 (31), 13723–13731. doi:10.1021/jp901309y.

Ostwald, W. 1897. Studien ber die Bildung und Umwandlung fester Körper. Zeitschrift für Physikalische Chemie 22, 289–330.

Pardavi-Horvath, M., and L. Takacs. 1995. Magnetic nanocomposites by reaction milling. *Scripta Metallurgica et Materialia* 33 (10/11), 1731–1740.

Park, J., K. An, Y. Hwang, J. G. Park, H. J. Noh, J. Y. Kim, J. H. Park, N. M. Hwang, and T. Hyeon. 2004. Ultra-large-scale syntheses of monodisperse nanocrystals. *Nature Materials* 3 (12), 891–895. doi: 10.1038/nmat1251.

Price, A. D., and D. L. Huber. 2013. Controlled polymer monolayer synthesis by radical transfer to surface immobilized transfer agents. *Polymer Chemistry* 4 (5), 1565–1574. doi:10.1039/C2PY20955A.

Prozorov, T., R. Prozorov, and K. S. Suslick. 2004. High velocity interparticle collisions driven by ultrasound. *Journal of the American Chemical Society* 126 (43), 13890–13891. doi:10.1021/ja049493o.

Reichert, J. M. 2014. Global antibody development trends. *mAbs* 1 (1), 86–87. doi:10.4161/mabs.1.1.7645.

Rodrigues, T. C., M. L. S. Oliveira, A. Soares-Schanoski, S. L. Chavez-Rico, D. B. Figueiredo, V. M.

Gonçalves, D. M. Ferreira, N. K. Kunda, I. Y. Saleem, and E. N. Miyaji. 2018. Mucosal immunization with PspA (Pneumococcal surface protein A)-adsorbed nanoparticles targeting the lungs for protection against pneumococcal infection. *PLoS One* 13 (1), e0191692. doi:10.1371/journal.pone.0191692.

Sahoo, Y., H. Pizem, T. Fried, D. Golodnitsky, L. Burstein, C. N. Sukenik, and G. Markovich. 2001. Alkyl phosphonate/phosphate coating on magnetite nanoparticles: A comparison with fatty acids. *Langmuir* 17 (25), 7907–7911. doi:10.1021/la010703+.

Satzer, P., F. Svec, G. Sekot, and A. Jungbauer. 2016. Protein adsorption onto nanoparticles induces conformational changes: Particle size dependency, kinetics, and mechanisms. *Engineering in Life Sciences* 16 (3), 238–246. doi:10.1002/elsc.201500059.

Saville, S. L., B. Qi, J. Baker, R. Stone, R. E. Camley, K. L. Livesey, L. F. Ye, T. M. Crawford, and O. T. Mefford. 2014. The formation of linear aggregates in magnetic hyperthermia: Implications on specific absorption rate and magnetic anisotropy. *Journal of Colloid and Interface Science* 424, 141–151. doi:10.1016/j.jcis.2014.03.007.

Smith, T. W., and D. Wychick. 1980. Colloidal iron dispersions prepared via the polymer-catalyzed decomposition of iron pentacarbonyl. *Journal of Physical Chemistry* 84 (12), 1621–1629.

Stone, R., S. Hipp, J. Barden, P. J. Brown, and O. T. Mefford. 2013. Highly scalable nanoparticle-polymer composite fiber via wet spinning. *Journal of Applied Polymer Science* 130 (3), 1975–1980. doi:10.1002/App.39408.

Sun, S. H., S. Anders, T. Thomson, J. E. E. Baglin, M. F. Toney, H. F. Hamann, C. B. Murray, and B. D. Terris. 2003. Controlled synthesis and assembly of FePt nanoparticles. *Journal of Physical Chemistry B* 107 (23), 5419–5425. doi:10.1021/Jp027314o.

Sun, S. H., C. B. Murray, D. Weller, L. Folks, and A. Moser. 2000. Monodisperse FePt nanoparticles and ferromagnetic FePt nanocrystal superlattices. *Science* 287 (5460), 1989–1992.

Sun, Y. P., X. Q. Li, J. Cao, W. X. Zhang, and H. P. Wang. 2006. Characterization of zero-valent iron nanoparticles. *Advances in Colloid and Interface Science* 120 (1–3), 47–56. doi:10.1016/j.cis.2006.03.001.

Suslick, K. S., S. B. Choe, A. A. Cichowlas, and M. W. Grinstaff. 1991. Sonochemical synthesis of amorphous iron. *Nature* 353 (6343), 414–416. doi:10.1038/353414a0.

Suslick, K. S., Y. Didenko, M. M. Fang, T. Hyeon, K. J. Kolbeck, W. B. McNamara, M. M. Mdleleni, and M. Wong. 1999. Acoustic cavitation and its chemical consequences. *Philosophical Transactions of the Royal Society a-Mathematical Physical and Engineering Sciences* 357 (1751), 335–353. doi:10.1098/rsta.1999.0330.

Suslick, K. S., and G. J. Price. 1999. Applications of ultrasound to materials chemistry. *Annual Review of Materials Science* 29 (1), 295–326. doi:10.1146/annurev.matsci.29.1.295.

Sweeney, S. F., G. H. Woehrle, and J. E. Hutchison. 2006. Rapid purification and size separation of gold nanoparticles via diafiltration. *Journal of the American Chemical Society* 128 (10), 3190–3197. doi:10.1021/ja0558241.

Torrice, M. 2016. Does nanomedicine have a delivery problem? *ACS Central Science* 2 (7), 434–437. doi:10.1021/acscentsci.6b00190.

Vijayendran, R. A., and D. E. Leckband. 2001. A quantitative assessment of heterogeneity for surface-immobilized proteins. *Analytical Chemistry* 73 (3), 471–480. doi:10.1021/ac000523p.

Vreeland, E. C., J. Watt, G. B. Schober, B. G. Hance, M. J. Austin, A. D. Price, B. D. Fellows, T. C. Monson, N. S. Hudak, L. Maldonado-Camargo, A. C. Bohorquez, C. Rinaldi, and D. L. Huber. 2015. Enhanced nanoparticle size control by extending LaMer's mechanism. *Chemistry of Materials* 27 (17), 6059–6066. doi:10.1021/acs.chemmater.5b02510.

Wang, Y. X. 2011. Superparamagnetic iron oxide based MRI contrast agents: Current status of clinical application. *Quantitative Imaging in Medicine and Surgery* 1 (1), 35–40. doi:10.3978/j.issn.2223-4292.2011.08.03.

Watt, J., M. J. Austin, C. K. Simocko, D. V. Pete, J. Chavez, L. M. Ammerman, and D. L. Huber. 2018a. Formation of metal nanoparticles directly from bulk sources using ultrasound and application to e-waste upcycling. *Small* 14 (17), e1703615. doi:10.1002/smll.201703615.

Watt, J., G. C. Bleier, M. J. Austin, S. A. Ivanov, and D. L. Huber. 2017. Non-volatile iron carbonyls as versatile precursors for the synthesis of iron-containing nanoparticles. *Nanoscale* 9 (20), 6632–6637. doi:10.1039/c7nr01028a.

Watt, J., G. C. Bleier, Z. W. Romero, B. G. Hance, J. A. Bierner, T. C. Monson, and D. L. Huber. 2018b. Gram scale synthesis of Fe/Fe_xO_y core–shell nanoparticles and their incorporation into matrix-free superparamagnetic nanocomposites. *Journal of Materials Research* 33 (15), 2156–2167. doi:10.1557/jmr.2018.139.

White, L. D., and C. P. Tripp. 2000. Reaction of (3-aminopropyl)dimethylethoxysilane with amine catalysts on silica surfaces. *Journal of Colloid and Interface Science* 232 (2), 400–407. doi:10.1006/jcis.2000.7224.

Whitesell, J. K., and H. K. Chang. 1993. Directionally aligned helical peptides on surfaces. *Science* 261 (5117), 73–76. doi:10.1126/science.261.5117.73.

Wilhelm, S., A. J. Tavares, Q. Dai, S. Ohta, J. Audet, H. F. Dvorak, and W. C. W. Chan. 2016. Analysis of nanoparticle delivery to tumours. *Nature Reviews Materials* 1 (5). doi:10.1038/natrevmats.2016.14.

Wu, L., A. Mendoza-Garcia, Q. Li, and S. Sun. 2016. Organic phase syntheses of magnetic nanoparticles and their applications. *Chemical Reviews.* doi:10.1021/acs.chemrev.5b00687.

Yee, C., G. Kataby, A. Ulman, T. Prozorov, H. White, A. King, M. Rafailovich, J. Sokolov, and A. Gedanken. 1999. Self-assembled monolayers of alkanesulfonic and -phosphonic acids on amorphous iron oxide nanoparticles. *Langmuir* 15 (21), 7111–7115. doi:10.1021/La990663y.

Yu, W. W., J. C. Falkner, C. T. Yavuz, and V. L. Colvin. 2004. Synthesis of monodisperse iron oxide nanocrystals by thermal decomposition of iron carboxylate salts. *Chemical Communications (Camb)* (20), 2306–2307. doi:10.1039/b409601k.

Zhang, B., X. Wang, F. Liu, Y. Cheng, and D. Shi. 2012. Effective reduction of nonspecific binding by surface engineering of quantum dots with bovine serum albumin for cell-targeted imaging. *Langmuir* 28 (48), 16605–16613. doi:10.1021/la302758g.

Zhang, Z., Z. Wang, S. He, C. Wang, M. Jin, and Y. Yin. 2015. Redox reaction induced Ostwald ripening for size- and shape-focusing of palladium nanocrystals. *Chemical Science* 6 (9), 5197–5203. doi:10.1039/c5sc01787d.

7 Nanomagnetic Actuation: Controlling Cell Behavior with Magnetic Nanoparticles

Jon Dobson and Sarah H. Cartmell

CONTENTS

7.1 INTRODUCTION

The manipulation and control of cell behavior through magnetic nanoparticle-based actuation is a relatively new technique, which has led to novel and exciting biomedical applications. From its genesis as a theoretical model developed to predict the response of magnetic iron compounds in the brain to environmental electromagnetic fields, it has evolved into an elegant method for examining cellular mechanics, ion channel activation kinetics, and tissue engineering and regenerative medicine applications. This review examines the field of nanomagnetic actuation as applied to the manipulation and control of cellular functions and the potential for the technique in biological and clinical applications.

7.2 BACKGROUND

Though the use of nanomagnetic actuation for controlling cell behavior is relatively new, magnetic micro- and nanoparticles have been used in the investigation of cellular properties for nearly a century now. In 1922, LV Heilbrunn, working at the University of Michigan, introduced iron particles into slime molds to probe the viscosity of the protoplasm. By measuring the speed of movement of the particles in response to an applied magnetic field, he was able to work out the viscosity (Heilbronn 1922).

In 1924, William Seifriz developed a device based on Heilbrunn's earlier experiments that could be used to measure the protoplasm of individual cells (Seifriz 1924). The device was basically a micro-dissection system, which was used to insert rather large (ca. 7–16 μm) nickel particles into the cell (Figure 7.1). An electromagnet attached to the dissection microscope was used to apply the force to the particles. Though this was a rather clever way of combining these techniques to investigate the rheological properties of the cell, the values obtained varied significantly and were at odds with those determined using other techniques; thus, use of the technique waned for several decades.

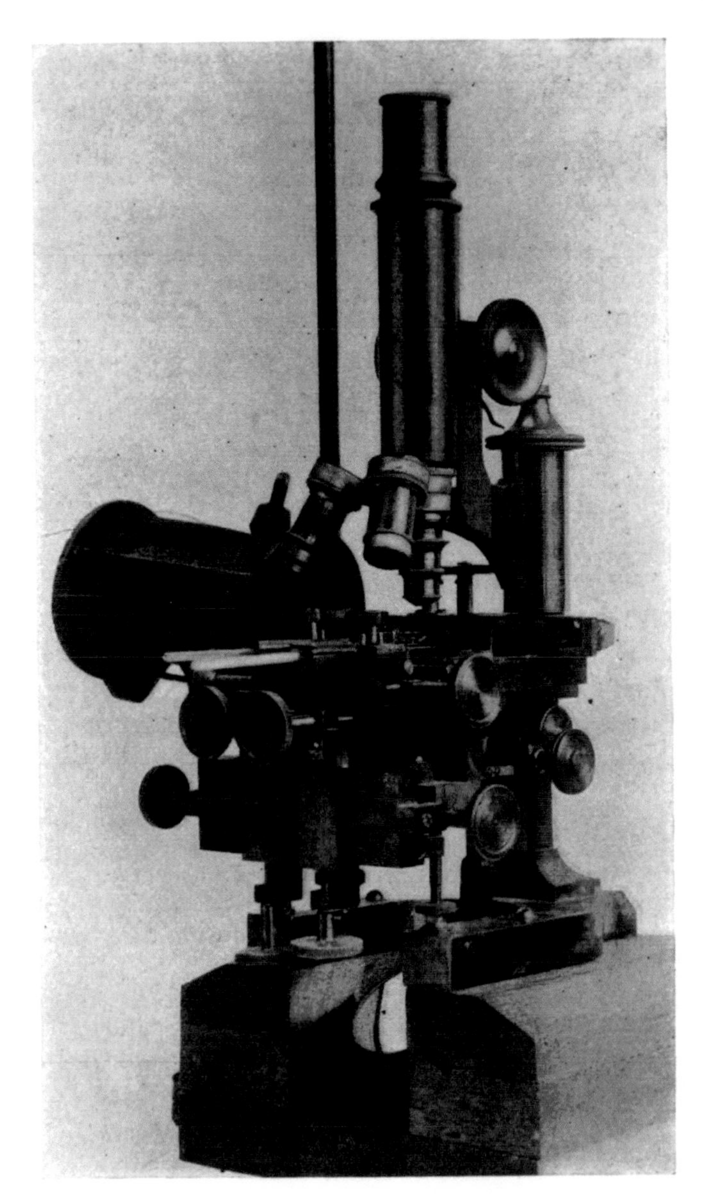

A Barber microdissection instrument attached to a microscope, with an electro-magnet in position for making elasticity measurements of living protoplasm. To the microdissection instrument are clamped two glass needles the right angle tips of which, together with the metal tip of the magnet core, project into a glass moist chamber under the microscope objective.

FIGURE 7.1 The original magnetic actuation device used by Seifriz in 1924. The system is a microdissection instrument attached to a microscope with an electromagnet for use in probing "living protoplasm," figure from (Seifriz 1924).

In 1949, Francis Crick and Arthur Hughes used magnetic microparticles to more accurately quantify the rheological properties of the cytoplasm of chick fibroblasts (Crick and Hughes 1950). In this case, the key to keeping the cells alive and healthy for the experiments was the fact that they added the particles to the growth medium and allowed the cells to phagocytose them over a period of several days. They examined particle motion under three sets of conditions—twisting, dragging, and prodding. The magnetite particles used in the experiments—between 2 μm and 10 μm—were well characterized and the paper even included a magnetization curve for the material. By quantifying these motions, they were able to get relatively accurate measures of the rheological properties of the cytoplasm in these cells.

In the 1980s, Valberg and others expanded upon the work of Crick and Hughes, introducing the concept of magnetic twisting of blocked magnetic micro- and nanoparticles in order to investigate the mechanical properties of the cell cytoplasm and organelles (Valberg and Albertini 1985, Valberg and Butler 1987, Valberg and Feldman 1987). Valberg's idea relied upon the fact that magnetically blocked particles within the cytoplasm and organelles could be physically "twisted" by applying a field at an angle to the magnetization vector of the particles. An initial, strong magnetizing field aligns the particles and then a smaller, twisting field (smaller than the saturating field) applied at 90° to the magnetizing field twists them out of alignment (Figure 7.2). The change in magnetization of the particles within the cells can then be measured via magnetometry. By comparing these measurements to measurements made in fluids and gels of known viscosity, they were able to determine the viscosity of the cellular components.

JP Butler, who worked with Valberg on some of these studies, along with Wang and Ingber, went on to develop an elegant variation on this concept in order to investigate the mechanical properties of the cell membrane and cytoskeleton (Wang, Butler, and Ingber 1993, Wang and Ingber 1995b, Meyer et al. 2000) The group coated relatively large (4.5 μm) magnetic particles with molecules (primarily arginine–glycine–aspartic RGD) that bind to integrin receptors on the cell surface. These receptors are essentially protrusions of the cell's cytoskeleton. By twisting the membrane-bound particles, they were able to investigate the mechanical properties of the membrane and cytoskeleton in a controlled fashion.

Interestingly, the origin of nanomagnetic actuation of cellular ion channels perhaps lies in a theoretical model developed to explain potential interactions of magnetic iron compounds in the brain with environmental electromagnetic fields. In 1992, Joseph Kirschvink, working at the California Institute of Technology, proposed a mechanism by which relatively weak magnetic fields from mains-powered electrical devices could activate mechanosensitive ion channels via actuation of nanoparticles of biogenic magnetite, which had recently been discovered in the human brain (Figure 7.3) (Kirschvink, Kobayashi-Kirschvink, and Woodford 1992, Adair 1991). The model demonstrated how a particle of magnetite with a stable magnetization (magnetically blocked) would "twist" in response to a magnetic field applied at an angle to the magnetization vector of the particle. The principle was similar to Valberg's earlier ideas of magnetic twisting cytometry but was more specific in that it investigated particles that could be coupled to the ion channels themselves. If such a particle were, in some way, coupled to a cellular ion channel, Kirschvink's calculations showed that the torque on the particle would be strong enough to force open the channel—activating and deactivating the channel in response to a sinusoidal magnetic field.

In 1995, a colleague, Tim St Pierre, and I were working on theoretical models of the interaction of biogenic magnetite in the brain with electromagnetic signals from cell

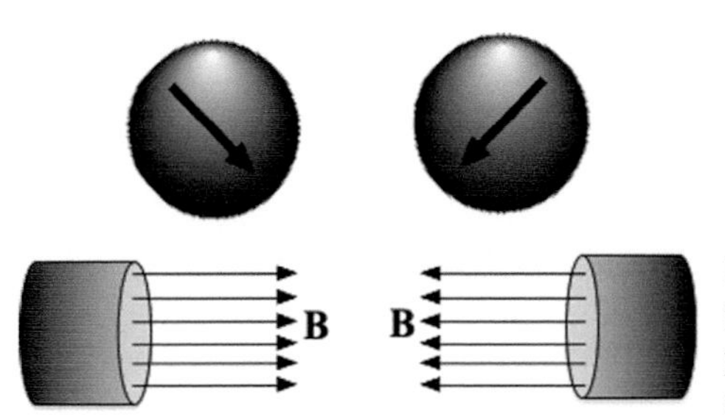

FIGURE 7.2 Schematic representation of magnetic twisting. A field applied at an angle to the blocked particle's internal magnetization vector will exert a torque on the particle as shown.

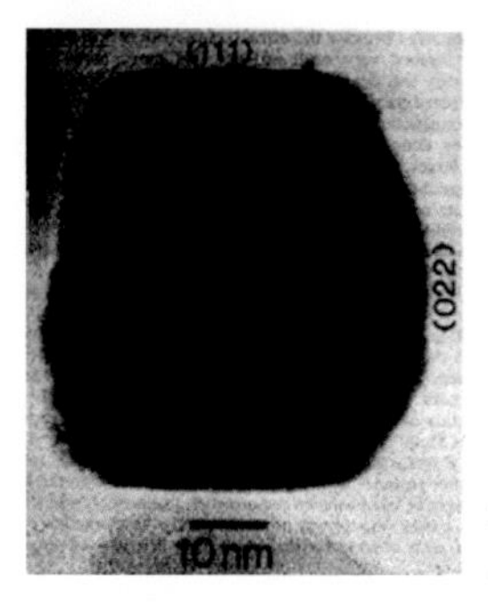

FIGURE 7.3 Transmission electron micrograph of biogenic magnetite extracted from the human cerebellum, figure with permission from Authors (Kirschvink, Kobayashi-Kirschvink, and Woodford 1992).

phones—in particular, fields associated with battery current bursts during discontinuous transmission (DTX), essentially a transponder signal the phone sends when the user is not speaking so that the location can be tracked and the signal transferred to the relevant cell when moving between coverage areas. We used these models to expand on Kirschvink's work and to show how it would be possible to open and close channels using magnetic pulses generated by these battery current bursts (Dobson and Pierre 1996). This idea eventually led to the development of a magnetic ion channel activation device for tissue engineering that will be discussed later.

In addition to twisting magnetically blocked nanoparticles, it is also possible to apply a translational force that can "pull" the particles toward a magnetic field source, provided there is a gradient to the field (Pankhurst et al. 2003). Magnetic nanoparticles will be attracted to such a field according to Equation 7.1:

$$F_{\mathrm{m}} = \frac{V_{\mathrm{m}}\Delta\chi}{\mu_0}(\boldsymbol{B}\cdot\nabla)\boldsymbol{B} \tag{7.1}$$

where F_{mag} is the force on the magnetic particle, μ_o is the magnetic permeability of free space, V_m is particle magnetic volume, $\Delta\chi = \chi_2 - \chi_1$ and χ_2 is the volume magnetic susceptibility of the magnetic particle, χ_1 is the volume magnetic susceptibility of the surrounding medium, **B** is the magnetic flux density in Tesla (T), $\nabla\mathbf{B}$ is field gradient and can be reduced to $\partial B/\partial x$, $\partial B/\partial y$, $\partial B/\partial z$. This attractive force (sometimes in combination with torque on blocked particles), when applied to magnetic nanoparticles, which are attached in some way to cell membrane receptors or cellular components, presents a unique opportunity to employ magnetic actuation in a way that may be used to control specific cellular processes and to interrogate the mechanical properties of cellular structures.

A short time later, several groups, led by Pommerenke, Sackmann, Glogauer, Ferrier and others, began to combine the ideas of attaching magnetic particles to cellular components and applying translational forces to them to investigate the activation kinetics of mechanosensitive (MS) cellular ion channels. By attaching magnetic particles to the cell membrane and applying a strong, high-gradient magnetic field, the particles could be pulled toward the field source, creating membrane

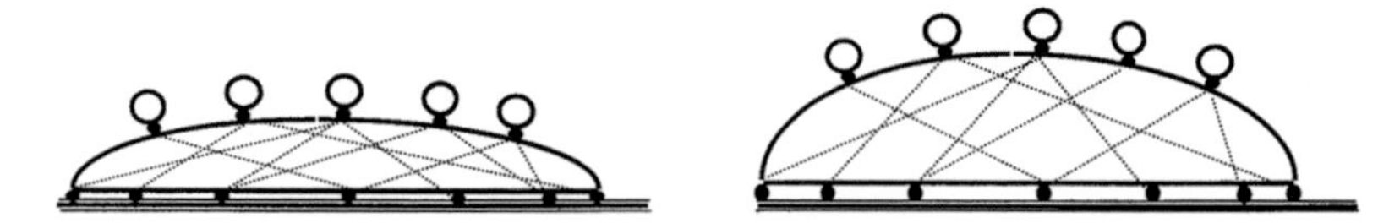

FIGURE 7.4 (Left) Schematic representation of magnetic particles attached to the dorsal surface of a substrate-attached cell, **(Right)** Application of a high-gradient magnetic field causes translational displacement of the particles and results in membrane deformation, image adapted from (Glogauer 1997).

stresses that are able to activate adjacent MS ion channels (Glogauer, Ferrier, and McCulloch 1995b, Glogauer and Ferrier 1997, Bausch, Möller, and Sackmann 1999, Pommerenke et al. 1996, Bausch et al. 2001). Attaching a variety of binding molecules to the magnetic micro- and nanoparticles makes it possible to investigate various biochemical reaction pathways and ion channel kinetics, essentially at the flip of a switch (Figure 7.4). (For a more detailed review see Hughes, El Haj, and Dobson (2005) and Waigh (2005)).

7.3 MECHANOTRANSDUCTION IN CELLS AND THE CELLULAR CYTOSKELETON

The cytoskeleton is a highly dynamic heterogeneous structure present within all cells. This protein structure provides the scaffolding within the cell and has multiple functions, including stabilizing cellular structure, regulation of cell division, organization of cellular traffic, and enabling of cell locomotion and migration (Sawada and Sheetz 2002, Ingber 1997, Wang, Butler, and Ingber 1993, Zhu, Bao, and Wang 2000, Janmey 1998). The cytoskeleton consists of complex networks of three main types of protein, the largest of which are microtubules. These are the most rigid of all cytoskeletal components; their long tubular structures extend throughout the cytoplasm forming a dynamic, rigid structure from which the remaining cytoskeletal components are anchored. The second significant structures are actin filaments. These are helical structures that extend into web-like protrusions traversing the spaces between microtubules. The third components of the cytoskeleton are intermediate filaments (Alberts et al. 2002).

It has long been accepted that the cytoskeleton plays a key role in a wide range of mechanotransductive processes. Disruption of these complex networks can have a significant influence and even eliminate mechanotransductive responses (Janmey 1998). The cytoskeleton also links the plasma membrane to the nuclear matrix (Traub 1995, Traub and Shoeman 1994) and may play a role in transmitting mechanical force from the plasma membrane to the nucleus (Hu et al. 2005). In support of this hypothesis, studies have shown that mechanical forces applied to the outer cell can be transmitted to the nuclear membrane (Guilak 1995, Maniotis, Chen, and Ingber 1997, Traub 1995). However, the mechanisms by which cytoskeletal deformation influences the promoter regions of mechanoresponsive genes are not currently known.

The cytoskeleton may also regulate mechanotransduction pathways by directly influencing the tension and deformability of the lipid membrane, which in turn may influence the behavior of membrane-bound mechanosensors such as integrins, stretch-activated enzymes, and mechanosensitive ion channels (Diamond, Sachs, and Sigurdson 1994, Sachs 1988). In addition to the indirect membrane tension effect, it has also been proposed that the cytoskeleton may interact directly with mechanosensitive ion channels.

Mechanotransduction has been described as "the process of converting physical forces into biochemical signals and integrating these signals into cellular responses" (Huang, Kamm, and Lee 2004). Mechanosensitive receptors and ion channels monitor these force transductions, activating cellular signaling events that regulate phenotypically specific biological processes. It should be noted that this is not the only means by which external mechanical cues are transduced to bone cells. The act of physically deforming tissues can induce movements in intercellular fluids, resulting in shear forces that can also influence cell activity (Netter 1987).

The precise methods by which the various cell types monitor and respond to mechanical stimuli are not fully understood. However, the methods by which cells attach and monitor their various extracellular matrices, share numerous commonalities. One such method is by attaching directly to the extracellular matrix via transmembrane glycoproteins known as cell surface integrins. These heterodimeric proteins consist of two distinct subunits known as α and β. There are currently 16α and 9β subunits known to exist, which can combine to form a variety of integrins with distinct binding specificities. The intracellular region of the α chain is physically connected to the cytoskeleton via interaction with talin, vinculin, and α-actinin. It is this connection that enables integrins to transduce mechanical stimuli from the extracellular matrix into the intracellular environment (Burridge et al. 1988).

The binding specificity of a particular integrin is dependant on the subunit composition. In bone, cells must be able to bind to a wide variety of amino acid sequences (binding motifs) present within the innate proteins of the bone matrix. Although the binding specificity varies between different heterodimers, there are some common motifs that adhere to a broad range of integrins. One such sequence was discovered in the study of laminin (Pierschbacher and Ruoslahti 1984) and has subsequently been identified in numerous matrix proteins including: fibronectin, thrombospondin, osteopontin, bone sialoprotein, type I collagen, and vitronectin. This moiety, known as Arg-Gly-Asp (RGD), represents a key mechanism by which bone cells are able to adhere to their surrounding environment (Grzesik and Robey 1994), and this sequence can therefore be utilized when targeting a broad range of integrin receptors (Fabry et al. 1999, Wang and Ingber 1995a, Bierbaum and Notbohm 1998, Chen et al. 2001, Puig-De-Morales et al. 2001, Wang, Butler, and Ingber 1993, Fodil et al. 2003).

Integrins have been known to play diverse roles in the regulation of numerous cell signaling pathways (Giancotti and Ruoslahti 1999, Clark and Brugge 1995, Hynes 1992). This is achieved by the action of the focal adhesion complex. This complex interacts with a wide range of signaling molecules, including G-protein, phosphotidylinositol, focal adhesion kinases, lipid kinases, serine/threonine kinases, and the Rho family of GTPases (Schwartz and Shattil 2000, Clark and Brugge 1995).

The direct adhesion of integrins to the extra-cellular matrix combined with their transmembrane domains enables direct transduction of mechanical forces across the cell membrane and into the cytoskeleton (Wang, Butler, and Ingber 1993), playing a key role in mechanotransductive signaling events (Pavalko et al. 1998,Goldschmidt, McLeod, and Taylor 2001) in numerous cell types (Wang, Butler, and Ingber 1993, Wang and Ingber 1995a, Glogauer, Ferrier, and McCulloch 1995a, Ingber 1997). It is also becoming increasingly clear that different classes of integrin receptors may be involved with the transduction of specific modes of mechanical stimulation (Salter, Robb, and Wright 1997, Salter et al. 2000, Pommerenke et al. 2002) and are responsible for distinct cellular responses (Yamada, Wirtz, and Kuo 2000).

Numerous cell types monitor and respond to mechanical forces by activating numerous cellular signaling cascades. One of the most widely studied signaling pathways is the mitogen-activated protein kinase (MAPK) pathway. This ubiquitous cascade is capable of regulating numerous cellular processes, including gene expression, mitosis, metabolism, motility, proliferation, apoptosis, and differentiation (Roux and Blenis 2004). To date, five distinct groups of MAPKs have been identified:

- Extracellular regulated kinases (ERKs)—1 and 2 (ERK1/2)
- C-Jun amino-terminal kinases (JNKs)—1, 2 and 3
- p38—isoforms α, β, γ, and δ
- ERKs 3 and 4 (Kyriakis and Avruch 2001, Chen et al. 2001)
- ERK5 (Chen et al. 2001, Kyriakis and Avruch 2001)

The most extensively characterised modules of the MAPK family are ERK1/2, JNK, and p38. Each module is activated by numerous different extracellular cues but in general mitogenic and developmental stimuli preferentially activate ERK1/2, whereas JNK and p38 are more responsive to stress stimuli (Roux and Blenis 2004). Each family of the MAPK pathway is composed of a set of three conserved sequentially activated kinases each becoming phosphorylated in a cascade effect, triggered in response to extracellular signals.

The MAPK signaling cascade follows a general format. MAPKKKs, or MAP3Ks, are often activated as a result of interactions with small GTP-binding proteins of the Ras/Rho family in response to extracellular stimuli (Kolch 2000, Dan, Watanabe, and Kusumi 2001). This then leads to activation of MAPKKs, or MAP2Ks, through phosphorylation on serine/threonine residues. Finally, MAPK activity is stimulated through dual phosphorylation on threonine and tyrosine residues located in the activation loop of kinase subdomain VIII. MAPK units are then able to act upon MAPK-activated protein kinases (MKs). These MKs mediate a wide range of cellular functions including mitogenic, developmental, and stress responses (Roux and Blenis 2004).

Specific MAPK components are capable of discriminating between proteins of a related wing of the cascade by differential substrate affinity, or binding to an appropriate residue, enabling its activation. However, it should be noted that pathway specificity is often not mediated by substrate affinity alone; many cascades involve complex interaction with multifaceted scaffolding proteins. These require the simultaneous interaction of multiple signaling proteins, often activated by numerous signaling pathways, ultimately resulting in the formation of an enzymatically active unit (Roux and Blenis 2004).

Studies have demonstrated that the MAPK signaling pathway plays a key role in the regulation of MSC differentiation into the osteogenic phenotype (Jaiswal et al. 2000, Jadlowiec et al. 2004). Jaiswal and colleagues (2000) demonstrated a correlation between the long-term activation of ERK2 and MSC osteogenesis, inhibition of ERK2 activation by MEK1-eliminated osteogenic differentiation. They demonstrated a possible role of JNK and p38 in the regulation of the latter stages of osteogenesis.

Activation of MAPK cascades is mediated by a wide range of extracellular cues, and mechanically triggered activation remains largely undefined, particularly within the MSC phenotype. However, numerous mechanical-force application techniques have been shown to activate MAPK pathways in osteoblasts (Inoue, Kido, and Matsumoto 2004, Boutahar et al. 2004, Jessop et al. 2002, Martineau and Gardiner 2001, Kapur, Baylink, and Lau 2003, You et al. 2001, Jansen et al. 2004). There have also been studies in which magnetic particles were used to apply force to human osteoblasts, assessing any changes in the levels of phosphorylated MAPK proteins. Pommerenke and colleagues (2002) demonstrated an activation of ERK1/2 in response to cyclic application (1 Hz) of a magnetic field to magnetic particles coated with anti-integrin β1 antibody. They also observed the activation of focal adhesion kinase (FAK) in response to the force-application technique. Another study used magnetic particles located within the cytoplasm of human osteoblasts and showed an increase in the levels of phosphorylated p38 and no change in the levels of phosphorylated ERK1/2 and JNK. The use of the p38 inhibitor SB 203580 to osteoblasts alone (no magnetic particles) completely inhibited alkaline phosphatase, bone nodule formation, and calcium deposition, indicating a possible key role of p38 in the regulation of the osteoblast phenotype (Yuge et al. 2003).

Clearly, the ability to manipulate, target, and control mechanical forces applied to sites on and within the cell would represent an important tool for activating and controlling these biochemical pathways. Fortunately, nanomagnetic actuation via biofunctionalized magnetic nanoparticles provides us with a remote control switch to initiate and control these processes. Before we explore this in more detail, however, a brief overview of the particles themselves is useful.

7.4 MAGNETIC PARTICLES

For a review of the chemical synthesis of particles, please refer back to Chapter 4. In addition, there are many good reviews of the chemical synthesis and surface modification and functionalization of magnetic micro- and nanoparticles used for these applications (Roca et al. 2009, Berry 2009), but a brief overview is probably useful here. Particles used for nanomagnetic actuation generally consist of an iron oxide core (usually magnetite, Fe_3O_4, or maghemite, γFe_2O_3, both of which have similar magnetic properties) with a polymer coating.

In some cases CrO_2-coated or iron oxide-coated polystyerene spheres are functionalized for twisting applications where remanent magnetization and larger forces are required (Wang, Butler, and Ingber 1993). The magnetic cores are coated with functionalizable (and, in many cases biocompatible) polymers such as PVA, silica, and dextran, to which binding molecules such as RGD or specific antibodies are attached.

Particles range in size from a few tens of nanometers to several microns. Generally, smaller particles (10–300 nm) are used in studies that focus on targeting specific ion channels and for tissue engineering applications, while larger particles are used for investigations of cellular and cytoskeletal mechanical properties due to the larger forces that can be generated. In the case of magnetic twisting cytometry, the particles must be magnetically blocked in order to maintain their magnetization following the initial magnetizing pulse. For magnetite/maghemite, this equates to a diameter of roughly 30 nm or greater (however, micron-sized particles are generally used). Below this threshold, the particles behave as superparamagnets and do not preserve a stable remanent magnetization or experience a torque in the twisting field.

7.5 NANOMAGNETIC ACTUATION FOR TISSUE-ENGINEERING APPLICATIONS

The past decade has seen rapid acceleration in the use of nanomagnetic actuation for studies of cell function (particularly membrane receptor activation), tissue engineering and regenerative medicine, and stem cell research (Dobson 2008).

In the early 2000s, our group began work on the development of magnetic force bioreactor systems for mechanical conditioning applications in tissue engineering (Cartmell et al. 2002, Dobson, Keramane, and El Haj 2002, Dobson et al. 2006). These magnetic ion channel activation (MICA) systems were based on the expansion of Kirschvink's cell actuation work that Tim St Pierre and I did in the mid-1990s. The rationale for the design was to allow in vivo mechanical stresses to be mimicked in a bioreactor environment or applied in the body for tissue engineering and regenerative medicine applications.

Bioreactors are primarily cell culture environments that deliver nutrients and gasses to cells while removing waste products. In early bioreactor systems for tissue engineering, cells were (and still are, to a large extent) seeded onto porous, biodegradeable polymer scaffolds. As the cells on the scaffold grow and produce tissue matrix, the polymer degrades and is replaced by the newly formed tissue. The goal of such devices is to be able to grow replacement tissue from a patient's own cells.

For bone and connective tissue engineering in particular, the cells must experience mechanical stimulation in order to produce tissue matrix with the correct mechanical properties. This, it turns out, is a tricky problem to solve for bioreactors. The reason it is difficult to mechanically stimulate cells and tissue constructs growing in bioreactors is several-fold. The primary reason, however, is that in order to make polymer scaffolds that are biodegradable/bioresorbable, the polymers used are inherently mechanically weak. Compression of these scaffolds in order to mechanically stimulate the cells growing within them usually results in the scaffolds breaking apart inside the bioreactor. This makes delivery of the required mechanical cues to the cells virtually impossible.

Fortunately, the force required to activate ion channels and mechanically stimulate these cells is very small—in the picoNewton range (Howard and Hudspeth 1988, Walker et al. 1999). By attaching magnetic nanoparticles to the cell membrane via RGD, or other more specifically targeted linkers, it is possible to provide the required mechanical stimulation directly to the cells rather than relying on applying a force to the scaffold and having the scaffold transmit that force to the cells. This principle has been demonstrated in our MICA magnetic force bioreactor systems over the past several years (Dobson et al. 2006). General deformation of the cell membrane and cytoskeleton via nanomagnetic actuation in these systems has been shown to increase the speed of bone matrix mineralization in bone cells (osteoblasts) growing in culture, to effect changes in cell membrane potential, and to upregulate genes related to bone and cartilage production (Cartmell et al. 2002, Kirkham et al. 2010) (Figure 7.5).

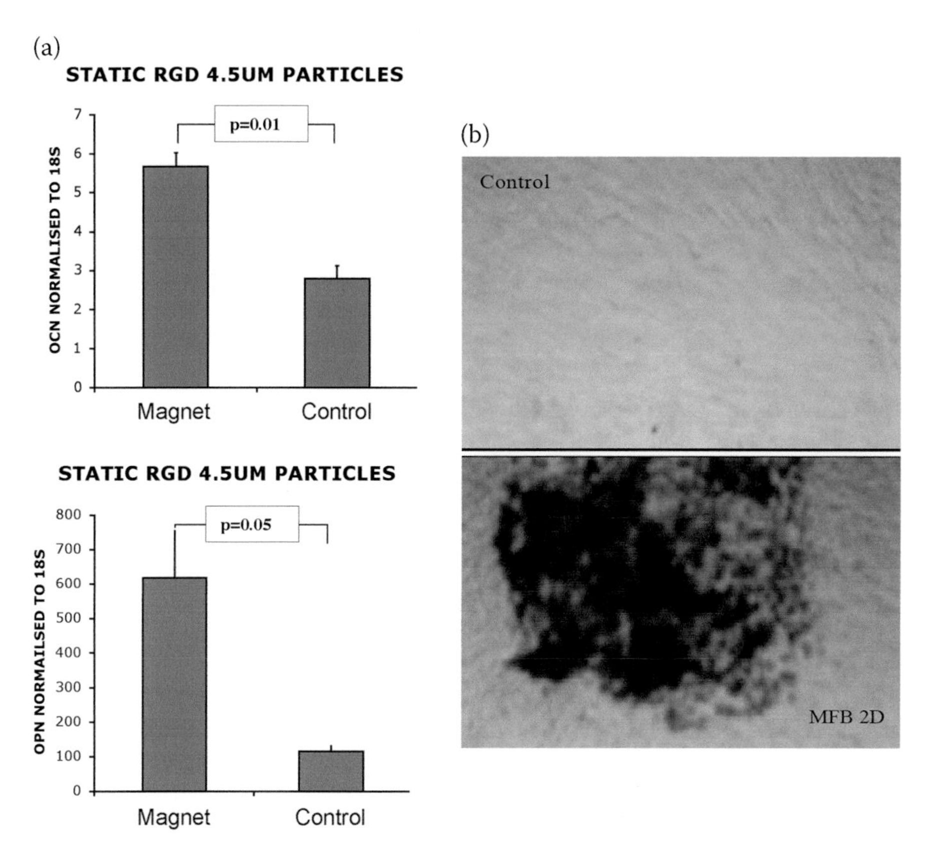

FIGURE 7.5 **(a)** RT-PCR data showing upregulation of osteocalcin and osteopontin (bone matrix genes) in response to nanomagnetic actuation of human osteoblasts within a 2D culture, **(b)** Light microscopy image of von Kossa stain showing bone matrix mineralization in human osteoblasts in response to nanomagnetic actuation, figure from (Dobson et al. 2006).

In related work, Ito, Honda, and colleagues in Japan have used nanoscale magnetoliposomes (liposomal spheres containing a magnetite/maghemite core) for other tissue-engineering applications. They have been able to construct three-dimensional (3D) tubular structures with potential applications in urinary and vascular tissue engineering (Ito et al. 2005a, b, c). They did this by attaching the magnetoliposomes to the appropriate cell type and then growing these cells in sheets on ultra-low attachment plates with magnets placed beneath. As the cells grow and produce the tissue required for the specific structure, the magnet can be removed and then a cylindrical magnet

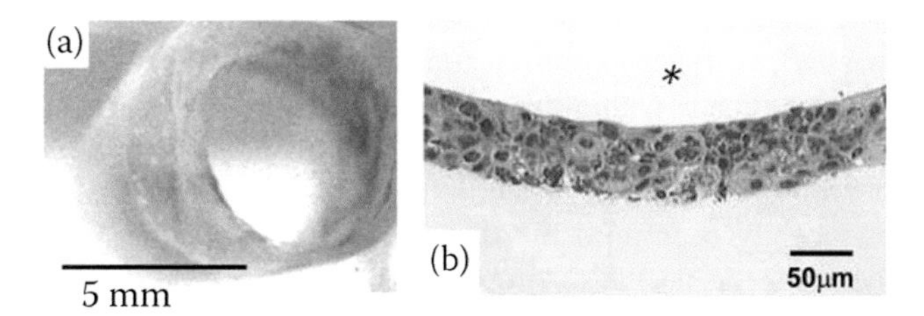

FIGURE 7.6 Light microscope images of **(a)** A tubular construct of urothelial tissue and **(b)** Hematoxylin-eosin stained cross section of the same tissue type. Both constructs were produced using nanomagnetic actuation techniques applied to cell culture/tissue engineering as described in the text, figure from (Ito et al. 2005b).

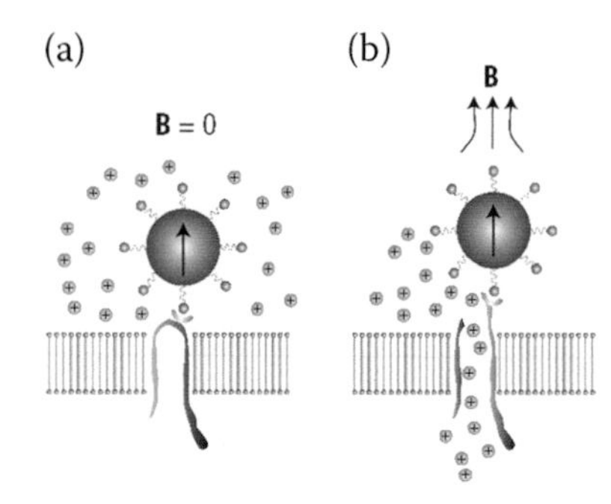

FIGURE 7.7 Schematic representation of targeted nanomagnetic ion channel actuation. A magnetic nanoparticle is attached to the ion channel via a linker molecule. **(a)** When the field is off, the channel functions normally (in this case, the channel is closed), **(b)** Actuation occurs when the field is switched on, opening the channel and allowing ions to flow into the cell. **B** represents the applied magnetic field vector, adapted from (Dobson 2008).

can be used to simply "roll up" the sheet into a tubular structure (Figure 7.6). With scale-up, this technique has the potential to find clinical application in a variety of diseases.

7.6 MAGNETIC ACTIVATION OF RECEPTOR SIGNALING (MARS)

More recent work has focused on using nanomagnetic actuation to target and activate specific ion channels or cell surface receptors on the cell membrane. Such targeted activation could be useful for investigating biochemical signaling cascades, regulating cell signaling-mediated processes such as tissue matrix production and, as is now becoming clear, directing the differentiation of stem cells for tissue engineering and regenerative medicine applications.

While the studies discussed in the introduction section above relied primarily on particle internalization, relatively non-specific binding (such as via RGD), or no binding, the objective in this case is to attach magnetic nanoparticles directly to a specific type of ion channel or membrane receptor in order to control it without necessarily interfering with the normal functioning of other channels in the cell's membrane (Figure 7.7).

In order to activate ion channels via general membrane or cytoskeletal deformation for applications such as mechanical stimulation for bone and connective tissue engineering, previous studies have focused initially on non-specific RGD tagging to integrin receptors on the cell surface to induce deformation. However, in order to actuate and control very specific biochemical reaction cascades involved in cell functioning as well as to examine ion channel activation kinetics and function, specific, targeted actuation techniques are being developed. These specific targeting and binding techniques provide a method of independently actuating ion channels with a high degree of specificity that is not possible with any other technique.

An important enabling technology being developed to exploit the potential of selective ion channel activation is the electromagnetic micro-needle actuator, developed by Ingber's group at Harvard/MIT. This device can be used to focus magnetic forces to small areas for highly targeted nanomagnetic actuation of cells growing in culture (Matthews et al. 2004). These needles have the advantage of not only focusing the field to a very specific target at high resolution but also of creating a very high field gradient in order to produce the activating force. The needle is positioned near a cell to which magnetic particles have been attached and, by activating the field, it is possible to actuate individual, targeted cells within the culture. The high gradient produced by the needle will enable actuation of both individual ion channels via direct attachment as well as deformation of the membrane and cytoskeleton via non-specific binding.

This technique has recently been used by Mannix, Ingber, and colleagues in an elegant experiment that used the magnetic needle to promote membrane receptor clustering in RBL-2H3 mast cells (Mannix et al. 2008). By coating magnetic nanoparticles with dinitrophenol-lysine ligands and attaching them to IgE-FcεRI receptor complexes on the cell surface, they were able to use the focused field produced by the needle to induce receptor clustering (Figure 7.8). Clustering of these receptors activated intracellular calcium signaling, demonstrating the potential of this technique for use as a biosensor.

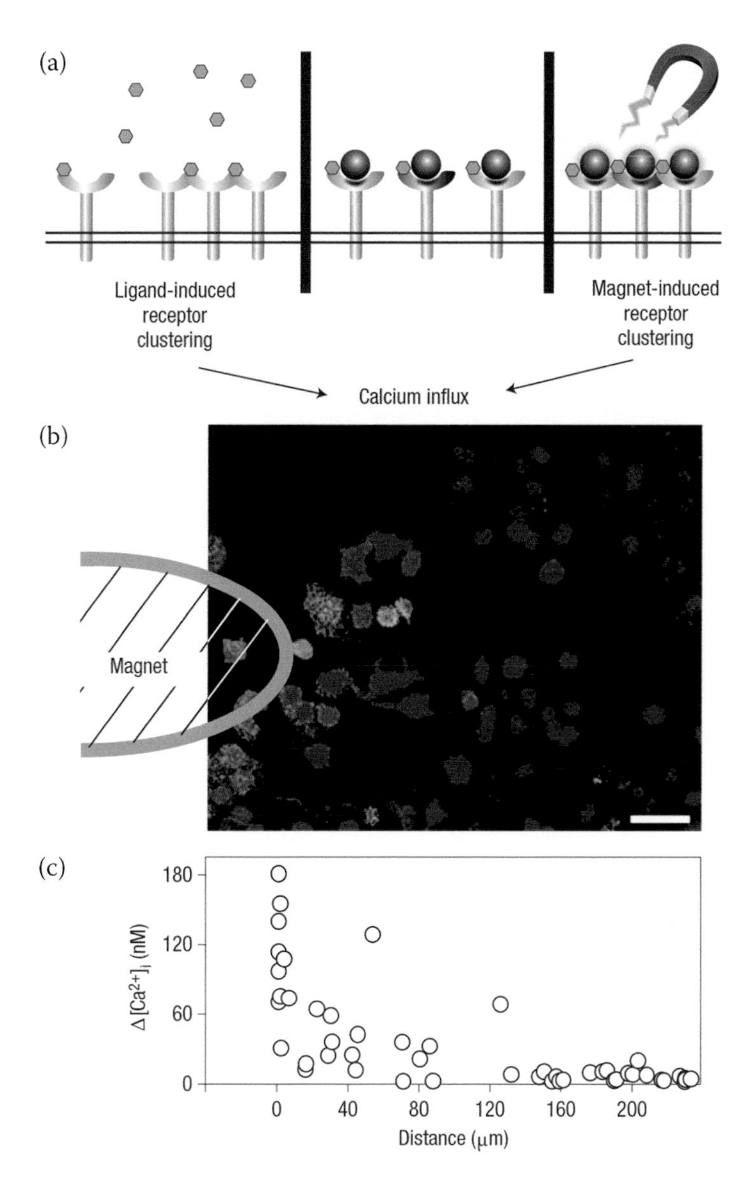

FIGURE 7.8 (a) Schematic representation of nanomagnetic actuation of receptor clustering, **(b)** Fluorescent microscope image of clustering-induced calcium influx actuated by magnetic forces acting on particles bound to the receptors, **(c)** Calcium flux vs distance from the magnet tip, figure from (Mannix et al. 2008).

Over the past several years, our group has been working on experiments aimed at demonstrating the concept of selective ion channel activation for generating cartilage matrix and controlling stem cell differentiation both in vitro and in vivo. In initial experiments, a histidine tag was inserted into the external loop of the TREK-1 mechanosensitive potassium channel, which was then transfected into COS-7 cells. Such channels are capable of initiating biochemical reaction cascades that result in the production of functional tissue (Mullender et al. 2004). Coated magnetic nanoparticles (magnetite, Fe_3O_4) ranging in size from 130 nm to 2.7 μm were attached to the histidine tag either via a Ni-NTA or anti-HIS linker. By applying both static and 1 Hz high-gradient magnetic fields generated by computer-controlled, moveable arrays of NdFeB magnets, TREK-1 activation was observed electrophysiologically by patch clamp in cells with

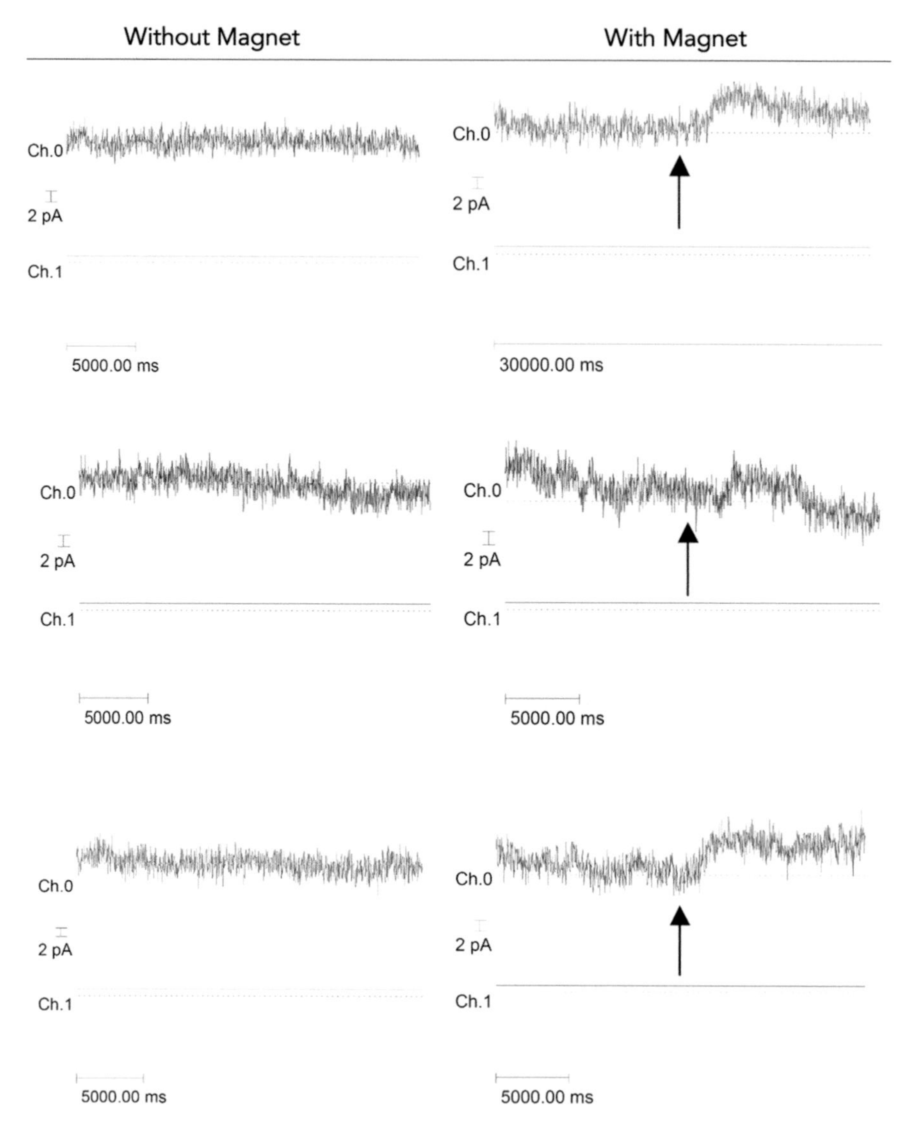

FIGURE 7.9 Patch clamp recordings of TREK-1 ion channel currents activated via the application of a magnetic force to magnetic nanoparticles attached to the channel. The arrow indicates application of the actuating field. The opening of the channel is indicated by a rise in the current (trace), redrawn from (Hughes et al. 2008).

both anti-HIS tagged and in Ni-NTA tagged particles down to 250 nm in size (Figure 7.9) (Hughes et al. 2008, Kirkham et al. 2010).

Though these results are encouraging, demonstrating that it is possible to activate and observe specific ion channels in cells, native TREK-1 channels do not possess the histidine tag to which we can link, limiting the use of the technique in vivo. In order to address this problem, over the past several years we have focused our efforts on targeting and activating the native TREK-1 channel by using anti-TREK antibodies in human mesenchymal stem cells (MSCs).

In this work, magnetic nanoparticles coated with anti-TREK antibodies were targeted to the channel in these cells. In vitro actuation resulted in significant upregulation of the cartilage-related genes osteopontin, CBFA-1, and SOX-9, in comparison to controls that included particles targeted to calcium channels using a Ca antibody coating. Taking the work a step further, we used the

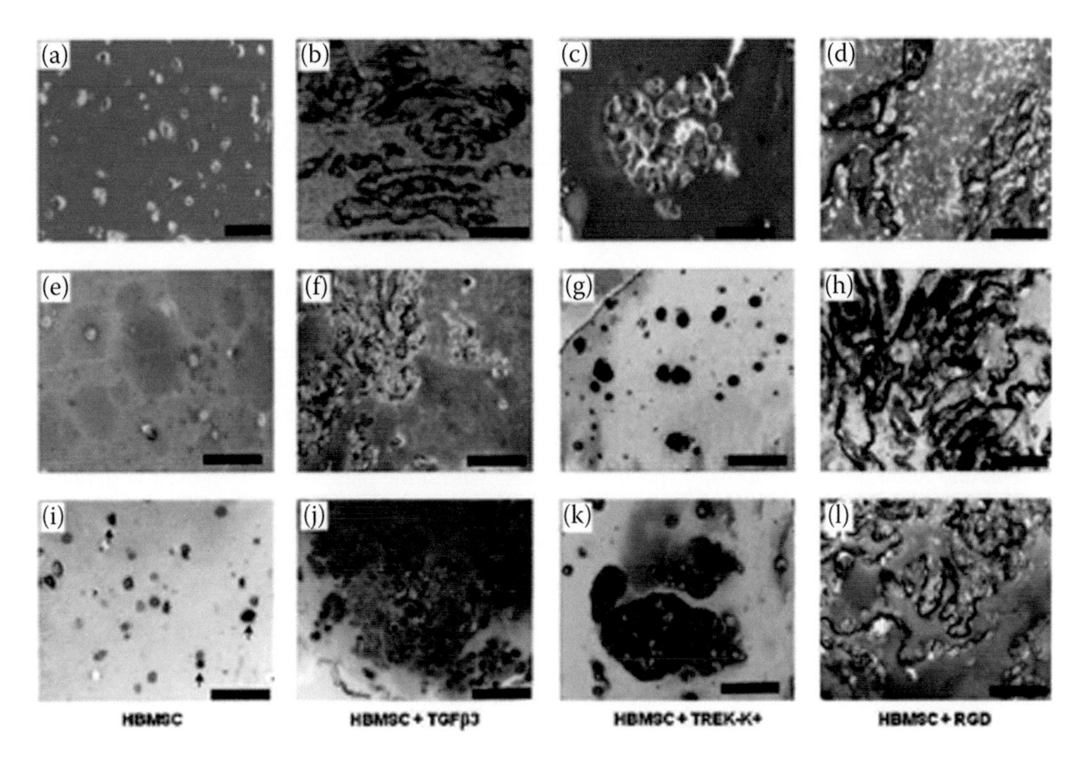

FIGURE 7.10 Histology and immunohistochemistry of human bone marrow stem cells (HBMSC) only; HBMSC & TGFβ3; HBMSC labelled with TREK-K^+ or HBMSC labelled with RGD particles encapsulated in alginate/chitosan capsules implanted subcutaneously in MF-1 nu/nu mice and exposed to a magnetic field for 21 days at 1 hr/day. Representative 6 μm tissue sections stained for alcian blue/Sirius red (**a–d**); Type-1 collagen (**e–h**); and Type-2 collagen (**i–l**). Arrows=positively stained HBMSC for Type 1 collagen. Scale bar=100 μm, figure from (Kanczler et al. 2010).

technique in vivo to control stem cell differentiation. Human MSCs were tagged with anti-TREK-coated magnetic nanoparticles, encapsulated in alginate, and implanted into the backs of nude mice (these are mice that have their immune system genetically knocked out so that they do not reject implanted human cells). The mice were exposed to magnetic fields during one hour per day spent in special magnetic cages where their normal movement through the magnetic field gradients would cyclically actuate the channels. After 21 days, the magnetically tagged MSCs were producing collagen matrix while the controls were not (in fact, many of the control cells had migrated out of the capsule) (Figure 7.10) (Kanczler et al. 2010). This work demonstrates the potential for this technique in stem cell therapy and regenerative medicine.

In the past few years, we have demonstrated the activation of platelet-derived growth factor receptors (PDGFR) using MARS technology to control both MSC differentiation and up-regulation of smooth muscle actin. Activation of PDGRF-α was shown to up-regulate osteogenic gene expression (collagen type 1 and Runx2) in MSCs over the first seven days in culture when compared to MARS targeting of PDGFR-β and cell surface integrins (Hu, El Haj, and Dobson 2013). MARS targeting of PDGRF-α was also shown to upregulate smooth muscle actin compared to activation of PDGFR-β or cell surface integrins after three hours of magnetomechanical stimulation (Hu, Dobson, and El Haj 2014).

Most recently, our group has begun the development of nanomagnetic techniques for the control of growth factor activation for regenerative medicine applications. Specifically, a bio-inspired conjugation strategy was used to graft transforming growth factor-beta (TGF-β) latent complex to magnetic nanoparticles. With the application of radiofrequency magnetic fields and the transfer of energy from the field to the particles, a conformation change was induced in the latent complex, resulting in the triggered release of active TGF-β (Monsalve et al. 2015). Though this technique is in its infancy, it has the potential to provide both remote spatial and temporal control over the activity of growth factors in the body, reducing potentially dangerous off-target side effects.

7.7 MAGNETICALLY ACTUATED BIOCHIPS AND CELL PATTERNING

Following on from their pioneering work on the development of the electromagnetic micro-needle and receptor clustering studies, the Ingber group recently developed a technique for the fabrication of magnetically actuated cellular microchips. By laser etching a patterned array of magnetizable nanospikes ("magnetic field gradient concentrators") onto a biocompatible substrate, they produced cell seeding patterns via attachment of RGD-coated magnetic microparticles to human umbilical vein endothelial (HUVE) cells. These cells are dependent on substrate adhesion for survival and removing the field from specific "islands" within the pattern has been used to actuate rapid cell detachment and apoptosis (Polte et al. 2007). The system can be configured for simultaneous investigation of multiple substrate ligands or soluble stimuli on multiple cell types on a single chip.

7.8 A FINAL THOUGHT

Nanomagnetic actuation is more than simply a tool for studying cell function—it is opening up a new field of research with applications that go beyond cell biology and, hopefully, one day into the clinic. In the future, it may be possible to treat disease or grow new tissue by remote activation and manipulation of specific receptors in the body. Until that time, the ability to target and manipulate cellular structures with such elegant precision has already opened up exciting new possibilities in regenerative medicine and the investigation of cell function.

REFERENCES

Adair, R. K. 1991. Constraints on biological effects of weak extremely-low-frequency electromagnetic fields. *Physical Review A* 43 (2), 1039.

Alberts, B., A. Johnson, J. Lewis, M. Raff, K. Roberts, and P. Walter. 2002. *Molecular Biology of the Cell*. 4th ed. New York and London: Garland.

Bausch, A. R., U. Hellerer, M. Essler, M. Aepfelbacher, and E. Sackmann. 2001. Rapid stiffening of integrin receptor-actin linkages in endothelial cells stimulated with thrombin: A magnetic bead microrheology study. *Biophysical Journal* 80 (6), 2649–2657.

Bausch, A. R., W. Möller, and E. Sackmann. 1999. Measurement of local viscoelasticity and forces in living cells by magnetic tweezers. *Biophysical Journal* 76 (1), 573–579.

Berry, C. C. 2009. Progress in functionalization of magnetic nanoparticles for applications in biomedicine. *Journal of Physics D: Applied Physics* 42 (22), 224003. doi:10.1088/0022-3727/42/22/224003.

Bierbaum, S., and H. Notbohm. 1998. Tyrosine phosphorylation of 40 kDa proteins in osteoblastic cells after mechanical stimulation of beta1-integrins. *European Journal of Cell Biology* 77 (1), 60–67.

Boutahar, N., A. Guignandon, L. Vico, and M. H. Lafage-Proust. 2004. Mechanical strain on osteoblasts activates autophosphorylation of focal adhesion kinase and proline-rich tyrosine kinase 2 tyrosine sites involved in ERK activation. *Journal of Biological Chemistry* 279 (29), 30588–30599.

Burridge, K., K. Fath, T. Kelly, G. Nuckolls, and C. Turner. 1988. Focal adhesions: Transmembrane junctions between the extracellular matrix and the cytoskeleton. *Annual Review of Cell and Developmental Biology* 4, 487–525.

Cartmell, S. H., J. Dobson, S. B. Verschueren, and A. J. El Haj. 2002. Development of magnetic particle techniques for long-term culture of bone cells with intermittent mechanical activation. *IEEE Transactions on Nanobioscience* 99 (2), 92–97.

Chen, J., B. Fabry, E. L. Schiffrin, and N. Wang. 2001. Twisting integrin receptors increases endothelin-1 gene expression in endothelial cells. *American Journal of Physiology-Cell Physiology* 280 (6), C1475–C1484.

Clark, E. A., and J. S. Brugge. 1995. Integrins and signal transduction pathways: The road taken. *Science* 268 (5208), 233–239.
Crick, F. H. C., and A. Hughes. 1950. The physical properties of cytoplasm. A study by means of the magnetic particle method. Part II. Theoretical treatment. *Experimental Cell Research* 1 (4), 505–533.
Dan, I., N. M. Watanabe, and A. Kusumi. 2001. The Ste20 group kinases as regulators of MAP kinase cascades. *Trends in Cell Biology* 11 (5), 220–230.
Diamond, S. L., F. Sachs, and W. J. Sigurdson. 1994. Mechanically induced calcium mobilization in cultured endothelial cells is dependent on actin and phospholipase. *Arteriosclerosis Thrombosis* 14 (12), 2000–2006.
Dobson, J., S. H. Cartmell, A. Keramane, and A. J. El Haj. 2006. A magnetic force mechanical conditioning bioreactor for tissue engineering, stem cell conditioning and dynamic in vitro screening. *IEEE Transactions on Nanobioscience* 5, 173–177.
Dobson, J., A. Keramane, and A. J. El Haj. 2002. Theory and applications of a magnetic force bioreactor. *European Cells and Materials* 4, 130–131.
Dobson, J. 2008. Remote control of cellular behaviour with magnetic nanoparticles. *Nature Nanotechnology* 3 (3), 139–143.
Dobson, J., and T. St Pierre. 1996. Application of the ferromagnetic transduction model to DC and pulsed magnetic fields: Effects on epileptogenic tissue and implications for cellular phone safety. *Biochemical and Biophysical Research Communications* 227 (3), 718–723.
Fabry, B., G. N. Maksym, R. D. Hubmayr, J. P. Butler, and J. J. Fredberg. 1999. Implications of heterogeneous bead behavior on cell mechanical properties measured with magnetic twisting cytometry. *Journal of Magnetism and Magnetic Materials* 194, 120–125.
Fodil, R., V. Laurent, E. Planus, and D. Isabey. 2003. Characterization of cytoskeleton mechanical properties and 3D-actin structure in twisted adherent epithelial cells. *Biorheology* 40 (1–3), 241–245.
Giancotti, F. G., and E. Ruoslahti. 1999. Integrin signaling. *Science* 285 (5430), 1028–1032.
Glogauer, M., and J. Ferrier. 1997. A new method for application of force to cells via ferric oxide beads. *Pflügers Archiv* 435 (2), 320–327.
Glogauer, M., J. Ferrier, and C. A. McCulloch. 1995a. Magnetic fields applied to collagen-coated ferric oxide beads induce stretch-activated Ca^{2+} flux in fibroblasts. *American Journal of Physiology* 269 (5 Pt 1), C1093–C1104.
Glogauer, M., J. Ferrier, and C. A. McCulloch. 1995b. Magnetic fields applied to collagen-coated ferric oxide beads induce stretch-activated Ca^{2+} flux in fibroblasts. *American Journal of Physiology-Cell Physiology* 269 (5), C1093–C1104.
Goldschmidt, M. E., K. J. McLeod, and W. R. Taylor. 2001. Integrin-mediated mechanotransduction in vascular smooth muscle cells: Frequency and force response characteristics. *Circulation Research* 88 (7), 674–680.
Grzesik, W. J., and P. G. Robey. 1994. Bone matrix RGD glycoproteins: Immunolocalization and interaction with human primary osteoblastic bone cells in vitro. *Journal of Bone and Mineral Research* 9 (4), 487–496.
Guilak, F. 1995. Compression-induced changes in the shape and volume of the chondrocyte nucleus. *Journal of Biomechanics* 28 (12), 1529–1541.
Heilbronn, A. 1922. Eine neue methode zur bestimmung der viskosität lebender protoplasten. *Jahrbücher für Wissenschaftliche Botanik* 61, 284.
Howard, J., and A. J. Hudspeth. 1988. Compliance of the hair bundle associated with gating of mechanoelectrical transduction channels in the bullfrog's saccular hair cell. *Neuron* 1 (3), 189–199.
Hu, B., J. Dobson, and A. J. El Haj. 2014. Control of smooth muscle α-actin (SMA) up-regulation in HBMSCs using remote magnetic particle mechano-activation. *Nanomedicine: Nanotechnology, Biology and Medicine* 10 (1), 45–55.
Hu, B., A. J. El Haj, and J. Dobson. 2013. Receptor-targeted, magneto-mechanical stimulation of osteogenic differentiation of human bone marrow-derived mesenchymal stem cells. *International Journal of Molecular Sciences* 14 (9), 19276–19293.
Hu, S., J. Chen, J. P. Butler, and N. Wang. 2005. Prestress mediates force propagation into the nucleus. *Biochemical and Biophysical Research Communication* 329 (2), 423–428.
Huang, H., R. D. Kamm, and R. T. Lee. 2004. Cell mechanics and mechanotransduction: Pathways, probes, and physiology. *American Journal of Physiology-Cell Physiology* 287 (1), C1–C11.
Hughes, S., A. J. El Haj, and J. Dobson. 2005. Magnetic micro-and nanoparticle mediated activation of mechanosensitive ion channels. *Medical Engineering & Physics* 27 (9), 754–762.

Hughes, S., S. McBain, J. Dobson, and A. J. El Haj. 2008. Selective activation of mechanosensitive ion channels using magnetic particles. *Journal of The Royal Society Interface* 5 (25), 855–863.

Hynes, R. O. 1992. Integrins: Versatility, modulation, and signaling in cell adhesion. *Cell* 69 (1), 11–25.

Ingber, D. E. 1997. Tensegrity: The architectural basis of cellular mechanotransduction. *Annual Review of Physiology* 59, 575–599.

Inoue, D., S. Kido, and T. Matsumoto. 2004. Transcriptional induction of FosB/DeltaFosB gene by mechanical stress in osteoblasts. *Journal of Biological Chemistry* 279 (48), 49795–49803.

Ito, A., E. Hibino, C. Kobayashi, H. Terasaki, H. Kagami, M. Ueda, T. Kobayashi, and H. Honda. 2005a. Construction and delivery of tissue-engineered human retinal pigment epithelial cell sheets, using magnetite nanoparticles and magnetic force. *Tissue Engineering* 11 (3–4), 489–496.

Ito, A., K. Ino, M. Hayashida, T. Kobayashi, H. Matsunuma, H. Kagami, M. Ueda, and H. Honda. 2005b. Novel methodology for fabrication of tissue-engineered tubular constructs using magnetite nanoparticles and magnetic force. *Tissue Engineering* 11 (9–10), 1553–1561.

Ito, A., K. Ino, T. Kobayashi, and H. Honda. 2005c. The effect of RGD peptide-conjugated magnetite cationic liposomes on cell growth and cell sheet harvesting. *Biomaterials* 26 (31), 6185–6193.

Jadlowiec, J., H. Koch, X. Zhang, P. G. Campbell, M. Seyedain, and C. Sfeir. 2004. Phosphophoryn regulates the gene expression and differentiation of NIH_3T_3, MC_3T_3-E_1, and human mesenchymal stem cells via the integrin/MAPK signaling pathway. *Journal of Biological Chemistry* 279 (51), 53323–53330.

Jaiswal, R. K., N. Jaiswal, S. P. Bruder, G. Mbalaviele, D. R. Marshak, and M. F. Pittenger. 2000. Adult human mesenchymal stem cell differentiation to the osteogenic or adipogenic lineage is regulated by mitogen-activated protein kinase. *Journal of Biological Chemistry* 275 (13), 9645–9652.

Janmey, P. A. 1998. The cytoskeleton and cell signaling: Component localization and mechanical coupling. *Physiological Reviews* 78 (3), 763–781.

Jansen, J. H., F. A. Weyts, I. Westbroek, H. Jahr, H. Chiba, H. A. Pols, J. A. Verhaar, J. P. van Leeuwen, and H. Weinans. 2004. Stretch-induced phosphorylation of ERK1/2 depends on differentiation stage of osteoblasts. *Journal of Cellular Biochemistry* 93 (3), 542–551.

Jessop, H. L., S. C. Rawlinson, A. A. Pitsillides, and L. E. Lanyon. 2002. Mechanical strain and fluid movement both activate extracellular regulated kinase (ERK) in osteoblast-like cells but via different signaling pathways. *Bone* 31 (1), 186–194.

Kanczler, J. M., H. S. Sura, J. Magnay, D. Green, R. O. C. Oreffo, J. P. Dobson, and A. J. El Haj. 2010. Controlled differentiation of human bone marrow stromal cells using magnetic nanoparticle technology. *Tissue Engineering Part A* 16 (10), 3241–3250.

Kapur, S., D. J. Baylink, and K. H. Lau. 2003. Fluid flow shear stress stimulates human osteoblast proliferation and differentiation through multiple interacting and competing signal transduction pathways. *Bone* 32 (3), 241–251.

Kirkham, G. R., K. J. Elliot, A. Keramane, D. M. Salter, J. P. Dobson, A. J. El Haj, and S. H. Cartmell. 2010. Hyperpolarization of human mesenchymal stem cells in response to magnetic force. *IEEE Transactions on Nanobioscience* 9 (1), 71–74.

Kirschvink, J. L., A. Kobayashi-Kirschvink, and B. J. Woodford. 1992. Magnetite biomineralization in the human brain. *Proceedings of the National Academy of Sciences* 89 (16), 7683–7687.

Kolch, W. 2000. Meaningful relationships: The regulation of the Ras/Raf/MEK/ERK pathway by protein interactions. *Biochemical Journal* 351 (Pt 2), 289–305.

Kyriakis, J. M., and J. Avruch. 2001. Mammalian mitogen-activated protein kinase signal transduction pathways activated by stress and inflammation. *Physiological Reviews* 81 (2), 807–869.

Maniotis, A. J., C. S. Chen, and D. E. Ingber. 1997. Demonstration of mechanical connections between integrins, cytoskeletal filaments, and nucleoplasm that stabilize nuclear structure. *Proceedings of the National Academy of Sciences of the United States of America* 94 (3), 849–854.

Mannix, R. J., S. Kumar, F. Cassiola, M. Montoya-Zavala, E. Feinstein, M. Prentiss, and D. E. Ingber. 2008. Nanomagnetic actuation of receptor-mediated signal transduction. *Nature Nanotechnology* 3 (1), 36.

Martineau, L. C., and P. F. Gardiner. 2001. Insight into skeletal muscle mechanotransduction: MAPK activation is quantitatively related to tension. *Journal of Applied Physiology* 91 (2), 693–702.

Matthews, B. D., D. A. LaVan, D. R. Overby, J. Karavitis, and D. E. Ingber. 2004. Electromagnetic needles with submicron pole tip radii for nanomanipulation of biomolecules and living cells. *Applied Physics Letters* 85 (14), 2968–2970.

Meyer, C. J., F. J. Alenghat, P. Rim, J. H.-J. Fong, B. Fabry, and D. E. Ingber. 2000. Mechanical control of cyclic AMP signalling and gene transcription through integrins. *Nature Cell Biology* 2 (9), 666–668.

Monsalve, A., A. C. Bohórquez, C. Rinaldi, and J. Dobson. 2015. Remotely triggered activation of TGF with magnetic nanoparticles. *IEEE Magnetics Letters* 6, 1–4.

Mullender, M., A. J. El Haj, Y. Yang, M. A. Van Duin, E. H. Burger, and J. Klein-Nulend. 2004. Mechanotransduction of bone cells in vitro: Mechanobiology of bone tissue. *Medical and Biological Engineering and Computing* 42 (1), 14–21.

Netter, F. H. 1987. Musculoskeletal system: Anatomy, physiology, metabolic disorders, Vol. 8. In *Netter Collection of Medical Illustrations*. New Jersey: Saunders.

Pankhurst, Q. A., J. Connolly, S. K. Jones, and J. Dobson. 2003. Applications of magnetic nanoparticles in biomedicine. *Journal of Physics D: Applied Physics* 36, R167.

Pavalko, F. M., N. X. Chen, C. H. Turner, D. B. Burr, S. Atkinson, Y. F. Hsieh, J. Qiu, and R. L. Duncan. 1998. Fluid shear-induced mechanical signaling in MC_3T_3-E_1 osteoblasts requires cytoskeleton-integrin interactions. *American Journal of Physiology* 275 (6 Pt 1), C1591–C1601.

Pierschbacher, M. D., and E. Ruoslahti. 1984. Cell attachment activity of fibronectin can be duplicated by small synthetic fragments of the molecule. *Nature* 309 (5963), 30–33.

Polte, T. R., M. Shen, J. Karavitis, M. Montoya, J. Pendse, S. Xia, E. Mazur, and D. E. Ingber. 2007. Nanostructured magnetizable materials that switch cells between life and death. *Biomaterials* 28 (17), 2783–2790.

Pommerenke, H., E. Schreiber, F. Dürr, B. Nebe, C. Hahnel, W. Möller, and J. Rychly. 1996. Stimulation of integrin receptors using a magnetic drag force device induces an intracellular free calcium response. *European Journal of Cell Biology* 70 (2), 157–164.

Pommerenke, H., C. Schmidt, F. Durr, B. Nebe, F. Luthen, P. Muller, and J. Rychly. 2002. The mode of mechanical integrin stressing controls intracellular signaling in osteoblasts. *Journal of Bone and Mineral Research* 17 (4), 603–611.

Puig-De-Morales, M., M. Grabulosa, J. Alcaraz, J. Mullol, G. N. Maksym, J. J. Fredberg, and D. Navajas. 2001. Measurement of cell microrheology by magnetic twisting cytometry with frequency domain demodulation. *Journal of Applied Physiology* 91 (3), 1152–1159.

Roca, A. G., S. Veintemillas-Verdaguer, M. Port, C. Robic, C. J. Serna, and M. P. Morales. 2009. Effect of nanoparticle and aggregate size on the relaxometric properties of MR contrast agents based on high quality magnetite nanoparticles. *The Journal of Physical Chemistry B* 113 (19), 7033–7039. doi:10.1021/jp807820s.

Roux, P. P., and J. Blenis. 2004. ERK and p38 MAPK-activated protein kinases: A family of protein kinases with diverse biological functions. *Microbiology and Molecular Biology Reviews* 68 (2), 320–344.

Sachs, F. 1988. Mechanical transduction in biological systems. *Critical Reviews in Biomedical Engineering* 16 (2), 141–169.

Salter, D. M., J. E. Robb, and M. O. Wright. 1997. Electrophysiological responses of human bone cells to mechanical stimulation: Evidence for specific integrin function in mechanotransduction. *Journal of Bone and Mineral Research* 12 (7), 1133–1141.

Salter, D. M., W. H. Wallace, J. E. Robb, H. Caldwell, and M. O. Wright. 2000. Human bone cell hyperpolarization response to cyclical mechanical strain is mediated by an interleukin-1beta autocrine/paracrine loop. *Journal of Bone and Mineral Research* 15 (9), 1746–1755.

Sawada, Y., and M. P. Sheetz. 2002. Force transduction by Triton cytoskeletons. *Journal of Cell Biology* 156 (4), 609–615.

Schwartz, M. A., and S. J. Shattil. 2000. Signaling networks linking integrins and rho family GTPases. *Trends in Biochemical Sciences* 25 (8), 388–391.

Seifriz, W. 1924. An elastic value of protoplasm, with further observations on the viscosity of protoplasm. *Journal of Experimental Biology* 2 (1), 1–11.

Traub, P. 1995. Intermediate filaments and gene regulation. *Physiological Chemistry and Physics and Medical NMR* 27 (4), 377–400.

Traub, P., and R. L. Shoeman. 1994. Intermediate filament proteins: Cytoskeletal elements with gene-regulatory function? *International Review of Cytology* 154, 1–103.

Valberg, P. A., and J. P. Butler. 1987. Magnetic particle motions within living cells. Physical theory and techniques. *Biophysical Journal* 52 (4), 537–550.

Valberg, P. A., and D. F. Albertini. 1985. Cytoplasmic motions, rheology, and structure probed by a novel magnetic particle method. *The Journal of Cell Biology* 101 (1), 130–140.

Valberg, P. A., and H. A. Feldman. 1987. Magnetic particle motions within living cells. Measurement of cytoplasmic viscosity and motile activity. *Biophysical Journal* 52 (4), 551–561.
Waigh, T. A. 2005. Microrheology of complex fluids. *Reports on Progress in Physics* 68 (3), 685.
Walker, L. M., Å. Holm, L. Cooling, L. Maxwell, Å. Öberg, T. Sundqvist, and A. J. El Haj. 1999. Mechanical manipulation of bone and cartilage cells with 'optical tweezers.' *FEBS Letters* 459 (1), 39–42.
Wang, N., J. P. Butler, and D. E. Ingber. 1993. Mechanotransduction across the cell surface and through the cytoskeleton. *Science* 260 (5111), 1124–1127.
Wang, N., and D. E. Ingber. 1995a. Probing transmembrane mechanical coupling and cytomechanics using magnetic twisting cytometry. *Biochemistry and Cell Biology* 73 (7–8), 327–335.
Wang, N., and D. E. Ingber. 1995b. Probing transmembrane mechanical coupling and cytomechanics using magnetic twisting cytometry. *Biochemistry and Cell Biology* 73 (7–8), 327–335.
Yamada, S., D. Wirtz, and S. C. Kuo. 2000. Mechanics of living cells measured by laser tracking microrheology. *Biophysical Journal* 78 (4), 1736–1747.
You, J., G. C. Reilly, X. Zhen, C. E. Yellowley, Q. Chen, H. J. Donahue, and C. R. Jacobs. 2001. Osteopontin gene regulation by oscillatory fluid flow via intracellular calcium mobilization and activation of mitogen-activated protein kinase in MC_3T_3-E_1 osteoblasts. *Journal of Biological Chemistry* 276 (16), 13365–13371.
Yuge, L., A. Okubo, T. Miyashita, T. Kumagai, T. Nikawa, S. Takeda, M. Kanno, Y. Urabe, M. Sugiyama, and K. Kataoka. 2003. Physical stress by magnetic force accelerates differentiation of human osteoblasts. *Biochemical and Biophysical Research Communications* 311 (1), 32–38.
Zhu, C., G. Bao, and N. Wang. 2000. Cell mechanics: Mechanical response, cell adhesion, and molecular deformation. *Annual Review of Biomedical Engineering* 2, 189–226.

8 Magnetic Nanoparticles: Challenges and Opportunities in Drug Delivery

Allan E. David, Mahaveer S. Bhojani, and Adam J. Cole

CONTENTS

8.1 INTRODUCTION

Drug discovery and development remains a formidable challenge. Despite advances in synthetic chemistry and in techniques such as high-throughput screening, the search for drugs to treat human diseases remains in many instances an unfruitful endeavor. Even after successful identification of a drug candidate, the development of a formulation that provides sufficient delivery of the drug to its target site in the body is a hurdle that is often difficult to overcome. Toxicity that may arise due to the non-specific effect of a drug on healthy tissue is another issue that must be addressed. Indeed, side effects can often limit safe dosing to levels too low to achieve therapeutic concentrations at a disease site. Because of these potential roadblocks, drug development generally requires a significant commitment of both time and resources, which can prevent many otherwise promising drugs from achieving clinical translation (i.e. approval for use in human patients). Nanoparticle-based drug delivery systems developed over the last decade or so, however, have generated much enthusiasm for their potential to overcome the difficulties faced in drug delivery. Additionally, the incorporation of imaging agents into nanoparticles or the ability to image a nanoparticle intrinsically has further generated interest in nanomedicine. Indeed, the potential for simultaneous therapy and diagnosis of disease with a single agent has led to the classification of some nanoparticulate delivery systems as "theranostic." Due to their unique properties, nanoparticles offer the potential to: (1) more accurately target diseased sites in the body, (2) reduce the dose required to achieve a therapeutic concentration of an agent at its target site, and (3) alleviate side effects associated with off-target accumulation of drugs in normal tissues (Arruebo, Fernández-Pacheco, Ibarra, & Santamaría, 2007).

Nanotechnology, in its broadest definition, is the engineering and utilization of materials having at least one dimension between 1 and 1000 nm. In a more stringent case, it is generally accepted that the upper limit of nanotechnology is about 100 nm. This size range lies between the "molecular" (~0.1 nm) and "bulk" (hundreds of nanometers) thresholds. High surface energies and quantum effects characteristic of materials of this size give rise to unique chemical, physical, and optical properties not observed in materials at other length scales. With large surface-to-volume ratios, nanomaterials enable the high loading of different functional ligands on to a single platform. Biomedical research, particularly focused on cancer detection and therapy, has shown great interest in utilizing these distinct properties of nanoparticles.

Recently, magnetic nanoparticles (MNPs) have drawn interest because of their potential for both diagnostic imaging and drug delivery. MNPs utilized generally consist of a magnetic core encapsulated in a shell coating (e.g. dextran, poly(lactic-*co*-glycolic) [PLGA], starch), as shown in Figure 8.1. This structure provides a scaffold onto which other components can be attached, including drugs, imaging agents, and targeting ligands. The magnetic core enables the nanoparticles to be manipulated with external magnetic fields for magnetic targeting (Chertok et al. 2008), magnetically responsive drug release (Dobson 2006), and magnetic imaging (Ferrari, Lee, Lee, & Josephson, 2006).

MNPs have been studied over the past two decades primarily as contrast agents in magnetic resonance imaging (MRI) and for magnetic cell sorting (Ferrari et al. 2006, Weissleder et al. 1990, Sun et al. 1998). In fact, two superparamagnetic iron oxide agents (i.e. Feridex and Resovist) have been approved for clinical use as MR contrast agents. The demonstrated ability of MNPs to enhance MRI of tumors (see Figure 8.2), which also indicates their accumulation in tumors, has generated considerable excitement for their potential to deliver anticancer drugs.

8.2 MAGNETIC NANOPARTICLES (MNPS)

The coating surrounding the magnetic core of an MNP functions to enhance both biocompatibility and particle stability, and also provide a substrate for the coupling of various ligands (see Figure 8.1). Magnetic materials can be classified as ferrimagnetic, ferromagnetic, antiferromagnetic, paramagnetic, or diamagnetic. High-coercivity ferromagnets are "permanent

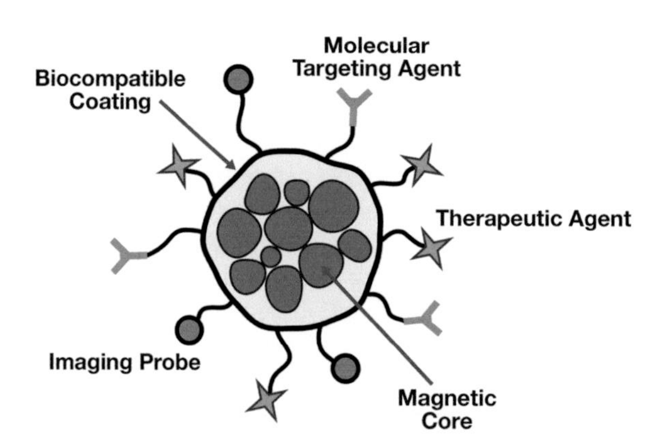

FIGURE 8.1 Illustration of the core-shell structure typical of magnetic nanoparticles utilized for drug delivery. The magnetic core is composed of multiple, small iron oxide crystals that are then encapsulated by a biocompatible coat. The coating stabilizes the particle against aggregation and also provides anchor points for the attachment of targeting agents, imaging probes, and drugs.

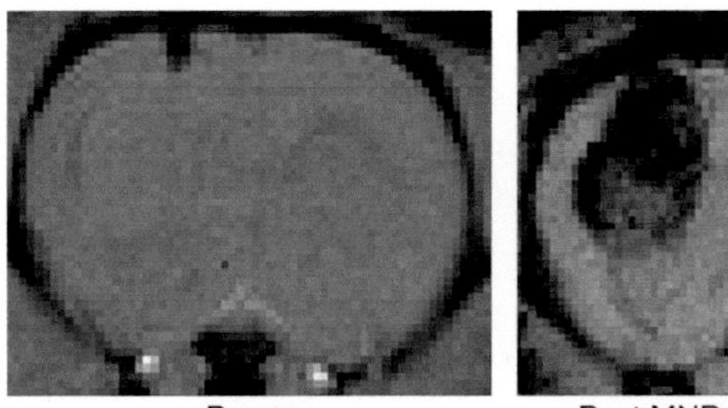

FIGURE 8.2 Magnetic resonance imaging (MRI) acquired of the brains of Fisher 344 rats, taken before (prescan) and after the administration of magnetic iron-oxide nanoparticles. The animal was exposed to a magnetic field for one hour (magnetic targeting) after tail vein injection of MNP prior to MRI. Clear delineation of the tumor margins is observed by the accumulation of MNP, while this was difficult to see in the prescan image.

magnets" that retain their magnetic properties even in the absence of an applied magnetic field. These materials are typically not considered for drug delivery applications because of their tendency to stick together due to interparticle magnetic attraction. Aggregation of nanoparticles into larger structures could lead to an embolism in blood vessels of the brain, lungs, and other tissues. Impeded bloodflow to the heart or brain often leads to debilitating consequences such as cardiac arrest or stroke, respectively.

Paramagnetic materials, on the other hand, possess no net magnetic moment unless under the influence of an applied magnetic field; in other words, their magnetization must be induced. In addition, the magnetic susceptibility of paramagnetic particles is relatively small, such that paramagnetic particles have relatively small magnetic moments absent strong external magnetic fields. Superparamagnetic materials have properties similar to both ferromagnetic and paramagnetic materials. Individual superparamagnetic particles possess a strong magnetic moment at any given time. Yet, in the absence of an external field, this magnetic moment freely rotates in time such that the time-averaged field over a period of magnetic measurement is zero. Thus, superparamagnetic materials are not characterized with remanent magnetization (i.e. hysteresis) in the absence of an applied field. Rotation of the magnetic moment for a superparamagnetic particle is due to Brownian and Néel relaxation processes. In order to minimize magnetically induced aggregation of MNPs, superparamagnetic MNPs are the most commonly utilized in biomedical applications, including for drug delivery.

While MNPs comprised of cobalt (Scarberry et al. 2008), iron alloy (Pal and Bahadur 2010), or nickel cores (Scarberry, Dickerson, McDonald, & Zhang, 2008) have been investigated, iron oxide-based MNPs (e.g. magnetite, Fe_3O_4, or maghemite, γ-Fe_2O_3) are used most frequently because of their better biocompatibility (Jain, Reddy, Morales, Leslie-Pelecky, & Labhasetwar, 2008). Iron oxides display superparamagnetism when the particles have a single magnetic domain, corresponding to a single crystal size of below approximately 20–30 nm (Sjogren et al. 1997). Most MNPs utilized for drug delivery, therefore, either consist of single crystals below this size threshold, or have a clustered core consisting of small single crystals physically bound together within the nanoparticle coating. Iron oxide MNPs have been shown to have limited toxicities at even high doses. One explanation for their low toxicity could be the fact that iron from the nanoparticle eventually joins the body's own iron pool, being stored in iron-containing proteins such as hemoglobin and ferritin. As iron oxide MNPs are fairly inert, they generally require the incorporation of additional, bioactive agents to exert a therapeutic effect on diseased tissues or cells.

The focus in this chapter is on use of MNPs as drug delivery vehicles for cancer therapy. While cancer is our focus, it should be noted that MNPs can be used for treatment of many other diseases. For example, MNPs can be used to deliver drugs to the inner ear via magnetic forces

(Nacev et al. 2012), to treat retinal diseases (Yanai et al. 2012), and to treat magnetic stents (Chorny et al. 2010, Polyak et al. 2008). MNPs can also be conjugated to viruses to enhance genetic transfection in vitro and in vivo (McBain et al. 2008). MNPs may also be used to make genetically modified cells and stem cells magnetic in order to guide them to specific organs in vivo (Cheng et al. 2013).

8.3 CONJUGATION AND RELEASE OF DRUGS FROM MNPs

With the availability of a wide range of coating materials (e.g. polysaccharides, fatty acids, proteins, synthetic polymers), therapeutic compounds can be attached to nanoparticles via chemical conjugation, by electrostatic or hydrophobic interactions, or by encapsulation within the coating. Extensive pre-clinical work, using animal cancer models, has been done to evaluate MNPs that incorporate small-molecule drugs (e.g. paclitaxel, doxorubicin), radiotherapeutics, and photoactivated agents (e.g. Photofrin) (Veiseh et al. 2010, Bhojani, Van Dort, Rehemtulla, & Ross, 2010). While displaying some therapeutic efficacy, these results have yet to translate into clinical studies, likely because most studies done with animal models show only marginal improvements in survival.

Recently some of the focus has shifted toward potentially therapeutic macromolecular agents (e.g. proteins and polynucleotides) due to their enhanced potency and specificity in cancer therapy when compared to small molecules—promising more effective therapy with decreased side effects. One measure of a substance's potency, or toxicity, is the concentration at which it inhibits the growth of cells by 50%—the inhibitory concentration (IC50). The IC50 of cytotoxic proteins (e.g. gelonin) can be in the nanomolar (10^{-9} M) to picomolar (10^{-12} M) range, compared to micromolar (10^{-6} M) range for most small-molecule drugs (e.g. carmustine > 100 μM). The clinical utility of macromolecular agents, however, has been hindered by their poor plasma pharmacokinetics, their tendency to induce an immune response, poor tissue selectivity, and their inability to penetrate cellular membranes and reach the interior regions of cells, which may be required for agents that exert their therapeutic effect intracellularly. The size and surface properties of MNPs can be designed to improve plasma pharmacokinetics, to achieve target selectivity, and to also provide a means for nanoparticle internalization into cells, and, thus, any conjugated macromolecular agent. The few reports detailing the successful in vivo delivery of macromolecules entail the successful delivery of survivin small interfering RNA (siRNA) to human colorectal cancer xenografts and the delivery of BIRC5 siRNA to human breast cancer xenografts (Medarova, Pham, Farrar, Petkova, & Moore, 2007, Kumar, Yigit, Dai, Moore, & Medarova, 2010).

The goal of drug delivery is to achieve targeted delivery of therapeutic molecules to diseased cells in the body, while limiting the exposure of normal cells. Many drugs are inactivated when attached to MNPs, which helps to minimize their toxicity against healthy tissue. Following accumulation of drug-loaded MNPs in the target tumor, however, it is necessary for the attached drug to be released in order for it to exert its therapeutic effect. Ideally, this is achieved by attaching a drug molecule to an MNP in such a manner that its release can be "triggered" by a process or condition that is unique to tumor tissue and not found in healthy tissues. In practice, however, the differences between tumor cells and normal cells are subtle. Therefore, most techniques under development attempt to utilize slight variations in physiological conditions or the overexpression of particular enzymes. In targeting cancer, for example, it has been shown that the microenvironment of most tumors is more acidic than normal physiological pH (pH ~7.4). The release of conjugated anticancer drugs can therefore be triggered by selecting chemical bonds that are broken under acidic conditions, see Figure 8.3. Examples of such covalent bonds include the esters, acetals, and hydrazones, among others. While a drug would remain attached to the MNP during circulation in the blood (pH ~7.4), it would be released once the MNP enters the acidic tumor microenvironment (pH < 7). Another

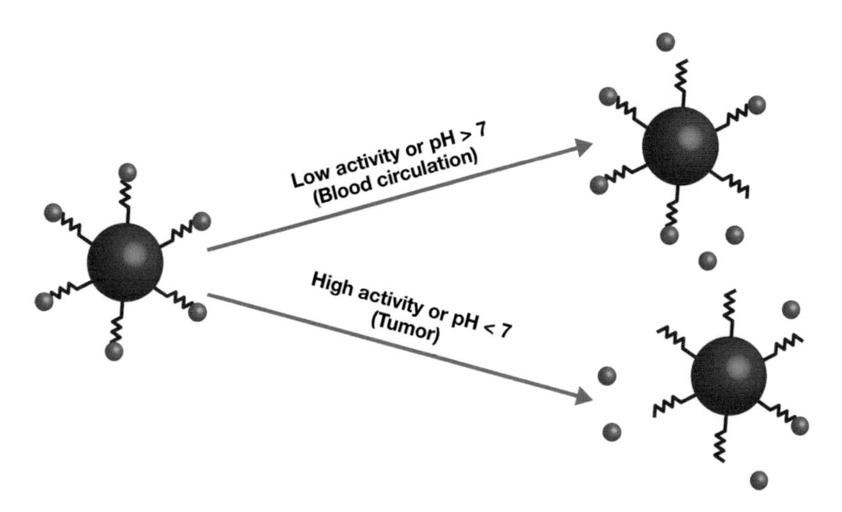

FIGURE 8.3 Illustration of the tumor-selective release of drug (smaller dots) attached to an MNP due to a more acidic pH or the presence of high protease activity in the tumor compared to systemic conditions.

approach is to utilize a coating, or drug, that undergoes a pH-mediated change in charge (i.e. negative-to-positive, or vice versa) or in hydrophilicity (i.e. hydrophobic-to-hydrophilic, or vice versa) in order to alter the interaction between an adsorbed drug and the MNP coating. In the ideal case, a drug would strongly adhere to the MNP while in systemic circulation, avoiding premature release, and then be released within the tumor due to reduced affinity of the drug for the MNP surface.

Still another approach is to utilize enzymes that are overexpressed in tumors, when compared to normal tissue, to initiate release of therapeutic molecules from an MNP. Several enzymes, found in greater concentrations within tumors, have been explored for tumor-specific drug release, including matrix metalloproteinase-7 (MMP-7), collagenase, and legumain.

8.4 STEPS TO DRUG DELIVERY

8.4.1 Routes of Administration

A goal of using MNPs for cancer therapy is to maximize drug delivery to the tumor while minimizing side effects. We will survey methods of drug delivery, describe conditions that affect circulation half-life of an MNP in the blood, and discuss methods to increase specific delivery to the tumor. Of course, the MNP must first be introduced into the body before any drug delivery could take place. There are many possible routes of administration for a therapeutic agent, including oral administration, inhalation, transdermal administration, and through direct injection (see Figure 8.4). Each administration route has its own set of advantages and drawbacks, which are also briefly discussed.

8.4.1.1 Oral Delivery

Oral delivery is a preferred route for the majority of therapeutic agents because it is non-invasive, is generally accepted by patients, and, thus, promotes a high level of patient compliance. The gastrointestinal (GI) tract is comprised of the oral cavity, the stomach, and the small and large intestines, which together provide a large surface area for adsorption. The microenvironment along this route can vary greatly with regards to pH, water content, proteolytic activity, and flora. For example, the normal stomach has an acidic environment with pH as low as 1.2, while other sections of the GI tract are typically neutral to basic with pH as high as 8.5. While this diversity provides a potential means for microenvironment-sensitive

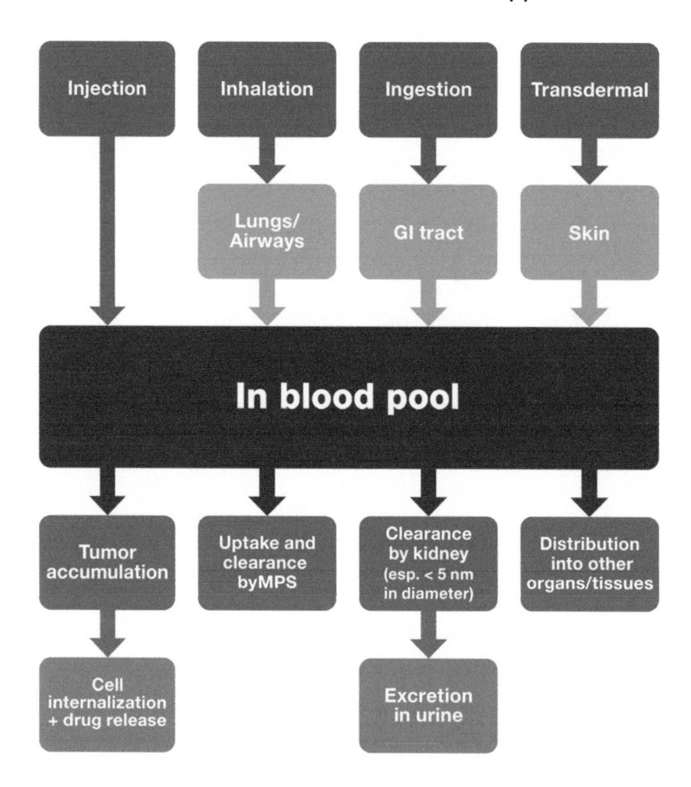

FIGURE 8.4 Pathways for drug administration, biodistribution, and clearance from the body.

targeting, it also raises a host of challenges for systemic drug delivery. In general, nanoparticles are thermodynamically unstable; that is, they have a tendency to aggregate into larger structures. The kinetics of this aggregation process can be very slow, and this allows for stable dispersions under certain conditions. A significant, rapid change in microenvironment, however, can induce instability and lead to the aggregation of MNPs. Moreover, because retention in the stomach can be up to several hours, any MNP that is susceptible to acidic degradation, as is generally the case for iron oxides, is unlikely to remain intact unless it is able to rapidly escape the stomach and pass into the intestine or be absorbed into the blood pool. Because all compounds absorbed into the circulation from the digestive tract enter the hepatic portal system, clearance of MNPs absorbed from the digestion may be subject to significant liver clearance.

While the oral route is preferred, there has been, to our knowledge, no report of systemic delivery of MNPs into the body via the GI tract. Nevertheless, the oral route of administration is mentioned here to bring awareness of its general importance in drug delivery. With ever increasing reports of methods for nanoparticle synthesis and functionalization, there is certainly future potential for the use of oral administration routes for systemic delivery, which may provide a substantial leap in the clinical viability of an MNP drug delivery vehicle. At present, though, published reports indicate that MNPs administered orally and retained in the GI tract with an external magnetic field, though not necessarily taken up systemically, can enhance the bioavailability of protein drugs (e.g. insulin) (Cheng et al. 2006, Laulicht, Gidmark, Tripathi, & Mathiowitz, 2011). Moreover, orally administered MNPs have also been utilized as MRI contrast agents to study the bowels (Fidler, Guimaraes, & Einstein, 2009).

8.4.1.2 Inhalation

With a surface area of approximately 70 m^2 in adult humans, the lungs offer a large surface area for potential MNP adsorption. The pulmonary route of delivery, however, demands considerable effort for the optimization of formulations, whether dry or suspended forms, to insure deep penetration into the lungs. There is also considerable concern for lung-associated toxicities (e.g. inflammation or blockage of airways) with this administration route. This would be especially true for nanoparticles that degrade slowly, as is typically the case with MNPs. The fact that deposition of particles in the lungs can cause adverse health effects is supported by both epidemiological and toxicity studies (Schmid et al. 2009). While larger particles remain in the lungs for extended periods, it has been shown that particles of size less than about 100 nm are able to translocate into the blood vessels and are found in secondary organs (e.g. brain and heart), potentially increasing toxicity concerns. Some studies, however, have indicated that lung toxicity may not be an insurmountable challenge. In one study, toxicity observed in mice after inhalation of fluorescent MNPs for four weeks was systemic in nature (i.e. weight loss, extramedullary hematopoiesis in the spleen, and changes in white blood cell level), yet without obvious pulmonary effects (Kwon et al. 2009). While this route of administration would seem an unlikely choice for repetitive dosing of MNPs, which may be required for treating chronic conditions, it may be feasible if only a single, small dose is required or if the lungs are the target tissue. Indeed, variations in the alveolar clearance of magnetic particles in patients with diseased lungs have been used to study the effects of smoking, sarcoidosis, idiopathic pulmonary fibrosis, and chronic obstructive bronchitis, along with the effects of age. Inhaled MNPs, therefore, also provide the potential for non-invasive diagnosis of lung-associated diseases (Moller et al. 2001).

8.4.1.3 Transdermal Delivery

Transdermal delivery of nanoparticles can be challenging due to the natural barrier properties of the skin. The skin is divided into four separate layers: the stratum corneum, epidermis, dermis, and the subcutaneous connective tissue. Making up the outermost layer, the stratum corneum is composed of dead and dehydrated cells that form a protective barrier. Below the stratum corneum lies the epidermis, which contains several layers of primarily keratinocytes along with several other cell types. In the third layer, the dermis, we find the presence of blood capillaries, lymphatic vessels, and nerve endings. For any systemic delivery of MNPs to occur through the transdermal route, the nanoparticles would have to navigate into the dermis and then penetrate into the residing blood vessels. Another route for penetration is through the orifices of the hair follicles. It has been demonstrated that MNPs smaller than 10 nm can penetrate through the stratum corneum and reach the epidermis, but that they aggregate prior to reaching the dermis layer (Baroli et al. 2007). The addition of an applied external field to provide magnetic force on MNPs could also enhance the penetration of nanoparticles into layers of the skin. For now, though, MNPs appear to be limited to topical delivery by this route of administration and do not seem ready for systemic delivery through the transdermal route. This may change in the future with greater understanding of particle/biological system interactions and improved design of MNP surface properties.

8.4.1.4 Direct Injection

Injection of nanoparticles directly into the body is the most effective means of their systemic delivery. Direct injection of MNP into the system provides the greatest control of dosing, avoiding the uptake barriers present with other routes, and also gives the highest bioavailability since the carrier is introduced either directly into, or where it can more readily access, the blood flow. The injection site could be, among others, intravenous, intra-arterial, intraperitoneal, intrathecally, subcutaneous, intramuscular, or directly into selected tissues (e.g. intratumoral). The intra-arterial route is often selected when it is desired to perfuse a particular tissue with a high dose and/or avoid first-pass clearance. A number of studies evaluating MNPs for treatment of tumors have also utilized direct intratumoral injections. This method, however, is not universally feasible for all tumors as many are deep-seated within the body and difficult to target externally. Among all these routes of administration, intravenous injection is the most frequently utilized, for, among other reasons, its validated use in clinical settings.

8.5 CELLULAR INTERNALIZATION OF MNPS

In some cases, once the MNP reaches the target tissue, it is desirable for the MNP to internalize within target cells, either into the cytoplasm or even within the nuclear envelope. This is often true when the attached drug cargo acts on intracellular processes and, either the drug itself is unable to cross the cellular membrane, or the target cell has developed resistance against the drug and rapid delivery of a high dose is required. Molecular targeting ligands, such as antibodies, can often also be used to facilitate internalization of MNPs into cells, primarily via endocytosis (Veiseh *et al.* 2010).

Endocytosis, meaning "within" (endo) "cell" (cytosis), is a process by which larger chemical species like macromolecules or nanoparticles gain access to intracellular compartments of cells without traversing through the cell membrane (Mellman 1996). This process plays a critical role in the retention of nanoparticles in diseased tissues. Generally, endocytosis is subdivided into three categories: pinocytosis, phagocytosis, or receptor-mediated phagocytosis. In each of the three cases, an intracellular vesicle is generated by invagination of the plasma membrane; they differ, however, in the exact mechanism by which the process occurs. Pinocytosis (meaning "cell drinking") is a process by which the plasma membrane forms an invagination and takes up any water-soluble substance found in the vicinity. During phagocytosis, the cell changes shape by jetting out projections called pseudopodia that engulf nearby macromolecules or nanoparticles. In receptor-mediated endocytosis, specific macromolecules are taken up by receptor-peptide, protein-protein, antigen-antibody, or other receptor-ligand interactions. Receptors on the targeting protein specifically bind their ligand on the extracellular surface of cells and trigger their endocytic internalization. Once within a vesicle, the trapped nanoparticles can be released from the vesicle and exposed to the intracellular milieu by a few methods. One of the most common mechanisms utilized involves increasing the acidic environment generated within vesicles after internalization, leading to rupture of the vesicles and release of their contents into the cell. This strategy is often found in nature, as with the pathogen listeria, which uses this method to gain entry into cells. In another approach, multifunctional nanoparticles have exploited the interaction between F3-peptide and nucleolin to achieve targeting to the nucleus of cells (Reddy et al., 2006). An alternative approach to cellular internalization of MNPs is to use highly cationic polymers, such as cell-penetrating peptides (e.g. HIV-TAT; low-molecular-weight protamine [LMWP]) or synthetic polymers (e.g. polyethylenimine). The cationic nature of these agents has been shown especially useful in ferrying MNPs into cells (Suh et al. 2009, Torchilin 2008, Medarova et al. 2007, Veiseh et al. 2010, Chertok, David, and Yang 2010). It should be noted, however, that while highly efficient at transporting MNPs across cell membranes, these cationic agents do not act with any specificity for cell types (Schwarze et al. 1999). In addition, due to their rapid clearance from blood circulation, the exposure of target tissue to MNPs having a high surface charge can be low, as discussed in the next section.

8.6 PHARMACOKINETICS AND THE MONONUCLEAR PHAGOCYTE SYSTEM

In many cases, MNPs rely on the circulatory system (i.e. blood flow) in order to reach the target tissue. Nanoparticles, and indeed any other substance introduced into the body, have several possible fates; they may: (1) enter the blood circulation, (2) distribute into body tissues, (3) be degraded, or (4) be excreted from the body. This is often termed ADME, for absorption, distribution, metabolism, and excretion. The rate and extent to which each of these processes occurs determines the pharmacokinetics of the particle. Pharmacokinetics, a quantitative description of how a substance travels through the body, how the body acts upon it, and how it is eliminated from the body, are therefore an important consideration for design of MNP-based drug delivery systems. Surface properties and size of the MNP have the greatest impact on the particle's pharmacokinetics (Yoo, Chambers, and Mitragotri 2010). These properties determine the extent of MNP interaction with the mononuclear phagocyte system (MPS), which is a part of the immune system comprised of a group of cells whose primary function is to eliminate foreign particles from the body. Note that in some texts, the MPS is referred to as the "reticuloendothelial system (RES)," but this is an older term that is no longer in favor as it gives incorrect prominence to endothelial cells, which do not play a major role in this process.

The primary components of the MPS include opsonin proteins circulating in the blood, macrophages, and cells in the liver, spleen, and bone marrow. Opsonins are a group of proteins (including certain antibodies) circulating in the blood stream that make foreign bodies more susceptible to phagocytosis by macrophages. Macrophages, derived from Greek words meaning "big eaters," are a type of white blood cell that ingest and then digest foreign particles and microorganisms. MNPs that enter the blood stream are coated by opsonin proteins and then taken up by activated macrophages. In general, particles with highly charged surfaces are more quickly adsorbed from plasma and, thus, display substantially shorter circulation times when compared to neutrally charged MNPs (Chertok, David, and Yang 2010).

Modification of the nanoparticle surface can be used to exert resistance to MPS interactions. Surface modification of MNPs with poly(ethylene glycol) (PEG), or "PEGylation," is a commonly utilized method that provides steric resistance to opsonization and macrophage uptake. PEG is, generally, a linear molecule that interferes with interactions between an MNP and both opsonins and macrophages. One commonly used measure of this interaction between MNPs and the body system is the plasma half-life. The plasma half-life is defined as the time required for the concentration of MNP in blood plasma to drop to half of its initial equilibrium value. MNPs that are rapidly opsonized and cleared have a short half-life, while MNPs that remain in circulation for long periods have a long half-life. As an example, the circulation half-life of starch-coated MNPs (170 nm hydrodynamic diameter), measured in rats, was increased from less than 10 minutes to almost 12 hours after PEGylation (Cole et al. 2011b).

With respect to size, MNPs of hydrodynamic diameter in the range of 10–100 nm seem to provide the ideal pharmacokinetics for drug delivery (Gupta and Wells 2004). Nanoparticles smaller than 10 nm are able to cross the endothelial linings that define the blood vessels and extravasate into normal tissues, and are also subject to rapid renal (kidney) clearance. On the other hand, MNPs larger than 100 nm tend to be more rapidly eliminated from blood via the MPS, although this depends upon several factors such as particle elasticity and shape. Microparticles with low elastic modulus can mimic red blood cells (RBC) and have much longer circulation times than stiffer particles. Indeed DeSimone's group showed that RBC-shaped polymer microparticles display a two-component pharmacokinetic profile with an elimination half-life of three hours for hydrogels with a modulus of 64 kPa, and a half-life of 93 hours for hydrogels with a modulus of 8 kPa, achieved using less crosslinker (Merkel et al. 2011). In addition, particles can be encapsulated in RBC ghosts to prolong their circulation (Cinti, Taranta, Naldi, & Grimaldi, 2011). Independent variation of MNP size and surface properties can be utilized to achieve the desired MNP pharmacokinetics. MNPs with longer circulation half-lives have greater exposure to target tissues, potentially improving targeting efficiency and, thereby, also enhancing therapeutic efficacy. The rate of drug release from the MNP, however, must also be considered to avoid release into general blood circulation by long-circulating MNPs.

8.7 BIODISTRIBUTION OF MNPS

In addition to the pharmacokinetics, it is also very important to consider how the MNPs distribute throughout the body. Blood in the body circulates from arteries, which carry oxygen-rich blood away from the heart, into arterioles, which branch into the capillaries, and then returns toward the heart by passing through venules that lead into the veins. Because of their much greater surface area and relatively thin capillary walls, most of the exchange of chemical species, including MNPs, from blood into perfused tissues occurs at the capillaries. Capillaries are divided into three main groups: (1) Continuous capillaries, found in the central nervous system (e.g. blood–brain barrier) and skeletal muscles, are characterized by tight junctions between cells that limit diffusion to small molecules; (2) Fenestrated capillaries (Gerd, 1971), which contain small pores that provide limited permeation of proteins, and can be found in tissues such as the intestine and kidney; and (3) Sinusoidal capillaries, which have openings large enough to allow the passage of cells, including macrophages, and are typically found in the liver and spleen. The relative abundance of these different types of capillaries in various tissues and the physical properties of an MNP determines the distribution of MNP throughout the body. Often called biodistribution, this distribution of MNPs is an important indicator in evaluating the target selectivity and efficiency of an MNP-targeting approach. Moreover, biodistribution may also give clues to the potential of an MNP and/or targeting strategy for off-target toxicities (Chertok, David, & Yang, 2010).

A large fraction of an administered dose of MNPs is taken up in tissues of the MPS—mainly in the liver and spleen (Jain et al. 2008, Lee et al. 2010, Medarova et al. 2007, Kunzmann, Andersson, Thurnherr, Krug, Scheynius, and Fadeel 2011, Chertok et al. 2010). While liver uptake occurs by action of the Kupffer macrophage cells located in sinusoids, both macrophage and mechanical filtration processes occur in the spleen (Gupta and Wells 2004, Kunzmann et al. 2011). Significant accumulation of MNPs in any non-targeted tissue raises concerns about the potential for toxicity. Many MNPs on their own, however, do not show a significant median lethal dose (LD_{50}) (Gupta, Naregalkar, Vaidya, & Gupta, 2007). The LD_{50} is the dose at which a substance causes the death of half of the population to which it has been administered. This is a typical measure of toxicity determined in preclinical animal trials prior to introduction of a potential therapeutic compound to human patients. A lower LD_{50} indicates higher toxicity and places greater constraints on safe dose levels. As MNPs, specifically those derived from iron oxide, are generally accepted as safe, the toxicity concerns for MNP drug delivery systems typically arise from the agents attached to the nanoparticles.

The overall toxicity of drug-loaded MNPs can depend on a number of factors, including: synthesis procedures; purity; properties of any coatings; potency and specificity of attached drugs; drug release kinetics; other ligands (e.g. agents used for cellular internalization of the MNP); route of administration; and the biodistribution (Gupta et al. 2007, Kunzmann et al. 2011, Veiseh, Gunn, & Zhang, 2010). A thorough assessment of toxicity requires the use of several tests, including liver functionality, detection of signaling proteins (e.g. cytokines), histology, measurement of tissue oxidative stress, and blood counts (Jain et al. 2008, Lee et al. 2010, Medarova et al. 2007, Sun et al. 2010). The ratio of the concentration of a pharmaceutical agent that produces a therapeutic effect to the amount that generates toxicity is called the therapeutic index. Theoretically, an optimum therapeutic index ratio of infinity is achieved when the drug-loaded MNPs all reach the target tissue with none remaining in the normal tissue. In practice, the fraction of MNPs that reaches the target tissue is only a few percent, at best, and more typically well below 1% of the injected dose.

8.8 TARGETING MNPS TO TUMORS

8.8.1 Passive Targeting: The Aberrant Physiology of Tumors

The supply of oxygen, nutrients, growth factors, and the removal of cellular waste are all critical for the growth and survival of cells. While normal tissues are penetrated by blood vessels to provide access to these nutrients, this is not the case with newly formed solid tumors. These tumors, without blood supply, are restricted in their ability to grow to a diffusion-limited size of about 2 mm (Jain 1987). Further growth of cancerous tissue is enabled when tumor cells activate an angiogenic switch, whereby they overexpress proangiogenic factors to promote the formation of new blood vessels. These newly formed blood vessels deliver nutrients and other factors to the growing tumor mass and also enable the spread of cancer cells to distant sites in the body (metastasis). The process of new blood vessel formation is called angiogenesis (or tumor angiogenesis in the case of cancer). Tumor angiogenesis is significantly different from angiogenesis seen in normal tissue (such as occurs during the repair of injured skin tissue) and is considered abnormal. In normal tissue, the endothelial cells that form the walls of blood vessels are closely packed and present a barrier to the extravascular diffusion of MNPs. However, a major manifestation of the abnormality of tumor angiogenesis is the formation of blood vessels that have poorly formed walls, having gaps between the endothelial cells. Additionally, the diameter of most tumor vessels is irregular with an atypical pattern of branching that does not fit well into any usual classification of arterioles, capillaries, or venules. Even large-caliber vessels in tumors have thin, leaky walls that allow the extravascular movement of MNPs across the blood vessel walls. Combined with the typically observed limited draining of fluids from tumors, which limits MNP wash out, this provides a means for accumulation of MNPs within the tumor milieu (see Figure 8.5). Indeed, particles of up to 1 µm in size have been shown to penetrate certain types of solid tumors (Maeda 2010). The combination of increased retention due to a leaky vasculature and poor drainage of fluids is often referred to as enhanced permeation and retention (EPR). Moreover, the EPR effect is generally used as a synonym for passive targeting to tumor cells. The difference in vascular permeability between tumor and normal tissue provides a useful approach for targeted drug delivery using MNPs, as they are selectively retained within this environment.

Not only is the blood vessel formation abnormal in tumors, but the composition of the basement membrane of tumor vessels is also distinct from their normal counterpart. The basement membrane is formed primarily of collagen and other glycoproteins, and serves to envelop the endothelial cells, pericytes, and smooth muscle cells. This, together with the interstitial matrix, forms the extracellular matrix (ECM), which provides mechanical support to the cells. Compared to normal tissue, the ECM in tumors has an aberrantly higher density and stiffness (Jain 1987). It has also been demonstrated that tumors have a relatively higher interstitial fluid pressure (IFP). The dense ECM and high IFP in tumors can both serve as barriers that inhibit the free diffusion of MNPs into the

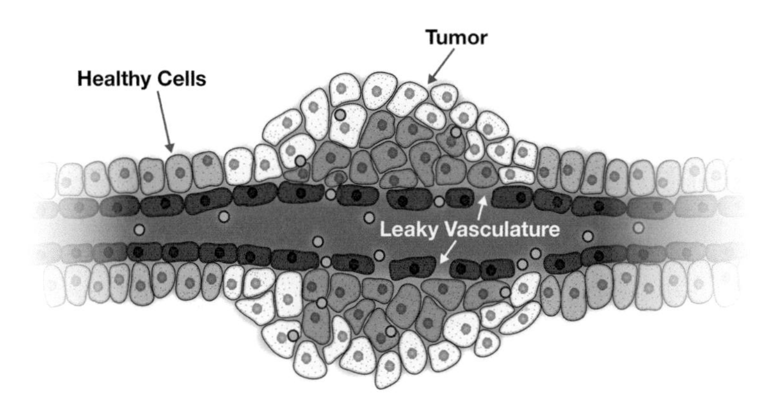

FIGURE 8.5 Illustration of passive targeting of MNP to tumors by the enhanced permeability and retention (EPR) effect. Abnormal growth in tumors results in fenestrations between the endothelial cells that form the walls of blood vessels. This enables MNPs to passively diffuse out of the blood vessel and into the tumor, but not into normal tissues that maintain an intact vessel wall.

tumor mass, thus limiting the penetration of most therapeutic molecules, and nanoparticles, to the periphery of the tumor volume close to the vasculature. In general, MNPs of smaller hydrodynamic diameter can be expected to penetrate the extracellular space more rapidly than similar larger particles. Surface properties and the strength of interaction between MNPs and cells and extracellular matrix are also important determinants of tumor penetration. MNPs that bind strongly to cells, or the matrix, tend to get "stuck" on the perimeter of the tumor, while those with a weaker interaction could penetrate deeper, but may also offer limited residence time for drug release within the tumor.

8.8.2 Active Targeting of Tumor-Specific Biomarkers

In addition to passive targeting, a number of active tumor targeting/retaining approaches, which take advantage of specific interactions between MNPs other materials or applied forces, are currently being investigated—including molecular targeting. Molecular targeting is achieved by conjugation of specific ligands that recognize and interact with unique molecular signatures on tumor cells—leading to increased accumulation of MNPs within the tumor microenvironment. Molecular targeting has been greatly assisted by sequencing of the human genome and the associated development of paradigm-changing expression-profiling technologies such as microarray, serial analysis of gene expression, subtractive proteomic mapping, and in vivo phage display. Using these modalities, libraries of tumor-specific biomarkers have been identified that can be exploited as tumor-specific targeting/retention moieties. In this approach, the MNP surface is modified with targeting ligands that specifically bind biomarkers that are, ideally, uniquely present on the surface of cancer cells or in the tumor microenvironment (see Figure 8.6); more commonly, the biomarker is simply expressed to a greater extent on cancer cells compared to normal cells. The large surface areas of MNPs enable the attachment of multiple targeting ligands that can significantly increase the binding strength with target cells, through multivalent binding (Veiseh *et al.* 2010). A wide variety of targeting ligands have been studied and are available for tumor targeting, including peptides, e.g. chlorotoxin (Veiseh et al. 2009), RGD (Montet, Montet-Abou, Reynolds, Weissleder, & Josephson, 2006), lung cancer targeting peptide (Guthi et al. 2010), CREKA (Simberg et al. 2007), bombesin (Martin et al. 2010), F3 (Zhang et al. 2009, Reddy et al., 2006a), A54 (Jiang et al. 2009), LHRH, antibodies (e.g. Anti-HER2 [John et al. 2010], Anti-EGFR/EGFRvIII [Hadjipanayis et al. 2010, Cho et al. 2010]), and small molecules (e.g. folate). Several antibodies approved as clinical drugs and which may also be capable of targeting MNPs are listed in Table 8.1.

TABLE 8.1
Targeting Antibodies Currently in Clinical Use

Targeting Antibody	Trade Name	Mechanism of Action
CD20, HER2/Neuregulin EGF receptor	Herceptin, Rituxan, HuMax-CD20 and HuMax-EGFr	Cell-mediated cytotoxicity
CD33	Myotarg	Toxin-mediated killing
CD40, CD137	Various	Agonist activity
CTLA4	MDX-010	Antagonist activity
EGF receptor	Erbitux	Blockage ligand binding
HER2/Neuregulin	Pertuzumab	Disruption signaling
VEGF	Avistatin	Angiogenesis inhibition

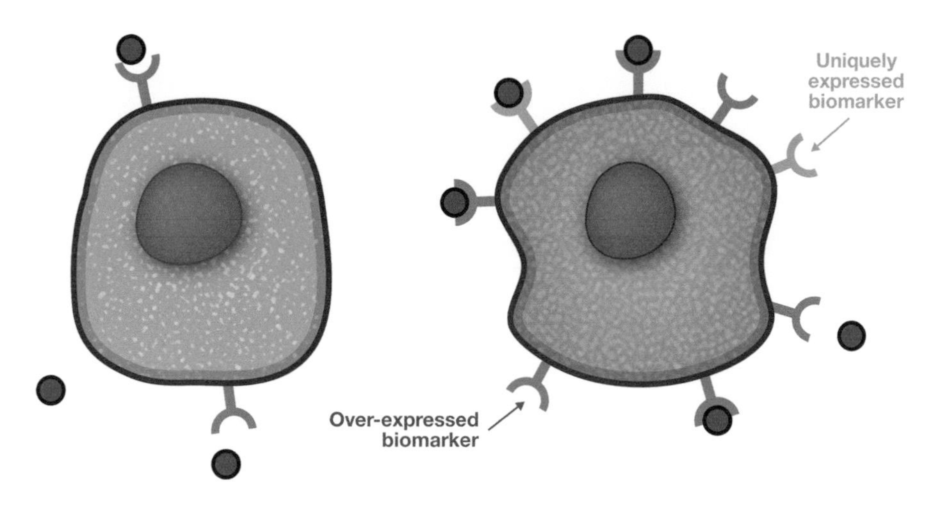

FIGURE 8.6 Molecular targeting of cancer cells. Molecular targeting of cancer takes advantage of biomarkers over-expressed or, preferentially, uniquely expressed on cancer cells compared to normal cells. MNPs are surface-decorated with targeting ligands (e.g. antibodies) that recognize and bind to these biomarkers.

8.8.3 Active Targeting with Magnetic Fields

The attraction of MNPs to a magnetic field offers a unique method for active targeting to tumors. Referred to as *magnetic targeting,* an externally applied magnetic field gradient is used to counteract the hydrodynamic drag force exerted on the particle by blood flow and to retain the MNP within a target site, see Figure 8.7. The magnetic force exerted on an MNP by an externally applied magnetic field (B) and its gradient (∇B) can be calculated according to the equation (Dobson 2006):

$$F_{magnetic} = \frac{\Delta\chi V_c}{2\mu_0} B(\nabla B) \tag{8.1}$$

where $\nabla\chi$ is the difference in magnetic susceptibility of the magnetic core and the medium, V_c is the volume of the magnetic core, and μ_0 is the magnetic permeability of free space. The retention of MNP occurs when the magnetic force is greater than the hydrodynamic drag force (Chertok, David, Huang, & Yang, 2007), which is described by Stoke's law:

$$\overrightarrow{F_{drag}} = 6\pi\eta r_p(\overrightarrow{v_f} - \overrightarrow{v_p}) \tag{8.2}$$

where η is the fluid viscosity, r_p is the particle radius, and v_f and v_p are the fluid and particle velocities, respectively.

A closer examination of these two equations allows us to realize a few important points regarding magnetic targeting. First, looking at Equation 8.1, we observe that not only is the magnetic force proportional to the strength of the applied field (greater magnetic field produces a larger magnetic force), but that a magnetic field gradient (∇B) is also necessary to generate a magnetic force. In order words, magnetic fields that are uniformly homogeneous in magnitude, although of high field strength, cannot be used to magnetically target MNPs to tissues in the body. In general, MNPs under the influence of a magnetic field gradient are deflected in the direction of increasing magnetic flux gradient. It is also apparent that the magnetic susceptibility of the MNP, which roughly is an indication of its response to the applied field, is an important parameter that affects magnetic targeting. This has led to considerable interest in

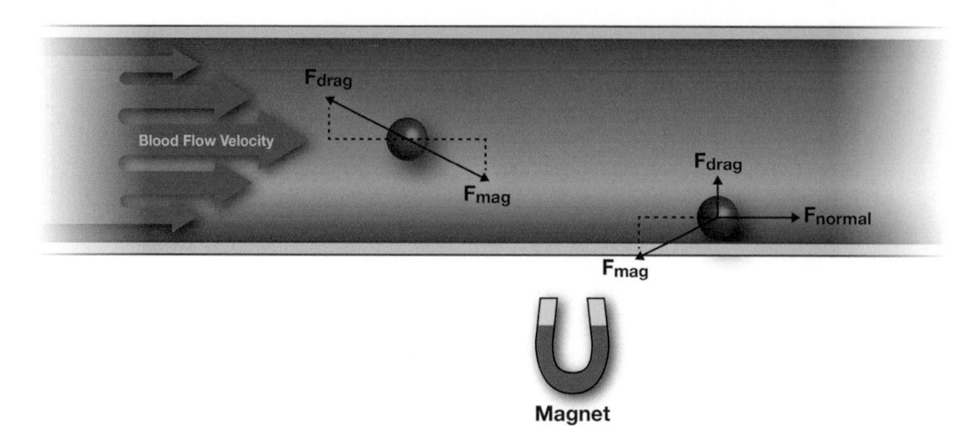

FIGURE 8.7 Schematic for the magnetic targeting of MNPs. The two primary forces that must be accounted for in magnetic targeting of MNPs are the drag force, exerted on the MNP by blood flow, and the magnetic force. Magnetic capture of the MNP occurs when the magnetic force is sufficiently strong to deflect the MNP to the capillary wall and retain the particle against the flow.

developing MNPs having greater susceptibilities to facilitate their capture *in vivo*. We also see that the magnetic force exerted on an MNP is proportional to the volume ($\sim r^3$) of its magnetic core, specified by V_c. MNP size, r_p, is also found in Equation 8.2 and indicates that the drag force experienced by the MNP also increases with size. However, because the magnetic force increases according to r^3 while the drag force increases only in proportion to r, MNPs with larger magnetic cores can typically be expected to accumulate to a greater degree than those with smaller ones. In general, MNPs optimal for magnetic targeting consist of a core diameter greater than about 100 nm (Goya, Grazu, & Ibarra, 2008).

The final parameter that must be considered from these equations is the fluid velocity (v_f), which determines the drag force exerted on the MNP. As previously mentioned, tumors possess a leaky vasculature due to their abnormal growth. In addition to their hyperpermeability, blood vessels in tumors also tend to have larger diameters (i.e. greater cross-sectional areas) than the vasculature in normal tissue. It has also been shown that the blood perfusion, or volumetric blood flow, in tumors is lower than that in normal tissues. The linear blood flow velocity, v, which is equal to the volumetric flow rate divided by the vessel cross-sectional area, in tumors is therefore found to be, on average, lower than the flow velocity in normal tissue. Because the linear blood velocity determines the drag force, according to Equation 8.2, the MNP can be more easily retained by a given magnetic field in the tumor compared to normal tissue, see Figure 8.8. It is this difference in MNP retention between tumor and normal tissues that enables the selective capture of MNP in tumors by magnetic targeting (Chertok et al. 2007, David, Cole, Chertok, Park, & Yang., 2011). It should be noted that as MNPs accumulate in a blood vessel there is a corresponding reduction in the cross-sectional area available for blood flow. This would result in an increase in blood flow velocity through this region of the vessel and thereby increasing drag on MNP, which could place a limit on total MNP accumulation.

While magnetic targeting has been applied to a number of animal tumor models (Dobson 2006), much of the recent work is related to brain cancer (Chertok et al. 2007, 2009, 2008, Chertok, David, and Yang 2010, David et al. 2011). Brain tumors can be especially challenging to treat by traditional surgical methods due to the risk of disrupting vital brain function in surrounding normal tissues and due to the presence of the blood–brain barrier (BBB). Magnetic targeting has been shown to enhance MNP accumulation in 9L glioma brain tumors in rats by five-fold compared to passive targeting by the EPR effect (Cole, David, Wang, Galbán, &

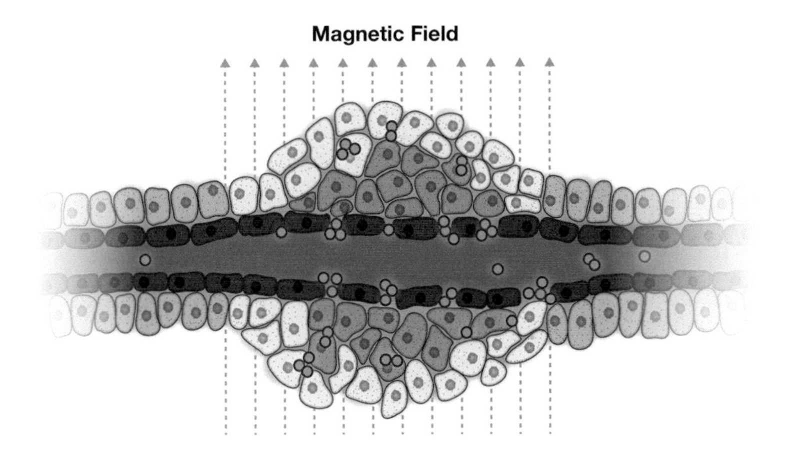

FIGURE 8.8 Active targeting of MNP to a tumor with magnetic targeting. MNPs, carried by blood flow into an applied magnetic field gradient, experience a magnetic force that retains them within the tumor lesion. Blood vessels in tumors, on average, have a significantly greater diameter and lower volumetric blood flow compared to those found in normal tissues—a difference that is critical for magnetic targeting. The corresponding higher drag force in normal tissue prevents MNP accumulation, while a sufficiently strong magnetic force would trap MNP within tumor lesions.

Yang, 2011a). Magnetic targeting has also been combined with focus ultrasound (to disrupt the BBB) in order to achieve greater targeting (Liu et al. 2010). The deep seating of tumors within the body can make them especially difficult to target magnetically, because magnetic flux density falls sharply with increasing distance for its source (Liu et al. 2010). Optimization of the magnetic field gradient for targeting deep-seated tissues within the human body is one of the challenges that must be addressed to enable clinical adoption of magnetic targeting approaches for drug delivery.

8.9 CONSIDERATIONS FOR THE FUTURE

Despite their promising diagnostic and therapeutic potential, clinical adoption of MNPs has been limited by a number of hurdles. Obtaining FDA approval for nanoparticle-based drug carriers, for example, can be difficult because administration route, biodistribution, drug dosing, size, shape, surface properties, etc. all factor into achieving successful treatment efficacy and toxicity. Polydispersity (e.g. a measure of the variation of nanoparticle size in a population of nanoparticles) in the size and shape of MNPs that result from current MNP manufacturing techniques, yields a heterogeneous product with poor batch-to-batch reproducibility. Such variability affects the predictability of each batch's dose-to-dose and patient-to-patient behavior. Indeed, predictability is an important factor in gaining regulatory approval of an agent. Additionally, with magnetic targeting of MNPs, the effects of magnetic fields on particle stability and biodistribution must also be considered, which can vary based on the applied field geometry and MNP surface and magnetic properties. Clinical translation will continue to be a significant challenge for nanoparticle-based drug delivery vehicles until we are able to reasonably predict their behavior in a patient.

The development of magnetic field gradients sufficient for magnetic targeting is another important challenge for successful magnetic targeting of MNPs. While the magnetic forces generated by current technology (e.g. 3 Tesla clinical MRI scanners) is sufficient to magnetically capture MNPs at deep-seated tissues within the human body, the challenge is to limit MNP accumulation in surrounding tissues; in other words, to achieve sufficient focusing of the magnetic force within the given target tissue volume. Because the strength of magnetic fields decreases rapidly with distance, most magnetic targeting applications have been limited to short distances from the magnet sources (e.g. subcutaneous targeting). While the task appears daunting, progress over the past few years has offered some hope for eventual therapeutic translation of this technology, as evidenced by the clinical trials listed in Table 8.2.

TABLE 8.2
Recently Initiated Clinical Trials Evaluating the Utility of MNPs in Diagnostic Applications*

Status	Start Date	Study Title	Condition	Sponsor
Recruiting	Dec. 2011	Plasmonic Photothermal and Stem Cell Therapy of Atherosclerosis With The Use of Gold Nanoparticles With Iron Oxide-Silica Shells Versus Biodegradable Stenting	Atherosclerosis	Ural State Medical Academy
Recruiting	Nov. 2010	Ferumoxytol for Magnetic Resonance Imaging of Myocardial Infarction	Myocardial Infarction	University of Edinburgh
Enrolling	Aug. 2010	Study of the Detection of Lymphoblasts by a Novel Magnetic Needle and Nanoparticles in Patients With Leukemia	Leukemia	University of New Mexico
Ongoing	Apr. 2010	Silica-Gold Nanoparticles and Mesenchymal Stem Cells Versus Composite Ferro-Magnetic Approach For Management of Atherosclerotic Plaque and Artery Remodeling	Atherosclerosis	Ural State Medical Academy
Recruiting	Jan. 2010	Inflammatory Cell Labelling and Tracking With Magnetic Resonance Imaging After Myocardial Infarction	Myocardial Infarction; Inflammation	University of Edinburgh
Recruiting	Oct. 2008	Assessing Dynamic Magnetic Resonance (MR) Imaging in Patients With Recurrent High Grade Glioma Receiving Chemotherapy	Brain Neoplasms	OHSU Knight Cancer Institute
Ongoing	Jul. 2008	Evaluation of Magnetic Nanoparticle Enhanced Imaging in Autoimmune Diabetes	Diabetes Mellitus, Type 1	Joslin Diabetes Center
Recruiting	Jul. 2008	Improved Pre-Operative Staging of Pancreatic Cancer Using Superparamagnetic Iron Oxide Magnetic Resonance Imaging	Pancreatic Cancer	Massachusetts General Hospital
Completed	Sept. 2007	Evaluation of Magnetic Nanoparticle Enhanced Imaging in Autoimmune Diabetes	Diabetes	Joslin Diabetes Center
Terminated	Jul. 2005	A Validation Study of MR Lymphangiography Using SPIO, a New Lymphotropic Superparamagnetic Nanoparticle Contrast	Bladder, Genitourinary, and Prostate Cancers	M.D. Anderson Cancer Center

Note:
* Data accessed from www.clinicaltrial.gov on March 2012.

REFERENCES

Arruebo, M., R. Fernández-Pacheco, M. R. Ibarra, and J. Santamaría. 2007. Magnetic nanoparticles for drug delivery. *Nano Today* 2 (3), 22–32.

Baroli, B., M. G. Ennas, F. Loffredo, M. Isola, R. Pinna, and M. A. Lopez-Quintela. 2007. Penetration of metallic nanoparticles in human full-thickness skin. *Journal of Investigative Dermatology* 127 (7), 1701–1712.

Bhojani, M. S., M. Van Dort, A. Rehemtulla, and B. D. Ross. 2010. Targeted imaging and therapy of brain cancer using theranostic nanoparticles. *Molecular ikPharmaceutics* 7 (6), 1921–1929. doi:10.1021/mp100298r.

Cheng, J., B. A. Teply, S. Y. Jeong, C. H. Yim, D. Ho, I. Sherifi, S. Jon, O. C. Farokhzad, A. Khademhosseini, and R. S. Langer. 2006. Magnetically responsive polymeric microparticles for oral delivery of protein drugs. *Pharmaceutical Research* 23 (3), 557–564. doi:10.1007/s11095-005-9444-5.

Cheng, L., C. Wang, X. Ma, Q. Wang, Y. Cheng, H. Wang, Y. Li, and Z. Liu. 2013. Multifunctional upconversion nanoparticles for dual-modal imaging-guided stem cell therapy under remote magnetic control. *Advanced Functional Materials* 23, 272–280. doi:10.1002/adfm.201201733.

Chertok, B., A. E. David, Y. Z. Huang, and V. C. Yang. 2007. Glioma selectivity of magnetically targeted nanoparticles: A role of abnormal tumor hydrodynamics. *Journal of Controlled Release* 122 (3), 315–323. doi:10.1016/j.jconrel.2007.05.030.

Chertok, B., B. A. Moffat, A. E. David, F. Q. Yu, C. Bergemann, B. D. Ross, and V. C. Yang. 2008. Iron oxide nanoparticles as a drug delivery vehicle for MRI monitored magnetic targeting of brain tumors. *Biomaterials* 29 (4), 487–496. doi:10.1016/j.biomaterials.2007.08.050.

Chertok, B., A. E. David, B. A. Moffat, and V. C. Yang. 2009. Substantiating in vivo magnetic brain tumor targeting of cationic iron oxide nanocarriers via adsorptive surface masking. *Biomaterials* 30 (35), 6780–6787. doi:10.1016/j.biomaterials.2009.08.040.

Chertok, B., A. J. Cole, A. E. David, and V. C. Yang. 2010. Comparison of electron spin resonance spectroscopy and inductively-coupled plasma optical emission spectroscopy for biodistribution analysis of iron-oxide nanoparticles. *Molecular Pharmaceutics* 7 (2), 375–385. doi:10.1021/mp900161h.

Chertok, B., A. E. David, and V. C. Yang. 2010. Polyethyleneimine-modified iron oxide nanoparticles for brain tumor drug delivery using magnetic targeting and intra-carotid administration. *Biomaterials* 31 (24), 6317–6324.

Cho, Y.-S., T.-J. Yoon, E.-S. Jang, K. S. Hong, S. Y. Lee, O. R. Kim, C. Park, Y.-J. Kim, G.-C. Yi, and K. Chang. 2010. Cetuximab-conjugated magneto-fluorescent silica nanoparticles for in vivo colon cancer targeting and imaging. *Cancer Letters* 299 (1), 63–71.

Chorny, M., I. Fishbein, B. B. Yellen, I. S. Alferiev, M. Bakay, S. Ganta, R. Adamo, M. Amiji, G. Friedman, and R. J. Levy. 2010. Targeting stents with local delivery of paclitaxel-loaded magnetic nanoparticles using uniform fields. *Proceedings of the National Academy of Sciences* 107 (18), 8346–8351. doi:10.1073/pnas.0909506107.

Cinti, C., M. Taranta, I. Naldi, and S. Grimaldi. 2011. Newly engineered magnetic erythrocytes for sustained and targeted delivery of anti-cancer therapeutic compounds. *PLoS One* 6 (2), 1-9.

Cole, A. J., A. E. David, J. Wang, C. J. Galbán, and V. C. Yang. 2011a. Magnetic brain tumor targeting and biodistribution of long-circulating PEG-modified, cross-linked starch-coated iron oxide nanoparticles. *Biomaterials* 32 (26), 6291–6301.

Cole, A. J., A. E. David, J. Wang, C. J. Galban, H. L. Hill, and V. C. Yang. 2011b. Polyethylene glycol modified, cross-linked starch-coated iron oxide nanoparticles for enhanced magnetic tumor targeting. *Biomaterials* 32 (8), 2183–2193.

David, A. E., A. J. Cole, B. Chertok, Y. S. Park, and V. C. Yang. 2011. A combined theoretical and in vitro modeling approach for predicting the magnetic capture and retention of magnetic nanoparticles in vivo. *Journal of Controlled Release* 152 (1), 67–75.

Dobson, J. 2006. Magnetic nanoparticles for drug delivery. *Drug Development Research* 67 (1), 55–60. doi:10.1002/ddr.20067.

Ferrari, M., A. P. Lee, L. J. Lee, and L. Josephson. 2006. Magnetic nanoparticles for MR imaging. In *BioMEMS and Biomedical Nanotechnology*, 227–237. Springer New York.

Fidler, J. L., L. Guimaraes, and D. M. Einstein. 2009. MR imaging of the small bowel. *Radiographics* 29 (6), 1811–1825. doi:10.1148/rg.296095507.

Gerd G. M. 1971. Structure and formation of pores in fenestrated capillaries. *Journal of Ultrastructure Research* 36 (5-6), 768–782.

Goya, G. F., V. Grazu, and M. R. Ibarra. 2008. Magnetic nanoparticles for cancer therapy. *Current Nanoscience* 4 (1), 1–16.

Gupta, A. K., and S. Wells. 2004. Surface-modified superparamagnetic nanoparticles for drug delivery: Preparation, characterization, and cytotoxicity studies. *IEEE Transactions on NanoBioscience* 3 (1), 66–73.

Gupta, A. K., R. R. Naregalkar, V. D. Vaidya, and M. Gupta. 2007. Recent advances on surface engineering of magnetic iron oxide nanoparticles and their biomedical applications. *Nanomedicine* 2 (1), 23–39. doi:10.2217/17435889.2.1.23.

Guthi, J. S., S. G. Yang, G. Huang, S. Z. Li, C. Khemtong, C. W. Kessinger, M. Peyton, J. D. Minna, K. C. Brown, and J. M. Gao. 2010. MRI-visible micellar nanomedicine for targeted drug delivery to lung cancer cells. *Molecular Pharmaceutics* 7 (1), 32–40. doi:10.1021/mp9001393.

Hadjipanayis, C. G., R. Machaidze, M. Kaluzova, L. Y. Wang, A. J. Schuette, H. W. Chen, X. Y. Wu, and H. Mao. 2010. EGFRvIII antibody-conjugated iron oxide nanoparticles for magnetic resonance imaging-guided convection-enhanced delivery and targeted therapy of glioblastoma. *Cancer Research* 70 (15), 6303–6312. doi:10.1158/0008-5472.can-10-1022.

Jain, R. K. 1987. Transport of molecules across tumor vasculature. *Cancer and Metastasis Reviews* 6 (4), 559–593. doi:10.1007/bf00047468.

Jain, T. K., M. K. Reddy, M. A. Morales, D. L. Leslie-Pelecky, and V. Labhasetwar. 2008. Biodistribution, clearance, and biocompatibility of iron oxide magnetic nanoparticles in rats. *Molecular Pharmaceutics* 5 (2), 316–327. doi:10.1021/mp7001285.

Jiang, J. S., Z. F. Gan, Y. Yang, B. Du, M. Qian, and P. Zhang. 2009. A novel magnetic fluid based on starch-coated magnetite nanoparticles functionalized with homing peptide. *Journal of Nanoparticle Research* 11 (6), 1321–1330. doi:10.1007/s11051-008-9534-5.

John, R., R. Rezaeipoor, S. G. Adie, E. J. Chaney, A. L. Oldenburg, M. Marjanovic, J. P. Haldar, B. P. Sutton, and S. A. Boppart. 2010. In vivo magnetomotive optical molecular imaging using targeted magnetic nanoprobes. *Proceedings of the National Academy of Sciences of the USA* 107 (18), 8085–8090. doi:10.1073/pnas.0913679107.

Kumar, M., M. Yigit, G. Dai, A. Moore, and Z. Medarova. 2010. Image-guided breast tumor therapy using a small interfering RNA nanodrug. *Cancer Research* 70 (19), 7553–7561. doi:10.1158/0008-5472.can-10-2070.

Kunzmann, A., B. Andersson, T. Thurnherr, H. Krug, A. Scheynius, and B. Fadeel. 2011. Toxicology of engineered nanomaterials: Focus on biocompatibility, biodistribution and biodegradation. *Biochimica et Biophysica Acta (BBA)-General Subjects*1810 (3), 361–373.

Kwon, J.-T., D.-S. Kim, A. Minai-Tehrani, S.-K. Hwang, S.-H. Chang, E.-S. Lee, C.-X. Xu, H. T. Lim, J.-E. Kim, B.-I. Yoon, G.-H. An, K.-H. Lee, J.-K. Lee, and M.-H. Cho. 2009. Inhaled fluorescent magnetic nanoparticles induced extramedullary hematopoiesis in the spleen of mice. *Journal of Occupational Health* 51 (5), 423–431.

Laulicht, B., N. J. Gidmark, A. Tripathi, and E. Mathiowitz. 2011. Localization of magnetic pills. *Proceedings of the National Academy of Sciences* 108 (6), 2252–2257. doi:10.1073/pnas.1016367108.

Lee, M. J.-E., O. Veiseh, N. Bhattarai, C. Sun, S. J. Hansen, S. Ditzler, S. Knoblaugh, D. Lee, R. Ellenbogen, M. Zhang, and J. M. Olson. 2010. Rapid pharmacokinetic and biodistribution studies using cholorotoxin-conjugated iron oxide nanoparticles: A novel non-radioactive method. *PLoS ONE* 5 (3), e9536.

Liu, H. L., M. Y. Hua, H. W. Yang, C. Y. Huang, P. C. Chu, J. S. Wu, I. C. Tseng, J. J. Wang, T. C. Yen, P. Y. Chen, and K. C. Wei. 2010. Magnetic resonance monitoring of focused ultrasound/magnetic nanoparticle targeting delivery of therapeutic agents to the brain. *Proceedings of the National Academy of Sciences of the USA* 107 (34), 15205–15210. doi:10.1073/pnas.1003388107.

Maeda, H. 2010. Tumor-selective delivery of macromolecular drugs via the EPR effect: Background and future prospects. *Bioconjugate Chemistry* 21 (5), 797–802. doi:10.1021/bc100070g.

Martin, A. L., J. L. Hickey, A. L. Ablack, J. D. Lewis, L. G. Luyt, and E. R. Gillies. 2010. Synthesis of bombesin-functionalized iron oxide nanoparticles and their specific uptake in prostate cancer cells. *Journal of Nanoparticle Research* 12 (5), 1599–1608. doi:10.1007/s11051-009-9681-3.

McBain, S. C., U. Griesenbach, S. Xenariou, A. Keramane, C. D. Batich, E. W. F. W. Alton, and J. Dobson. 2008. Magnetic nanoparticles as gene delivery agents: Enhanced transfection in the presence of oscillating magnet arrays. *Nanotechnology* 19 (40), 405102.

Medarova, Z., W. Pham, C. Farrar, V. Petkova, and A. Moore. 2007. In vivo imaging of siRNA delivery and silencing in tumors. *Nature Medicine* 13 (3), 372–377.

Mellman, I. 1996. Endocytosis and molecular sorting. *Annual Review of Cell and Developmental Biology* 12 (1), 575–625. doi:10.1146/annurev.cellbio.12.1.575.

Merkel, T. J., S. W. Jones, K. P. Herlihy, F. R. Kersey, A. R. Shields, M. Napier, J. C. Luft, H. Wu, W. C. Zamboni, A. Z. Wang, J. E. Bear, and J. M. DeSimone. 2011. Using mechanobiological mimicry of red blood cells to extend circulation times of hydrogel microparticles. *Proceedings of the National Academy of Sciences*. doi:10.1073/pnas.1010013108.

Moller, W., W. Barth, M. Kohlhaufl, K. Haussinger, W. Stahlhofen, and J. Heyder. 2001. Human alveolar long-term clearance of ferromagnetic iron oxide microparticles in healthy and diseased subjects. *Experimental Lung Research* 27 (7), 547–568. doi:10.1080/019021401753181827.

Montet, X., K. Montet-Abou, F. Reynolds, R. Weissleder, and L. Josephson. 2006. Nanoparticle imaging of integrins on tumor cells. *Neoplasia* 8 (3), 214–222. doi:10.1593/neo.05769.

Nacev, A., A. Komaee, A. Sarwar, R. Probst, S. H. Kim, M. Emmert-Buck, and B. Shapiro. 2012. Towards control of magnetic fluids in patients: Directing therapeutic nanoparticles to disease locations. *IEEE Control Systems* 32 (3), 32–74. doi:10.1109/mcs.2012.2189052.

Pal, S. K., and D. Bahadur. 2010. Shape controlled synthesis of iron-cobalt alloy magnetic nanoparticles using soft template method. *Materials Letters* 64 (10), 1127–1129. doi:10.1016/j.matlet.2010.01.086.

Polyak, B., I. Fishbein, M. Chorny, I. Alferiev, D. Williams, B. Yellen, G. Friedman, and R. J. Levy.. 2008. High field gradient targeting of magnetic nanoparticle-loaded endothelial cells to the surfaces of steel stents. *Proceedings of the National Academy of Sciences* 105 (2), 698–703. doi:10.1073/pnas.0708338105.

Reddy, G. R., M. S. Bhojani, P. McConville, J. Moody, B. A. Moffat, D. E. Hall, G. Kim, Y. E. L. Koo, M. J. Woolliscroft, J. V. Sugai, T. D. Johnson, M. A. Philbert, R. Kopelman, A. Rehemtulla, and B. D. Ross. 2006. Vascular targeted nanoparticles for imaging and treatment of brain tumors. *Clinical Cancer Research* 12 (22), 6677–6686. doi:10.1158/1078-0432.ccr-06-0946.

Rodríguez-Llamazares, S., J. Merchán, I. Olmedo, H. P. Marambio, J. P. Muñoz, P. Jara, J. C. Sturm, B. Chornik, O. Peña, N. Yutronic, and M. J. Kogan. 2008. Ni/Ni oxides nanoparticles with potential biomedical applications obtained by displacement of a nickel-organometallic complex. *Journal of Nanoscience and Nanotechnology* 8 (8), 3820–3827.

Scarberry, K. E., E. B. Dickerson, J. F. McDonald, and Z. J. Zhang. 2008. Magnetic nanoparticle-peptide conjugates for in vitro and in vivo targeting and extraction of cancer cells. *Journal of the American Chemical Society* 130 (31), 10258–10262. doi:10.1021/ja801969b.

Schmid, O., W. Möller, M. Semmler-Behnke, G. A. Ferron, E. Karg, J. Lipka, H. Schulz, W.G. Kreyling, and T. Stoeger. 2009. Dosimetry and toxicology of inhaled ultrafine particles. *Biomarkers* 14 (s1), 67–73. doi:10.1080/13547500902965617.

Schwarze, S. R., A. Ho, A. Vocero-Akbani, and S. F. Dowdy. 1999. In vivo protein transduction: Delivery of a biologically active protein into the mouse. *Science* 285 (5433), 1569–1572.

Simberg, D., T. Duza, J. H. Park, M. Essler, J. Pilch, L. L. Zhang, A. M. Derfus, M. Yang, R. M. Hoffman, S. Bhatia, M. J. Sailor, and E. Ruoslahti. 2007. Biomimetic amplification of nanoparticle homing to tumors. *Proceedings of the National Academy of Sciences of the USA* 104 (3), 932–936. doi:10.1073/pnas.0610298104.

Sjogren, C. E., C. Johansson, A. Naevestad, P. C. Sontum, K. Briley-Saebo, and A. K. Fahlvik. 1997. Crystal size and properties of superparamagnetic iron oxide (SPIO) particles. *Magnetic Resonance Imaging* 15 (1), 55–67. doi:10.1016/S0730-725X(96)00335-9.

Suh, J. S., J. Y. Lee, Y. S. Choi, F. Yu, V. Yang, S. J. Lee, C. P. Chung, and Y. J. Park. 2009. Efficient labeling of mesenchymal stem cells using cell permeable magnetic nanoparticles. *Biochemical and Biophysical Research Communications* 379 (3), 669–675. doi:10.1016/j.bbrc.2008.12.041.

Sun, C., K. Du, C. Fang, N. Bhattarai, O. Veiseh, F. Kievit, Z. Stephen, D. Lee, R. G. Ellenbogen, B. Ratner, and M. Zhang. 2010. PEG-mediated synthesis of highly dispersive multifunctional superparamagnetic nanoparticles: Their physicochemical properties and function in vivo. *ACS Nano* 4 (4), 2402–2410. doi:10.1021/nn100190v.

Sun, L., M. Zborowski, L. R. Moore, and J. J. Chalmers. 1998. Continuous, flow-through immunomagnetic cell sorting in a quadrupole field. *Cytometry* 33 (4), 469–475. doi:10.1002/(sici)1097-0320(19981201)33:4<469::aid-cyto11>3.0.co;2-6.

Torchilin, V. P. 2008. Cell penetrating peptide-modified pharmaceutical nanocarriers for intracellular drug and gene delivery. *Biopolymers* 90 (5), 604–610. doi:10.1002/bip.20989.

Tran, N., and T. J. Webster. 2010. Magnetic nanoparticles: Biomedical applications and challenges. *Journal of Materials Chemistry* 20 (40), 8760–8767. doi:10.1039/c0jm00994f.

Veiseh, O., C. Sun, C. Fang, N. Bhattarai, J. Gunn, F. Kievit, K. Du, B. Pullar, D. Lee, R. G. Ellenbogen, J. Olson, and M. Zhang. 2009. Specific targeting of brain tumors with an optical/magnetic resonance imaging nanoprobe across the blood–brain barrier. *Cancer Research* 69 (15), 6200–6207. doi:10.1158/0008-5472.can-09-1157.

Veiseh, O., F. M. Kievit, C. Fang, N. Mu, S. Jana, M. C. Leung, H. Mok, R. G. Ellenbogen, J. O. Park, and M. Zhang. 2010. Chlorotoxin bound magnetic nanovector tailored for cancer cell targeting, imaging, and siRNA delivery. *Biomaterials* 31 (31), 8032–8042.

Veiseh, O., J. W. Gunn, and M. Q. Zhang. 2010. Design and fabrication of magnetic nanoparticles for targeted drug delivery and imaging. *Advanced Drug Delivery Reviews* 62 (3), 284–304. doi:10.1016/j.addr.2009.11.002.

Weissleder, R., G. Elizondo, J. Wittenberg, C. A. Rabito, H. H. Bengele, and L. Josephson. 1990. Ultrasmall superparamagnetic iron oxide: Characterization of a new class of contrast agents for MR imaging. *Radiology* 175 (2), 489–493.

Yanai, A., U. O. feli, A. L. Metcalfe, P. Soema, L. Addo, C. Y. Gregory-Evans, K. Po, X. Shan, O. L. Moritz, and K. Gregory-Evans. 2012. Focused magnetic stem cell targeting to the retina using superparamagnetic iron oxide nanoparticles. *Cell Transplantation* 21 (6), 1137–1148.

Yoo, J. W., E. Chambers, and S. Mitragotri. 2010. Factors that control the circulation time of nanoparticles in blood: Challenges, solutions and future prospects. *Current Pharmaceutical Design* 16 (21), 2298–2307.

Zhang, Y., M. Yang, J. H. Park, J. Singelyn, H. Q. Ma, M. J. Sailor, E. Ruoslahti, M. Ozkan, and C. Ozkan. 2009. A surface-charge study on cellular-uptake behavior of F3-peptide-conjugated iron oxide nanoparticles. *Small* 5 (17), 1990–1996. doi:10.1002/smll.200900520.

9 Magnetic Particle Biosensors

Yunzi Li, Paivo Kinnunen, Alexander Hrin,
Mark A. Burns, and Raoul Kopelman

CONTENTS

9.1 INTRODUCTION

Biosensors are a set of relatively new technologies that have been used to detect the presence of various biological agents. Development of more effective biosensors has been a major area of research with the goal of obtaining methods for early detection of cancers, infections, and other diseases we are confronted with today. A few important parameters to consider when designing a biosensor are: speed (how quickly the biosensor assay can be performed), sensitivity (the minimum amount of biological target that can be detected), specificity (how well the biosensor detects only the target analyte), simplicity (the ease with which the assay can be performed), dynamic range (the range of concentrations of the target over which the biosensor is capable of reliably detecting the target), robustness (how well the assay can be reliably performed under varying conditions), amount of sample required (how much of the biological sample of interest is required for use with the assay), and multiplexing (the ability to detect multiple targets with a single assay).

In past decades, the research involving micro- and nanoparticle biosensors has grown by leaps and bounds. As understanding of these systems and the methods of synthesizing magnetic particles have steadily improved, magnetic particles have become more common as a tool in many avenues of this research. Magnetic particles offer a way to increase efficacy of biosensors by increasing the specificity and sensitivity of an assay through binding and enrichment of a target analyte. The magnetic particles are able to act as mobile capture vehicles that can bind the targets of interest and be quickly separated or manipulated with the use of an external magnetic field. It has been shown that at low concentration of biological analytes the sensitivity of a biosensor is limited by the binding affinity and binding rates (Sheehan and Whitman 2005). Using a large number of magnetic particles can increase the binding rates and total number of molecules bound to the sensors. When more targets are captured more quickly, it becomes simpler for the biosensor to exceed a signal-to-noise ratio (S/N) of three, which is generally considered to be the limit of detection for biosensors.

Iron oxide magnetic particles—either magnetite (Fe_3O_4) or maghemite (γ-Fe_2O_3)—are among the most prolific in areas of biological research due to their stability and biocompatibility

(Tamanaha et al. 2008). These iron oxides display a ferrimagnetic character, meaning that they display a domain structure, but with lower saturation magnetization than related ferromagnetic materials, due to the anti-parallel alignment of adjacent moments. The domain structure of ferrimagnetic materials results in superparamagnetic behavior when particle size is smaller than a single domain. A great deal of the work done with magnetic particles takes advantage of the superparamagnetic behavior of very small particles.

Initially, the most prevalent biological application of magnetic particles was in the isolation and separation of desired cellular components (Hsing, Ying, and Wenting 2007). Magnetic separation has been used to isolate nucleic acids (Obata et al. 2002), proteins, enzymes (Safarik and Safarikova 2004), and entire cells (Safarık and Safarıkova 1999) for study. Given that magnetic particles have been successful as a tool for isolating various systems of interest, there has been a recent push to discover ways in which the magnetic particles can be used in a way that also facilitates making some kind of measurement of the isolated system of interest. It is in this context that the use of magnetic particles as biosensors has arisen.

9.2 SURFACE MODIFICATION

Before we discuss magnetic particle biosensors, it is worth briefly mentioning the basic ways in which such synthetic particles are modified to interact with biological systems. In order to have specific, directed interactions with biological components of interest, the surfaces of magnetic particles are modified to facilitate the desired interactions. The first step in modifying the surfaces of magnetic particles is generally to encapsulate the magnetic material in some kind of coating. This encapsulation serves several purposes, the first of which is to increase the biocompatibility of the particles (Gupta and Gupta 2005). Without a coating of some kind, the particles will often form clusters due to hydrophobic interactions. Once these clusters are formed, they may start exhibiting dipole–dipole interactions, which in turn cause the clusters to become larger, which can be problematic if the particles are to be used as a biosensor. A hydrophilic or amphiphilic coating on the particles will help to prevent this kind of agglomeration. Another potential benefit of using biocompatible shells to encapsulate magnetic particles is the capacity to form micron-sized beads consisting of many superparamagnetic nanoparticles embedded in a polymer matrix. The popular Dynabeads from Invitrogen are precisely this kind of particle. The advantage of these magnetic beads is the fact that they contain a much higher density of magnetic material (and hence they have a faster response to external magnetic fields) while retaining the superparamagnetic character of the nanoparticles (Lea et al. 1988).

Many different materials have been used to provide surface coatings and matrices for magnetic particles. Both synthetic and biopolymers have been commonly used for this purpose. Among the most popular polymeric coatings have been polyethylene glycol (PEG), dextran, polyvinyl alcohol (PVA), and polyacrylamide; also, numerous non-polymer coatings, such as stearic acid, silica, and gold, have also been used to modify the surfaces of magnetic particles (Gupta et al. 2007). In some cases the encapsulating agents have been directly used as targets for biological interactions. Dextran coatings have been used to bind the biomolecule Concanavalin A in several different experiments, demonstrating the efficacy of magnetic particles as biosensors (Kriz, Radevik, and Kriz 1996). Rather than try to find a coating that can also be used as a targeting mechanism, most magnetic particles used in biological systems have the desired targeting ligands or receptor proteins directly attached to the surface of the magnetic particles themselves (Gupta et al. 2007). Among the many different proteins and ligands bound to the surface, two of the most common are biotin and avidin variants. The biotin–avidin bond is among the strongest of biological systems (Weber et al. 1989). Coating the particles with one of these two molecules allows further surface modification of generic sets of particles, for instance, magnetic beads coated with streptavidin can be incubated with any number of biotinylated molecules (antibodies, oligonucleotides) to set the desired targeting mechanism of the magnetic beads. This method of modifying magnetic particles has made for great versatility in their use as biosensors.

9.3 MAGNETIC PARTICLE BIOSENSORS

As previously mentioned, magnetic particles have been employed as key components in numerous diverse biosensors. While the way in which such particles have been used varies a great deal, magnetic particles have turned into successful biosensors because biological systems of interest do not contain any significant magnetic background that would be affected by the relatively small magnetic fields associated with magnetic particle biosensors (Haun et al. 2010). This means that magnetic particles can easily isolate targeted systems or biomolecules, but more importantly, it means that if measurements associated with the magnetic particles can be made reliably, there is no magnetic signal associated with the biological system that could contaminate the information from the magnetic particles. For this reason, even small signals associated with magnetic particles can reliably detect instances of biological interaction.

Since their first use as biosensors in the mid-1990s, magnetic particles have been used for biosensing in a wide variety of detection methods and platforms (Graham, Ferreira, and Freitas 2004, Haun et al. 2010, Hsing, Ying, and Wenting 2007, Kasatkin, Vasil'eva, and Murav'ev 2010, Katz, Willner, and Wang 2004, Kuramitz 2009, Megens and Prins 2005, Tamanaha et al. 2008). While the techniques have been numerous, magnetic particle biosensors can be divided into three categories by the functionality of the magnetic particles in these biosensors: magnetic particles as carriers for signal intensification, magnetic particles as labels to detect magnetic contents, and biosensing based on the dynamics of magnetic particles. While not every example fits perfectly into one category, these are the major ways in which magnetic particles are used as biosensors.

The first category uses the magnetic particles as a solid phase for capturing the target of interest and increasing the concentration of the target via magnetic separation. Some kind of label, generally non-magnetic, is used to report the presence of the target. For this category, the aspect that changes is the reporter—the system or method that is used to detect the presence of the target. The magnetic label method uses the magnetic particles themselves to report the presence of the target of interest. In these systems, it is the method of magnetic particle detection that changes throughout the various techniques. More recently, the dynamics of magnetic particles in a magnetic field have been widely studied and used to infer different physical and biological properties of biological samples as well as to detect the presence of the target, adding more dimensions to magnetic particle biosensors. Generally, the rotational motion of magnetic particles is measured optically, and properties/targets of interest can be quantified from the dynamic theory. The dynamic theory, thus the method of quantification, varies with the shape and quantity of magnetic particles used in these systems. Next we will discuss specific systems in each of the three categories. While several review articles have thoroughly discussed different aspects of theory and technology behind magnetic particle biosensors (Schrittwieser et al. 2016, van Reenen et al. 2014, Issadore et al. 2014, Rocha-Santos 2014), we focus on introduction of the technologies and their successful implementation in various applications.

9.3.1 MAGNETIC PARTICLES AS CARRIERS

In systems where magnetic particles are used as carriers, the magnetic particles do not play a direct role in the detection of the target. The essentially common part of all these systems is that the magnetic particles are first used to isolate the target of interest from a background of unwanted components. This might consist of a particular sequence of DNA, a particular protein or antibody, or even entire cells. After binding to the magnetic particles, the target can be isolated by magnetic separation of the particles from the other parts that would otherwise make it difficult to detect the target. What differentiates the various systems of detection is the label that is used to report the presence of the target. The reporter aspect of the sensor is crucial because it is the reporter element that allows the presence of the target of interest to be readily detected. The most common reporting methods are: electrochemical detection, chemiluminescent detection, fluorescent detection, bio-barcode detection, and agglutination assay. We'll take a brief look at each of these methods in turn. Table 9.1 provides an overview of the different kinds of reporter-label techniques, magnetic particles, targets, and reported sensitivities from a wide array of literature sources.

9.3.1.1 Electrochemical Detection

Electrochemical detection is among the broadest methods of reporting the presence of a biological target. The essential idea is that the system is set up in such a way that when the magnetic particles bind the target and are brought to an electrode, some kind of electrochemical reaction occurs that would not have happened if no target were present. The systems that employ this scheme are numerous and oftentimes seem to have little in common.

One method of electrochemical detection is by direct oxidation of the target (Erdem et al. 2006, Palecek, Fojta, and Jelen 2002). Figure 9.1 provides a scheme to demonstrate the direct oxidation process. Generally this method has been used to detect target nucleic acids. The first step is to isolate the target with the magnetic particles. Once the target has been secured, it can be removed from the magnetic particles or left attached, depending on the system. The target nucleic acids are placed into conditions that will facilitate the electrochemical oxidation of specific bases (purines usually). In this case, the magnitude of the signal from the oxidation will correspond to the amount of target present. This method is often described as "label free" because no other agents are bound to the target to detect the oxidation signal. See Figure 9.1 for the basic layout of this detection method.

Some other systems take advantages of the steric hindrance of antibody–antigen complex to block the electrochemical reaction of a redox probe, usually $K_3Fe(CN)_6$ (He et al. 2009, Zamfir et al. 2011, Yang et al. 2014). In these systems, the target binds to the surface of magnetic beads via antibody–antigen interaction. The complex blocks the active surface of the electrodes and inhibits the electron transfer between the anionic redox species and the electrode. As a result, the electric signal due to the electrochemical reaction decreases and its magnitude correlates with the concentration of the target.

Another method of detection is the electrochemical stripping of metal labels that bind to the target and the magnetic beads (Wang et al. 2001, Li et al. 2010). In this case, the target in fact binds metal labels to the magnetic beads before separation. The number of labels is proportional to the amount of target present. Next the metal label is "stripped" by an acid into metal ions. The stripped metals result in a detectable change in the signal, measured by the electrodes of the system. In addition to simple gold nanoparticles as metal tags, this technique has also been used in conjunction with iron-core gold nanoparticles (Wang 2003), semiconductor nanoparticles (Zhu et al. 2004), and gold nanoparticles enhanced by silver deposition (Wang, Polsky, and Xu 2001, Wang, Xu, and Polsky 2002). This method has also been used to detect multiple targets from a single sample, by relying on the fact that different labels result in different electrochemical signals (Liu et al. 2004). Metal stripping has been used to detect the presence of both nucleic acids and proteins.

A final way of using electrochemical signals to detect the presence of a target nucleic acid or protein is to utilize enzyme labels to start the electrochemical reaction (Gehring et al. 1996). Two common strategies to ensure high specificity and sensitivity are sandwich and competitive immunoassays. In sandwiched assays, we see that the target will attach some sort of active protein

TABLE 9.1
Systems Using Magnetic Particles as Carriers to Detect Biological Targets

Method	Magnetic Particles	Target	Reported Sensitivity	References
Electrochemical label-free	Dynabeads M-280 2.8 μm	DNA oligonucleotide	10 fM	Erdem et al. (2006)
Electrochemical probe ($K_3Fe(CN)_6$)	Ademtech S.A. 200 nm	Ochratoxin A	10 pg/ml	Zamfir et al. (2011)
Electrochemical probe ($K_3Fe(CN)_6$)	Xi'an Goldmag Nanobiotech 20 nm	Clenbuterol	0.2 pg/ml	Yang et al. (2014)
Electrochemical stripping (Au nanoparticle tags)	Bangs 0.8 μm	DNA oligonucleotide	2 nM	Wang et al. (2001)
Electrochemical stripping (Ag enhanced Au tag)	N/A	DNA oligonucleotide	32 pM	Wang, Polsky, and Xu (2001)
Electrochemical stripping (Ag enhanced Au tag)	N/A	DNA oligonucleotide	50 fM	Wang, Xu, and Polsky (2002)
Electrochemical stripping (Au coated Fe particles)	Bangs 10 μm	DNA oligonucleotide	50 ng/ml	Wang (2003)
Electrochemical stripping (ZnS label)	N/A	DNA oligonucleotide	0.3 pM	Zhu et al. (2004)
Electrochemical stripping (MW CNT w/alkaline phosphatase tags)	N/A	DNA oligonucleotide IgG	54 aM 6 fM	Wang, Liu, and Jan (2004)
Electrochemical label (β-galactoseidase)	Invitrogen 2.8 μm	IgG	4 fmol	Thomas et al. (2004)
Electrochemical labels (multiplexed)	Bangs 0.83 μm	B2-microglobulin/IgG/BSA/CRP	0.1 μg/μl	Liu et al. (2004)
Electrochemical label (alkaline phosphatase)	Dynabeads M-280 2.8 μm	IgG	16 ng/ml	Do and Ahn (2008)
Electrochemical label (lipase/ferrocene)	Invitrogen 1 μm	DNA oligonucleotide	20 amol	Ferapontova et al. (2010)
Electrochemical label (horseradish peroxidase)	29 nm (made in-house)	Carcinoembryonic antigen	1 μg/ml	Li et al. (2010)
Electrochemical label (horseradish peroxidase)	Streptavidin magnesphere 1 μm	Ochratoxin A	70 pg/ml	Bonel et al. (2011)
Electrochemical label (horseradish peroxidase)	N/A	Thrombin	30 fM	Zhao et al. (2011)
Electrochemical label (horseradish peroxidase)	Polyscience 1 μm	Interleukin-8	0.1 fM	Munge et al. (2011)
Electrochemical label (horseradish peroxidase and glucose oxidase)	360 nm (made in-house)	Glucose	10 μM	Chen et al. (2011)
Electrochemical label (invertase)	Bangs 1 μm	Cocaine Adenosine Interferon-γ Uranium	3 μM 18 μM 3 nM 9 nM	Li, Ji, and Liu (2011)

(Continued)

TABLE 9.1 (Continued)

Method	Magnetic Particles	Target	Reported Sensitivity	References
Electrochemical (horseradish peroxidase)	Xi'an Goldmag Nanobiotech 200–500 nm	H_2O_2	12 μM	Xin et al. (2013)
Microelectrode capacitance measurement	Invitrogen 2.8 μm	Salmonella typhimurium cells	10 CFU/ml	Yang and Li (2006)
Chemiluminescence (Ru based tag)	Dynabeads M-450	Carcinoembryonic antigen	0.2–0.4 μg/l	Blackburn et al. (1991)
		α-fetoprotein	0.4 μg/;	
		Digoxin	0.1 ng/ml	
		Thyrotropin	40 μIU/l	
		DNA oligonucleotide	10 copies	
Chemiluminescence (Ru based tag)	Invitrogen 2.8 μm	*Escherichia coli* O157 cells	10 cells	Shelton and Karns (2001)
Chemiluminescence (luminol)	CPG, Inc. 5 μm	DNA oligonucleotide	10 fM	Weizmann et al. (2003)
Chemiluminescence (luminol)	500 nm (made in-house)	DNA oligonucleotide	7 aM	Bi et al. (2010)
		Ramos cells	56 cells/ml	
Chemiluminescence (unspecified)	Polysciences 1.5 μm	Adenosine	5 nM	Yan et al. (2010)
		Cocaine	3 nM	
Chemiluminescence (luminol)	Polysciences (*unspecified size*)	DNA oligonucleotide	50 aM	Cai et al. (2010)
Chemiluminescence (luminol)	Tianjin BaseLine ChroTech Research Center, 3–4 μm	Cocaine	0.5 nM	Li et al. (2011)
Chemiluminescence (luminol)	N/A	Thyroid-stimulating hormone	0.8 μIU/l	Ng et al. (2012)
		17β-estradiol	21 pg/ml	
Chemiluminescence (luminol)	25–30 nm (made in-house)	Cry1Ac	0.3 pg/ml	Li et al. (2013)
Chemiluminescence (Ru based tag)	2.8 μm	Carcino-embryonic antigen	0.4 ng/ml	Zhou et al. (2014)
		DNA oligonucleotide	16 pM	
Fluorescent label (quantum dots)	Invitrogen 2.8 μm	TNF-α	6 fM	Agrawal, Sathe, and Nie (2007)
Fluorescent label (fluorescent dyes)	Invitrogen MyOne	Protein biomarkers (Interleukin 17 and 9 others)	10–100 pg/l	Todd et al. (2007)
Fluorescent label (quantum dots)	Invitrogen 2.8 μm	DNA oligonucleotide	250 zM	Liu et al. (2008)
Fluorescent label (fluorescein)	Dynabeads M-280	TNF-α	3 pM	Herrmann et al. (2008)
Fluorescent label (quantum dots)	Bangs 8.27 μm	DNA oligonucleotide	50 nM	Lim et al. (2009)
Fluorescent label (fluorescein)	Dynabeads M-270 2.8 μm	IgG	~0.1 μg/ml	Peyman et al. (2009)

(*Continued*)

TABLE 9.1 (Continued)

Method	Magnetic Particles	Target	Reported Sensitivity	References
Fluorescent label (quantum dots)	2.8 μm	C-reactive protein	10 pM	Zhu, Duan, and Publicover (2010)
Fluorescent label (quantum dots)	Invitrogen 2.8 μm	DNA oligonucleotide	890 zM	Kim and Son (2010)
Fluorescent (OliGreen Dye)	Polysciences (*unspecified size*)	Genomic DNA	180 fM	Wang et al. (2010)
Fluorescent label (FAM fluorophore)	Dynabeads M-270 2.8 μm	Adenosine	170 μM	Huang and Liu (2010)
Fluorescent label (β-galactosidase catalyzed fluorogenic reaction)	Varian 2.7 μm	PSA TNF-α	~200 aM ~600 aM	Rissin et al. (2010)
Fluorescent label (target catalyzed fluorogenic reaction)	Dynabeads MyOne 1 μm	Human neutrophil elastase Thrombin	100 fM 2 fM	Zhao et al. (2011)
Bio-barcode	Iron oxide 1 μm	PSA	3 aM	Nam, Thaxton, and Mirkin (2003)
Bio-barcode	Polysciences (*unspecified size*)	DNA oligonucleotide	500 zM	Nam, Stoeva, and Mirkin (2004)
Bio-barcode	Invitrogen (*unspecified size*)	PSA	10 fg/ml	Bao et al. (2006)
Bio-barcode (fluorescence)	Polysciences 1 μm	PSA	300 aM	Oh et al. (2006)
Bio-barcode	Invitrogen M-280	DNA oligonucleotide (multiplexed)	500 fM	Stoeva et al. (2006)
Bio-barcode	Invitrogen 1 μm	PSA/HCG/AFP (multiplexed)	170 fM	Stoeva et al. (2006)
Bio-barcode	Invitrogen 2.8 μm	Genomic DNA	3 fM	Hill, Vega, and Mirkin (2007)
Bio-barcode	Invitrogen 1 μm	PSA	330 fg/ml	Thaxton et al. (2009)

label to the magnetic particles (again in a quantity proportional to the amount of target present). Next, the enzymes will be brought to the electrode, where their activity will—either directly or indirectly—cause a change in the detectable electrochemical signal. Figure 9.2 shows an example of a sandwich immunoassay system in which an electrochemical label is used to detect the presence of a target. In competitive assays, the enzyme is displaced by the target and released to the solution to catalyze reactions of a substrate for electrochemical detection. Figure 9.3 shows an example of a system using competitive assay. Many different kinds of enzymes have been used, including: β-galactosidase (Thomas et al. 2004), alkaline phosphatase (Wang, Liu, and Jan 2004, Do and Ahn 2008), horseradish peroxidase (Li et al. 2010, Bonel et al. 2011, Munge et al. 2011, Chen et al. 2011, Xin et al. 2013), glucose oxidase (Chen et al. 2011), invertase (Xiang and Lu 2011) and a lipase (Ferapontova et al. 2010). The experimental techniques for electrochemical label detection are similar no matter what label is used.

Cellular growth has been detected electrochemically by monitoring the changes in impedance of the media as the cells grow and divide (Yang and Li 2006). After isolation and separation of the target cells by magnetic beads, the growth area is essentially inserted between two plates of a

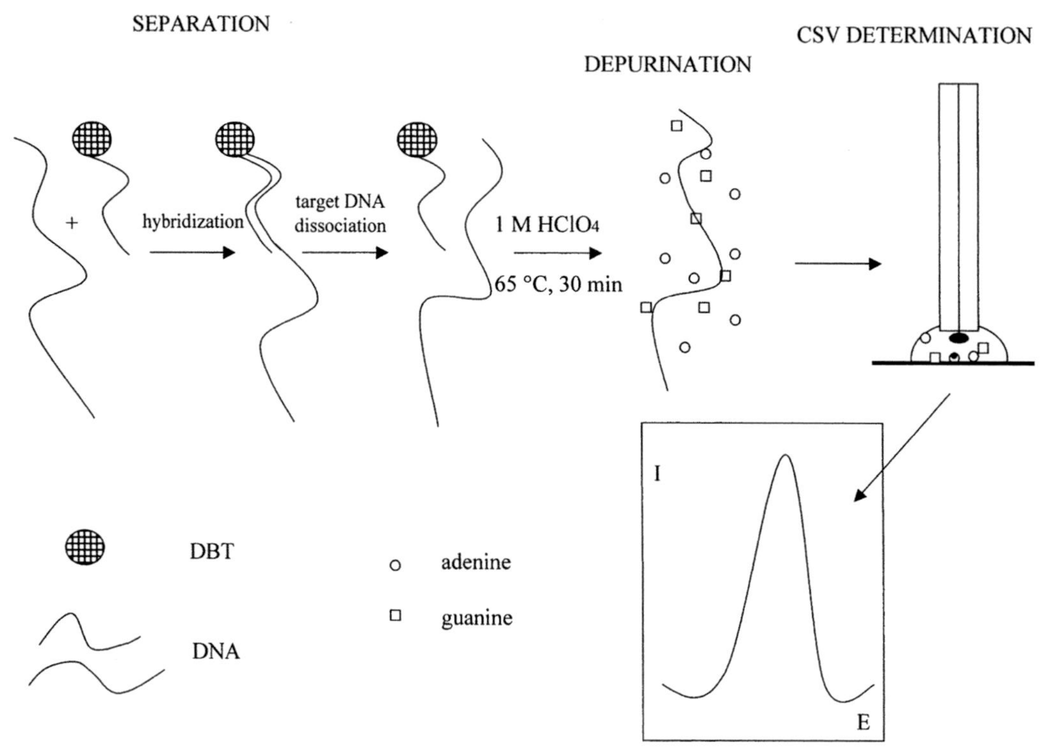

FIGURE 9.1 A basic schematic for the direct oxidation process. The magnetic particles are used to capture the DNA. The purine bases are stripped and the resulting signal can be detected electrochemically, reprinted with permission from (Palecek et al. 2002).

capacitor, and the cells are allowed to grow. The growth of the cells causes a change in the dielectric constant, in between the two plates of the capacitor. This change can be detected as a change in the impedance of the connected circuit.

9.3.1.2 Chemiluminescent Detection

A method somewhat related to electrochemical detection is chemiluminescent detection. Chemiluminescent detection utilizes an appropriate agent (usually luminol or Ru-chelate derivatives) to detect the presence of the target of interest. In this case, it is a chemiluminescent glow that is detected when the target of interest is present. However, there are generally intermediate steps required in order to generate the conditions under which the luminol will exhibit luminescence. It is in this fashion that chemiluminescent detection methods are similar to electrochemical detection methods. For instance, gold nanoparticles are sometimes used as

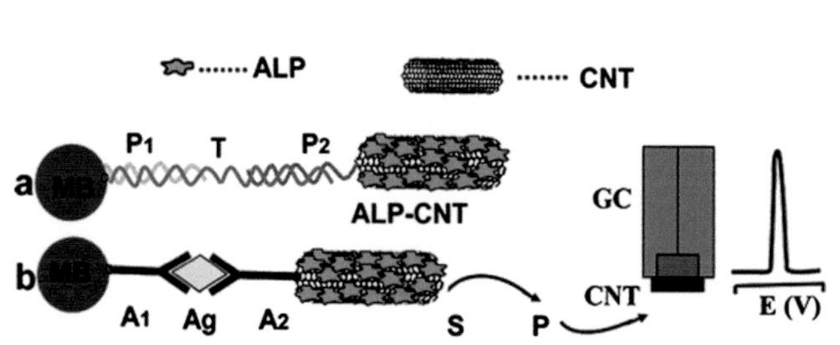

FIGURE 9.2 A sandwich immunoassay, where the target binds an enzyme-labeled carbon nanotube to the magnetic particle. The enzyme alkaline phosphatase catalyzes the electrochemical reaction of a substrate and the signal can be detected. The process by which electrochemical label detection occurs is essentially similar, no matter the label being used, reprinted with permission from (Wang et al. 2004).

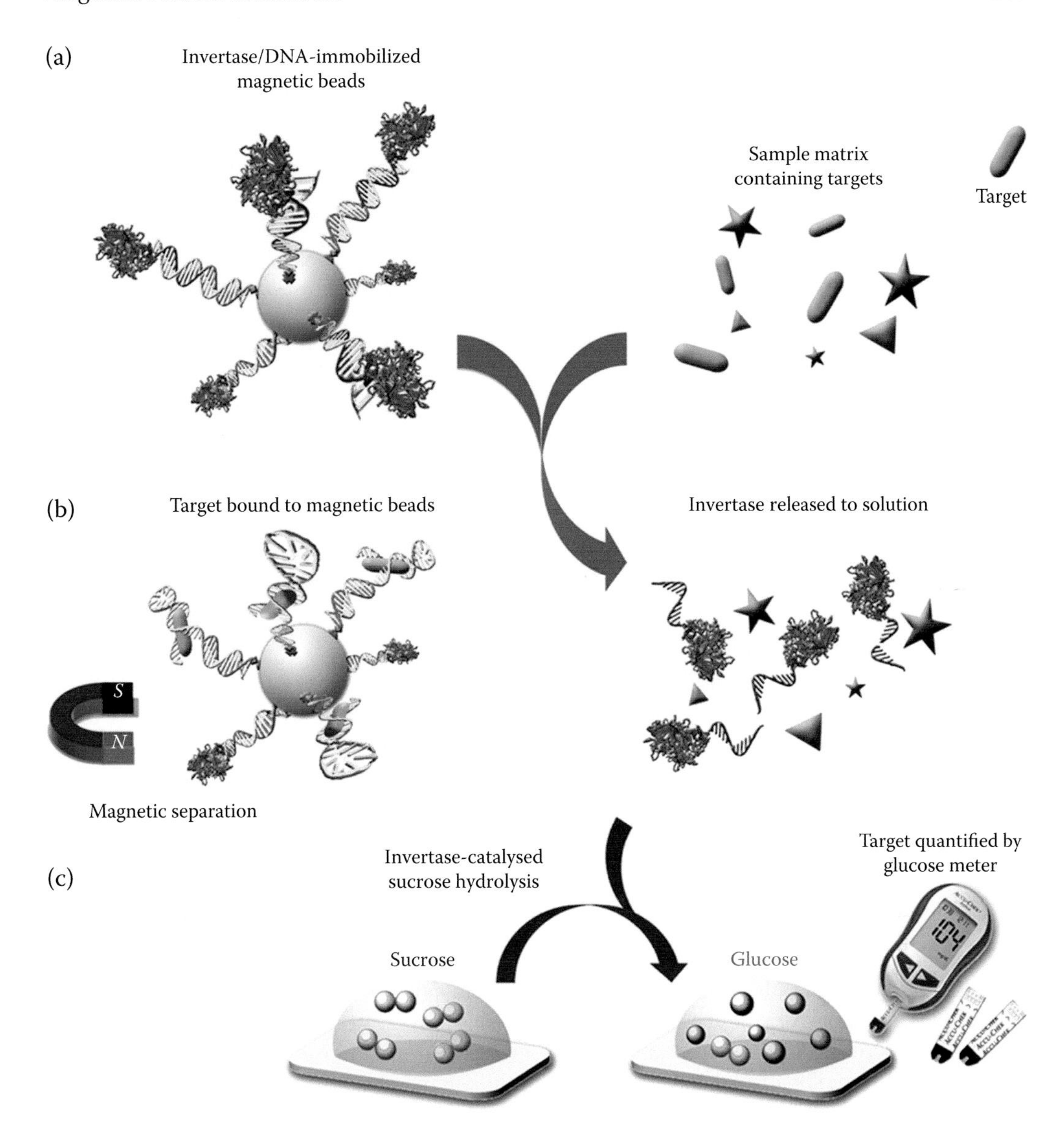

FIGURE 9.3 In this schematic, the target competes with enzyme invertase to bind to the magnetic beads. The enzyme releases and catalyzes the formation of glucose, the latter of which is then detected by an electrochemical glucose sensor, reprinted with permission from Springer Customer Service Centre GmbH (Xiang and Lu 2011).

tags that attach to the magnetic beads, in the presence of the target; they are then detected via a chemiluminescent reaction of luminol catalyzed by metal ions (Cai et al. 2010, Bi, Zhou, and Zhang 2010). Other methods have used enzymes as tags in order to generate the needed oxidizing reactants for luminol to emit its blue glow (Weizmann et al. 2003, Yan et al. 2010, Li, Ji, and Liu 2011, Ng et al. 2012). In some cases, it's an electrochemical reaction that enables the luminescence (Blackburn et al. 1991, Shelton and Karns 2001, Li et al. 2013, Zhou et al. 2014), and in this way, chemiluminescence is similar to electrochemical detection in that many similar designs and experimental setups can be used for each. Figure 9.4 shows the basic setup for a reaction chemiluminescent detection method.

9.3.1.3 Fluorescent Detection

Detection of biological analytes by means of fluorescence is one of the most commonly used techniques in the field today (D'Orazio 2003). The use of fluorescence in conjunction with magnetic particle isolation has potential in the sense that the magnetic particles could be used to isolate a small number of biomolecules on their surface, which could then be detected by means of fluorescence. Again, the idea is to use the target of interest to attach a fluorescent agent or an enzyme of a fluorogenic reaction to the magnetic particles. Effectively, the magnetic particles are locally enriching the concentration of the target such that the fluorescence is detectable. The target concentration can be quantified because the amount of fluorescence is proportional to the concentration of the target. Many different fluorescent labels have been used in conjunction with magnetic particle detection, including quantum dots (Liu et al. 2008), fluorescein (Huang and Liu 2010) and DNA-staining dyes (Wang et al. 2010). In some other systems, enzyme labels have been used to catalyze the reaction of a fluorogenic substrate in the solution (Rissin et al. 2010, Zhao, Li, and Le 2011). Figure 9.5 provides an example of how fluorescent labels have been used to detect target proteins (Agrawal, Sathe, and Nie 2007, Zhu, Duan, and Publicover 2010, Todd et al. 2007, Herrmann, Veres, and Tabrizian 2008, Peyman, Iles, and Pamme 2009, Rissin et al. 2010, Zhao, Li, and Le 2011), but the technique has also been used to detect the presence of nucleic acids (Kim and Son 2010, Lim et al. 2009, Huang and Liu. 2010).

9.3.1.4 Bio-barcode Detection

Bio-barcode detection is a technique developed by the Mirkin group (Hill and Mirkin 2006, Nam, Jang, and Groves 2007). The essential idea behind bio-barcode sensors is that rather than try to detect the target directly, one instead detects a pre-prepared oligonucleotide sequence corresponding to that target. The actual choice of sequence in this case is arbitrary, so long as the barcode can fulfill all the steps needed in the process. The initial bio-barcode research did not use magnetic particles, and relied on the variations in the C-G content of the barcode oligonucleotides to identify different binding analytes (Nam, Park, and Mirkin 2002). The use of magnetic particles made the bio-barcode technology far more effective. The magnetic particles are covered with the appropriate detection elements for the target of interest. After capturing and isolating the target with the magnetic particles, the system could be incubated with gold nanoparticles that also have the appropriate recognition element attached to the surface, which would form a sandwich complex with the total number of gold nanoparticles captured, in proportion to the amount of target present. In addition to the target recognition element, the gold nanoparticles have a large quantity—generally around 100× the number of target recognition elements—of barcode DNA on their surface. This means that every time a gold nanoparticle encounters the target, there will be a large increase in the amount of bio-barcode DNA. Figure 9.6 shows broadly how the bio-barcode method can be used to detect many relevant biological targets.

The barcode DNAs are generally detected using a scanometric method, also developed by the Mirkin group (Taton, Mirkin, and Letsinger 2000). The basic idea is that because the sequence of the barcode DNA is a portion of the complementary sequence, it can be bound to a surface, and gold nanoparticles can be attached to each of the DNA barcodes using the other portion of the complementary sequence. The final step is to catalytically deposit silver onto the gold

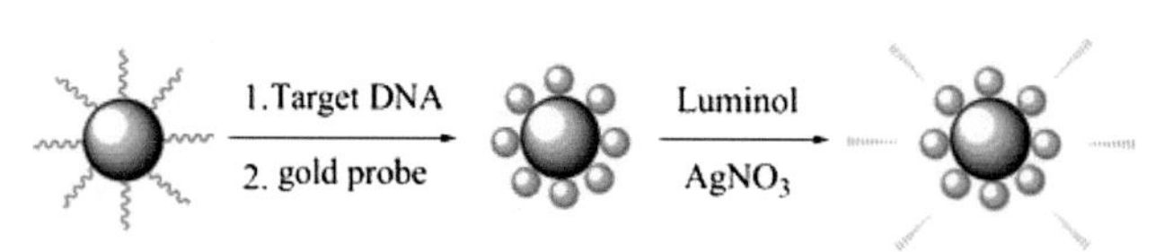

FIGURE 9.4 The basic setup for a chemiluminescent label is to use the target to bind a label that can facilitate a chemiluminescent reaction. Without the target there will be no luminescence, reprinted with permission from (Cai et al. 2010).

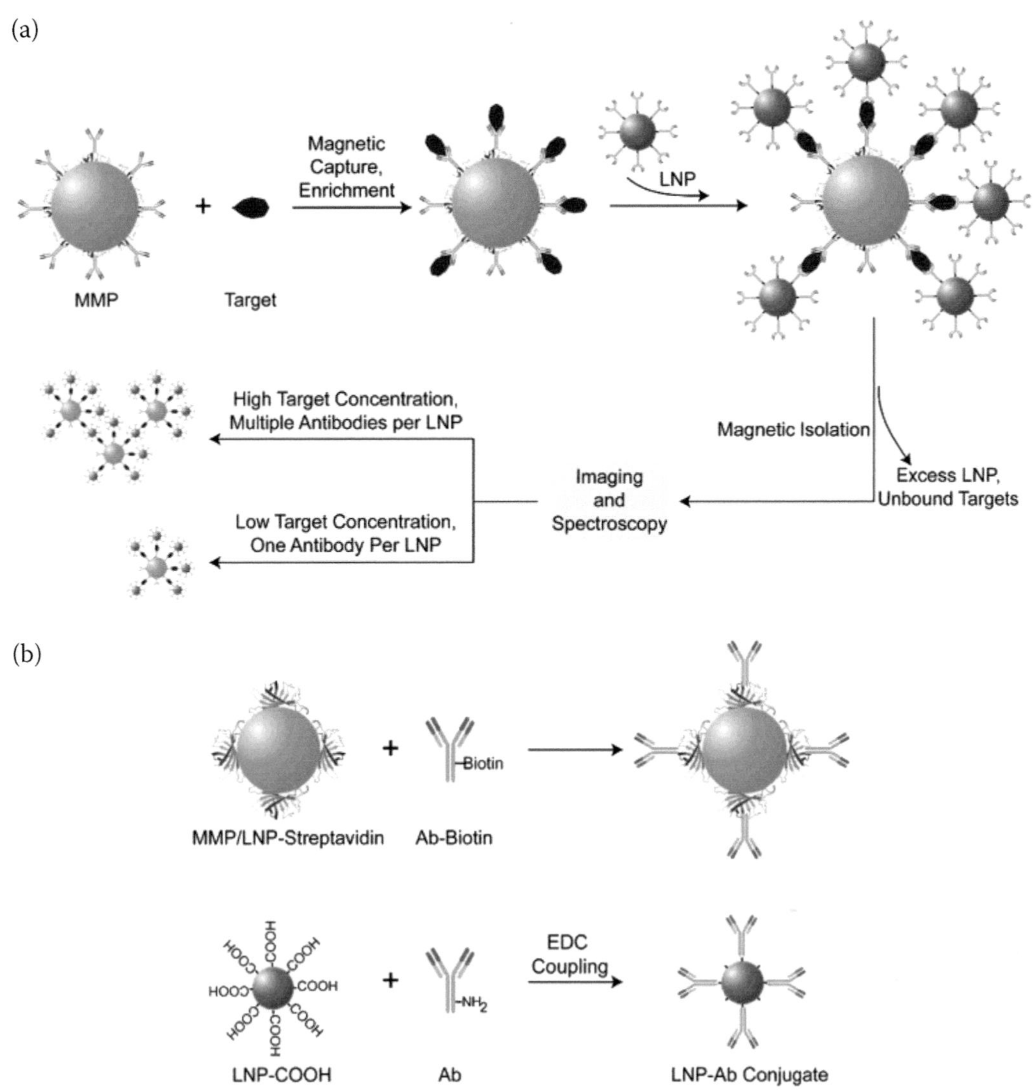

FIGURE 9.5 The most common fluorescent labels used to detect biological targets in conjunction with magnetic particles are quantum dots, reprinted with permission from (Agrawal et al. 2007).

nanoparticles attached to the surface through the barcode DNA. Once this has been done, the surface will be effectively stained a darker color, depending on the concentration of bio-barcode DNA. The extent of this staining corresponds to the amount of target in the initial system of interest.

Another method of detection of bio-barcodes is with fluorescence (Oh et al. 2006). In this setup, the barcodes are labeled with fluorophores before they are ever attached to the gold nanoparticle probes. Once the target has been isolated by the magnetic particles, the barcodes can be removed and the overall fluorescence of the solution will correspond to the concentration of the barcodes and, by extension, to the concentration of the target.

The bio-barcode method has demonstrated extreme sensitivity, on the order of 10 molecules for some targets (Nam, Thaxton, and Mirkin 2003). It has been successful in detecting low levels of biologically relevant proteins (Bao et al. 2006, Nam, Jang, and Groves 2007, Thaxton et al. 2009), oligonucleotides (Nam, Stoeva, and Mirkin 2004), and genomic DNA (Hill, Vega, and Mirkin

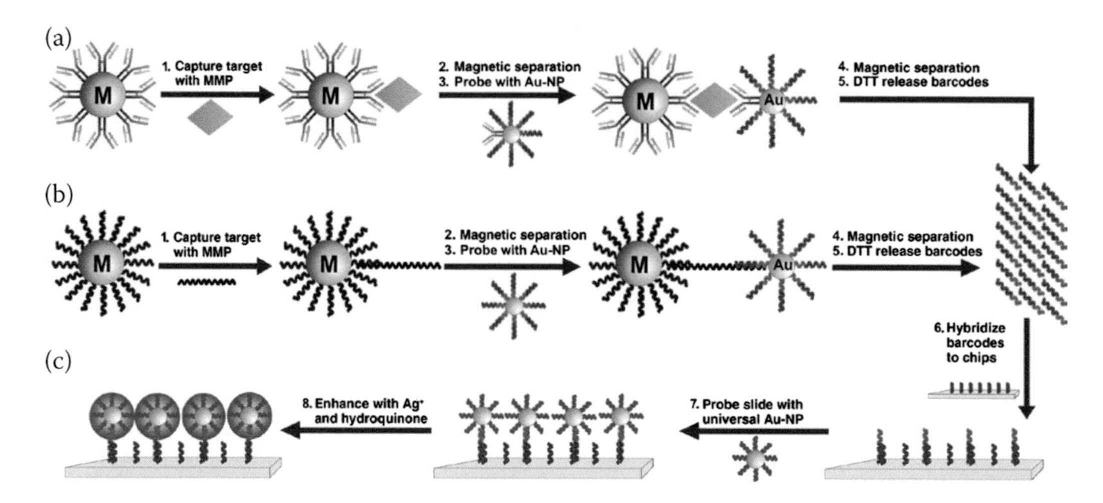

FIGURE 9.6 The bio-barcode method has proven both versatile in its ability to detect very minute amounts of both nucleic acids and proteins. The detection method is the same whether or not the setup is detecting DNA or a specific protein. Schematic representation of (a) protein, (b) nucleic acid and (c) scanometric detection using the bio-barcode assay. Au-NP, gold nanoparticles; MMP, magnetic microparticles, reprinted with permission from Springer Customer Service Centre GmbH (Hill and Mirkin 2006).

2007). Additionally, bio-barcodes have proven amenable to multiplexing, with the ability to detect multiple targets of both proteins (Stoeva et al. 2006) and DNA (Stoeva et al. 2006).

9.3.2 Magnetic Particles as Labels

The magnetic label methods of biosensing continue to rely on the capacity of magnetic particles to capture and isolate targets of interest from a mixture of unwanted elements. However, these types of sensors attempt to directly detect the magnetic particles that have bound the target to their surface. Magnetic label detection methods utilize systems in which the effects of the magnetic particles themselves are measured, rather than the effects of a related reporter. Like with methods where magnetic particles are used as carriers, the number of individual techniques reported in the literature is quite numerous, but in a broad sense, the methods of detecting magnetic particles fall into one of the following categories: anisotropic magnetic resistance (AMR); induction coils; superconducting quantum interference device (SQUID); and magnetic relaxation using nuclear magnetic resonance (NMR). In addition, in some of the literature, a simple CCD camera is used to detect the magnetic particles if they are detectable by such means (Morozov et al. 2007, Morozova and Morozov 2008, Mulvaney et al. 2009, Shlyapnikov et al. 2010). For many of these "magnetic label" detection methods, the ability to manipulate the magnetic particles is vital, as oftentimes the detection of particles must be performed on a substrate, and the ability to quickly and easily bring magnetic particles to a desired substrate with an external magnet is a major advantage.

Table 9.2 lists numerous examples of the different techniques used for magnetic particle detection, the kinds of magnetic particles used, the biological targets, and the sensitivity reported in the literature today.

9.3.2.1 Giant Magnetoresistance Detection

Giant magnetoresistance (GMR) is the first of the anisotropic magnetic resistance (AMR) phenomena we will explore. GMR is one of the more common methods of detecting magnetic particle labels in biosensors. Essentially, the observed phenomenon of GMR is a measurable change in the

TABLE 9.2
Systems Using Magnetic Particles as Labels to Detect Biological Targets

Method	Magnetic Particles	Target	Reported Sensitivity	References
Giant magnetoresistance	Bangs 350 nm	DNA oligonucleotide	16 pg/μl (25 pM)	Schotter (2004)
Giant magnetoresistance/CCD	Invitrogen 2.8 μm	DNA oligonucleotide	10 fM	Mulvaney et al. (2007)
		Ricin A chain	300 fM	
		Staphylococcal enterotoxin B	300 fM	
Giant magnetoresistance (spin valve)	MACs 50 nm	CEA	5 fM	Osterfeld et al. (2008)
		G-CSF	119 fM	
		IL-1α	53 fM	
		IL-10	56 fM	
		IFN-γ	59 fM	
		Lactoferrin	13 fM	
		TNF-α	57 fM	
Giant magnetoresistance (spin-valve type)	Miltenyi 50 nm	CEA	50 aM	Gaster et al. (2009)
Giant magnetoresistance	FeCo 13 nm	IL-6	125 fM	Li et al. (2010)
Giant magnetoresistance	FeCo 13 nm	endoglin	83 fM	Srinivasan et al. (2011)
Giant magnetoresistance (spin-valve type)	MACS 46 nm	Epithelial cell adhesion molecule	20 zmol	Gaster et al. (2011)
Giant magnetoresistance (spin-valve type)	50 nm	Secretory leukocyte peptidase inhibitor	10 fM	Hall et al. (2013)
Giant magnetoresistance	Miltenyi Biotec 50 nm	DNA oligonucleotide	39 pM	Rizzi et al. (2016)
Spin valve	Nanomag-D 250 nm	DNA oligonucleotide	10 pM	Graham et al. (2005)
Hall probe	Polysciences 2–20 nm	IgG	100 pg/ml	Aytur et al. (2006)
Hall probe	$MnFe_2O_4$ 10, 12, 16 nm	Rare cells	<100 cells/ml whole blood	Issadore et al. (2012)
MTJ	Fe_3O_4 16 nm	DNA oligonucleotide	2.5 μM	Shen et al. (2008)
Coil/Maxwell Bridge	Dextran ferrofluid	Glucose	40 mM	Kriz, Radevik, and Kriz (1996)
Coil/Maxwell Bridge	Iron oxide 20–200 nm	Concanavalin A	1.5 μM	Kriz, Gehrke, and Kriz (1998)
Coil/resonance	Invitrogen 2.8 μm	Human transferrin	260 fM	Richardson, Hawkins, and Luxton (2001)
Coil/permeability	Fe_3O_4 dextran coating 40–110 nm	C-reactive protein	0.2 μg/ml	Kriz et al. (2005)
Coil/resonance	Seradyne 0.824 μm	Troponin I	0.5 ng/ml	Kiely et al. (2007)
Cantilever	Clontech borosilicate 5 μm	Genomic DNA	71 zM	Weizmann et al. (2004)
		Telomerase (activity from cancer cells)	100 cells	
SQUID (substrate)	N/A	Collagen	100 pM	Kotitz et al. (1997)

(*Continued*)

TABLE 9.2 (Continued)

Method	Magnetic Particles	Target	Reported Sensitivity	References
SQUID (substrate)	Fe_3O_4 25 nm	IgE	2 amol (10 fM)	Enpuku et al. (2005)
SQUID (substrate)	Fe_3O_4 25 nm	IgE	15 amol	Enpuku et al. (2007)
SQUID (no substrate)	Nanomag 40, 130, 250 nm	DNA oligonucleotide	11 pM	Strömberg et al. (2008b)
SQUID (no substrate)	130, 250 nm	DNA oligonucleotide	40 pM	Strömberg et al. (2009)
SQUID (substrate)	Fe_3O_4 25 nm	IL-8/IgE *Candida albicans* cells	4 amol 300 cells	Kuma et al. (2010)
NMR relaxation	Iron oxide (40–50 nm)	DNA oligonucleotide	10 pM	Perez et al. (2002)
NMR relaxation	Iron oxide 40 nm	Telomere DNA (telomerase activity)	300 amol	Grimm et al. (2004)
NMR relaxation	Iron oxide 38 nm	Avidin *Staphylococcus aureus* cells	3 nM 10 bacteria	Lee et al. (2008)
NMR relaxation	Iron oxide 30 nm	Avidin	20 fmol	Sun et al. (2009)
NMR relaxation	Mn^{2+}-doped iron oxide 10, 12, 16 nm	SkBr3 cells	2 cells	Lee et al. (2009)
NMR relaxation	Iron oxide 16 nm	Bacillus Calmette-Guérin cells	20 CFU	Lee, Yoon, and Weissleder (2009)
NMR relaxation	Iron oxide (>100 nm)	A549 cells (mammalian)/MAP cells (bacterial)	10 cells	Kaittanis, Santra, and Perez (2009)
NMR relaxation	Iron oxide 38 nm	Avidin *S. aureus* cells	1 nM 10 CFU	Issadore et al. (2011)
NMR relaxation	Iron oxide 21 nm	*S. aureus* cells *Staphylococcus epidermidis* cells	1300 CFU 35,100 CFU	Chung et al. (2011)
NMR relaxation	Iron oxide 21 nm	DNA oligonucleotide	0.5 pM	Chung et al. (2013)
NMR relaxation	Iron oxide 30 nm	DNA oligonucleotide Genomic DNA	1 nM 1–5 pieces	Liong et al. (2013)
NMR relaxation	Iron oxide 7 nm	Microvesicles	~10^4 counts	Shao et al. (2012)
NMR relaxation	Iron oxide 21 nm	Microvesicles	2×10^6 counts/μl	Rho et al. (2013)
CCD	Invitrogen 1 μm	Streptavidin/IgG/West Nile Virus particles	400–800 molecules	Morozov et al. (2007)
CCD	Invitrogen 2.8 μm	Genomic DNA	10 fM	Mulvaney et al. (2009)
CCD	Invitrogen 1 μm	DNA oligonucleotide	100 aM	Shlyapnikov et al. (2010)

resistance of a material when a small magnetic field is applied. For GMR sensors, the change in resistance due to a fairly small number of particles is detectable, and the effect increases with the number of magnetic particles resting on the sensor. The idea for the setup of a GMR biosensor is to attach a binding ligand to the substrate directly above the GMR sensors, and another binding ligand to the magnetic particles. When the target is present, it binds the magnetic particles to the substrate in an amount proportional to the amount of target. Finally, the presence of those magnetic particles results in a detectable change in the resistance of the GMR sensor, a change proportional to the amount of particles attached to the surface. Figure 9.7 shows fabricated GMR sensors and the essential process for how the GMR phenomenon can be used to detect the presence of biological agents.

In 1998, an idea for an integrated system of biosensors relying on this method of detection was published (Baselt et al. 1998). The name they gave to their system was the Bead ARray Counter (BARC). The setup for the BARC used a 64-sensor array of 80 × 5 μm GMR sensors for the detection of magnetic beads. Each of the GMR sensors was incorporated into a Wheatstone bridge circuit and amplified for the detection of the smallest voltage variations. The initial sensitivity of the system was around 100 particles. In order to remove beads nonspecifically bound to the surface, the device was capable of deploying a magnetic field that created a vertical force on the beads and removed the vast majority of beads that had nonspecifically stuck to the surface, but left the beads bound by the target attached to the sensor surface. One of the great advantages of using the BARC setup is the great capacity for parallelization, as a large array of GMR sensors might be able to detect ~1000 analytes for a given sample. The BARC concept went through many iterations and development after the original paper was published (Miller et al. 2001, Rife 2003). In recent years, the BARC setup has been successful in detecting a great range of proteins (Gaster et al. 2009, Osterfeld et al. 2008, Hall et al. 2013, Li et al., 2010Marquina et al. 2012), nucleic acids (Schotter 2004, Ferreira et al. 2005, Mulvaney et al. 2007, Rizzi et al. 2016) and cells, and quantifying the kinetics of antigen–antibody binding (Gaster et al. 2011). Its multiplexing capacity and great potential as an integrated point-of-care solution, as envisioned by its creators, have also been demonstrated (Gaster, Hall, and Wang 2011, Wang et al. 2014, Koets et al. 2009).

Other iterations of the BARC design have used the fluid force discrimination (FFD) technique to remove nonspecifically bound particles from the sensor surface (Baselt et al. 1998, Edelstein et al. 2000). Here, the idea is that the sensor surfaces are used to capture the target and then are placed in a fluid cell undergoing laminar flow and the magnetic beads functionalized to bind the target are washed over the sensor substrate, often with an external magnet ensuring that the particles stay in close proximity to the surface. If the magnetic beads encounter the target bound to the substrate, they will quickly bind to the surface; otherwise they will simply be ushered along by the flow of the cell. Over a small amount of time, the sensors with capture targets will capture far more magnetic beads that can, in turn, be detected by the GMR sensor. FFD assays have also detected the magnetic beads by using a simple CCD camera (Morozova and Morozov 2008, Mulvaney et al. 2009). Figure 9.8 shows a setup from Morozov et al. (2007), which outlines a magnetic particle biosensor using FFD and detection with a CCD. The use of a CCD makes for simpler experimental detection, but doesn't possess the scope for parallelization of the GMR sensors.

A limitation of GMR sensors is the essential give-and-take between the different dimensions of the biosensor. In order to make an effective biosensor, the area of the sensor substrate should be as large as possible, in order to interact and bind with a greater number of targets. However, sensitivity of the sensor increases as the magnetoresistive strip becomes thinner, with a smaller surface area (Baselt et al. 1998). This means that a balance must be struck between the area available for analyte detection and the sensitivity of the GMR sensor itself.

9.3.2.2 Single-Particle Detection Methods

As GMR sensors have gained some recognition in biosensor platforms, several new techniques capable of detecting single particles have arisen, all of which rely on some kind of anisotropic magnetic resistance (AMR) phenomenon. The motivation behind such techniques is that they offer the capacity for extremely sensitive target detection. However, many of these techniques are only

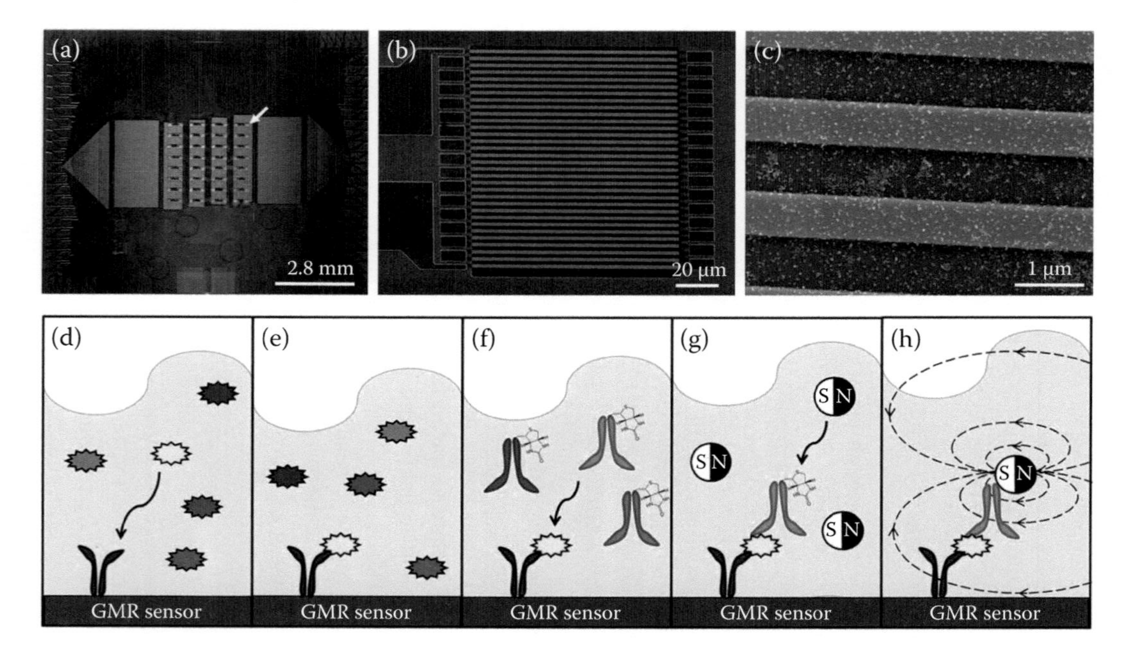

FIGURE 9.7 GMR sensors are small fabricated **(a–c)** devices that change their resistance under the influence of magnetic fields. The use of GMR as a biosensor **(d–h)** involves using a biological target to attach magnetic particles to the surface such that their presence can be detected by the GMR sensor, reprinted with permission from Springer Customer Service Centre GmbH (Gaster et al. 2009).

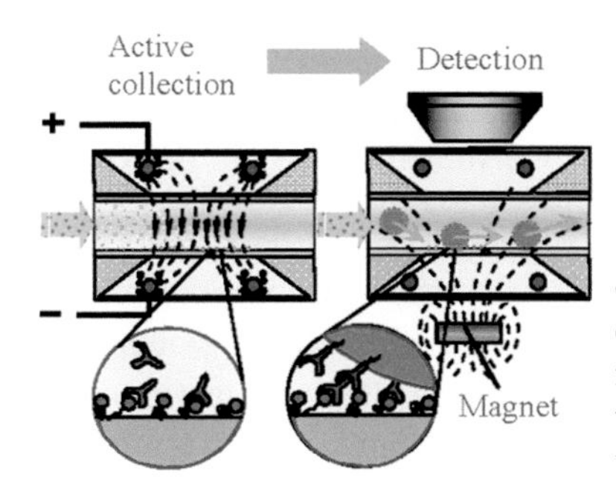

FIGURE 9.8 Fluid force discrimination in conjunction with optical detection methods uses similar principles as the BARC design, but does away with the magnetoresistive elements. Though the magnetic fields of the particles are not detected, the ability to force them to surface is vital to optimal function of this biosensor, reprinted with permission from (Morozov et al. 2007).

just beginning to advance into actual detection of biological targets, because of the small detection area associated with such sensors.

The first of the single-particle detection technologies is the sub-micron GMR sensor (Wood et al. 2005). The essential mechanism is the same as for larger GMR sensors but the dimensions are drastically reduced to around 2 μm × 100 nm. The reduced size of the sensors makes it possible for the magnetic field generated from a single 100 nm magnetic particle to cause a detectable voltage difference across the GMR sensor.

Spin valves are another method of single-particle detection that has been widely investigated (Ferreira et al. 2003, Heim et al. 1994, Li et al. 2003, 2006, Loureiro et al. 2009, Jeong Dae, Sang Don, and Myung Ae 2009, Lagae et al. 2002). Spin valves rely on an anisotropic magnetic resistance (AMR) in order to detect very small magnetic fields. The essence of spin-valve technology is a coupling of ferromagnetic thin films across a non-magnetic metal layer (Dieny et al. 1991, Ferreira et al. 2003, Speriosu et al. 1991). In spin valves, the resistance of the material changes depends on the angle between the magnetization directions of the two ferromagnetic layers. Most spin valves today also use an additional ferrimagnetic layer to "pin" the orientation of the magnetic moment of one of the ferromagnetic layers. Figure 9.9 shows the basic setup of the sensor. The single magnetic particle can then be detected by the way it affects the orientation of the magnetic moment of the other ferromagnetic layer. Of all the single-particle detection mechanisms, spin valves are among the most researched and furthest along in detection of biological agents. Spin valves have been both employed in microarrays of sensors (Wang et al. 2005), detection of DNA oligonucleotides (Graham et al. 2005, Ferreira et al. 2005), and cell counting (Loureiro et al. 2009, 2011).

The magnetic tunnel junction (MTJ) is another type of single-particle sensor with a design similar to that of spin valves (Grancharov et al. 2005, Shen et al. 2005). The difference in structure between the two types of sensor is that the non-ferromagnetic layer of the spin valve is replaced with a tunnel barrier in the MTJ (Brzeska et al. 2004, Moodera et al. 1995). In these sensors the current cannot conduct across the intermediate layer; the electrons must effectively tunnel through the barrier. Similar to spin valves, the amount of current that gets through the intermediate layer is dependent on the relative orientations of the magnetizations on either side

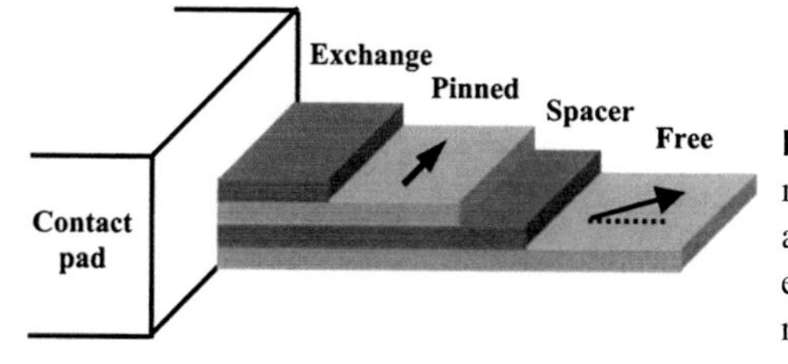

FIGURE 9.9 A spin valve consists of a free ferromagnetic layer, a non-ferromagnetic metal spacer layer, another pinned ferromagnetic layer, and a ferrimagnetic exchange layer that pins the previous ferromagnetic layer, reprinted with permission from (Ferreira et al. 2003.)

of the tunnel barrier. The presence of magnetic particles changes the tunneling efficiency and consequently the conduction current through the sensor. Like the related spin valves, MTJs have also been employed in microarrays (Wang et al. 2005) and demonstrated success in detecting DNA oligonucleotides (Shen et al. 2008).

The Hall effect has also been shown to have the capacity to detect the magnetic fields from single particles (Besse et al. 2002, Ejsing et al. 2004, Mihajlović et al. 2005, Skucha et al. 2011, Gabureac et al. 2013). The Hall effect, in essence, is the generation of a transverse voltage across a conducting material in the presence of a magnetic field. The Hall effect results from the magnetic force acting upon moving charge carriers in a direction perpendicular to the direction of current flow. The magnetic force on the charge carriers causes them to tend to stack on one side of the conductor as they flow. The uneven distribution of charge carriers in the conductor results in a measurable voltage perpendicular to the direction of current flow. The field from a single magnetic particle is sufficient to generate a transverse voltage across specially designed Hall probes. Often, Hall sensors are used in conjunction with an applied external field in order to magnetize the particles sufficiently that they can be detected. Figure 9.10 shows an example of how magnetic particles can be magnetized by a DC field in the vertical direction. Arrays of Hall probe sensors have shown effectiveness in detecting some biological targets (Aytur et al. 2006) and rare cells (Issadore et al. 2012).

The final technique of single-particle detection that employs an AMR effect is the AMR ring sensor. Essentially, these sensors are an AMR material deposited in a ring pattern such that a single particle could fit within the ring's inner radius (Jiang et al. 2006, Miller et al. 2002). Ring sensors take advantage of the axial symmetry of the fields of magnetic particles. The rings are constructed such that when no particle is present, the ring sensor is in a high-resistance state, because the magnetization is parallel to the direction of current flow. When a magnetic particle is present, the radial field of the particle modulates the magnetization such that it becomes perpendicular to the direction of current flow, putting the sensor into a low-resistance state. Figure 9.11 shows a schematic of the circuit and setup used in this method.

9.3.2.3 Magnetic Coil Detection

Magnetic coils set up in certain kinds of circuits have successfully been used to detect the presence of magnetic particles bound to a substrate. Magnetic particles have a magnetic permeability different from that of air and hence change the inductance of a solenoid coil when they are placed within the interior of the coil (Kriz, Radevik, and Kriz 1996). The setup for this method of detection involves placing target ligands on a simple substrate—generally plastic or glass, but silica microparticles have also been used (Kriz, Gehrke, and Kriz 1998)—and binding the magnetic particles to the substrate with the target. The substrates can then be placed into the coil and the

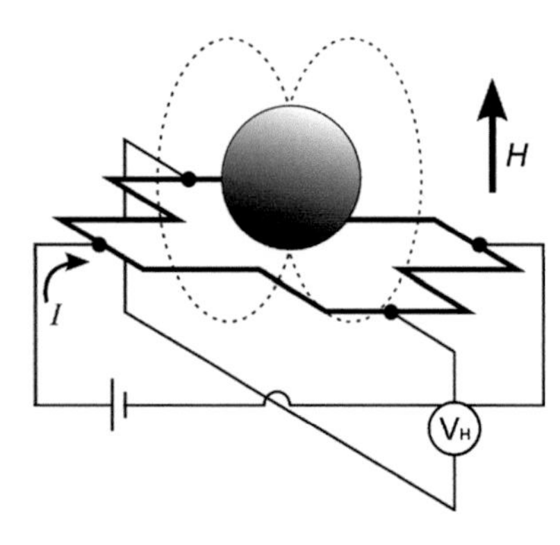

FIGURE 9.10 Sensors based upon the Hall effect take advantage of the magnetic force exerted on charge carriers in flowing current. This magnetic force causes a detectable transverse voltage to be formed across the Hall sensor. An external magnetic field is usually required in order to magnetize the particle sufficiently so that its magnetic field can be detected, reprinted with permission from (Tamanaha et al. 2008).

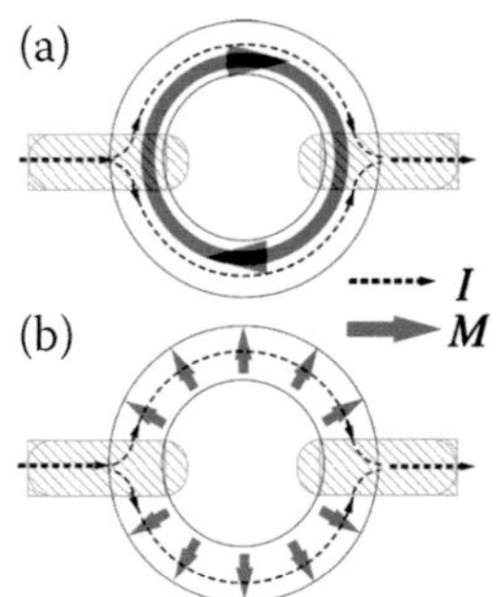

FIGURE 9.11 Ring AMR sensors are made from magnetoresistive material deposited in a ring shape. The geometry of the ring sensor makes it more sensitive to the magnetic fields generated by single particles. **(a)** The magnetization of the ring when no particles are present (high resistance), **(b)** The magnetization when a particle rests in the middle of the sensor (low resistance), reprinted with permission from (Miller et al. 2002).

number of magnetic particles can be detected by observing the changes in the inductance of the solenoid coil.

There are multiple ways of detecting changes in the inductance of the detection coil. The first is through the use of a Maxwell bridge circuit (Kriz, Radevik, and Kriz 1996, Kriz, Gehrke, and Kriz 1998). Essentially, this setup is an RLC circuit with several variable resistances, being driven with an alternating current source. The resistors can be adjusted such that the output of the circuit is zero before the sample is inserted into the coil. Placing magnetic material into the coil results in a voltage change at the output of the circuit, a change that is proportional to the amount of magnetic material placed in the interior of the coil. Figure 9.12 shows the basic circuit schematic used in this method of detection.

The other method of detecting the change of inductance in a solenoid coil is by setting up an LC resonant circuit (Richardson et al. 2001). In this case, the change in the inductance of the coil, resulting from the presence of the magnetic particles, corresponds to a change in the resonant frequency of the circuit. The change of the circuit's resonance frequency can be detected and corresponds to the number of magnetic particles in the sample. Figure 9.13 shows the circuit schematic for this detection mechanism.

These detection methods have been used to detect many biologically relevant molecules such as C-reactive protein (Kriz et al. 2005), human transferrin (Richardson, Hawkins, and Luxton 2001), and toponin I (Kiely et al. 2007). This technique is limited by the large number of magnetic particles required for detection (compared to other detection techniques). Initial investigations reported that at least 10^5 magnetic particles are required for detection (Richardson et al. 2001).

9.3.2.4 Cantilever Detection

Micro-cantilevers as biosensors have been widely researched in the past several years. Cantilevers have been used to detect the presence of single cells, and even single-cell growth (Boisen and Thundat 2009, Fritz 2008, Ziegler 2004). Recent work has used magnetic particles in conjunction

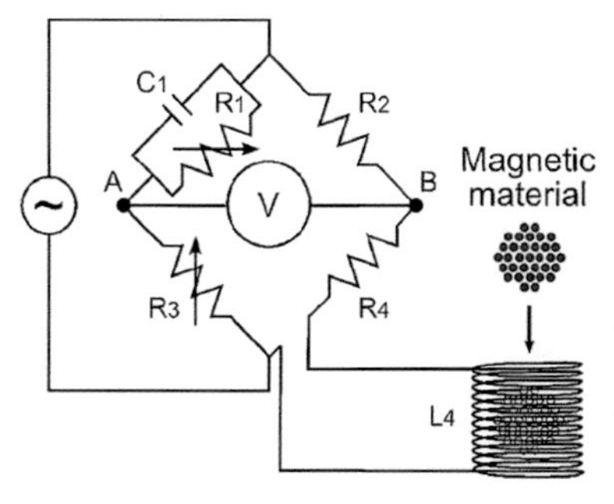

FIGURE 9.12 One setup of using inductance coils to detect magnetic particle sensors uses a Maxwell bridge. In this circuit, a change in the inductance of the coil caused by the addition of the magnetic particles results in a voltage difference between points A and B, reprinted with permission from (Tamanaha et al. 2008).

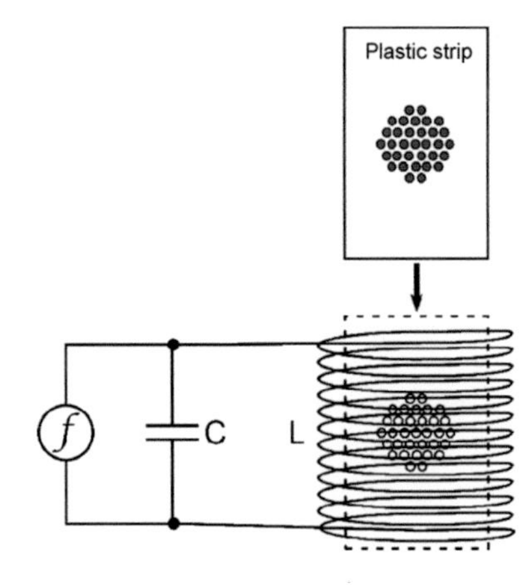

FIGURE 9.13 An alternative method to detecting inductance change in a magnetic coil is with a resonant LC circuit. A change is observed in the resonance frequency of the circuit upon the addition of the magnetic particles, reprinted with permission from (Tamanaha et al. 2008).

with cantilever technology as a biosensor (Weizmann et al. 2005, 2004). The apparatus for this detection method consists of using the target to bind magnetic particles to the surface of a cantilever. An external magnet is placed near the cantilever and the deflection of the cantilever can be measured by bouncing a laser beam off the surface of the cantilever and measuring its displacement on the photodiode array. The cantilever will be deflected in proportion to the number of magnetic particles attached to the surface, and the number of magnetic particles is proportional to the amount of target present in the incubation. Figure 9.14 shows the basic setup by which cantilever detection functions.

9.3.2.5 SQUID Detection

Superconducting quantum interference devices (SQUIDs) are used to measure the magnetic relaxation of materials. In other words, they measure the time it takes for a given material to lose its magnetization after being saturated by an external magnetic field, or if it does not lose its magnetization, what is the remanent magnetization of the sample. For situations involving magnetic particles diffusing through a solution, the magnetization of the system will eventually return to zero. Magnetic remanent relaxation occurs in one of two forms: Brownian and Néel relaxation

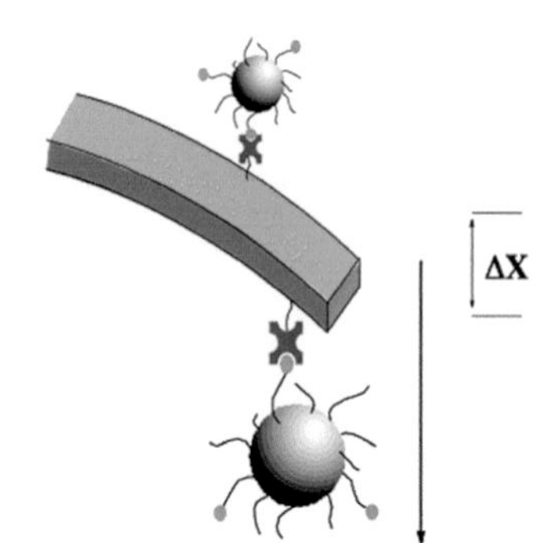

FIGURE 9.14 Magnetic particles can be used to detect biological targets in conjunction with cantilevers by binding the magnetic particles to the cantilever with the target and measuring the cantilever deflection when an external magnetic field acts on the magnetic particles attached to the cantilever, reprinted with permission from (Weizmann et al. 2004).

(Fannin et al. 2006). Brownian relaxation is the rearrangement of the magnetic moments that occurs due to the diffusion of the particles. Because the particles are free to diffuse, their magnetic moments will randomly reorient themselves as the particles diffuse. Néel relaxation occurs when the particles (or material) are not free to diffuse and hence the internal magnetization of the material must realign itself due to ambient thermal energy. Néel relaxation is generally slower than Brownian relaxation for the systems of interest.

Knowledge of relaxation times can help detect the presence of biological targets in different ways. The first method of detection involves using the biological target to bind the magnetic particles to a substrate (Kotitz et al. 1997, Tsukamoto et al. 2005). In this case, the particles that are bound will have to relax via the Néel mechanism, while unbound particles can relax using the much faster Brownian mechanism. By detecting the extent of Néel relaxation against Brownian, SQUID measurements can detect the amount of target analyte present. The basic scheme for this detection process can be seen in Figure 9.15.

Another method of detecting the presence of biological targets with SQUID is by using the binding analytes to form clusters of particles (Strömberg et al. 2008a, b). While these clusters will still be free to reorient due to diffusion, they will do so more slowly than individual, non-clustered, particles. The detectable change in relaxation time can be analyzed in such a way that it corresponds to differences in the amount of biological target.

SQUID techniques have been successful in detecting proteins (Chemla et al. 2000, Enpuku et al. 2005, 2007, Kotitz et al. 1997, Kuma et al. 2010), oligonucleotides (Strömberg et al. 2009, Zardán Gómez de la Torre et al. 2010), and cells (Hathaway et al. 2011). The substrate-free version of SQUID detection has relied on a DNA amplification technique, called rolling circle amplification (RCA) (Gusev et al. 2001), in order to form the clusters of magnetic particles needed for this method of detection. Figure 9.16 shows how RCA can be used to detect DNA in conjunction with SQUID.

9.3.2.6 Spin Relaxation Detection with NMR

Nuclear magnetic resonance (NMR) is another technique that has been used to detect the presence of biological targets bound to magnetic particles without a substrate (Grimm et al. 2004, Perez et al. 2002). In essence, the setup is similar to the substrate-free SQUID detection mechanism: targets bind the magnetic particles together into clusters and the presence of clusters can be measured by the apparatus. For NMR detection, though, the system takes advantage of the fact that the spin-spin relaxation times of the protons in the water are reduced by larger clusters of magnetic particles. Thus, when the binding target is present, the spin-spin relaxation will be lower than for samples without the binding analyte. Figure 9.17 gives a basic schematic for how biological targets can be detected with the spin relaxation method. The differences in spin relaxation can be easily detected with simple benchtop relaxometer devices (Perez et al. 2002). It is also possible that this method of detection could be performed in vivo using larger magnetic resonance imaging (MRI) machines.

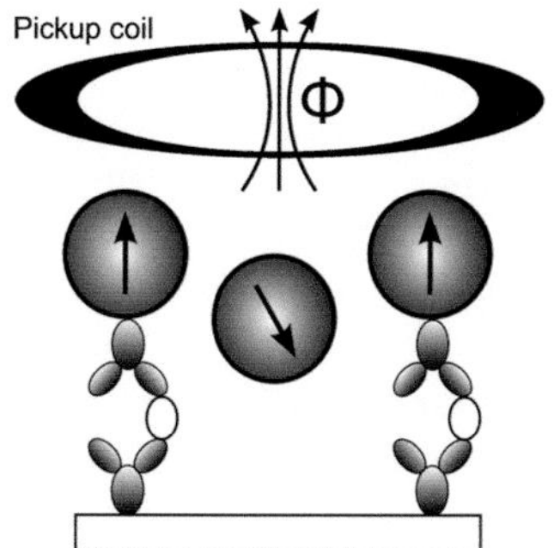

FIGURE 9.15 Detection of bound magnetic particles with SQUID exploits the fact that bound particles are not able to disperse themselves through diffusion and their remanent magnetic relaxation will be much slower than for unbound particles, reprinted with permission from (Tamanaha et al. 2008).

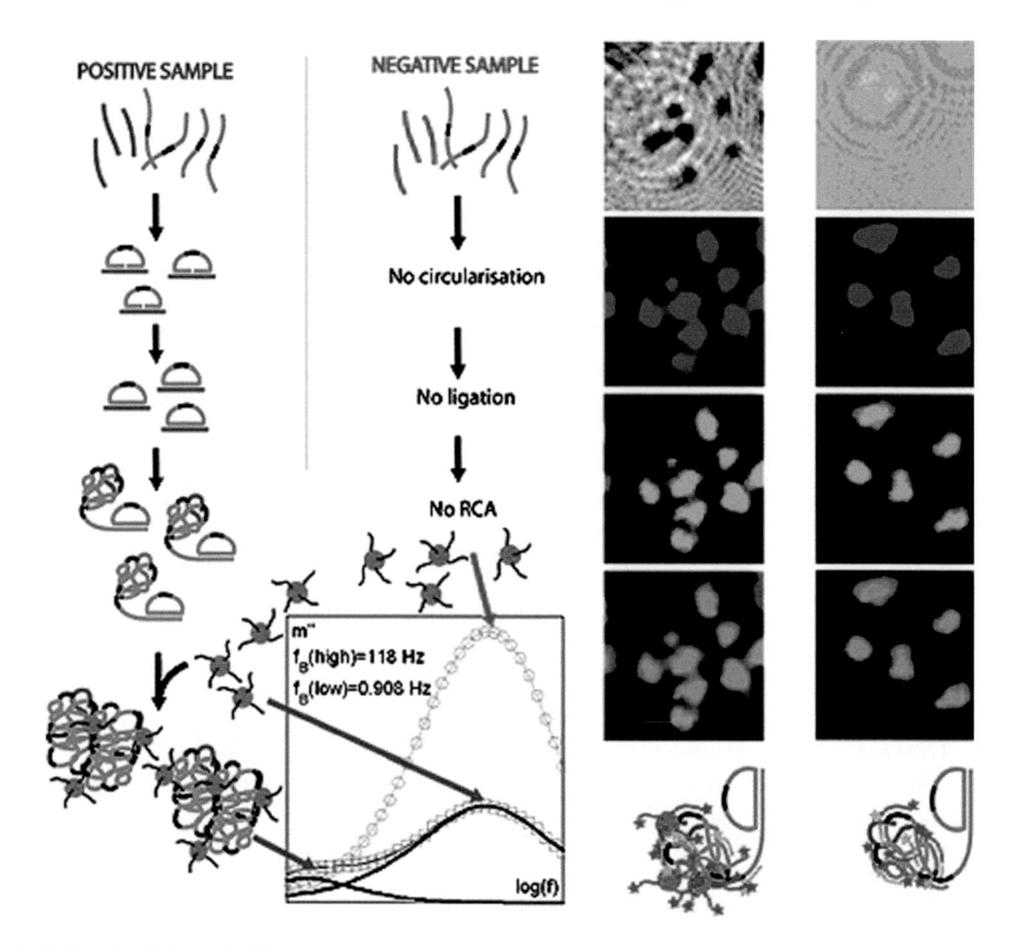

FIGURE 9.16 Schematic illustration of a magnetic nanobead detection method, where detection of DNA using SQUID has been demonstrated without the use of a substrate to bind the particles. For this detection method, the target DNA is amplified into large coils that then bind the magnetic particles together in clumps resulting in a change in the Brownian relaxation frequency. The positive sample and the negative sample show two very distinct magnetization curves and give different peak frequency values when fitting with Cole-Cole procedure. Microscopic images in the presence or absence of functionalized nanobeads are also shown.

Many biological systems have been investigated using spin relaxation. DNA oligonucleotides (Perez et al. 2002, Chung et al. 2013, Liong et al. 2013), proteins (Grimm et al. 2004, Lee et al. 2008, Sun et al. 2009, Bamrungsap et al. 2011), microvesicles (Shao et al. 2012, Rho et al. 2013), whole cells (Kaittanis, Santra, and Perez 2009, Lee et al. 2008, 2009, Lee, Yoon, and Weissleder 2009, Issadore et al. 2011, Chung et al. 2011), and cellular growth (Kaittanis, Nath, and Perez 2008) have all been detected using the spin relaxation NMR method. More recently, an integrated point-of-care μNMR device for multiplexed tumor analysis in clinical settings has been developed and received more attention (Haun et al. 2011, Ghazani et al. 2014).

This technology has also been widely used in three-dimensional imaging in vitro (cells under microscope) or in vivo (non-invasively, in animals and humans) (Panagiotopoulos et al. 2015, Buzug 2012, Gleich 2013, Bulte et al. 2011, Pablico-Lansigan, Situ, and Samia 2013, Ahrens and Bulte 2013). One of the earliest usages is via MRI, where superparamagnetic iron oxide nanoparticles (SPIONs) are loaded into cells and provide an MRI contrast effect under magnetic fields (Gallagher et al. 2008, Chen et al. 2013, Hu et al. 2013, Longo et al. 2014, de Vries et al. 2005).

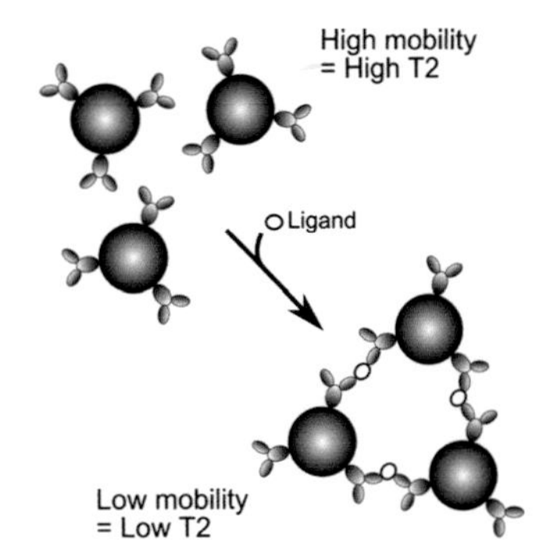

FIGURE 9.17 Magnetic particles clump together in the presence of the biological target. This clumping effect causes the spin-spin relaxation times to change in a detectable fashion, reprinted with permission from (Tamanaha et al. 2008).

More recently, another method, magnetic particle imaging (MPI) has been developed, offering higher sensitivity and quantification capability over MRI. The SPIONs, previously used as contrast agents in MRI, are used as MPI tracers. First proposed by Gleich and Weizenecker (2005), MPI directly maps the magnetic tracers, taking advantage of the nonlinearity of magnetization curves and magnetic saturation capability of SPIONs. This new technique has demonstrated great potential for applications that require high temporal and spatial resolutions or quantitative analysis (Zheng et al. 2015, 2016, Murase, Song, and Hiratsuka 2014, Rahmer et al. 2013, Weizenecker et al. 2009).

9.3.3 Dynamics-Based Magnetic Particle Biosensors

Dynamics of magnetic particles, including translational, rotational, and oscillatory motions, have been studied and used for biosensing. Early applications of the dynamics of magnetic particles in biosensing analyzed creep response and relaxation curves of magnetic particles so as to measure viscoelastic properties of cytoplasm (Crick and Hughes 1950, Valberg and Feldman 1987, Zaner and Valberg 1989, Ziemann, Rädler, and Sackmann 1994, Amblard et al. 1996, Bausch, Möller, and Sackmann 1999, Keller, Schilling, and Sackmann 2001). The majority of these systems track motions of magnetic particles with video microscopy and complex imaging processing. Others measure the decay of remanent magnetic fields as an indicator of the particle relaxation. More recent research has been focused on rotational motions of magnetic particles, in a rotating magnetic field, owing to the simplicity of this method and the wide variety of potential applications. These studies measure the periodicity of the rotational motion of magnetic particles, using a microscope or a photodiode, to infer different biomedical as well as biophysical properties (McNaughton et al. 2007, 2009). Here, we will focus on discussing the technologies based on the rotational dynamics of magnetic particles.

Magnetic particles driven with rotating magnetic fields have interesting qualities that have found numerous applications. The applications include more traditional ones, such as magnetic microdrills (Honda et al. 1996, McNaughton, Anker, and Kopelman 2005), micromixing (Biswal and Gast 2004), and artificial bacterial flagella (Ghosh and Fischer 2009, Zhang et al. 2009), as well as less straightforward applications, such as magnetic separation (Yellen et al. 2007), background extraction (Anker, Behrend, and Kopelman 2003), biomolecule detection (Hecht et al. 2011, Park, Handa, and Sandhu 2010, Ranzoni et al. 2011), viscosity measurement (Gaub and McConnell 1986, Sinn et al. 2012, Li et al. 2014, Berret 2016), and cell growth detection and antimicrobial susceptibility testing (Elbez et al. 2010, Kinnunen et al. 2010, McNaughton et al. 2007, 2006, 2009, Kinnunen et al. 2011, Sinn et al. 2011).

The underlying theory that relates the rotational periodicity to the measured properties differentiates among the previously mentioned systems. The first study of ferrohydrodynamics was conducted by Moskowitz and Rosensweig in 1967, where ferrofluids were placed in a rotating magnetic field and their rotation rates were observed (Moskowitz and Rosensweig 1967). It was soon followed by a theoretical study on single ferromagnetic (FM) and superparamagnetic (SPM) particles in rotating magnetic fields, so as to answer some of the questions that were raised (Caroli and Pincus 1969). Caroli and Pincus theoretically predicted that there exists a critical frequency for FM particles, only below which the particles rotate synchronously with the rotating field, anticipating more recent work (McNaughton et al. 2006). Two decades later Popplewell et al., using numerical integrations of the equations of motion, found that above the critical driving frequency the FM particle rotation rate is steady, but with a superimposed oscillatory component of relatively small magnitude; furthermore, the mean particle rotation rate decreases as the driving frequency increases (Popplewell, Rosensweig, and Johnston 1990). Recent dynamics-based magnetic particle biosensors further develop and apply the theory of rotational dynamics of magnetic particles, including both synchronous and asynchronous rotation of magnetic particles. The theoretical relationships used in these systems are largely dependent on the shape/assembly of the magnetic particle(s). The most common shapes/assemblies of magnetic particle(s) include: sphere, rod, chain, and cluster. Table 9.3 shows various rotational-dynamic-based biosensing systems using different shapes/assemblies of magnetic particle(s) and their applications reported in the literature.

9.3.3.1 Single Spherical Magnetic Particles

The application of rotational dynamics of single spherical magnetic particles for biosensing is one of the earliest and most diverse methods in the literature. In such systems, magnetic particles (i.e. ferromagnetic, superparamagnetic, or paramagnetic particles) are placed in a rotating magnetic field generated by perpendicular electromagnetic coils or a rotating magnet. The rotational motion of the magnetic particles in the synchronous or the asynchronous regime is recorded, and the periodicity of the rotation of the spherical magnetic particles is correlated with the measured properties. An example of a ferromagnetic particle rotating synchronously (below the critical driving frequency) and asynchronously (above the critical driving frequency) with the external rotating magnetic field can be seen in Figure 9.18 (McNaughton et al. 2006). The merit of the use of spherical magnetic particles lies in the geometric symmetry of the particles, which greatly simplifies the theoretically derived relationship.

The rotation of magnetic particles in the synchronous regime is used in the technology of magnetically modulated optical nanoprobes (MagMOONs) (Anker and Kopelman 2003, Anker et al. 2005). In some cases, the MagMOONs have fluorescent dyes visible on only one side of the sphere and the periodic magnetic modulation of the MagMOONs is used to separate a weak fluorescent signal from the background fluorescence (Figure 9.19). This simple procedure increases the signal-to-background ratio by 3–4 orders of magnitude (Anker and Kopelman 2003), enabling the use of MagMOONs as real-time fluorescent imaging agents in samples with highly scattering or fluorescent backgrounds. In other cases, fluorescent labels are attached to one hemisphere of the MagMOONs in the presence of analytes, thus blinking with fluorescent signals upon rotational actuation. This technology can potentially be used in vitro or in vivo for fluorescent imaging, as well as for spatially resolved sensing of chemical and physical parameters, such as pH and viscosities (Roberts, Anker, and Kopelman 2005, Anker et al. 2005, Behrend et al. 2005, Lee and Kopelman 2009).

The anomalous rotational phenomenon, where the magnetic particle rotates asynchronously with the driving field, became the basis of the method of asynchronous magnetic bead rotation (AMBR), for biosensors, developed by the Kopelman group. The governing theory underlying the asynchronous rotation phenomenon varies with the magnetic property (e.g. ferromagnetic, superparamagnetic, or paramagnetic) of the spherical magnetic particles used. For ferromagnetic particles, the apparent rotational frequency $\langle\dot{\theta}\rangle$ is a time-averaged frequency, and its value decreases

TABLE 9.3
Biosensing Systems Based on Rotational Dynamics of Magnetic Particles

Method	Magnetic Particles	Applications	References
Single spherical magnetic particles	Spherotech 4.4 μm ferromagnetic	Chemically responsive fluorescent imaging; biomolecule detection	Anker and Kopelman (2003)
Single spherical magnetic particles	Spherotech 4–5 μm ferromagnetic	Chemically responsive fluorescent imaging; biomolecule detection	Anker et al. (2005)
Single spherical magnetic particles	1–10 μm ferromagnetic (made in-house)	pH sensing; viscosity measurement	McNaughton et al. (2006)
Single spherical magnetic particles	2 μm ferromagnetic	Single-cell detection (*E. coli*)	McNaughton et al. (2007)
Single spherical magnetic particles	Spherotech 4.6 μm ferromagnetic	Measurement of magnetic field strength; measurement of inter-particle magnetic moment uniformity; measurement of viscosity; measurement of binding dynamics; estimation of shape factor	McNaughton et al. (2007)
Single spherical magnetic particles	Spherotech 40 μm ferromagnetic	Single-cell detection; bacteria growth monitoring	McNaughton et al. (2009)
Single spherical magnetic particles	Invitrogen M-280 2.8 μm superparamagnetic	Bacteria growth monitoring; antimicrobial susceptibility testing of microorganisms	Kinnunen et al. (2011)
Single spherical magnetic particles	Ocean Nanotech 30 nm superparamagnetic	Cell size and morphology monitoring (Hela cells); cytotoxicity and drug sensitivity assay	Elbez et al. (2011)
Single spherical magnetic particles	Spherotech 8.8 μm superparamagnetic; Invitrogen M-280 2.8 μm superparamagnetic	Bacteria growth monitoring; antimicrobial susceptibility testing of microorganisms	Sinn et al. (2011)
Single spherical magnetic particles	Invitrogen Dynal MyOne 1 μm superparamagnetic	Biomolecule detection (thrombin)	Hecht, Kumar, and Kopelman (2011)
Single spherical magnetic particles	Spherotech 16 μm superparamagnetic; Spherotech 8.8 μm; 10 μm (made in-house)	Viscosity measurement; bacteria growth monitoring; antimicrobial susceptibility testing of microorganisms	Sinn et al. (2012)
Single spherical magnetic particles	Spherotech 45 μm paramagnetic	Viscosity measurement; DNA detection and quantification	Li et al. (2014)
Single rod-shaped magnetic particles	Diameter 0.2 μm, length 20 μm (made in-house)	Viscosity measurement (butterfly saliva)	Tokarev et al. (2013)

(Continued)

TABLE 9.3 (Continued)

Method	Magnetic Particles	Applications	References
Single rod-shaped magnetic particles	Diameter 0.5 μm, length 12 μm superparamagnetic (made in-house)	Viscoelasticity measurement (cytoplasm)	Berret (2016)
Magnetic particle chains	Spherotech 0.86 μm	Chemically responsive fluorescent imaging; biomolecule detection	Anker et al. (2005)
Magnetic particle chains	Masterbeads 500 nm; Bio-Adembeads 300 nm	Biomolecule detection (biotinylated BSA)	Ranzoni et al. (2011)
Magnetic particle chains	Masterbeads 500 nm	Biomolecule detection (PSA)	Ranzoni et al. (2010)
Magnetic particle clusters	Invitrogen M-280 2.8 μm superparamagnetic	Bacteria growth monitoring; antimicrobial susceptibility testing of microorganisms	Kinnunen et al. (2012)
Magnetic particle clusters	Invitrogen Dynabead 1 μm superparamagnetic	Biomolecule detection (thrombin)	Hecht et al. (2013)
Magnetic particle clusters	1 μm	Biomolecule detection (thrombin)	Hecht et al. (2014)
Magnetic particle clusters	Invitrogen 2.8 μm	Bacteria growth monitoring (*E. coli* and *S. aureus*)	Kinnunen et al. (2014)

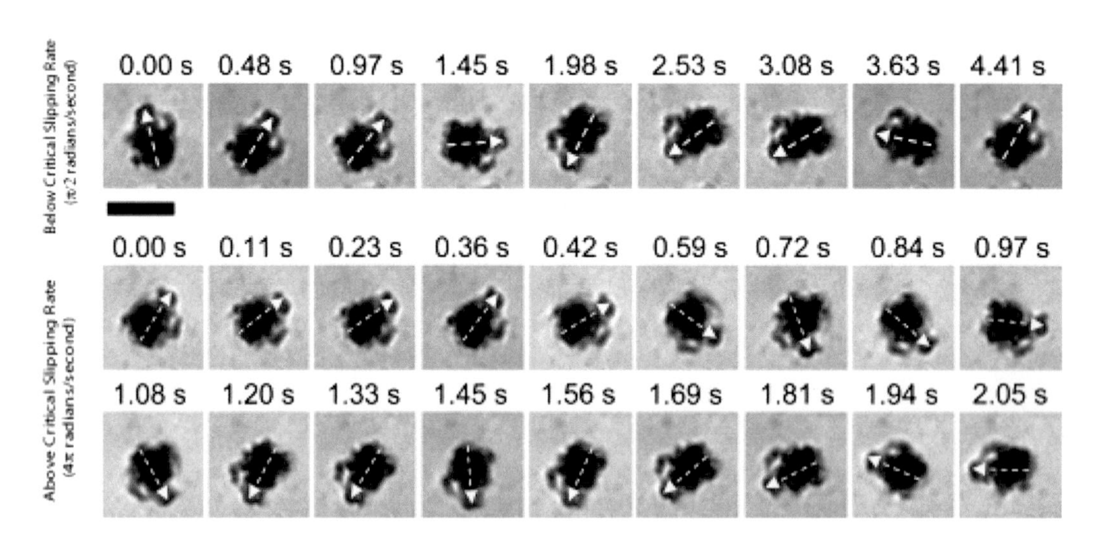

FIGURE 9.18 Sequence of bright-field microscopy images of a single magnetic particle, where the black scale bar is 7 μm. Top row: The particle is driven below the critical driving frequency, rotating in the clockwise direction. Bottom two rows: The particle is driven above the critical driving frequency. The particle begins in the clockwise direction, at t = 0.00, 0.11 and 0.23 seconds, but then rotates in a direction that is opposite of the external driving field, the counterclockwise direction, at t = 0.36, 0.84, 0.97,1.45, and 2.05 seconds, reprinted with permission from (McNaughton et al. 2006).

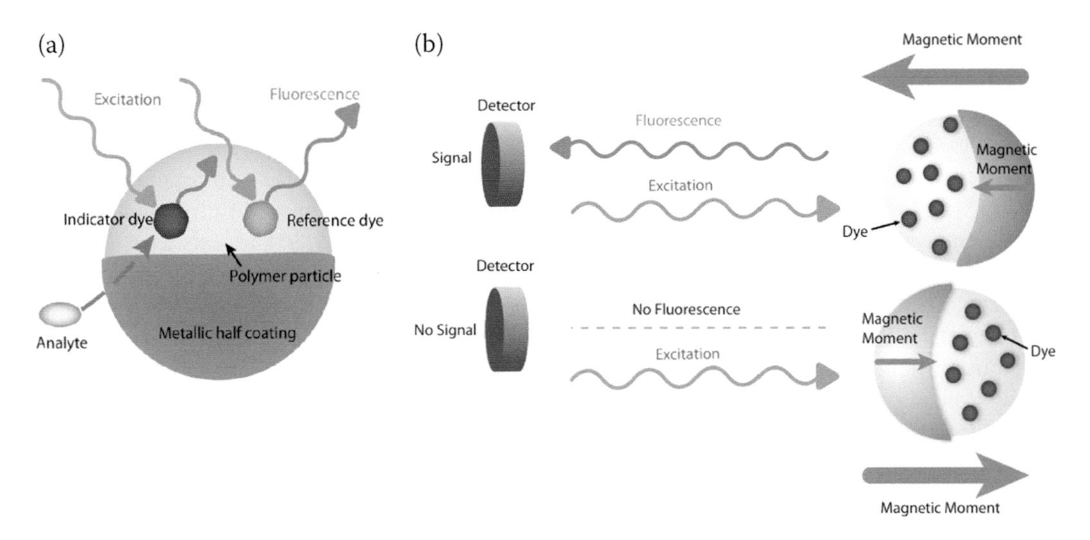

FIGURE 9.19 **(a)** Modulated optical nanoprobe sensor, **(b)** Background-free measurement taken by a magnetically modulated optical nanoprobe (MagMOON). An external magnetic field orients the MagMOON, causing its fluorescent excitation and observed emission to blink on and off as it rotates. Note that the background fluorescence does not blink, reprinted with permission from (Lee, Kopelman and Smith et al. 2009).

with driving frequency in the asynchronous regime due to a drag-induced wobbling. Theoretical analysis on the equation of motion for a ferromagnetic particle gives Equation 9.1 for the rotation frequency of the particle (Helgesen, Pieranski, and Skjeltorp 1990, McNaughton et al. 2007):

$$\langle \dot{\theta} \rangle = \begin{cases} \Omega, & \Omega < \Omega_c \\ \Omega - \sqrt{\Omega^2 - \Omega_c^2}, & \Omega > \Omega_c \end{cases} \tag{9.1}$$

where Ω is the field driving frequency, and Ω_c is the critical frequency. The critical frequency is correlated with various properties (McNaughton et al. 2009),

$$\Omega_c = \frac{mB}{\kappa \eta V} \tag{9.2}$$

Here m is the magnetic moment of the particle, B is strength of the magnetic field, κ is the shape factor of the particle (e.g. $\kappa = 6$ for a sphere), η is the viscosity of the surrounding fluid, and V is the volume of the particle. Substituting Equations 9.2 into 9.1, we see that various properties could be determined by measuring the rotational frequency/period of the magnetic particles. A similar analysis has been conducted for magnetic particles with induced magnetic dipoles rotating in the asynchronous regime. The rotational frequency $\dot{\theta}$ of paramagnetic or superparamagnetic particles at high driving frequency follows the relationship (Janssen et al. 2009, Kinnunen et al. 2011):

$$\dot{\theta} = \frac{x'' V_m B^2}{\kappa \eta V \mu_o} \tag{9.3}$$

where x'' is the imaginary part of the magnetic susceptibility (which is frequency-dependent), V_m is the volume of the magnetic content, and μ_0 is the permeability of free space. Therefore, it is evident from Equation 9.3 that the periodicity of the rotational motion of superparamagnetic particles can also be used to measure different properties, in a similar fashion as for ferromagnetic particles.

One of the most common AMBR biosensors detects analyte-induced volume changes of magnetic particles by measuring the rotational periodicity of the magnetic particles. The volume of magnetic particles (i.e. V in Equations 9.2 and 9.3) increases upon binding of particle-labeled biomolecules or cells, thus decreasing the rotational frequency of the magnetic particles. A sensitivity of as low as a59 nm change in the bead diameter, or a 4 fl change in volume, has been achieved (Kinnunen et al. 2010). This method has been applied to biomolecule detection (Hecht et al. 2011, Hecht, Kumar, and Kopelman 2011), to single-cell detection (McNaughton et al. 2007, 2009), and to cell growth and drug susceptibility monitoring (Kinnunen et al. 2011, Sinn et al. 2011, Elbez et al. 2011). Figure 9.20 presents a schematic representation of how a magnetic particle detects a single cell and monitors its growth by measuring changes in its rotational frequency. A slight variation of this method is to label a non-magnetic particle or cell of sphere like shape with superparamagnetic nano-particles . In these circumstances, the non-magnetic objects are granted superparamagnetic properties and follow Equation 9.3 in the asynchronous regime (Elbez et al. 2011, Hecht, Kumar, and Kopelman 2011, Hecht et al. 2011).

Other AMBR systems measure physical and biomedical properties of biological fluids. As shown in Equations 9.1–9.3, various physical properties, such as the viscosity η, are related to the periodicity of rotation. For ferromagnetic particles, an alternative to measure the periodicity of rotation is to measure the critical driving frequency by scanning various driving frequencies (McNaughton et al. 2006). The AMBR method has thus been used to measure physical properties, such as viscosity, magnetic properties and morphological properties (McNaughton et al. 2007), including cancer cell morphodynamics (Elbez et al. 2011). The information on such physical properties has also been used for monitoring the progression of biochemical reactions (Li et al. 2014), monitoring bacterial growth, and testing antimicrobial susceptibility (Sinn et al. 2012).

9.3.3.2 Single Rod-Shaped Magnetic Particles

Single rod-shaped magnetic particles are mainly used in viscoelastic measurements of biological fluids. This method is similar to the methods using single spherical magnetic particles in that the same equation (i.e. Equation 9.1) is used to distinguish the synchronous and asynchronous regimes of the magnetic rotation. However, the expressions for the critical driving frequency are different due to the geometric and magnetic asymmetry of magnetic nanorods. For ferromagnetic nanorods (Tokarev et al. 2012, 2013, Aprelev et al. 2015),

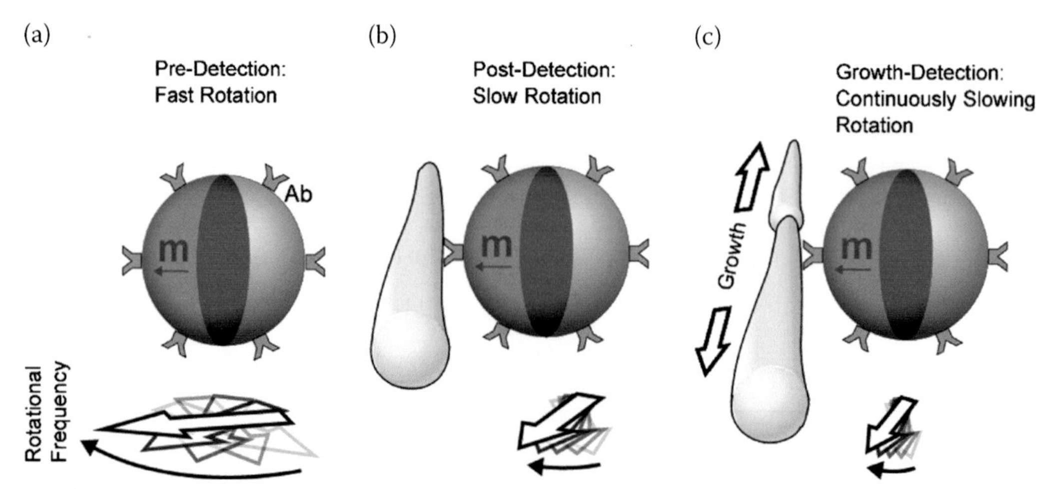

FIGURE 9.20 Magnetic microsphere functionalized with an antibody **(a)** with no attached bacteria, **(b)** with one bacterium, **(c)** after bacterial growth. The rotation rate slows the bacteria grows on the surface of the particle, reprinted with permission from (McNaughton et al. 2009).

$$\Omega_c = \frac{mB\left(3\text{In}\left(\frac{l}{d}\right) - 2.4\right)}{\pi\eta l^3} \tag{9.4}$$

where l and d are the length and diameter of the rod, respectively. For paramagnetic and superparamagnetic nanorods (Frka-Petesic et al. 2011, Berret 2016),

$$\Omega_c = \frac{3\mu_0 \Delta\chi B^2}{\pi\eta l^{*2}} \tag{9.5}$$

where $\Delta\chi = \chi^2/(2+\chi)$ and $l^* = (l/d)\sqrt{\text{In}(l/d) - 0.662 + 0.917(d/l) - 0.050(d/l)^2}$. Note that the aspect ratio of the nanorods (i.e. l/d), rather than the specific dimensions of the nanorods, determines the critical driving frequency in both cases. In magnetic nanorod-based systems, the critical driving frequency is measured optically by observing the onset of the wobbling motion, which is considered a signature of the asynchronous regime, in a magnetic field with an increasing/decreasing frequency. The value of the critical driving frequency is then substituted into Equation 9.4 or 9.5 so as to calculate viscosities of biological fluids, such as saliva (Tokarev et al. 2013) or cytoplasm (Berret 2016). The amplitude of the wobbling motion (i.e. angle of the backward motion) in the asynchronous regime is used to calculate the elastic modulus of cytoplasm (Berret 2016). Figure 9.21 shows an example of single rod-shaped magnetic particle biosensors.

9.3.3.3 Chains of Magnetic Particles

Magnetic particles have also been assembled into chain shape and used for biomolecule detection (Anker, Behrend, and Kopelman 2003, Ranzoni et al. 2011). The rotational motion of magnetic particle chains in the synchronous regime has been utilized (Ranzoni et al. 2010). In some systems, chains of magnetic particles are synthesized through chemical linking or heating above the glass transition temperature (e.g. 94°C for polystyrene) before using them as biosensors. Upon binding

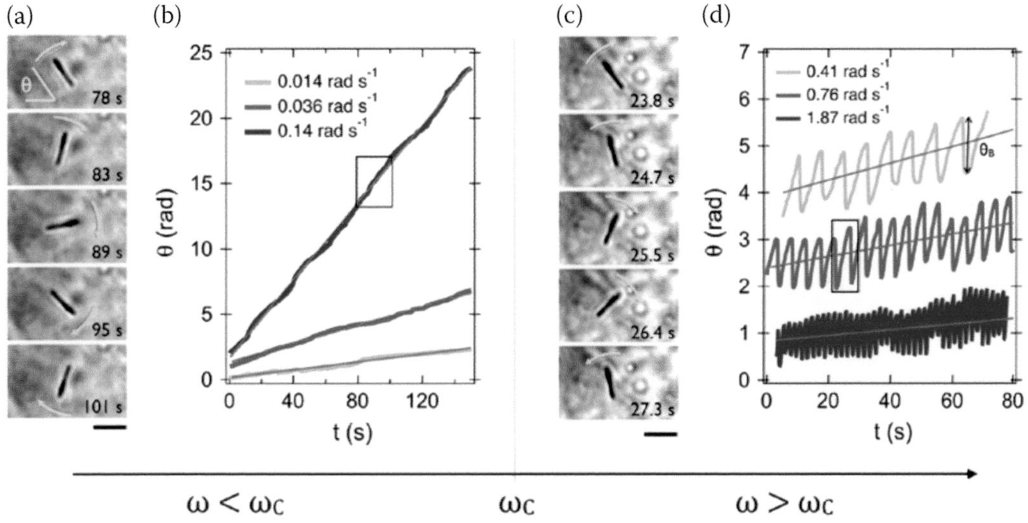

FIGURE 9.21 Synchronous and asynchronous magnetic nanorod rotations. **(a)** and **(b)** show the magnetic nanorod rotations in the synchronous regime, and **(c)** and **(d)** illustrate the rotations in the asynchronous regime. The critical driving frequency Ω_c is optically determined upon the onsite of backward motions. θ_B is the amplitude of the wobbling motion, reprinted with permission from (Berret 2016).

of targets to the magnetic particle chains, the magnetic particle chains are fluorescently labeled and start blinking with fluorescent signals under a rotating magnetic field (Anker, Behrend, and Kopelman 2003). In other systems, magnetic particles undergo target-assisted chaining reactions under a rotating magnetic field, and the rotational motion of the particle chains in the synchronous regime is measured by optical scattering, with the optical beam nearly aligned to the plane of rotation (Ranzoni et al. 2011). The amplitude of the periodic optical signal is linearly proportional to the number of particle chains, thus the target concentration, within a dynamic range of two orders of magnitude. A graphical illustration of the working principle of this system is provided in Figure 9.22.

9.3.3.4 Clusters of Magnetic Particles

The use of clusters of magnetic particles has also been explored in various studies, owing to the ease of device operation and signal measurement. The clusters are self-assembled under a rotating magnetic field and adopt amorphous shapes, which are well-maintained throughout their rotation. The cluster formation and rotational dynamics are governed by the properties of individual magnetic particles, thus following similar equations as for single magnetic particles (Kinnunen et al. 2012). The morphology and rotational dynamics of magnetic particle clusters have been analyzed and applied for biosensing. One approach monitors the periodicity of the asynchronous rotation of a self-assembled cluster in a bacterial growth media (Kinnunen et al. 2012, 2014). As the bacteria grow, the bacteria attach to the magnetic particles and the added volume from the bacterial growth can be calculated from the rotational periodicity. In another system, magnetic particles are self-assembled into clusters via bridges of target molecules, thus increasing the size of otherwise tightly packed clusters (Hecht et al. 2013, 2014). A higher concentration of target

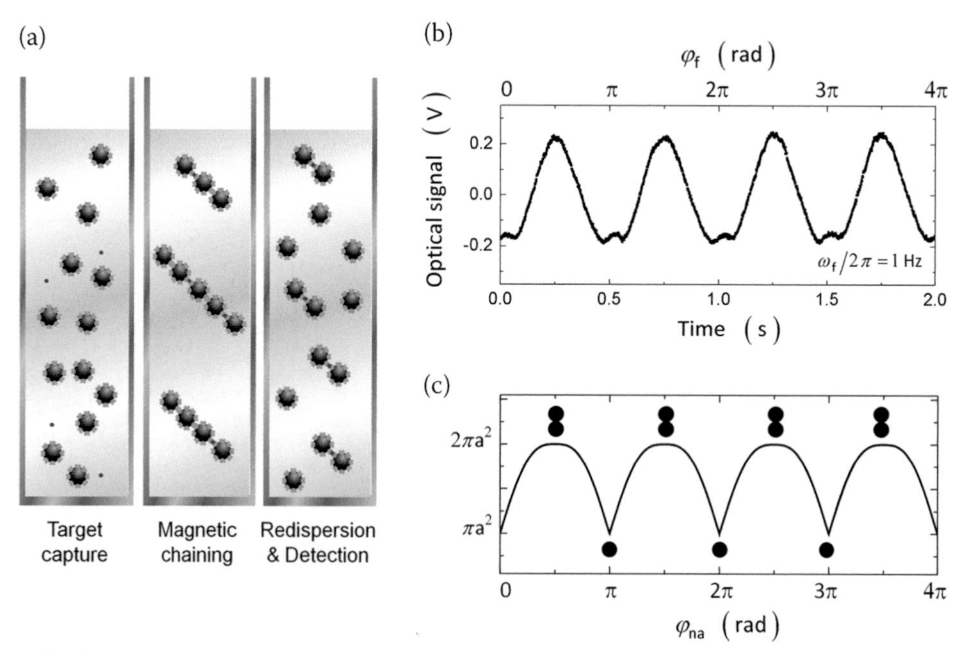

FIGURE 9.22 Biomolecule detection by measuring periodic optical signals due to rotation of magnetic particle chains in the synchronous regime. **(a)** Magnetic particles form chains upon biomolecule binding, **(b)** The measured signal of optical scattering shows periodicity, **(c)** Expected cross-sectional area of a single magnetic particle dimer detected by the photodetector, reprinted with permission from (Ranzoni et al. 2011).

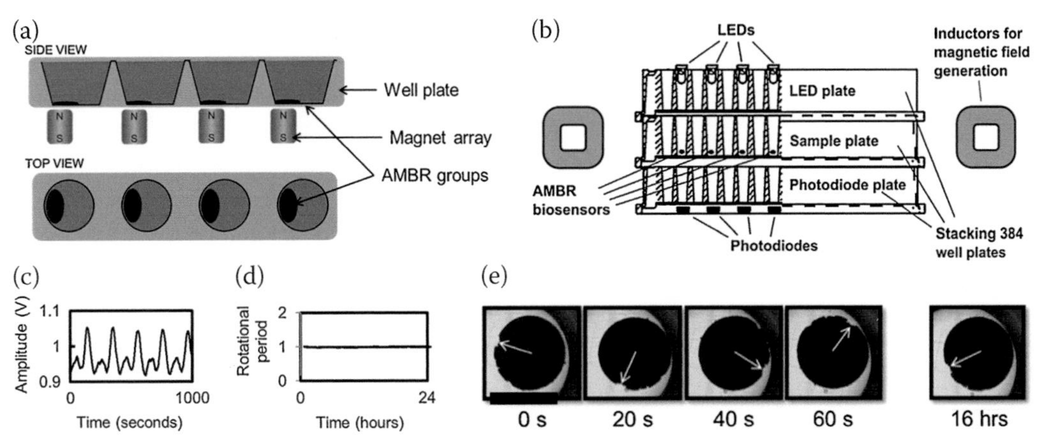

FIGURE 9.23 A prototype of a magnetic particle cluster biosensor with microscope-free operation. **(a)** Side and top view, **(b)** Cross-sectional side view of the prototype, **(c)** Raw intensity data from the photodiode, **(d)** Stability of measured rotational period over time, **(e)** Microscope image of the rotating cluster, reprinted with permission from (Kinnunen 2014).

molecules results in a higher rotational period, a larger cluster size, a lower fractal dimension, and a higher lacunarity ("gappiness"). Each of these variables can be used independently to quantify the concentration of biomolecules. The large size of magnetic particle clusters significantly improves the performance of microscope-free cluster detection, thus allowing a simple and cost-effective operation. A prototype of a photodiode-based biosensor using magnetic particle cluster is demonstrated in Figure 9.23.

ACKNOWLEDGEMENT

We would like to thank Chang Heon Lee for his help with literature update and we acknowledge support from NIH/NCI grant R01 CA250499 (RK).

REFERENCES

Agrawal, A., T. Sathe, and S. Nie. 2007. Single-bead immunoassays using magnetic microparticles and spectral-shifting quantum dots. *Journal of Agricultural and Food Chemistry* 55 (10), 3778–3782. doi:10.1021/jf0635006.

Ahrens, E. T., and J. W. Bulte. 2013. Tracking immune cells in vivo using magnetic resonance imaging. *Nature Reviews Immunology* 13 (10), 755–763. doi:10.1038/nri3531.

Amblard, F., B. Yurke, A. Pargellis, and S. Leibler. 1996. A magnetic manipulator for studying local rheology and micromechanical properties of biological systems. *Review of Scientific Instruments* 67 (3), 818. doi:10.1063/1.1146816.

Anker, J. N., C. J. Behrend, H. Huang, and R. Kopelman. 2005. Magnetically-modulated optical nanoprobes (MagMOONs) and systems. *Journal of Magnetism and Magnetic Materials* 293 (1), 655–662. doi:10.1016/j.jmmm.2005.01.031.

Anker, J. N., C. Behrend, and R. Kopelman. 2003. Aspherical magnetically modulated optical nanoprobes (MagMOONs). *Journal of Applied Physics* 93, 6698.

Anker, J. N., and R. Kopelman. 2003. Magnetically modulated optical nanoprobes. *Applied Physics Letters* 82 (7), 1102. http://link.aip.org/link/APPLAB/v82/i7/p1102/s1&Agg=doi.

Aprelev, P., Y. Gu, R. Burtovyy, I. Luzinov, and K. G. Kornev. 2015. Synthesis and characterization of nanorods for magnetic rotational spectroscopy. *Journal of Applied Physics* 118 (7), 074901. doi: 10.1063/1.4928401.

Aytur, T., J. Foley, M. Anwar, B. Boser, E. Harris, and P. R. Beatty. 2006. A novel magnetic bead bioassay platform using a microchip-based sensor for infectious disease diagnosis. *Journal of Immunological*

Methods 314 (1–2), 21–29. http://www.sciencedirect.com/science/article/B6T2Y-4KB107N-1/2/1e915e52002942b0d7a8fbbb699886ae.

Bamrungsap, S., M. I. Shukoor, T. Chen, K. Sefah, and W. Tan. 2011. Detection of lysozyme magnetic relaxation switches based on aptamer-functionalized superparamagnetic nanoparticles. *Analytical Chemistry* 83 (20), 7795–7799. doi:10.1021/ac201442a.

Bao, Y. P., T. F. Wei, P. A. Lefebvre, H. An, L. He, G. T. Kunkel, and U. R. Müller. 2006. Detection of protein analytes via nanoparticle-based bio bar code technology. *Analytical Chemistry* 78 (6), 2055–2059.

Baselt, D. R., G. U. Lee, M. Natesan, S. W. Metzger, P. E. Sheehan, and R. J. Colton. 1998. A biosensor based on magnetoresistance technology. *Biosensors and Bioelectronics* 13 (7–8), 731–739. http://www.sciencedirect.com/science/article/B6TFC-3V7468X-2/2/2062aec052caa896da42da32714f13ee.

Bausch, A. R., W. Möller, and E. Sackmann. 1999. Measurement of local viscoelasticity and forces in living cells by magnetic tweezers. *Biophysical Journal* 76 (1), 573–579.

Behrend, C. J., J. N. Anker, B. H. McNaughton, and R. Kopelman. 2005. Microrheology with modulated optical nanoprobes (MOONs). *Journal of Magnetism and Magnetic Materials* 293 (1), 663–670. doi:10.1016/j.jmmm.2005.02.072.

Berret, J. F. 2016. Local viscoelasticity of living cells measured by rotational magnetic spectroscopy. *Nature Communications* 7, 10134. doi:10.1038/ncomms10134.

Besse, P. A., G. Boero, M. Demierre, V. Pott, and R. Popovic. 2002. Detection of a single magnetic microbead using a miniaturized silicon Hall sensor. *Applied Physics Letters* 80 (22), 4199. http://link.aip.org/link/APPLAB/v80/i22/p4199/s1&Agg=doi.

Bi, S., H. Zhou, and S. Zhang. 2010. A novel synergistic enhanced chemiluminescence achieved by a multiplex nanoprobe for biological applications combined with dual-amplification of magnetic nanoparticles. *Chemical Science* 1 (6), 681. doi:10.1039/c0sc00341g.

Biswal, S. L., and A. P. Gast. 2004. Micromixing with linked chains of paramagnetic particles. *Analytical Chemistry* 76 (21), 6448.

Blackburn, G. F., H. P. Shah, J. H. Kenten, J. Leland, R. A. Kamin, J. Link, J. Peterman, M. J. Powell, A. Shah, and D. B. Talley. 1991. Electrochemiluminescence detection for development of immunoassays and DNA probe assays for clinical diagnostics. *Clinical Chemistry* 37 (9), 1534–1539.

Boisen, A., and T. Thundat. 2009. Design & fabrication of cantilever array biosensors. *Materials Today* 12 (9), 32–38.

Bonel, L., J. C. Vidal, P. Duato, and J. R. Castillo. 2011. An electrochemical competitive biosensor for ochratoxin A based on a DNA biotinylated aptamer. *Biosensors Bioelectronics* 26 (7), 3254–3259. doi:10.1016/j.bios.2010.12.036.

Brzeska, M., M. Panhorst, P. B. Kamp, J. Schotter, G. Reiss, A. Pühler, A. Becker, and H. Brückl. 2004. Detection and manipulation of biomolecules by magnetic carriers. *Journal of Biotechnology* 112 (1–2), 25–33.

Bulte, J. W., P. Walczak, B. Gleich, J. Weizenecker, D. E. Markov, H. C. Aerts, H. Boeve, J. Borgert, and M. Kuhn. 2011. MPI cell tracking: What can we learn from MRI? *Proceedings of SPIE—The International Society for Optical Engineering* 7965, 79650z. doi:10.1117/12.879844.

Buzug, T. M. 2012. *Magnetic Particle Imaging: An Introduction to Imaging Principles and Scanner Instrumentation*. Berlin Heidelberg:Springer.

Cai, S., L. Xin, C. Lau, and J. Lu. 2010. Highly sensitive non-stripping gold nanoparticles-based chemiluminescent detection of DNA hybridization coupled to magnetic beads. *The Analyst* 135 (3), 615. http://pubs.rsc.org/en/Content/ArticleLanding/2010/AN/b927359j.

Caroli, C., and P. Pincus. 1969. Response of an isolated magnetic grain suspended in a liquid to a rotating field. *Physik der Kondensierten Materie* 9 (4), 311–319. http://www.springerlink.com/content/a767l2t6h6620w37/.

Chemla, Y. R., H. L. Grossman, Y. Poon, R. McDermott, R. Stevens, M. D. Alper, and J. Clarke. 2000. Ultrasensitive magnetic biosensor for homogeneous immunoassay. *Proceedings of the National Academy of Sciences of the USA* 97 (26), 14268–14272. http://www.pnas.org/content/97/26/14268.abstract.

Chen, J., H. Pan, G. M. Lanza, and S. A. Wickline. 2013. Perfluorocarbon nanoparticles for physiological and molecular imaging and therapy. *Advances in Chronic Kidney Disease* 20 (6), 466–478. doi:10.1053/j.ackd.2013.08.004.

Chen, X., J. Zhu, Z. Chen, C. Xu, Y. Wang, and C. Yao. 2011. A novel bienzyme glucose biosensor based on three-layer Au–Fe_3O_4@SiO_2 magnetic nanocomposite. *Sensors and Actuators B: Chemical* 159 (1), 220–228. doi:10.1016/j.snb.2011.06.076.

Chung, H. J., C. M. Castro, H. Im, H. Lee, and R. Weissleder. 2013. A magneto-DNA nanoparticle system for rapid detection and phenotyping of bacteria. *Nature Nanotechnology* 8 (5), 369–375. doi:10.1038/nnano.2013.70.

Chung, H. J., T. Reiner, G. Budin, C. Min, M. Liong, D. Issadore, H. Lee, and R. Weissleder. 2011. Ubiquitous detection of Gram-positive bacteria with bioorthogonal magnetofluorescent nanoparticles. *ACS Nano* 5 (11), 8834–8841.

Crick, F. H. C., and A. F. W. Hughes. 1950. The physical properties of cytoplasm. *Experimental Cell Research* 1 (1), 37–80. doi:10.1016/0014-4827(50)90048-6.

D'Orazio, P. 2003. Biosensors in clinical chemistry. *Clinica Chimica Acta* 334 (1–2), 41–69.

de Vries, I. J., W. J. Lesterhuis, J. O. Barentsz, P. Verdijk, J. H. van Krieken, O. C. Boerman, W. J. Oyen, J. J. Bonenkamp, J. B. Boezeman, G. J. Adema, J. W. Bulte, T. W. Scheenen, C. J. Punt, A. Heerschap, and C. G. Figdor. 2005. Magnetic resonance tracking of dendritic cells in melanoma patients for monitoring of cellular therapy. *Nature Biotechnology* 23 (11), 1407–1413. doi:10.1038/nbt1154.

Dieny, B., V. S. Speriosu, S. Metin, S. S. P. Parkin, B. A. Gurney, P. Baumgart, and D. R. Wilhoit. 1991. Magnetotransport properties of magnetically soft spin-valve structures (invited). *Journal of Applied Physics* 69 (8), 4774. http://link.aip.org/link/JAPIAU/v69/i8/p4774/s1&Agg=doi.

Do, J., and C. H. Ahn. 2008. A polymer lab-on-a-chip for magnetic immunoassay with on-chip sampling and detection capabilities. *Lab Chip* 8 (4), 542–549. doi:10.1039/b715569g.

Edelstein, R. L., C. R. Tamanaha, P. E. Sheehan, M. M. Miller, D. R. Baselt, L. J. Whitman, and R. J. Colton. 2000. The BARC biosensor applied to the detection of biological warfare agents. *Biosensors and Bioelectronics* 14 (10–11), 805–813.

Ejsing, L., M. F. Hansen, A. K. Menon, H. A. Ferreira, D. L. Graham, and P. P. Freitas. 2004. Planar Hall effect sensor for magnetic micro-and nanobead detection. *Applied Physics Letters* 84, 4729.

Elbez, R., B. H. McNaughton, L. Patel, K. J. Pienta, and R. Kopelman. 2011. Nanoparticle induced cell magneto-rotation: Monitoring morphology, stress and drug sensitivity of a suspended single cancer cell. *PLoS One* 6 (12), e28475. doi:10.1371/journal.pone.0028475.

Enpuku, K., K. Inoue, K. Soejima, K. Yoshinaga, H. Kuma, and N. Hamasaki. 2005. Magnetic immunoassays utilizing magnetic markers and a high-Tc SQUID. *IEEE Transactions on Applied Superconductivity* 15 (2), 660–663.

Enpuku, K., K. Soejima, T. Nishimoto, T. Matsuda, H. Tokumitsu, T. Tanaka, K. Yoshinaga, H. Kuma, and N. Hamasaki. 2007. Biological immunoassays without bound/free separation utilizing magnetic marker and HTS SQUID. *IEEE Transactions on Applied Superconductivity* 17 (2), 816–819.

Erdem, A., M. I. Pividori, A. Lermo, A. Bonanni, M. del Valle, and S. Alegret. 2006. Genomagnetic assay based on label-free electrochemical detection using magneto-composite electrodes. *Sensors and Actuators B: Chemical* 114 (2), 591–598. http://apps.isiknowledge.com/CitedFullRecord.do?product=UA&db_id=WOS&SID=4BL5L1G2C26cdpkcA@9&search_mode=CitedFullRecord&isickref=149670028.

Fannin, P. C., L. Cohen-Tannoudji, E. Bertrand, A. T. Giannitsis, C. Mac Oireachtaigh, and J. Bibette. 2006. Investigation of the complex susceptibility of magnetic beads containing maghemite nanoparticles. *Journal of Magnetism and Magnetic Materials* 303 (1), 147–152. http://www.sciencedirect.com/science/article/B6TJJ-4HRN3H6-1/2/700aba01bb615b272c06e0a46e40b101.

Ferapontova, E. E., M. N. Hansen, A. M. Saunders, S. Shipovskov, D. S. Sutherland, and K. V. Gothelf. 2010. Electrochemical DNA sandwich assay with a lipase label for attomole detection of DNA. *Chemical Communications* 46 (11), 1836. http://xlink.rsc.org/?DOI=b924627d.

Ferreira, H. A., D. L. Graham, N. Feliciano, L. A. Clarke, M. D. Amaral, and P. P. Freitas. 2005. Detection of cystic fibrosis related DNA targets using AC field focusing of magnetic labels and spin-valve sensors. *IEEE Transactions on Magnetics* 41 (10), 4140–4142. doi:10.1109/tmag.2005.855340.

Ferreira, H. A., D. L. Graham, P. P. Freitas, and J. M. S. Cabral. 2003. Biodetection using magnetically labeled biomolecules and arrays of spin valve sensors (invited). *Journal of Applied Physics* 93, 7281.

Fritz, J. 2008. Cantilever biosensors. *The Analyst* 133 (7), 855. http://apps.isiknowledge.com/full_record.do?product=UA&search_mode=Refine&qid=58&SID=4Cae37FbIfNh@p1kGij&page=2&doc=14.

Frka-Petesic, B., K. Erglis, J. F. Berret, A. Cebers, V. Dupuis, J. Fresnais, O. Sandre, and R. Perzynski. 2011. Dynamics of paramagnetic nanostructured rods under rotating field. *Journal of Magnetism and Magnetic Materials* 323 (10), 1309–1313. doi:10.1016/j.jmmm.2010.11.036.

Gabureac, M. S., L. Bernau, G. Boero, and I. Utke. 2013. Single superparamagnetic bead detection and direct tracing of bead position using novel nanocomposite nano-hall sensors. *IEEE Transactions on Nanotechnology* 12 (5), 668–673.

Gallagher, F. A., M. I. Kettunen, S. E. Day, D. E. Hu, J. H. Ardenkjaer-Larsen, R. Zandt, P. R. Jensen, M. Karlsson, K. Golman, M. H. Lerche, and K. M. Brindle. 2008. Magnetic resonance imaging of pH in vivo using hyperpolarized 13C-labelled bicarbonate. *Nature* 453 (7197), 940–943. doi:10.1038/nature07017.

Gaster, R. S., D. A. Hall, C. H. Nielsen, S. J. Osterfeld, H. Yu, K. E. Mach, R. J. Wilson, B. Murmann, J. C. Liao, S. S. Gambhir, S. X. Wang. 2009. Matrix-insensitive protein assays push the limits of biosensors in medicine. *Nature Medicine* 1327–1332.

Gaster, R. S., D. A. Hall, and S. X. Wang. 2011. NanoLAB: An ultraportable, handheld diagnostic laboratory for global health. *Lab Chip* 11 (5), 950–956. doi:10.1039/c0lc00534g.

Gaster, R. S., L. Xu, S. J. Han, R. J. Wilson, D. A. Hall, S. J. Osterfeld, H. Yu, and S. X. Wang. 2011. Quantification of protein interactions and solution transport using high-density GMR sensor arrays. *Nature Nanotechnology* 6 (5), 314–320. doi:10.1038/nnano.2011.45.

Gaub, H. E., and H. M. McConnell. 1986. Shear viscosity of monolayers at the air-water interface. *The Journal of Physical Chemistry* 90 (26), 6830–6832. doi:10.1021/j100284a024.

Gehring, A. G., C. G. Crawford, R. S. Mazenko, L. J. Van Houten, and J. D. Brewster. 1996. Enzyme-linked immunomagnetic electrochemical detection of Salmonella typhimurium. *Journal of Immunological Methods* 195 (1–2), 15–25. http://www.sciencedirect.com/science/article/B6T2Y-3W3169C-3/2/08f8285b691b402c1ba46ed51403d678.

Ghazani, A. A., M. Pectasides, A. Sharma, C. M. Castro, M. Mino-Kenudson, H. Lee, J. A. Shepard, and R. Weissleder. 2014. Molecular characterization of scant lung tumor cells using iron-oxide nanoparticles and micro-nuclear magnetic resonance. *Nanomedicine* 10 (3), 661–668. doi:10.1016/j.nano.2013.10.008.

Ghosh, A., and P. Fischer. 2009. Controlled propulsion of artificial magnetic nanostructured propellers. *Nano Letters* 9 (6), 2243–2245.

Gleich, B. 2013. *Principles and Applications of Magnetic Particle Imaging.* Berlin/Heidelberg, Germany: Springer Science & Business Media.

Gleich, B., and J. Weizenecker. 2005. Tomographic imaging using the nonlinear response of magnetic particles. *Nature* 435 (7046), 1214–1217. doi:10.1038/nature03808.

Graham, D. L., H. A. Ferreira, and P. P. Freitas. 2004. Magnetoresistive-based biosensors and biochips. *TRENDS in Biotechnology* 22 (9), 455–462.

Graham, D. L., H. A. Ferreira, N. Feliciano, P. P. Freitas, L. A. Clarke, and M. D. Amaral. 2005. Magnetic field-assisted DNA hybridisation and simultaneous detection using micron-sized spin-valve sensors and magnetic nanoparticles. *Sensors and Actuators B: Chemical* 107 (2), 936–944.

Grancharov, S. G., H. Zeng, S. Sun, S. X. Wang, S. O'Brien, C. B. Murray, J. R. Kirtley, and G. A. Held. 2005. Bio-functionalization of monodisperse magnetic nanoparticles and their use as biomolecular labels in a magnetic tunnel junction based sensor. *Journal of Physical Chemical B* 109 (26), 13030–13035.

Grimm, J., J. M. Perez, L. Josephson, and R. Weissleder. 2004. Novel nanosensors for rapid analysis of telomerase activity. *Cancer Research* 64 (2), 639–643. http://cancerres.aacrjournals.org/content/64/2/639.abstract.

Gupta, A. K., and M. Gupta. 2005. Synthesis and surface engineering of iron oxide nanoparticles for biomedical applications. *Biomaterials* 26 (18), 3995–4021.

Gupta, A. K., R. R. Naregalkar, V. D. Vaidya, and M. Gupta. 2007. Recent advances on surface engineering of magnetic iron oxide nanoparticles and their biomedical applications. *Nanomedicine* 2 (1), 23–39.

Gusev, Y., J. Sparkowski, A. Raghunathan, H. Ferguson, J. Montano, N. Bogdan, B. Schweitzer, S. Wiltshire, S. F. Kingsmore, W. Maltzman, and V. Wheeler. 2001. Rolling circle amplification. *The American Journal of Pathology* 159 (1), 63–69.

Hall, D. A., R. S. Gaster, K. Makinwa, S. X. Wang, and B. Murmann. 2013. A 256 pixel magnetoresistive biosensor microarray in 0.18mum CMOS. *IEEE Journal of Solid-State Circuits* 48 (5), 1290–1301. doi:10.1109/JSSC.2013.2245058.

Hathaway, H. J., K. S. Butler, N. L. Adolphi, D. M. Lovato, R. Belfon, D. Fegan, T. C. Monson, J. E. Trujillo, T. E. Tessier, and H. C. Bryant. 2011. Detection of breast cancer cells using targeted magnetic nanoparticles and ultra-sensitive magnetic field sensors. *Breast Cancer Research* 13 (5), 1.

Haun, J. B., C. M. Castro, R. Wang, V. M. Peterson, B. S. Marinelli, H. Lee, and R. Weissleder. 2011. Micro-NMR for rapid molecular analysis of human tumor samples. *Science Translational Medicine* 3 (71), 71ra16.

Haun, J. B., T.-J. Yoon, H. Lee, and R. Weissleder. 2010. Magnetic nanoparticle biosensors. *Wiley Interdisciplinary Reviews: Nanomedicine and Nanobiotechnology.* http://onlinelibrary.wiley.com/doi/10.1002/wnan.84/pdf.

He, P., Z. Wang, L. Zhang, and W. Yang. 2009. Development of a label-free electrochemical immunosensor based on carbon nanotube for rapid determination of clenbuterol. *Food Chemistry* 112 (3), 707–714. doi:10.1016/j.foodchem.2008.05.116.

Hecht, A., P. Commiskey, F. Lazaridis, P. Argyrakis, and R. Kopelman. 2014. Fractal dimension of microbead assemblies used for protein detection. *ChemPhysChem* 15 (16), 3444–3446. doi:10.1002/cphc.201402048.

Hecht, A., P. Commiskey, N. Shah, and R. Kopelman. 2013. Bead assembly magnetorotation as a signal transduction method for protein detection. *Biosensors Bioelectronics* 48, 26–32. doi:10.1016/j.bios.2013.03.073.

Hecht, A., P. Kinnunen, B. H. McNaughton, and R. Kopelman. 2011. Label-acquired magnetorotation for biosensing: An asynchronous rotation assay. *Journal of Magnetism and Magnetic Materials* 323 (3–4), 272–278. http://www.sciencedirect.com/science/article/B6TJJ-512VTP7-5/2/63c6ad753d8fa40cf0d89ed032d87db2.

Hecht, A., A. A. Kumar, and R. Kopelman. 2011. Label-acquired magnetorotation as a signal transduction method for protein detection: Aptamer-based detection of thrombin. *Analytical Chemistry* 83 (18), 7123–7128. doi:10.1021/ac2014756.

Heim, D. E., R. E. Fontana, C. Tsang, V. S. Speriosu, B. A. Gurney, and M. L. Williams. 1994. Design and operation of spin valve sensors. *IEEE Transactions on Magnetics* 30 (2), 316–321.

Helgesen, G., P. Pieranski, and A. T. Skjeltorp. 1990. Nonlinear phenomena in systems of magnetic holes. *Physical Review Letters* 64 (12), 1425. http://link.aps.org/doi/10.1103/PhysRevLett.64.1425.

Herrmann, M., T. Veres, and M. Tabrizian. 2008. Quantification of low-picomolar concentrations of TNF-α in serum using the dual-network microfluidic ELISA platform. *Analytical Chemistry* 80 (13), 5160–5167.

Hill, H. D., and C. A. Mirkin. 2006. The bio-barcode assay for the detection of protein and nucleic acid targets using DTT-induced ligand exchange. *Nature Protocols* 1 (1), 324–336.

Hill, H. D., R. A. Vega, and C. A. Mirkin. 2007. Nonenzymatic detection of bacterial genomic DNA using the bio bar code assay. *Analytical Chemistry* 79 (23), 9218–9223.

Honda, T., K. I. Arai, K. Ishiyama, 1996. Micro swimming mechanisms propelled by external magnetic fields. *IEEE Transactions on Magnetics* 32 (5), 5085–5087.

Hsing, I., X. Ying, and Z. Wenting. 2007. Micro- and nano-magnetic particles for applications in biosensing. *Electroanalysis* 19 (7-8), 755–768.

Hu, L., J. Chen, X. Yang, S. D. Caruthers, G. M. Lanza, and S. A. Wickline. 2013. Rapid quantification of oxygen tension in blood flow with a fluorine nanoparticle reporter and a novel blood flow-enhanced-saturation-recovery sequence. *Magnetic Resonance in Medicine* 70 (1), 176–183. doi:10.1002/mrm.24436.

Huang, P.-J. J., and J. Liu. 2010. Flow cytometry-assisted detection of adenosine in serum with an immobilized aptamer sensor. *Analytical Chemistry* 82 (10), 4020–4026.

Issadore, D., C. Min, M. Liong, J. Chung, R. Weissleder, and H. Lee. 2011. Miniature magnetic resonance system for point-of-care diagnostics. *Lab Chip* 11 (13), 2282–2287. doi:10.1039/c1lc20177h.

Issadore, D., J. Chung, H. Shao, M. Liong, A. A. Ghazani, C. M. Castro, R. Weissleder, and H. Lee. 2012. Ultrasensitive clinical enumeration of rare cells ex vivo using a micro-hall detector. *Science Translational Medicine* 4 (141), 141ra92–141ra92.

Issadore, D., Y. I. Park, H. Shao, C. Min, K. Lee, M. Liong, R. Weissleder, and H. Lee. 2014. Magnetic sensing technology for molecular analyses. *Lab Chip* 14 (14), 2385–2397. doi:10.1039/c4lc00314d.

Janssen, X. J. A., A. J. Schellekens, K. van Ommering, L. J. van Ijzendoorn, and M. W. J. Prins. 2009. Controlled torque on superparamagnetic beads for functional biosensors. *Biosensors and Bioelectronics* 24 (7), 1937–1941.

Jeong Dae, S., J. Sang Don, and C. M. Ae. 2009. Spin valve ring sensors for superparamagnetic bead detections. *IEEE Transactions on Magnetics* 45 (6), 2730–2732. doi:10.1109/tmag.2009.2020539.

Jiang, Z., J. Llandro, T. Mitrelias, and J. A. C. Bland. 2006. An integrated microfluidic cell for detection, manipulation, and sorting of single micron-sized magnetic beads. *Journal of Applied Physics* 99, 08S105.

Kaittanis, C., S. Nath, and J. M. Perez. 2008. Rapid nanoparticle-mediated monitoring of bacterial metabolic activity and assessment of antimicrobial susceptibility in blood with magnetic relaxation. *PLoS One* 3 (9), e3253. doi:10.1371/journal.pone.0003253.

Kaittanis, C., S. Santra, and J. M. Perez. 2009. Role of nanoparticle valency in the nondestructive magnetic-relaxation-mediated detection and magnetic isolation of cells in complex media. *Journal of the American Chemical Society* 131 (35), 12780–12791. http://apps.isiknowledge.com/full_record.do?

product=UA&colname=WOS&search_mode=CitingArticles&qid=11&SID=4BL5L1G2C26cdpkcA@9&page=1&doc=2.

Kasatkin, S. I., N. P. Vasil'eva, and A. M. Murav'ev. 2010. Biosensors based on the thin-film magnetoresistive sensors. *Automation and Remote Control* 71 (1), 156–166.

Katz, E., I. Willner, and J. Wang. 2004. Electroanalytical and bioelectroanalytical systems based on metal and semiconductor nanoparticles. *Electroanalysis* 16 (1–2), 19–44. doi:10.1002/elan.200302930.

Keller, M., J. Schilling, and E. Sackmann. 2001. Oscillatory magnetic bead rheometer for complex fluid microrheometry. *Review of Scientific Instruments* 72 (9), 3626. doi:10.1063/1.1394185.

Kiely, J., P. Hawkins, P. Wraith, and R. Luxton. 2007. Paramagnetic particle detection for use with an immunoassay based biosensor. *IET Science, Measurement & Technology* 1 (5), 270–275.

Kim, G.-Y., and A. Son. 2010. Development and characterization of a magnetic bead-quantum dot nanoparticles based assay capable of *Escherichia coli* O157:H7 quantification. *Analytica Chimica Acta* 677 (1), 90–96. http://www.sciencedirect.com/science/article/B6TF4-50PVG3B-3/2/c89c572c205641723e4b2683d6eaec77.

Kinnunen, P., B. H. McNaughton, T. Albertson, I. Sinn, S. Mofakham, R. Elbez, D. W. Newton, A. Hunt, and R. Kopelman. 2012. Self-assembled magnetic bead biosensor for measuring bacterial growth and antimicrobial susceptibility testing. *Small* 8 (16), 2477–2482. doi:10.1002/smll.201200110.

Kinnunen, P., I. Sinn, B. H. McNaughton, and R. Kopelman. 2010. High frequency asynchronous magnetic bead rotation for improved biosensors. *Applied Physics Letters* 97 (22), 223701. http://link.aip.org.proxy.lib.umich.edu/link/APPLAB/v97/i22/p223701/s1&Agg=doi.

Kinnunen, P., I. Sinn, B. H. McNaughton, D. W. Newton, M. A. Burns, and R. Kopelman. 2011. Monitoring the growth and drug susceptibility of individual bacteria using asynchronous magnetic bead rotation sensors. *Biosensors and Bioelectronics*26 (5), 2751–2755. doi:10.1016/j.bios.2010.10.010.

Kinnunen, P., M. E. Carey, E. Craig, S. N. Brahmasandra, and B. H. McNaughton. 2014. Rapid bacterial growth and antimicrobial response using self-assembled magnetic bead sensors. *Sensors and Actuators B: Chemical* 190, 265–269. doi:10.1016/j.snb.2013.08.070.

Koets, M., T. van der Wijk, J. T. van Eemeren, A. van Amerongen, and M. W. Prins. 2009. Rapid DNA multi-analyte immunoassay on a magneto-resistance biosensor. *Biosensors and Bioelectronics*24 (7): 1893–1898. doi:10.1016/j.bios.2008.09.023.

Kotitz, R., H. Matz, L. Trahms, H. Koch, W. Weitschies, T. Rheinlander, W. Semmler, and T. Bunte. 1997. SQUID based remanence measurements for immunoassays. *IEEE Transactions on Applied Superconductivity* 7 (2), 3678–3681.

Kriz, C. B., K. Radevik, and D. Kriz. 1996. Magnetic permeability measurements in bioanalysis and biosensors. *Analytical Chemistry* 68 (11), 1966–1970.

Kriz, K., J. Gehrke, and D. Kriz. 1998. Advancements toward magneto immunoassays. *Biosensors and Bioelectronics* 13 (7–8), 817–823.

Kriz, K., F. Ibraimi, M. Lu, L. O. Hansson, and D. Kriz. 2005. Detection of C-reactive protein utilizing magnetic permeability detection based immunoassays. *Analytical Chemistry* 77 (18), 5920–5924.

Kuma, H., H. Oyamada, A. Tsukamoto, T. Mizoguchi, A. Kandori, Y. Sugiura, K. Yoshinaga, K. Enpuku, and N. Hamasaki. 2010. Liquid phase immunoassays utilizing magnetic markers and SQUID magnetometer. *Clinical Chemistry and Laboratory Medicine* 48 (9), 1263–1269. http://find.galegroup.com/gtx/retrieve.do?contentSet=IAC-Documents&resultListType=RESULT_LIST&qrySerId=Locale%28en%2CUS%2C%29%3AFQE%3D%28sp%2CNone%2C4%291263%3AAnd%3AFQE%3D%28iu%2CNone%2C1%299%3AAnd%3AFQE%3D%28sn%2CNone%2C9%291434-6621%3AAnd%3AFQE%3D%28vo%2CNone%2C2%2948%24&sgHitCountType=None&inPS=true&sort=DateDescend&searchType=AdvancedSearchForm&tabID=T002&prodId=AONE&searchId=R1¤tPosition=1&userGroupName=lom_umichanna&docId=A238477513&docType=IAC.

Kuramitz, H. 2009. Magnetic microbead-based electrochemical immunoassays. *Analytical and Bioanalytical Chemistry* 394 (1), 61–69. doi.org/10.1007/s00216-009-2650-y.

Lagae, L., R. Wirix-Speetjens, J. Das, D. Graham, H. Ferreira, P. P. F. Freitas, G. Borghs, and J. De Boeck. 2002. On-chip manipulation and magnetization assessment of magnetic bead ensembles by integrated spin-valve sensors. *Journal of Applied Physics* 91 (10), 7445. doi:10.1063/1.1447288.

Lea, T., F. Vartdal, K. Nustad, S. Funderud, A. Berge, T. Ellingsen, R. Schmid, P. Stenstad, and J. Ugelstad. 1988. Monosized, magnetic polymer particles: Their use in separation of cells and subcellular components, and in the study of lymphocyte function in vitro. *Journal of Molecular Recognition* 1 (1), 9–18. doi:10.1002/jmr.300010104.

Lee, H., E. Sun, D. Ham, and R. Weissleder. 2008. Chip-NMR biosensor for detection and molecular analysis of cells. *Nature Medicine* 14 (8), 869–874. doi:10.1038/nm.1711.

Lee, H., T.-J. Yoon, J.-L. Figueiredo, F. K. Swirski, and R. Weissleder. 2009. Rapid detection and profiling of cancer cells in fine-needle aspirates. *Proceedings of the National Academy of Sciences* 106 (30), 12459–12464.

Lee, H., T.-J. Yoon, and R. Weissleder. 2009. Ultrasensitive detection of bacteria using core–shell nanoparticles and an NMR-filter system. *Angewandte Chemie International Edition* 48 (31), 5657–5660.

Lee, Y.-E. K., and R. Kopelman. 2009. Optical nanoparticle sensors for quantitative intracellular imaging. *Wiley Interdisciplinary Reviews: Nanomedicine and Nanobiotechnology* 1 (1), 98–110.

Lee, Y.-E. K., R. Kopelman, and R. Smith. 2009. Nanoparticle PEBBLE sensors in live cells and in vivo. *Annu Rev Anal Chem (Palo Alto Calif).* 2, 57–76. doi:10.1146/annurev.anchem. https://www.ncbi.nlm.nih.gov/pmc/articles/PMC2809932/pdf/nihms121844.pdf.

Li, G., V. Joshi, R. L. White, S. X. Wang, J. T. Kemp, C. Webb, R. W. Davis, and S. Sun. 2003. Detection of single micron-sized magnetic bead and magnetic nanoparticles using spin valve sensors for biological applications. *Journal of Applied Physics* 93, 7557.

Li, G., S. Sun, R. J. Wilson, R. L. White, N. Pourmand, and S. X. Wang. 2006. Spin valve sensors for ultrasensitive detection of superparamagnetic nanoparticles for biological applications. *Sensors and Actuators A: Physical* 126 (1), 98–106.

Li, J., H. Gao, Z. Chen, X. Wei, and C. F. Yang. 2010. An electrochemical immunosensor for carcinoembryonic antigen enhanced by self-assembled nanogold coatings on magnetic particles. *Analytical Chimica Acta* 665 (1), 98–104. doi:10.1016/j.aca.2010.03.020.

Li, J., Q. Xu, X. Wei, and Z. Hao. 2013. Electrogenerated chemiluminescence immunosensor for *Bacillus thuringiensis* Cry1Ac based on Fe_3O_4@Au nanoparticles. *Journal of Agricultural and Food Chemistry* 61 (7), 1435–1440. doi:10.1021/jf303774x.

Li, Y., D. T. Burke, R. Kopelman, and M. A. Burns. 2014. Asynchronous magnetic bead rotation (AMBR) microviscometer for label-free DNA analysis. *Biosensors (Basel)* 4 (1), 76–89. doi:10.3390/bios4010076.

Li, Y., X. Ji, and B. Liu. 2011. Chemiluminescence aptasensor for cocaine based on double-functionalized gold nanoprobes and functionalized magnetic microbeads. *Analytical and Bioanalytical Chemistry* 401 (1), 213–219. doi:10.1007/s00216-011-5064-6.

Li, Y., B. Srinivasan, Y. Jing, X. Yao, M. A. Hugger, J.-P. Wang, and C. Xing. 2010. Nanomagnetic competition assay for low-abundance protein biomarker quantification in unprocessed human sera. *Journal of the American Chemical Society* 132 (12), 4388–4392.

Lim, S. H., F. Bestvater, P. Buchy, S. Mardy, and A. D. C. Yu. 2009. Quantitative analysis of nucleic acid hybridization on magnetic particles and quantum dot-based probes. *Sensors* 9 (7), 5590–5599. http://apps.isiknowledge.com/full_record.do?product=UA&colname=WOS&search_mode=CitingArticles&qid=26&SID=4Cae37FbIfNh@p1kGij&page=1&doc=6.

Liong, M., A. N. Hoang, J. Chung, N. Gural, C. B. Ford, C. Min, R. R. Shah, R. Ahmad, M. Fernandez-Suarez, S. M. Fortune, M. Toner, H. Lee, and R. Weissleder. 2013. Magnetic barcode assay for genetic detection of pathogens. *Nature Communications* 4, 1752. doi:10.1038/ncomms2745.

Liu, G., J. Wang, J. Kim, M. R. Jan, and G. E. Collins. 2004. Electrochemical coding for multiplexed immunoassays of proteins. *Analytical Chemistry* 76 (23), 7126–7130. doi:10.1021/ac049107l.

Liu, Y., D. Yao, H. Chang, C. Liu, and C. Chen. 2008. Magnetic bead-based DNA detection with multi-layers quantum dots labeling for rapid detection of *Escherichia coli* O157:H7. *Biosensors and Bioelectronics* 24 (4), 558–565. http://apps.isiknowledge.com/full_record.do?product=UA&search_mode=GeneralSearch&qid=1&SID=4DCMgl@oDdE2BD24kJk&page=1&doc=1&colname=WOS.

Longo, D. L., P. Z. Sun, L. Consolino, F. C. Michelotti, F. Uggeri, and S. Aime. 2014. A general MRI-CEST ratiometric approach for pH imaging: Demonstration of in vivo pH mapping with iobitridol. *Journal of the American Chemical Society* 136 (41), 14333–14336. doi:10.1021/ja5059313.

Loureiro, J., P. Z. Andrade, S. Cardoso, C. L. da Silva, J. M. Cabral, and P. P. Freitas. 2011. Magnetoresistive chip cytometer. *Lab Chip* 11 (13), 2255–2261. doi:10.1039/c0lc00324g.

Loureiro, J., R. Ferreira, S. Cardoso, P. P. Freitas, J. Germano, C. Fermon, G. Arrias, M. Pannetier-Lecoeur, F. Rivadulla, and J. Rivas. 2009. Toward a magnetoresistive chip cytometer: Integrated detection of magnetic beads flowing at cm/s velocities in microfluidic channels. *Applied Physics Letters* 95, 034104.

Marquina, C., J. M. de Teresa, D. Serrate, J. Marzo, F. A. Cardoso, D. Saurel, S. Cardoso, P. P. Freitas, and M. R. Ibarra. 2012. GMR sensors and magnetic nanoparticles for immuno-chromatographic assays. *Journal of Magnetism and Magnetic Materials* 324 (21), 3495–3498. doi:10.1016/j.jmmm.2012.02.074.

McNaughton, B. H., R. R. Agayan, R. Clarke, R. G. Smith, and R. Kopelman. 2007. Single bacterial cell detection with nonlinear rotational frequency shifts of driven magnetic microspheres. *Applied Physics Letters* 91, 224105.

McNaughton, B. H., R. R. Agayan, J. X. Wang, and R. Kopelman. 2007. Physiochemical microparticle sensors based on nonlinear magnetic oscillations. *Sensors & Actuators: B. Chemical* 121 (1), 330–340.

McNaughton, B. H., J. N. Anker, and R. Kopelman. 2005. Magnetic microdrill as a modulated fluorescent pH sensor. *Journal of Magnetism and Magnetic Materials* 293 (1), 696–701. http://www.sciencedirect.com/science/article/B6TJJ-4FTS076-6/2/5392ee1f14f984193660a388cd24db45.

McNaughton, B. H., K. A. Kehbein, J. N. Anker, and R. Kopelman. 2006. Sudden breakdown in linear response of a rotationally driven magnetic microparticle and application to physical and chemical microsensing. *Journal of Physical Chemistry B* 110 (38), 18958.

McNaughton, B. H., P. Kinnunen, R. G. Smith, S. N. Pei, R. Torres-Isea, R. Kopelman, and R. Clarke. 2009. Compact sensor for measuring nonlinear rotational dynamics of driven magnetic microspheres with biomedical applications. *Journal of Magnetism and Magnetic Materials* 321 (10), 1648–1652. http://www.sciencedirect.com/science/article/B6TJJ-4VNKGYC-N/2/0a4b577717a9a7e2c9e779c7d5d93465.

Megens, M., and M. Prins. 2005. Magnetic biochips: A new option for sensitive diagnostics. *Journal of Magnetism and Magnetic Materials* 293 (1), 702–708.

Mihajlović, G., P. Xiong, S. von Molnár, K. Ohtani, H. Ohno, M. Field, and G. J. Sullivan. 2005. Detection of single magnetic bead for biological applications using an InAs quantum-well micro-Hall sensor. *Applied Physics Letters* 87 (11), 112502. http://link.aip.org/link/APPLAB/v87/i11/p112502/s1&Agg=doi.

Miller, M. M., G. A. Prinz, S. F. Cheng, and S. Bounnak. 2002. Detection of a micron-sized magnetic sphere using a ring-shaped anisotropic magnetoresistance-based sensor: A model for a magnetoresistance-based biosensor. *Applied Physics Letters* 81 (12), 2211. http://link.aip.org/link/APPLAB/v81/i12/p2211/s1&Agg=doi.

Miller, M. M., P. E. Sheehan, R. L. Edelstein, C. R. Tamanaha, L. Zhong, S. Bounnak, L. J. Whitman, and R. J. Colton. 2001. A DNA array sensor utilizing magnetic microbeads and magnetoelectronic detection. *Journal of Magnetism and Magnetic Materials* 225 (1–2), 138–144. http://www.sciencedirect.com/science/article/B6TJJ-42T4FB3-S/2/6d15a99389fa27c041ea5c5ff6d88b30.

Moodera, J. S., L. R. Kinder, T. M. Wong, and R. Meservey. 1995. Large magnetoresistance at room temperature in ferromagnetic thin film tunnel junctions. *Physical Review Letters* 74 (16), 3273. http://link.aps.org/doi/10.1103/PhysRevLett.74.3273.

Morozov, V. N., S. Groves, M. J. Turell, and C. Bailey. 2007. Three minutes-long electrophoretically assisted zeptomolar microfluidic immunoassay with magnetic-beads detection. *Journal of the American Chemical Society* 129 (42), 12628–12629. doi:10.1021/ja075069m.

Morozova, T. Y., and V. N. Morozov. 2008. Force differentiation in recognition of cross-reactive antigens by magnetic beads. *Analytical Biochemistry* 374 (2), 263–271. http://www.sciencedirect.com/science/article/B6W9V-4R9JTXB-2/2/29f23994f4d52579722ed41a9f21ee68.

Moskowitz, R., and R. E. Rosensweig. 1967. Nonmechanical torque-driven flow of a ferromagnetic fluid by an electromagnetic field. *Applied Physics Letters*.11pp.301

Mulvaney, S. P., C. L. Cole, M. D. Kniller, M. Malito, C. R. Tamanaha, J. C. Rife, M. W. Stanton, and L. J. Whitman. 2007. Rapid, femtomolar bioassays in complex matrices combining microfluidics and magnetoelectronics. *Biosensors and Bioelectronics* 23 (2), 191–200.

Mulvaney, S. P., C. N. Ibe, C. R. Tamanaha, and L. J. Whitman. 2009. Direct detection of genomic DNA with fluidic force discrimination assays. *Analytical Biochemistry* 392 (2), 139–144. http://www.sciencedirect.com/science/article/B6W9V-4WF4J57-5/2/d70c54d9fee98336fa2efdc517c9fa49.

Munge, B. S., A. L. Coffey, J. M. Doucette, B. K. Somba, R. Malhotra, V. Patel, J. S. Gutkind, and J. F. Rusling. 2011. Nanostructured immunosensor for attomolar detection of cancer biomarker interleukin-8 using massively labeled superparamagnetic particles. *Angewandte Chemie International Edition* 50 (34), 7915–7918. doi:10.1002/anie.201102941.

Murase, K., R. Song, and S. Hiratsuka. 2014. Magnetic particle imaging of blood coagulation. *Applied Physics Letters* 104 (25), 252409. doi:10.1063/1.4885146.

Nam, J. M., K. J. Jang, and J. T. Groves. 2007. Detection of proteins using a colorimetric bio-barcode assay. *Nature Protocols* 2 (6), 1438–1444.

Nam, J. M., S. J. Park, and C. A. Mirkin. 2002. Bio-barcodes based on oligonucleotide-modified nanoparticles. *Journal of the American Chemical Society* 124 (15), 3820–3821.

Nam, J. M., S. I. Stoeva, and C. A. Mirkin. 2004. Bio-bar-code-based DNA detection with PCR-like sensitivity. *Journal of the American Chemical Society* 126 (19), 5932–5933.

Nam, J. M., C. S. Thaxton, and C. A. Mirkin. 2003. Nanoparticle-based bio-bar codes for the ultrasensitive detection of proteins. *Science* 301 (5641), 1884.

Ng, A. H., K. Choi, R. P. Luoma, J. M. Robinson, and A. R. Wheeler. 2012. Digital microfluidic magnetic separation for particle-based immunoassays. *Analytical Chemistry* 84 (20), 8805–8812. doi:10.1021/ac3020627.

Obata, K., H. Tajima, M. Yohda, and T. Matsunaga. 2002. Recent developments in laboratory automation using magnetic particles for genome analysis. *Pharmacogenomics* 3 (5), 697–708.

Oh, B. K., J. M. Nam, S. W. Lee, and C. A. Mirkin. 2006. A fluorophore-based bio-barcode amplification assay for proteins. *Small* 2 (1), 103–108.

Osterfeld, S. J., H. Yu, R. S. Gaster, S. Caramuta, L. Xu, S.-J. Han, D. A. Hall, R. J. Wilson, S. Sun, R. L. White, R. W. Davis, N. Pourmand, and S. X. Wang. 2008. Multiplex protein assays based on real-time magnetic nanotag sensing. *Proceedings of the National Academy of Sciences* 105 (52), 20637–20640. http://www.pnas.org/content/105/52/20637.abstract.

Pablico-Lansigan, M. H., S. F. Situ, and A. C. Samia. 2013. Magnetic particle imaging: Advancements and perspectives for real-time in vivo monitoring and image-guided therapy. *Nanoscale* 5 (10), 4040–4055. doi:10.1039/c3nr00544e.

Palecek, E., M. Fojta, and F. Jelen. 2002. New approaches in the development of DNA sensors: Hybridization and electrochemical detection of DNA and RNA at two different surfaces. *Bioelectrochemistry* 56 (1–2), 85–90. http://www.sciencedirect.com/science/article/B6W72-454TMDK-3/2/e41fec3453fd75f07a5bb024c2f45aaa.

Panagiotopoulos, N., R. L. Duschka, M. Ahlborg, G. Bringout, C. Debbeler, M. Graeser, C. Kaethner, K. Ludtke-Buzug, H. Medimagh, J. Stelzner, T. M. Buzug, J. Barkhausen, F. M. Vogt, and J. Haegele. 2015. Magnetic particle imaging: Current developments and future directions. *International Journal of Nanomedicine* 10, 3097–3114. doi:10.2147/IJN.S70488.

Park, S. Y., H. Handa, and A. Sandhu. 2010. Magneto-optical biosensing platform based on light scattering from self-assembled chains of functionalized rotating magnetic beads. *Nano Letters* 10 (2), 446–451. doi:10.1021/nl9030488.

Perez, J. M., L. Josephson, T. O'Loughlin, D. Hogemann, and R. Weissleder. 2002. Magnetic relaxation switches capable of sensing molecular interactions. *Nature Biotechnology* 20 (8), 816–820. doi:10.1038/nbt720.

Peyman, S. A., A. Iles, and N. Pamme. 2009. Mobile magnetic particles as solid-supports for rapid surface-based bioanalysis in continuous flow. *Lab on a Chip* 9 (21), 3110–3117.

Popplewell, J., R. E. Rosensweig, and R. J. Johnston. 1990. Magnetic field induced rotations in ferrofluids. *IEEE Transactions on Magnetics* 26 (5), 1852–1854.

Rahmer, J., A. Antonelli, C. Sfara, B. Tiemann, B. Gleich, M. Magnani, J. Weizenecker, and J. Borgert. 2013. Nanoparticle encapsulation in red blood cells enables blood-pool magnetic particle imaging hours after injection. *Physics in Medicine and Biology* 58 (12), 3965–3977. doi:10.1088/0031-9155/58/12/3965.

Ranzoni, A., X. J. Janssen, M. Ovsyanko, I. J. L. J. van, and M. W. Prins. 2010. Magnetically controlled rotation and torque of uniaxial microactuators for lab-on-a-chip applications. *Lab Chip* 10 (2), 179–188. doi:10.1039/b909998k.

Ranzoni, A., J. J. Schleipen, L. J. van Ijzendoorn, and M. W. Prins. 2011. Frequency-selective rotation of two-particle nanoactuators for rapid and sensitive detection of biomolecules. *Nano Letters* 11 (5), 2017–2022. doi:10.1021/nl200384p.

Rho, J., J. Chung, H. Im, M. Liong, H. Shao, C. M. Castro, R. Weissleder, and H. Lee. 2013. Magnetic nanosensor for detection and profiling of erythrocyte-derived microvesicles. *ACS Nano* 7 (12), 11227–11233.

Richardson, J., P. Hawkins, and R. Luxton. 2001. The use of coated paramagnetic particles as a physical label in a magneto-immunoassay. *Biosensors and Bioelectronics* 16 (9-12), 989–993. http://www.sciencedirect.com/science/article/B6TFC-447MWX3-1K/2/f9a0cbcf3b2c406f0e6e0f875244e761.

Richardson, J., A. Hill, R. Luxton, and P. Hawkins. 2001. A novel measuring system for the determination of paramagnetic particle labels for use in magneto-immunoassays. *Biosensors and Bioelectronics* 16 (9–12), 1127–1132.

Rife, J. 2003. Design and performance of GMR sensors for the detection of magnetic microbeads in biosensors. *Sensors and Actuators A: Physical* 107 (3), 209–218. http://apps.isiknowledge.com/full_record.do?product=UA&colname=WOS&search_mode=CitingArticles&qid=33&SID=4Cae37FbIfNh@p1kGij&page=26&doc=252.

Rissin, D. M., C. W. Kan, T. G. Campbell, S. C. Howes, D. R. Fournier, L. Song, T. Piech, P. P. Patel, L. Chang, A. J. Rivnak, E. P. Ferrell, J. D. Randall, G. K. Provuncher, D. R. Walt, and D. C. Duffy. 2010. Single-molecule enzyme-linked immunosorbent assay detects serum proteins at subfemtomolar concentrations. *Nature Biotechnology* 28 (6), 595–599. doi:10.1038/nbt.1641.

Rizzi, G., J. R. Lee, P. Guldberg, M. Dufva, S. X. Wang, and M. F. Hansen. 2016. Denaturation strategies for detection of double stranded PCR products on GMR magnetic biosensor array. *Biosensors and Bioelectronics*. doi:10.1016/j.bios.2016.09.031.

Roberts, T. G., J. N. Anker, and R. Kopelman. 2005. Magnetically modulated optical nanoprobes (MagMOONs) for detection and measurement of biologically important ions against the natural background fluorescence of intracellular environments. *Journal of Magnetism and Magnetic Materials* 293 (1), 715–724. doi:10.1016/j.jmmm.2005.02.070.

Rocha-Santos, T. A. P. 2014. Sensors and biosensors based on magnetic nanoparticles. *TrAC Trends in Analytical Chemistry* 62, 28–36. doi:10.1016/j.trac.2014.06.016.

Safarık, I., and M. Safarıkova. 1999. Use of magnetic techniques for the isolation of cells. *Journal of Chromatography B* 722, 33–53.

Safarik, I., and M. Safarikova. 2004. Magnetic techniques for the isolation and purification of proteins and peptides. *BioMagnetic Research and Technology* 2 (1), 7.

Schotter, J. 2004. Comparison of a prototype magnetoresistive biosensor to standard fluorescent DNA detection. *Biosensors and Bioelectronics* 19 (10), 1149–1156. http://apps.isiknowledge.com/full_record.do?product=UA&colname=WOS&search_mode=CitingArticles&qid=33&SID=4Cae37FbIfNh@p1kGij&page=25&doc=244.

Schrittwieser, S., B. Pelaz, W. J. Parak, S. Lentijo-Mozo, K. Soulantica, J. Dieckhoff, F. Ludwig, A. Guenther, A. Tschope, and J. Schotter. 2016. Homogeneous biosensing based on magnetic particle labels. *Sensors (Basel)* 16 (6). doi:10.3390/s16060828.

Shao, H., J. Chung, L. Balaj, A. Charest, D. D. Bigner, B. S. Carter, F. H. Hochberg, X. O. Breakefield, R. Weissleder, and H. Lee. 2012. Protein typing of circulating microvesicles allows real-time monitoring of glioblastoma therapy. *Nature Medicine* 18 (12), 1835–1840. doi:10.1038/nm.2994.

Sheehan, P. E., and L. J. Whitman. 2005. Detection limits for nanoscale biosensors. *Nano Letters* 5 (4), 803–807. doi:10.1021/nl050298x.

Shelton, D. R., and J. S. Karns. 2001. Quantitative detection of *Escherichia coli* O157 in surface waters by using immunomagnetic electrochemiluminescence. *Applied and Environmental Microbiology*. 67 (7), 2908–2915. http://aem.asm.org/cgi/content/abstract/67/7/2908.

Shen, W., X. Liu, D. Mazumdar, and G. Xiao. 2005. In situ detection of single micron-sized magnetic beads using magnetic tunnel junction sensors. *Applied Physics Letters* 86, 253901.

Shen, W., B. D. Schrag, M. J. Carter, J. Xie, C. Xu, S. Sun, and G. Xiao. 2008. Detection of DNA labeled with magnetic nanoparticles using MgO-based magnetic tunnel junction sensors. *Journal of Applied Physics* 103, 07A306.

Shlyapnikov, Y. M., E. A. Shlyapnikova, T. Y. Morozova, I. P. Beletsky, and V. N. Morozov. 2010. Detection of microarray-hybridized oligonucleotides with magnetic beads. *Analytical Biochemistry* 399 (1), 125–131. http://www.sciencedirect.com/science/article/B6W9V-4Y1NV5X-2/2/b47fc339d67aaa422184ee2f3b952ea9.

Sinn, I., T. Albertson, P. Kinnunen, D. N. Breslauer, B. H. McNaughton, M. A. Burns, and R. Kopelman. 2012. Asynchronous magnetic bead rotation microviscometer for rapid, sensitive, and label-free studies of bacterial growth and drug sensitivity. *Analytical Chemistry* 84 (12), 5250–5256. doi:10.1021/ac300128p.

Sinn, I., P. Kinnunen, T. Albertson, B. H. McNaughton, D. W. Newton, M. A. Burns, and R. Kopelman. 2011. Asynchronous magnetic bead rotation (AMBR) biosensor in microfluidic droplets for rapid bacterial growth and susceptibility measurements. *Lab Chip* 11 (15), 2604–2611. doi:10.1039/c0lc00734j.

Skucha, K., P. Liu, M. Megens, J. Kim, and B. Boser. 2011. A compact Hall-effect sensor array for the detection and imaging of single magnetic beads in biomedical assays. *16th International Solid-State Sensors, Actuators and Microsystems Conference*, 1833–1836.

Speriosu, V. S., B. Dieny, P. Humbert, B. A. Gurney, and H. Lefakis. 1991. Nonoscillatory magnetoresistance in Co/Cu/Co layered structures with oscillatory coupling. *Physical Review B* 44 (10), 5358–5361.

Stoeva, S. I., J. S. Lee, J. E. Smith, S. T. Rosen, and C. A. Mirkin. 2006. Multiplexed detection of protein cancer markers with biobarcoded nanoparticle probes. *Journal of the American Chemistry Society* 128 (26), 8378–8379.

Stoeva, S. I., J. S. Lee, C. S. Thaxton, and C. A. Mirkin. 2006. Multiplexed DNA detection with biobarcoded nanoparticle probes. *Angewandte Chemie* 118 (20), 3381–3384.

Strömberg, M., J. Göransson, K. Gunnarsson, M. Nilsson, P. Svedlindh, and M. Strømme. 2008a. Sensitive molecular diagnostics using volume-amplified magnetic nanobeads. *Nano Letters* 8 (3), 816–821. doi:10.1021/nl072760e.

Strömberg, M., T. Zardán Gómez de la Torre, J. Göransson, K. Gunnarsson, M. Nilsson, M. Strømme, and P. Svedlindh. 2008b. Microscopic mechanisms influencing the volume amplified magnetic nanobead detection assay. *Biosensors and Bioelectronics* 24 (4), 696–703. http://www.sciencedirect.com/science/article/B6TFC-4SXH9X2-3/2/afd6bade2a464e0861cf6ca86c94ec22.

Strömberg, M., T. Z. Gómez de la Torre, J. Göransson, K. Gunnarsson, M. Nilsson, P. Svedlindh, and M. Strømme. 2009. Multiplex detection of DNA sequences using the volume-amplified magnetic nanobead detection assay. *Analytical Chemistry* 81 (9), 3398–3406. doi:10.1021/ac900561r.

Sun, N., Y. Liu, H. Lee, R. Weissleder, and D. Ham. 2009. CMOS RF biosensor utilizing nuclear magnetic resonance. *IEEE Journal of Solid-State Circuits* 44 (5), 1629–1643. doi:10.1109/jssc.2009.2017007.

Tamanaha, C. R., S. P. Mulvaney, J. C. Rife, and L. J. Whitman. 2008. Magnetic labeling, detection, and system integration. *Biosensors and Bioelectronics* 24 (1), 1–13.

Taton, T. A., C. A. Mirkin, and R. L. Letsinger. 2000. Scanometric DNA array detection with nanoparticle probes. *Science* 289 (5485), 1757.

Thaxton, C. S., R. Elghanian, A. D. Thomas, S. I. Stoeva, J. S. Lee, N. D. Smith, A. J. Schaeffer, H. Klocker, W. Horninger, G. Bartsch, and C. A. Mirkin. 2009. Nanoparticle-based bio-barcode assay redefines "undetectable" PSA and biochemical recurrence after radical prostatectomy. *Proceedings of the National Academy of Sciences of the USA* 106 (44), 18437–18442. doi:10.1073/pnas.0904719106.

Thomas, J., S. Kim, P. Hesketh, H. Halsall, and W. Heineman. 2004. Microbead-based electrochemical immunoassay with interdigitated array electrodes. *Analytical Biochemistry* 328 (2), 113–122. http://apps.isiknowledge.com/CitedFullRecord.do?product=UA&db_id=WOS&SID=3F9k9AG37dGoJlJka@d&search_mode=CitedFullRecord&isickref=134517330.

Todd, J., B. Freese, A. Lu, D. Held, J. Morey, R. Livingston, and P. Goix. 2007. Ultrasensitive flow-based immunoassays using single-molecule counting. *Clinical Chemistry* 53 (11), 1990–1995. doi:10.1373/clinchem.2007.091181.

Tokarev, A., I. Luzinov, J. R. Owens, and K. G. Kornev. 2012. Magnetic rotational spectroscopy with nanorods to probe time-dependent rheology of microdroplets. *Langmuir* 28 (26), 10064–10071. doi:10.1021/la3019474.

Tokarev, A., B. Kaufman, Y. Gu, T. Andrukh, P. H. Adler, and K. G. Kornev. 2013. Probing viscosity of nanoliter droplets of butterfly saliva by magnetic rotational spectroscopy. *Applied Physics Letters* 102 (3), 033701. doi:10.1063/1.4788927.

Tsukamoto, A., K. Saitoh, D. Suzuki, N. Sugita, Y. Seki, A. Kandori, K. Tsukada, Y. Sugiura, S. Hamaoka, H. Kuma, N. Hamasaki, and K. Enpuku. 2005. Development of multisample biological immunoassay system using HTSSQUID and magnetic nanoparticles. *IEEE Transactions on Applied Superconductivity* 15 (2), 656–659.

Valberg, P. A., and H. A. Feldman. 1987. Magnetic particle motions within living cells. Measurement of cytoplasmic viscosity and motile activity. *Biophysical Journal* 52 (4), 551.

van Reenen, A., A. M. de Jong, J. M. den Toonder, and M. W. Prins. 2014. Integrated lab-on-chip biosensing systems based on magnetic particle actuation—A comprehensive review. *Lab Chip* 14 (12), 1966–1986. doi:10.1039/c3lc51454d.

Wang, J. 2003. Particle-based detection of DNA hybridization using electrochemical stripping measurements of an iron tracer. *Analytica Chimica Acta* 482 (2), 149–155. http://apps.isiknowledge.com/CitedFullRecord.do?product=UA&db_id=WOS&SID=4BL5L1G2C26cdpkcA@9&search_mode=CitedFullRecord&isickref=129260972.

Wang, J., G. Liu, and M. R. Jan. 2004. Ultrasensitive electrical biosensing of proteins and DNA: carbon-nanotube derived amplification of the recognition and transduction events. *Journal of the American Chemical Society* 126 (10), 3010–3011. doi:10.1021/ja031723w.

Wang, J., R. Polsky, and D. Xu. 2001. Silver-enhanced colloidal gold electrochemical stripping detection of DNA hybridization. *Langmuir* 17 (19), 5739–5741 doi:10.1021/la011002f.

Wang, J., D. Xu, A.-N. Kawde, and R. Polsky. 2001. Metal nanoparticle-based electrochemical stripping potentiometric detection of DNA hybridization. *Analytical Chemistry* 73 (22), 5576–5581. http://apps.isiknowledge.com/CitedFullRecord.do?product=UA&db_id=WOS&SID=4BL5L1G2C26cdpkcA@9&search_mode=CitedFullRecord&isickref=122297636.

Wang, J., D. Xu, and R. Polsky. 2002. Magnetically-induced solid-state electrochemical detection of DNA hybridization. *Journal of the American Chemical Society* 124 (16), 4208–4209. doi:10.1021/ja0255709.

Wang, S. X., S. Y. Bae, G. Li, S. Sun, R. L. White, J. T. Kemp, and C. D. Webb. 2005. Towards a magnetic microarray for sensitive diagnostics. *Journal of Magnetism and Magnetic Materials* 293 (1), 731–736.

Wang, W., Y. Wang, L. Tu, Y. Feng, T. Klein, and J. P. Wang. 2014. Magnetoresistive performance and comparison of supermagnetic nanoparticles on giant magnetoresistive sensor-based detection system. *Scientific Reports* 4, 5716. doi:10.1038/srep05716.

Wang, Z., X. Wang, S. Liu, J. Yin, and H. Wang. 2010. Fluorescently imaged particle counting immunoassay for sensitive detection of DNA modifications. *Analytical Chemistry* 82 (23), 9901–9908. doi:10.1021/ac102416f.

Weber, P. C., D. H. Ohlendorf, J. J. Wendoloski, and F. R. Salemme. 1989. Structural origins of high-affinity biotin binding to streptavidin. *Science* 243 (4887), 85.

Weizenecker, J., B. Gleich, J. Rahmer, H. Dahnke, and J. Borgert. 2009. Three-dimensional real-time in vivo magnetic particle imaging. *Physics in Medicine and Biology* 54 (5), L1–L10. doi:10.1088/0031-9155/54/5/L01.

Weizmann, Y., R. Elnathan, O. Lioubashevski, and I. Willner. 2005. Magnetomechanical detection of the specific activities of endonucleases by cantilevers. *Nano Letters* 5 (4), 741–744. http://apps.isiknowledge.com/full_record.do?product=UA&colname=WOS&search_mode=CitingArticles&qid=45&SID=4Cae37FbIfNh@p1kGij&page=4&doc=31.

Weizmann, Y., F. Patolsky, E. Katz, and I. Willner. 2003. Amplified DNA sensing and immunosensing by the rotation of functional magnetic particles. *Journal of the American Chemical Society* 125 (12), 3452–3454 doi:10.1021/ja028850x.

Weizmann, Y., F. Patolsky, O. Lioubashevski, and I. Willner. 2004. Magneto-mechanical detection of nucleic acids and telomerase activity in cancer cells. *Journal of the American Chemical Society* 126 (4), 1073–1080. doi:10.1021/ja038257v.

Wood, D. K., K. K. Ni, D. R. Schmidt, and A. N. Cleland. 2005. Submicron giant magnetoresistive sensors for biological applications. *Sensors and Actuators A: Physical* 120 (1), 1–6.

Xiang, Y., and Y. Lu. 2011. Using personal glucose meters and functional DNA sensors to quantify a variety of analytical targets. *Nature Chemistry* 3 (9), 697–703. doi:10.1038/nchem.1092.

Xin, Y., X. Fu-bing, L. Hong-wei, W. Feng, C. Di-zhao, and W. Zhao-yang. 2013. A novel H_2O_2 biosensor based on Fe_3O_4–Au magnetic nanoparticles coated horseradish peroxidase and graphene sheets–Nafion film modified screen-printed carbon electrode. *Electrochimica Acta* 109, 750–755. doi:10.1016/j.electacta.2013.08.011.

Yan, X., Z. Cao, C. Lau, and J. Lu. 2010. DNA aptamer folding on magnetic beads for sequential detection of adenosine and cocaine by substrate-resolved chemiluminescence technology. *Analyst* 135 (9), 2400–2407. doi:10.1039/c0an00163e.

Yang, L., and Y. Li. 2006. Detection of viable Salmonella using microelectrode-based capacitance measurement coupled with immunomagnetic separation. *Journal of Microbiological Methods* 64 (1), 9–16. http://www.sciencedirect.com/science/article/B6T30-4G9GN1G-6/2/5c1ff4f5094305f7af4d8776fc49f966.

Yang, X., F. Wu, D.-Z. Chen, and H.-W. Lin. 2014. An electrochemical immunosensor for rapid determination of clenbuterol by using magnetic nanocomposites to modify screen printed carbon electrode based on competitive immunoassay mode. *Sensors and Actuators B: Chemical* 192, 529–535. doi:10.1016/j.snb.2013.11.011.

Yellen, B. B., R. M. Erb, H. S. Son, R. Hewlin, H. Shang, and G. U. Lee. 2007. Traveling wave magnetophoresis for high resolution chip based separations. *Lab on a Chip* 7 (12), 1681–1688.

Zamfir, L.-G., I. Geana, S. Bourigua, L. Rotariu, C. Bala, A. Errachid, and N. Jaffrezic-Renault. 2011. Highly sensitive label-free immunosensor for ochratoxin A based on functionalized magnetic nanoparticles and EIS/SPR detection. *Sensors and Actuators B: Chemical* 159 (1), 178–184. doi:10.1016/j.snb.2011.06.069.

Zaner, K. S., and P. A. Valberg. 1989. Viscoelasticity of F-actin measured with magnetic microparticles. *The Journal of Cell Biology* 109 (5), 2233–2243.

Zardán Gómez de la Torre, T., M. Strömberg, C. Russell, J. Göransson, M. Nilsson, P. Svedlindh, and M. Strømme. 2010. Investigation of immobilization of functionalized magnetic nanobeads in rolling circle amplified DNA coils. *The Journal of Physical Chemistry B* 114 (10), 3707–3713. doi:10.1021/jp911251k.

Zhang, L., J. J. Abbott, L. Dong, B. E. Kratochvil, D. Bell, and B. J. Nelson. 2009. Artificial bacterial flagella: Fabrication and magnetic control. *Applied Physics Letters* 94 (6).064107

Zhao, Q., X. F. Li, and X. C. Le. 2011. Aptamer capturing of enzymes on magnetic beads to enhance assay specificity and sensitivity. *Analytical Chemistry* 83 (24), 9234–9236. doi:10.1021/ac203063z.

Zheng, B., T. Vazin, P. W. Goodwill, A. Conway, A. Verma, E. U. Saritas, D. Schaffer, and S. M. Conolly. 2015. Magnetic particle imaging tracks the long-term fate of in vivo neural cell implants with high image contrast. *Scientific Reports* 5, 14055. doi:10.1038/srep14055.

Zheng, B., M. P. von See, E. Yu, B. Gunel, K. Lu, T. Vazin, D. V. Schaffer, P. W. Goodwill, and S. M. Conolly. 2016. Quantitative magnetic particle imaging monitors the transplantation, biodistribution, and clearance of stem cells in vivo. *Theranostics* 6 (3), 291–301. doi:10.7150/thno.13728.

Zhou, X., D. Zhu, Y. Liao, W. Liu, H. Liu, Z. Ma, and D. Xing. 2014. Synthesis, labeling and bioanalytical applications of a tris(2,2′-bipyridyl)ruthenium(II)-based electrochemiluminescence probe. *Nature Protocols* 9 (5), 1146–1159. doi:10.1038/nprot.2014.060.

Zhu, N., A. Zhang, P. He, and Y. Fang. 2004. DNA hybridization at magnetic nanoparticles with electrochemical stripping detection. *Electroanalysis* 16 (23), 1925–1930. http://onlinelibrary.wiley.com/doi/10.1002/elan.200303028/pdf.

Zhu, X., D. Duan, and N. G. Publicover. 2010. Magnetic bead based assay for C-reactive protein using quantum-dot fluorescence labeling and immunoaffinity separation. *The Analyst* 135 (2), 381. http://apps.isiknowledge.com/full_record.do?product=UA&colname=WOS&search_mode=CitingArticles&qid=23&SID=4Cae37FbIfNh@p1kGij&page=1&doc=4.

Ziegler, C. 2004. Cantilever-based biosensors. *Analytical and Bioanalytical Chemistry* 379 (7–8). http://apps.isiknowledge.com/full_record.do?product=UA&colname=WOS&search_mode=CitingArticles&qid=33&SID=4Cae37FbIfNh@p1kGij&page=23&doc=226.

Ziemann, F., J. Rädler, and E. Sackmann. 1994. Local measurements of viscoelastic moduli of entangled actin networks using an oscillating magnetic bead micro-rheometer. *Biophysical Journal* 66 (6), 2210.

10 Magnetic Contrast Imaging: Magnetic Nanoparticles as Probes in Living Systems

Robert C. Woodward, Matthew R. J. Carroll, Micheal J. House, and Timothy G. St Pierre

CONTENTS

10.1 INTRODUCTION

Continued progress in non-invasive diagnostic medicine relies on the constant improvement of imaging techniques such as x-rays, computed tomography (CT), magnetic resonance imaging (MRI), and positron emission tomography (PET) and the development of new imaging modalities such as optical coherence tomography (OCT) and magnetic particle imaging (MPI). These diagnostic tools are used by medical practitioners to identify anomalous conditions in patients and to plan and monitor their treatment. In addition, they are also used by medical researchers to improve our understanding of disease and improve our ability to detect and treat these diseases more effectively.

In order to improve the effectiveness of medical imaging, or to enhance the visualization of specific components of a living body, it is often necessary to administer chemical compounds, known as contrast agents or media, so that the contrast between different tissues (e.g. healthy against diseased) can be increased. These contrast agents can take a variety of forms depending on the imaging technique with which they will be used. In x-rays and CT scans, electron-dense compounds such as iopamidol are used, which absorb more x-rays. In PET, compounds are labeled with a positron-emitting isotope such as ^{18}F fluorodeoxyglucose, whereas for ultrasound, ligand-carrying micro-bubbles provide a strong reflection of ultrasound waves (Herranz and Sanchez-Garcia 2007, Sharma et al. 2006).

The major limitations in the clinical administration of contrast media relate to issues of efficacy, side effects, and toxicity. Ideally, the optimal contrast agent should create enhanced contrast at minimum dosage, so that unwanted side effects are minimized or avoided. Additionally, the contrast medium should be excreted from the body without any interference with endogenous species so that it does not accumulate within the body. For research-based applications, however, performance issues such as degree of enhancement, specificity, and sensitivity may be assigned higher priorities, meaning that contrast agents used in research environments may differ from those used clinically.

In some cases the contrast agents can be multimodal particles capable of multiple mechanisms of contrast enhancement. These multimodal systems are used because no single imaging modality is perfect and sufficient to gain all possible information that may be required. As such, multi-modality contrast agents offer an exciting way forward in diagnostic imaging (Jennings and Long 2009).

Contrast agents are widely used in clinical practice, with about one in three MRI scans being contrast-enhanced (Caravan et al. 2009). The global market for contrast agents in 2019 was $5 billion and is expected to grow to 6 billion by 2024 (GlobalData 2009). In a global survey by the European Science and Technology Observatory in 2006, more than 150 startups and small- to medium-enterprise companies that focused on nanomedicine research and development projects were identified, with contrast agents accounting for the third largest sector at 13% (Wagner et al. 2006).

10.2 ADVANCED IMAGING APPLICATIONS

Many current contrast agents use relatively simple methods for localization within the body, relying on blood circulation routes or natural biodistribution. However, scientists and clinicians are increasingly looking for more advanced contrast agents that can provide enhanced performance and functionality. These new contrast agents can then be used to provide improved performance or new imaging functionalities. Some of these are discussed below.

10.2.1 Targeted and Molecular Imaging

Many diseases result in either the expression of new compounds at the surface of the cells and affected tissues or over-expression of naturally occurring compounds, which act as markers for the disease. If a contrast agent is bound to a biochemical compound that binds either specifically or preferentially to these markers, it will preferentially localize at the site of the disease and improve our ability to detect the disease at an early stage. This is the concept of targeted contrast imaging.

Scientists in the laboratory have taken this concept further to develop the field of *molecular imaging* (Cai and Chen 2007, Debbage and Jaschke 2008). Originally developed to identify the localization of biochemical compounds in cells via optical microscopy in immunofluorescence (Coons 1961), molecular imaging is a powerful tool for the characterization of biological processes at a cellular or molecular level. It provides researchers with the ability to view the biological and biochemical processes in living organisms. It vastly improves our understanding of the biochemical factors that underline the disease and streamlines the drug development process. Applied in vivo in a clinical setting, the technique switches the diagnostic focus away from the abnormalities that occur in the later stages of the disease process to the biochemical events that precipitate the earliest stages of the disease (Waters and Wickline 2008, Jennings and Long 2009).

10.2.2 Quantitative Molecular Imaging

Recently it has been shown that contrast from naturally occurring iron oxide nanoparticles (ferritin) can be used to perform quantitative imaging where the MR images are used to measure the liver iron concentration in patients with iron overload diseases such as thalassemia and hemochromatosis (St Pierre et al. 2005). In a similar way, it may be possible to measure the level of a biochemical marker based on contrast from suitably targeted contrast agents. This ability would have significant implications, particularly in terms of assessing diseases and the personalization of treatments based on quantitative measurements.

10.2.3 Cellular Tracking

If a cell can be made to take up a contrast agent without affecting its function, as has been shown for some iron oxide nanoparticles (Elias et al. 2008), then it is possible to track the movement of these cells in vivo. Cell tracking can be used to follow the movement of diseased cells, for example metastasizing cancer cells, or the motion of labeled cells used in stem cell therapies (Rogers, Meyer, and Kramer 2006). In addition, red blood cells can be loaded with contrast agent to make a blood contrast agent that is virtually invisible to the immune system and hence has a high circulation time (Brahler et al. 2006).

10.2.4 Combined Therapeutic and Diagnostic (Theranostic) Agents

It is possible to design contrast agents to undertake multiple roles in both diagnosis and therapy. For example, a magnetic particle incorporating a drug could be used to provide (1) magnetically field-guided drug delivery, (2) images of particle distribution via MRI, and (3) hyperthermia treatment by exposing the particles to an alternating magnetic field. This combination of diagnostic and therapeutic functions in a single compound is called *theragnostics* (Shubayev, Pisanic, and Jin 2009). Such materials will provide clinicians with the ability to monitor disease, control the localized release of drugs or other therapies, and quickly access the efficacy of treatment.

10.3 MAGNETIC IMAGING TECHNOLOGIES

Generally in biological systems, and particularly in the human body, there is very little naturally occurring magnetic material. This means that magnetic nanoparticles make very good contrast agents for imaging modalities that rely on magnetically induced effects for generation of images. The only significant sources of magnetic signals in the human body are Fe-based compounds, particularly ferritin, which is the body's major form of Fe storage. However, the magnitude of the magnetic response from these compounds is about two orders of magnitude smaller than an equivalent amount of Fe in the form of magnetic nanoparticles (Lopez, Gutierrez, and Lazaro 2007). So magnetic nanoparticles, and even some magnetic compounds, can generate a significant improvement in contrast and image quality at relatively low concentrations.

10.3.1 Magnetic Resonance Imaging (MRI)

Of all the imaging techniques that rely on the magnetic properties of matter to derive images, the most common and clinically relevant is MRI. Based on the principles of nuclear magnetic resonance, MRI uses a combination of a strong magnetic field and radio frequency excitation to generate three-dimensional (3D) images without the use of ionizing radiation. The images are generated by manipulation and measurement of changes in the small magnetic moments associated with hydrogen atoms in the imaged object. In the human body, the majority of the signal is associated with the hydrogen atoms in water molecules (Kuperman 2000, Buxton 2003). The technique has inherently very good soft-tissue contrast and MRI is one of the most widespread techniques used in diagnostic medicine for detecting the presence or extent of pathologies, especially cancer. Contrast agents for MRI are based on magnetic compounds that create tissue differentiation, in the region of accumulation, by magnetically interfering with the proton relaxation rates of surrounding water molecules. The mechanism of contrast enhancement and how this affects the design of new contrast agents are reviewed later in this chapter.

In addition to MRI, there are several novel imaging techniques at various stages of development that can be used to image the presence of magnetic contrast agents.

10.3.2 Magnetomotive Optical Coherence Tomography

Optical coherence tomography (OCT) is a 3D imaging technique capable of imaging structures to a depth of 1–2 mm with micrometer resolution. The images are generated using a low-time-coherence light source and an interferometer (Fercher et al. 2003). Combined with advanced catheters and endoscopes, it provides a high-resolution imaging modality for internal organs, which is noninvasive or minimally invasive.

In order to achieve high noise rejection and improve the imaging quality, it is possible to use modulated OCT imaging, where the OCT signal is modulated by an external system (Oldenburg et al. 2008). One such technique is magnetomotive OCT (Oldenburg et al. 2005). Here an electromagnet is used to generate a magnetic field gradient, this field gradient causes a displacement of scattering objects, such as cells, to which magnetic nanoparticles are bound. This displacement causes a change in the OCT signal and through the use of a combination of multiple images and/or frequency filtering. it is possible to generate images where other scattering including physiological motion is effectively ignored. Magnetomotive OCT is capable of detection limits of less than 50 ppm of particles (Oldenburg et al. 2008).

10.3.3 Magnetic Particle Imaging (MPI)

Developed by Gleich and Weizenecker in 2005 (Gleich and Weizenecker 2005), magnetic particle imaging (MPI) makes use of the non-linear magnetization curve of magnetic nanoparticles to generate images of the particle concentration in a subject. In zero field, the response of a magnetic particle to a small alternating field is large. However, if a significant field is applied, the particle's magnetization saturates and the response to the alternating field diminishes. The technique uses a series of three field gradients to generate a single point in space with zero field. It then detects the magnetic response as a signal in the receiver coils at the drive frequency of the AC field and its higher harmonics. This signal is dominated by the particles in and around the zero field point. The zero field point is then scanned through space using a 3D Lissajous trajectory.[1] The main advantage of MPI is that it can generate 3D images of particle concentrations in real time (Weizenecker et al. 2009). To do this presently requires a calibration of the scanner at the voxel level in order to solve the inverse reconstruction problem. Several issues still need to be overcome prior to successful commercial development, including improving the signal-to-noise ratio, generating high field gradients in a human-sized scanner, and controlling patient heating due to the alternating magnetic field (Weizenecker et al. 2009).

10.3.4 Magnetomotive Ultrasound

Ultrasonic imaging is a relatively cheap and widely available technique that has the advantages of being portable, being able to image deep structures, producing real-time images, and not using ionizing radiation.

Magnetic nanoparticles can be used as ultrasonic contrast agents by looking for Doppler shifting of the ultrasound waves via the movement of tissue containing the magnetic nanoparticles in either a sinusoidal (Oh et al. 2006) or pulsed magnetic field (Mehrmohammadi et al. 2009). It is referred to as either magnetomotive ultrasound or magneto-acoustic imaging. Detection limits of around 50 ppm, similar to magnetomotive OCT, have been demonstrated in the lab, but these systems presently use large fields (0.5–2 T) and high field gradients.

10.3.5 Other Techniques

A number of in vivo and in vitro techniques use the properties of magnetic nanoparticles in diagnostic or biochemical tests, including:

- magnetotransduction, which uses magnetic particles to apply forces to cells to measure their mechanical properties (Wang, Tytell, and Ingber 2009),
- magnetopneumography, which looks at the time-dependent decay in a magnetic signal to probe intercellular motions in the lungs (Moller et al. 1997), and
- Brownian relaxometry, which looks at changes to the Brownian relaxation rates of magnetic nanoparticles as they bind to biomolecules, and hence change size (Chung et al. 2005).

For applications such as magnetomotive OCT where the ability to generate forces on cells and other tissue structures is the mechanism of contrast, then it is the moment[2] of the particle and the ability to attach the particle to the tissue of interest that is the critical factor affecting contrast. While in MPI, where the contrast is associated with the non-linear magnetization of the particles, it is both the magnetic susceptibility at low field and the field required to saturate the magnetization that will affect the contrast and resolution, respectively, of the imaging system. Despite these differences, many of the challenges of generating improved contrast agents are similar for all imaging modalities. In this chapter we will focus primarily on design considerations for MRI contrast agents.

10.4 BASIC PRINCIPLES OF MAGNETIC RESONANCE IMAGING

10.4.1 Protons

MRI is based upon the physical phenomenon of nuclear magnetic resonance (NMR), which arises from the interaction of a non-zero magnetic moment of an atomic nucleus with static and time-varying magnetic fields (Kuperman 2000, Levitt 2008). Hydrogen nuclei (and many other nuclei) possess the property of spin, an intrinsic form of angular momentum. In the case of hydrogen nuclei, two spin states are possible, +1/2 or −1/2. With no external magnetic field, the energy of these two states is the same, i.e. they are degenerate. While it is possible to use other elements, hydrogen is abundant within the body and so it can be easily exploited to render images of the soft tissue within the body. Each hydrogen nucleus within the body possesses a magnetic moment, μ_H (14.106×10^{-27} J/T). In the absence of a field and under normal conditions, thermal energy keeps the proton magnetic moments fluctuating resulting in a random orientation of moments throughout the body. Hence, the vector sum of all of these moments is zero and the bulk magnetization is zero.

10.4.2 Applying a Static Magnetic Field

When a static magnetic field is applied to a system, a preferential direction of magnetization is created. This preferential direction is in the same direction as the applied magnetic field, which by convention is termed the Z direction. The magnetization arises as a result of the lower energy configuration when a magnetic moment is parallel to the applied magnetic field. This difference in energy levels in the presence of an applied magnetic field is known as Zeeman splitting of the energy levels. The +1/2 spin state corresponds to a spin angular momentum that is parallel to the external magnetic field and therefore has a lower energy level than the −1/2 spin state (anti-parallel to the external field). This energy difference results in a slightly higher number of nuclei in the +1/2 spin state as compared with the −1/2 spin state. This difference in the number of positive and negative spin states results in a net magnetic moment μ along the field direction given by:

$$\mu = N\mu_H \exp^{-E/kT} \tag{10.1}$$

where N is the number of protons, μ_H is the nuclear magnetic moment of the proton, E is the energy difference between the two spin states, k is Boltzmann's constant, and T is the temperature. It is this bulk magnetic moment that is manipulated and measured during an MRI procedure.

In addition to the creation of a bulk magnetic moment, the external magnetic field causes the individual magnetic moments of the nuclei to precess around the field direction, as shown in Figure 10.1. The precession of the magnetic moment occurs at a given angular frequency, known as the Larmor frequency, ω, which is directly proportional to the net magnetic field experienced by the proton, B:

$$\omega = \gamma_H B \tag{10.2}$$

where γ_H is the proton gyromagnetic ratio.

10.4.3 Exciting the Magnetization

The application of the static external magnetic field creates a bulk magnetization vector, which the MRI manipulates to create images. This manipulation is achieved by inducing resonance absorption within the system. In order to produce resonance, an oscillating magnetic field must be applied that has a frequency equal to that of the Larmor frequency. For typical magnetic field strengths used in NMR and MRI, the Larmor frequency is in the radiofrequency region of the spectrum. Maximum energy absorption occurs when the RF frequency is equal to the Larmor frequency and thus proportional to the energy difference between the +1/2 and −1/2 states. The absorbed energy causes a change of some of the spins from the +1/2 state (parallel) to −1/2 states (antiparallel) and hence results in a change in the bulk magnetization vector away from its equilibrium position. It is this perturbation of the magnetization vector, and its subsequent relaxation back to its original position, that is measured during MRI.

There are basically two different ways to measure the relaxation of the bulk magnetization vector. The first is to measure its longitudinal relaxation, i.e. its relaxation in the direction of the applied external magnetic field. The second is transverse relaxation, i.e. relaxation perpendicular to

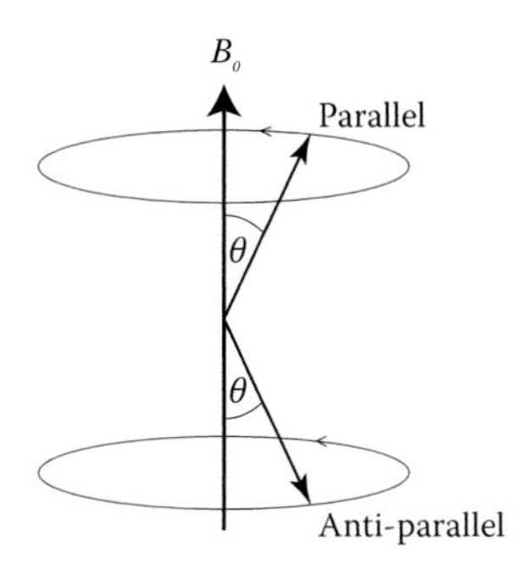

FIGURE 10.1 Application of an external magnetic field, B, causes the magnetic moments of the protons to precess about the field direction.

the applied external field. The rate at which the relaxation occurs determines the brightness of the image. With different relaxation rates in different parts of the body, it is possible to generate contrast between different tissue types, as described later.

10.4.3.1 Longitudinal Relaxation

If an RF pulse is applied at the Larmor frequency, it will cause spins in the lower energy (parallel) configuration to be excited to the higher energy (anti-parallel) configuration. As the number of spins in the parallel direction decreases, the magnitude of the bulk magnetization vector will gradually be reduced. Depending upon the duration of the RF pulse, this process can continue until there are more anti-parallel spins than parallel spins. At this instant the bulk magnetization vector will be in the opposite direction to the applied magnetic field. The final direction and magnitude of the bulk magnetization vector depends upon the duration of the excitation pulse and can vary from a small change to a full 180° rotation.

When the RF pulse is removed, the magnetization begins to return to its original equilibrium value as given by Equation 10.1. This process is known as longitudinal relaxation and the change in magnetization (for a 180° pulse) is shown in Figure 10.2. Excitation pulses that rotate the magnetization vector through a different angle will have the same form, but a different value immediately after the excitation pulse. This relaxation curve is characterized by the parameter R_1, known as the spin-lattice relaxation rate. Specifically R_1 is derived from the general equation[3]:

$$M_z(t) = M_0 + [M_z(0) - M_0]\exp^{(-tR_1)} \tag{10.3}$$

where $M_z(t)$ is the longitudinal magnetization at time, t, M_0 is the bulk saturation magnetization at equilibrium, and $M_z(0)$ is the longitudinal magnetization immediately after the excitation pulse.

Longitudinal relaxation is known as spin-lattice relaxation and is due to spins changing from the high-energy state to the low-energy state, causing a change in the bulk magnetization vector in the longitudinal direction. For this relaxation to occur, there must be both an agent to accept the energy from the spin (the "lattice") and a mechanism for the energy transfer. Although potentially this change in energy state can be either spontaneous or stimulated, in fact the probability of spontaneous transitions is very low and hence almost all transitions are stimulated (Wood and Hardy 1993). In order to stimulate a transition, a spin needs to experience an interaction with a fluctuating magnetic field at the appropriate frequency, the Larmor frequency. These fluctuating fields are generated by the random motion of magnetic moments associated with nuclei, electrons, and atoms within the surrounding medium that generate a broad spectrum of fluctuations that overlap the Larmor frequency of the proton spin.

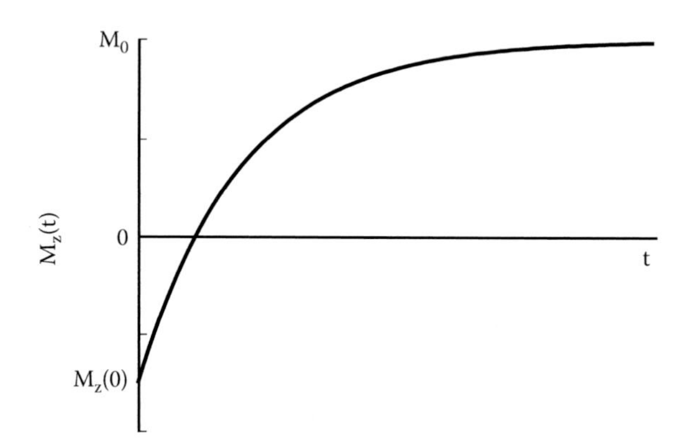

FIGURE 10.2 Relaxation of the longitudinal component of magnetization after the application of a radio frequency pulse. The relaxation is fitted with Equation 10.3 to determine the relaxation rate (R_1).

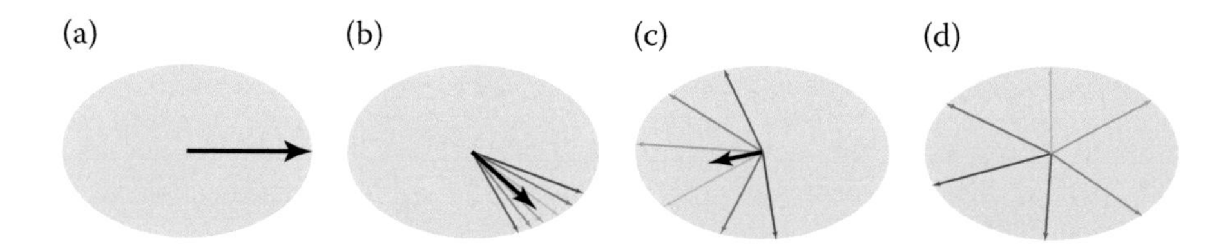

FIGURE 10.3 Dephasing of spins in the x–y plane after removal of the RF pulse, black arrow is the transverse signal, which is the vector sum of spins. Spins start in phase **(a)** and precess in a clockwise direction but lose phase coherence after the removal of the perturbing RF field due to local field inhomogeneities, which cause variations in their Larmor frequency **(b–d)**, until complete loss of the transverse signal **(d)**.

10.4.3.2 Transverse Relaxation

Transverse relaxation describes the reduction of the magnetization vector perpendicular to the applied magnetic field. Here an RF pulse is applied that forces the bulk nuclear magnetization vector from the z-direction (the direction of applied magnetic) through an angle that generates a component of magnetization in the x–y plane. Immediately after this pulse, the spins will be in phase with one another and the bulk magnetization vector will be at an angle with the applied field. However, as the spins continue to precess around the applied magnetic field, the spins move out of phase with one another (i.e. lose phase coherence), as shown in Figure 10.3.

This dephasing of the spins will manifest at the macroscopic level by a reduction in the transverse magnetization signal, as shown in Figure 10.4. Eventually this dephasing will result in a random arrangement of phases and a bulk magnetization of zero in the x–y plane. This decay in the signal is known as the free induction decay (FID) and is characterized by the relaxation rate, R_2*, which is derived from:

$$M_{xy}(t) = M_{xy}(0)\exp^{(-tR_2^*)} \quad (10.4)$$

where $M_{xy}(t)$ is the transverse magnetization at time t after the pulse, and $M_{xy}(0)$ is the initial maximum magnetization immediately after the pulse.

This dephasing of the spins is related to the slightly different local magnetic fields that each spin experiences. These different fields result in different Larmor frequencies so that when the RF pulse

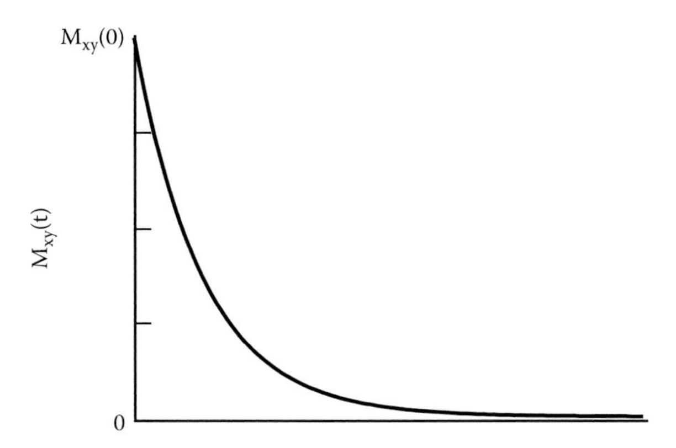

FIGURE 10.4 Relaxation of the transverse component of magnetization after the application of a radio frequency pulse. The relaxation is fitted with (10.4) to determine the relaxation rate (R_2^*).

is switched off, the proton spins are all rotating at slightly different frequencies. The variations in the local magnetic field are due to two major factors; the first is related to the bulk field inhomogeneities across the sample. As it is not possible to create a perfectly homogeneous static magnetic field, protons will experience a variation in field as a function of position in the sample due to this inhomogeneity, given by ΔB. This is effectively an extrinsic induced relaxation related to the field production system. The second cause of local field inhomogeneities is small variations in fields due to the random motion of spins in the sample, known as spin-spin interactions. These spin-spin interactions are an intrinsic property of the sample and produce a transverse relaxation rate called R_2. The free induction decay rate R_2^* is then given by: (Mansson and Bjornerud 2001)

$$R_2^* = R_2 + \gamma \Delta B \quad (10.5)$$

10.4.3.3 Spin Echo

The decay in the transverse signal is a result of both spin-spin interactions between the individual spins and inhomogeneities in the applied static magnetic field. To measure the decay in transverse magnetization solely as a result of the spin-spin interactions, a spin echo technique is employed. As in the previous procedure, this involves applying an RF pulse to move the bulk magnetization vector to an angle at which it has a component in the x–y plane. These spins are allowed to dephase over time similar to the free induction decay measurement of R_2^*. However, after a certain time, a 180° RF pulse is applied, which rotates all the individual magnetic moment vectors through 180° in the x–y plane. Following this pulse, instead of the moments dephasing further, they will begin to rephase, eventually reaching an instant when most of the moments are once again in phase with one another. This instant is known as the spin echo and is characterized by an increase in the transverse magnetization signal, see Figure 10.5.

Successive 180° radio frequency pulses can be applied, generating a series of spin echoes. With each successive pulse the maximum amplitude of the transverse magnetization decreases, see Figure 10.6. While R_2^* is related to the initial decay rate without any spin echoes, the decay in the spin echo peaks is only related to the spin-spin interactions and hence can be used to measure R_2.

10.4.4 Generation of Images in MRI

MRI uses longitudinal and transverse relaxation to generate images. Below is a brief simplified description of how these images are generated (Liang and Lauterbur 2000, Gibby 2005, Brown and

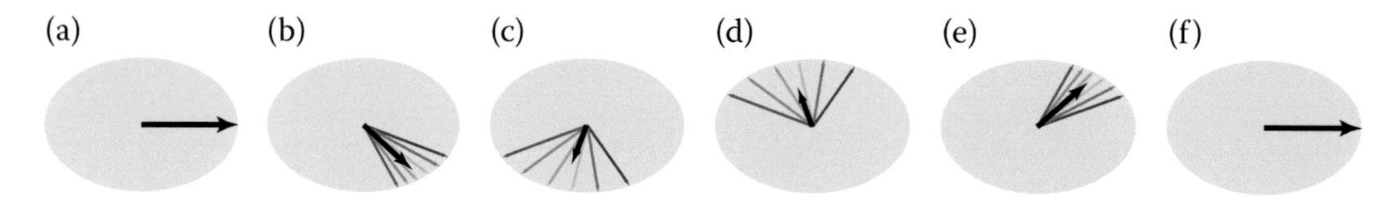

FIGURE 10.5 Dephasing and rephasing of spins in a spin echo measurement. Spins start in phase **(a)** but lose phase coherence due to local field inhomogeneities **(b–c)**. A 180° pulse is applied at **(c)**, which causes the spins to flip relative to the *y*-axis **(d)**. The spins then continue to precess clockwise but now the faster precessing spins (the ones to the left) catch up with the slower precessing spins (the ones on the right) causing an increase in phase coherence **(d–e)** until a spin echo is obtained **(f)** at which point they will start to dephase again.

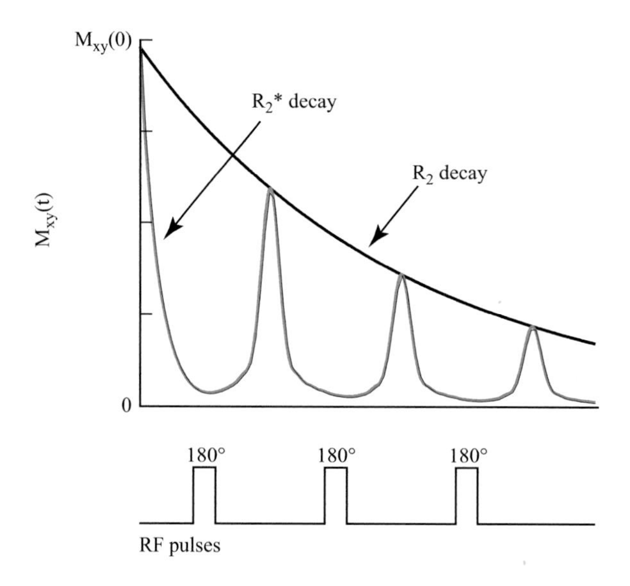

FIGURE 10.6 Spin echo measurement. Transverse magnetization decay with 180° radio frequency refocusing pulses leading to spin echoes. R_2 is determined by fitting (10.4) to the decay in the magnitude of the spin echoes.

Semelka 2010, Plewes and Kucharczyk 2012, Bushberg et al. 2002). A large uniform magnetic field, called the B_0 field, is applied to preferentially align the proton spins. This field accounts for the majority of the "B" field that spins experience in Equation 10.2. The spins are then excited using a resonant RF transmission generating longitudinal and/or transverse relaxation processes and an RF signal is acquired following excitation, which represents the data.

It is not possible to focus or collimate the RF transmission or the RF detection system so the RF signal that is detected is collected from all points in space simultaneously. In order to localize the signal and so generate useful images, a series of gradient coils is used to generate small variations in the magnetic field across the body. A schematic of the arrangement of the gradient coils and RF transceiver coils is shown in Figure 10.7.

The B_0 field is applied along the Z direction. The localization or encoding of position within the object requires three orthogonal magnetic field gradients. The resonant frequency of a proton is a function of the total magnetic field, as shown in Equation 10.2; hence small variations in the magnetic field provided by the gradient coils generate small changes in the resonant frequency of the protons.

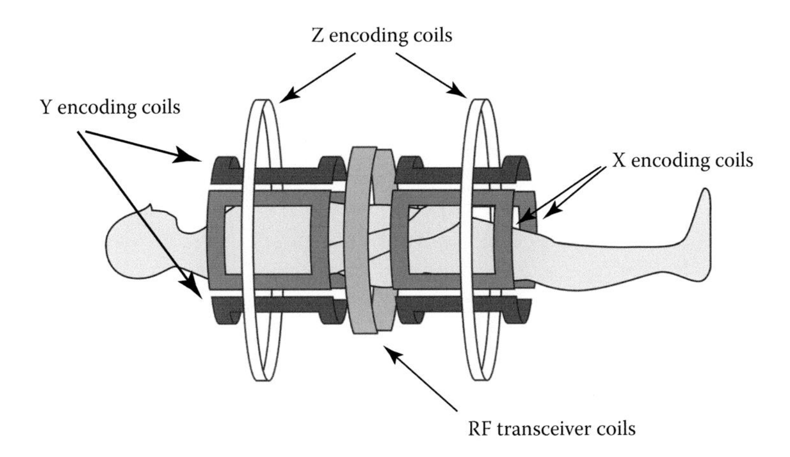

FIGURE 10.7 Arrangement of gradient and transceiver coils in MRI. Three sets of orthogonal gradient coils are used to encode the position information, X , Y and Z . The RF transceiver coils are used to both excite the protons and detect the signal.

- Z gradient coils provide a strong magnetic field gradient that selects a slice in the x–y plane. The thickness of the slice is determined by the strength of the gradient and bandwidth of the RF circuit. Protons outside the slice are no longer at resonance with the RF signal and are not significantly excited by it.
- X gradient coils generate a magnetic field gradient in the X direction, which is used to encode position in terms of the frequency within the slice. The resonant frequency of protons varies then as a function of position along the X direction (see Figure 10.8). The signal intensity at each frequency is extracted from the raw RF signal via a Fourier transform.
- Y gradient coils are use to generate phase encoding of the position within the slice. A pulse is applied so as to generate a spatially dependent phase shift in the resonant signal as shown in Figure 10.9. It is impossible to extract both phase and frequency information from a single measurement so a series of measurements with different strength pulse encoding gradients is used where the changes in the signal with pulse gradient strength uniquely describes the phase change.

The three gradient fields are applied as a series of pulses during the MR acquisition sequence; an example of a spin echo sequence is shown schematically in Figure 10.10. The slice encoding gradient (Z) is pulsed at the same time as the RF pulses so as to excite a specific slice through the subject. This must be applied for both the 90° pulse and 180° RF pulses. A phase encoding gradient pulse (Y) is applied between the 90° and 180° pulses to generate phase changes along the Y direction. The frequency encoding gradient pulse (X) is then designed to coincide with the spin echo in the MR signal. The frequency encoding gradient generates a spread of frequencies in the MR signal. A one-dimensional Fourier transform of the RF signal generates the signal intensity as a function of frequency. The pulse sequence is then repeated with different strength phase encoding pulses and eventually a complete 2D data set is obtained with the data frequency encoded in the X direction and phase encoded in the Y direction.

The raw MR data is collected and stored in a k-space matrix (frequency domain) with the frequency encoding generating data along the x-axis and phase encoding generating data along the y-axis (see Figures 10.10 and 10.11). Once a data set has been collected, an image can be reconstructed using a Fourier transformation (see Figure 10.11). Due to the nature of k-space in

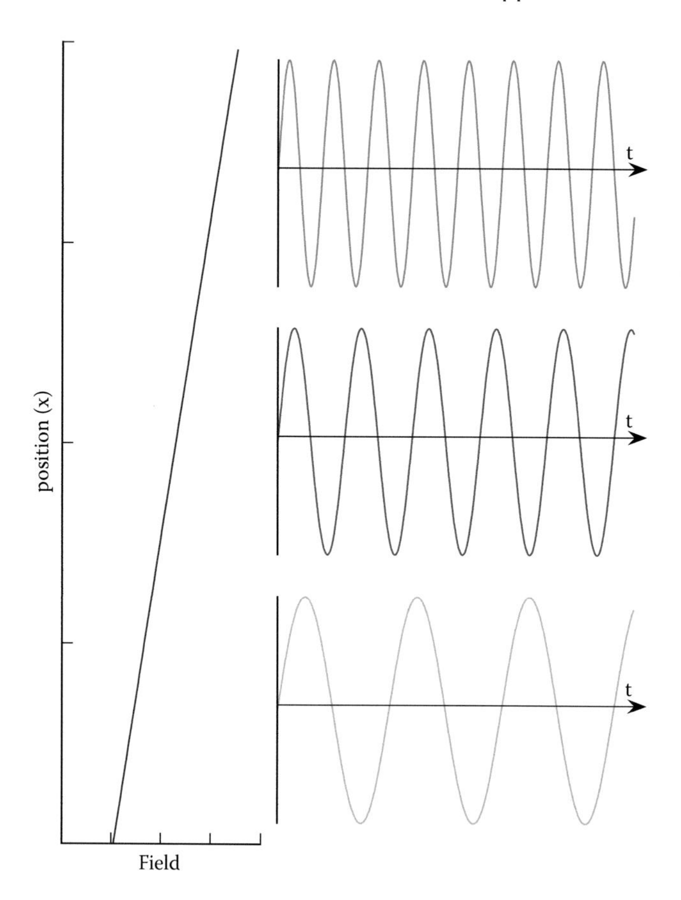

FIGURE 10.8 Illustration of frequency encoding using a gradient field applied along the X direction. The changes to the applied field result in variations in proton resonance frequency as function of position along the x-axis (Liang and Lauterbur 2000).

which the MR data is collected, not all points in the k-space matrix need to be collected in order to generate a reasonable representation of the image. This allows for faster acquisition times using different k-space sampling sequences at a cost of higher noise and reduced resolution in the associated image (Liang and Lauterbur 2000, Plewes and Kucharczyk 2012).

10.4.5 Pulse Sequences and Signal Intensity in MRI

A pulse sequence is the arrangement and timing of RF pulses and gradient field pulses used to produce the MR signal. A broad range of different pulse sequences are available for MRI, confused further by the fact that different manufacturers often use different names for what is effectively the same sequence. The most common sequences can be classified into three broad categories.

1. Spin echo sequences. In a spin echo sequence an excitation pulse is used to excite the proton spins and is then followed by one or more refocusing pulses to generate a spin echo(es). Spin echo sequences can be used to generate R_1, R_2, and proton density related contrast.
2. Inversion recovery sequences. Traditionally these pulse sequences emphasize R_1 related contrast. In this sequence, an initial RF pulse is used to flip the magnetization opposite to the field direction and is then followed by a 90° RF pulse to change the magnetization into the transverse direction. A third pulse is used to create an echo of the transverse signal.

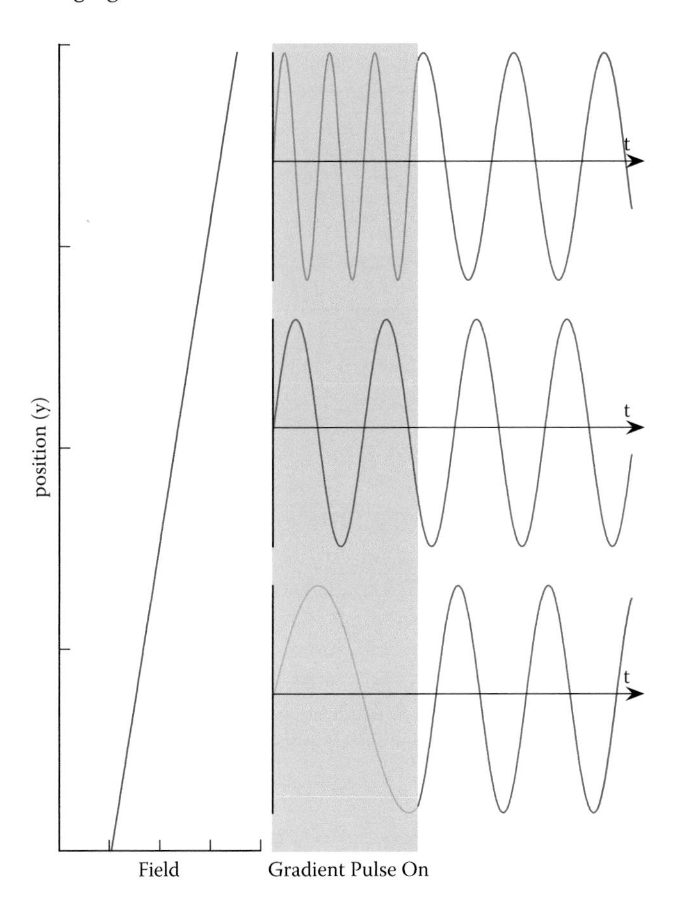

FIGURE 10.9 Illustration of phase encoding using a pulsed gradient field applied along Y direction. The pulsed gradient field results in a momentary change in the proton resonance frequency that generates a phase shift in the final signal as function of position along the y-axis (Liang and Lauterbur 2000).

3. Gradient echo sequences. Rather than use a refocusing RF pulse, this sequence uses magnetic field gradients to rephase the spins. Here two gradients of opposite polarity are applied to the sample to generate an echo. This technique is sensitive to field and susceptibility inhomogeneities and generates contrast related to R_2* rather than R_2.

As stated before, images are generated in an MRI through manipulation of the NMR of protons associated with water and fat in the body. Hence the signal is intrinsically associated with soft tissues. Fundamentally, the contrast in an MR image represents changes in R_1 and R_2 (or R_2*) proton relaxation rates and changes in proton density. However it is possible by manipulating the acquisition sequences and/or comparing multiple images to create contrast that reflects variations in a range of properties including magnetic susceptibility, diffusion, temperature, biomechanical properties, oxygen levels, and biochemical/chemical content (Plewes and Kucharczyk 2012).

The more common sequence parameters that allow the manipulation of the MR signal intensity are:

- T_R, the repetition time, is the time between successive applications of a pulse sequence to a given volume of space.
- T_E, the echo time, is the time between an excitation pulse and the maximum in the echo signal.

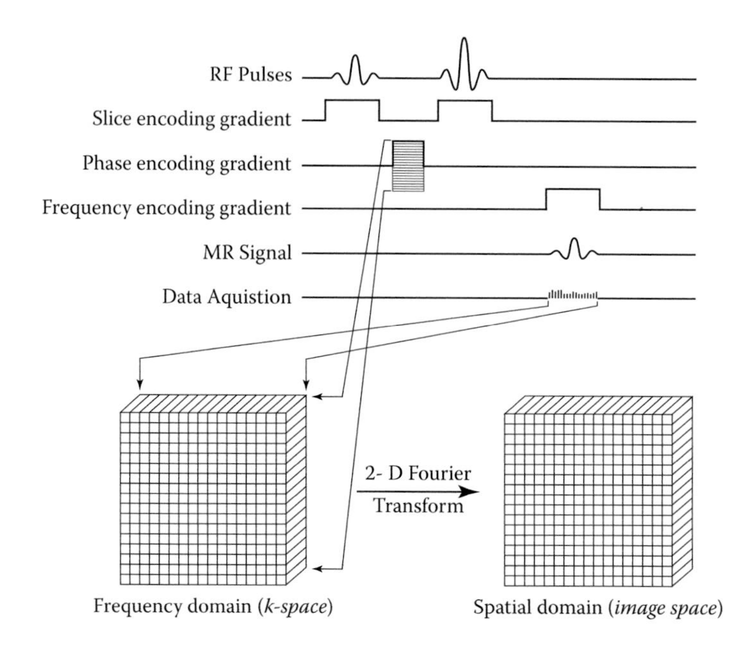

FIGURE 10.10 Schematic illustration of the sequence of RF and gradient field pulses used to collect the MR data set. A sequence must be repeated with phase encoding gradient pulses of different strengths in order to separate the MR signal in terms of frequency and phase in the final k-space data set (Bushberg et al. 2002).

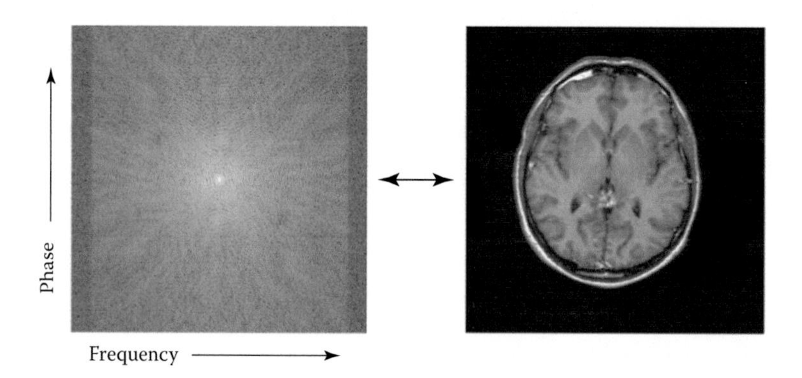

FIGURE 10.11 k-Space MR data **(left)** and associated MR image **(right)**. In the k-space image, the X direction signal is frequency encoded and the Y direction signal is phase encoded.

- T_I, the inversion time, is the time between a 180° inversion pulse and the imaging excitation pulse.
- α_E, the excitation angle or flip angle, is the amount the spins are rotated away from their equilibrium position by the RF pulse.

If we take as an example a simple *spin echo* sequence, then the MR signal intensity, I, is given by:

$$I = k\rho_P\left(1 - e^{-T_R R_1}\right)e^{-T_E R_2} \tag{10.6}$$

where k is a constant, ρ_P is the proton density, R_1 and R_2 are the longitudinal and transverse relaxation rates, and T_R and T_E are the repetition and echo times respectively. This signal intensity relationship illustrates that the MR signal intensity is a mix of the three fundamental parameters, ρ_P, R_1 and R_2. In general, tissues with higher proton density will appear brighter as there are more protons to generate the MR signal. Tissues with larger R_1 rates will recover their longitudinal magnetization more rapidly than tissues with smaller R_1 rates. As such, the tissues with larger R_1 will have a larger signal and appear brighter than the surrounding tissue. Conversely tissues with larger transverse relaxation rates (R_2) will appear darker in the MR image due to the more rapid dephasing of spins that leads to a decrease in signal.

It is possible by appropriate selection of the acquisition parameters to generate images that are weighted more toward changes in just one of these three parameters. For a single spin echo sequence these conditions would be:

$$T_1 \text{ weighted image–} T_R \leq 1/R_1 \text{ and } T_E \ll 1/R_2$$

$$T_2 \text{ weighted image–} T_R \gg 1/R_1 \text{ and } T_E \geq 1/R_2$$

$$\rho_P \text{ weighted image–} T_R \gg 1/R_1 \text{ and } T_E \ll 1/R_2$$

It should be noted that although such images are weighted to a particular parameter, the contrast is not exclusively related to changes in that parameter and large changes in the other parameters will affect the observed signal intensity.

10.5 CONTRAST ENHANCEMENT MECHANISMS

The mechanisms by which the bulk magnetization vector relaxes in the longitudinal and transverse directions are different and hence the mechanisms via which a contrast agent can increase this relaxation are also different. For this reason, there are two distinct forms of contrast agents. R_1 agents mainly increase longitudinal relaxation rates making parts of the image brighter and are hence called *positive* contrast agents. R_2 agents mainly increase the transverse relaxation rates making parts of the image darker and are therefore referred to as *negative* contrast agents.

The quality of a contrast agent is generally assessed in terms of the relaxivity, r_i ($i = 1, 2$), which is simply the rate of change of the relaxation rate with concentration of the contrast agent and is given by:

$$R_i(c) = R_i(0) + cr_i \tag{10.7}$$

where c is the concentration of the contrast agent, $R_i(c)$ is the relaxation rate at c and $R_i(0)$ is the relaxation rate in the absence of any contrast agent. The concentration of contrast agent is normally given in terms of mM of metal ion in the contrast agent, for example mM of Gd^{3+}, as any toxicity effects are expected to be related to the presence of these ions.

10.5.1 R_1 Contrast Agents

R_1 contrast agents work by increasing the longitudinal relaxation rate of water protons near them by enhancing the stimulation of spin-lattice relaxation processes. These materials effectively intensify the number and magnitude of fluctuating magnetic fields at or around the Larmor frequency of the protons. In the majority of cases, R_1 contrast agents are based around chelated paramagnetic ions such as Gd and Mn (Caravan et al. 1999, Lee and Koretsky 2004, Zhang, Nair, and McMurry 2005, Aime et al. 2006). However, recently there has been interest in the development of nanoparticulate systems based on very small (<5 nm) inorganic compounds of Gd, Mn, and Fe (Baek et al. 2010, Chambon et al. 1993, Na, Song, and Hyeon 2009, Park et al. 2009, Li et al. 2012, Wang et al. 2014).

The theory for longitudinal proton relaxation enhancement, based primarily on the affect of chelated paramagnetic ions, is relatively well-developed. A brief introduction to this theory is

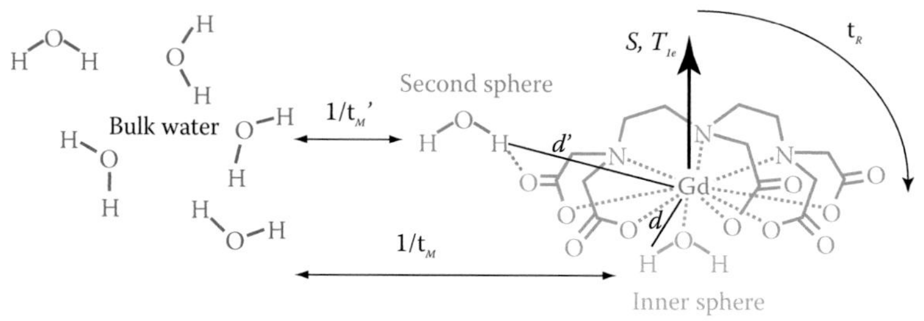

FIGURE 10.12 Some of the factors affecting r_1 relaxivity. The Gd has a magnetic moment related to S with an electronic relaxation time of T_{1e}. The Gd is complexed by DTPA with one water bound in the inner sphere and water bound to the second sphere at a greater distance from the Gd. The rate of exchange of the bound waters is given by the inverse of τ_m and τ'_m, based on Caravan et al. (2009).

included here. However, for a more comprehensive review, the reader is directed to one of a series of recent reviews on the subject (Peters, Huskens, and Raber 1996, Caravan et al. 1999, Toth, Helm, and Merbach 2002, Helm 2006).

The theory of longitudinal relaxivity is generally broken down into solvation spheres. In the *inner sphere,* water molecules are directly bound to a metal ion. The *second sphere* is less well-defined, but includes water molecules bound to the metal complex, possibly including those around counter-ions. Beyond the second sphere are water molecules freely diffusing around the complex, which make up the *outer sphere*. The total relaxivity (r_1^{obs}) is given by the sum of these components:

$$r_1^{obs} = r_1^{is} + r_1^{ss} + r_1^{os} \tag{10.8}$$

where r_1^{is} is the inner sphere contribution to the relaxivity, r_1^{ss} is the second sphere contribution, and r_1^{os} is the outer sphere contribution.

These solvation spheres and some of the factors affecting the observed relaxivity are shown schematically in Figure 10.12.

10.5.1.1 Inner-Sphere Relaxivity

The inner-sphere contribution to the relaxivity r_1^{is} arises from exchange of water bound to the paramagnetic ion with bulk water and is given by:

$$r_1^{is} = \frac{q/[H_2O]}{T_{1m} + \tau_m} \tag{10.9}$$

where q is the number of water molecules bound to the metal ion, H_2O is the molar concentration of water (0.556 M for pure water), $1/T_{1m}$ is the relaxation rate for a bound water molecule, and $1/\tau_m$ is the rate of exchange of the bound water with the bulk. q is generally 1 for Gd ion-based contrast agents as the other sites on the Gd ion are bound to chelating agents to ensure the stability, and thereby reduce the toxicity, of the chelated compound under physiological conditions (Hendrick and Haacke 1993, Aime et al. 2006, Toth, Helm, and Merbach 2002).

The relaxation rate for a bound water molecule is determined by the sum of the field-dependent relaxation rates from the dipole-dipole (DD), scalar (SC) and Curie (CS) interactions so that:

$$\frac{1}{T_{1m}} = \frac{1}{T_1^{DD}} + \frac{1}{T_1^{SC}} + \frac{1}{T_1^{CS}} \tag{10.10}$$

The dipole-dipole mechanism is an interaction through space in which the dipolar moment from the metal ion generates an alternating field at or near the Larmor frequency via tumbling of the compound. The scalar interaction requires contact between the protons and unpaired electrons to stimulate transition and depends on the moment of the ion and the electron spin relaxation time. The Curie interaction is related to the modulation of the magnetic moment of the ion. For fields and materials relevant to MRI the dominant mechanism is the dipole-dipole interaction and the other interactions can be generally ignored (Caravan et al. 2009, Toth, Helm, and Merbach 2002). The dipole-dipole relaxation rate $1/T_1^{DD}$ is given by:

$$\frac{1}{T_1^{DD}} = \frac{1}{120\pi^2}(\mu_0\gamma_n\gamma_e\hbar)^2\frac{S(S+1)}{d^6}\left[\frac{3\tau_{c1}}{1+\omega_n^2\tau_{c1}^2} + \frac{7\tau_{c2}}{1+\omega_e^2\tau_{c2}^2}\right] \tag{10.11}$$

where γ_n is the nuclear gyromagnetic ratio, γ_e is the electron gyromagnetic ratio, $\hbar$ is the reduced Planck's constant, μ_0 is the permeability of free space, S is the spin quantum number of the ion, d is the distance between the proton and the metal ion, ω_n is the nuclear Larmor frequency, ω_e is the electron Larmor frequency, and τ_{c1} and τ_{c2} are the longitudinal and transverse correlation times given by:

$$\frac{1}{\tau_{ci}} = \frac{1}{\tau_R} + \frac{1}{\tau_m} + \frac{1}{T_{ie\,(i=1,2)}} \tag{10.12}$$

where $1/\tau_R$ is the rotation rate of the compound, $1/\tau_m$ is the water exchange rate, and T_{1e} and T_{2e} are the longitudinal and transverse electron relaxation times respectively. For most paramagnetic ions either the electron relaxation times are very small due to orbital contributions to the magnetic moment or the moments themselves are small. Hence inner-sphere contributions to the relaxivity are small except for Gd^{3+}, Mn^{2+}, and Fe^{3+} ions (Hendrick and Haacke 1993, Peters, Huskens, and Raber 1996). The optimal relationship of time constants for relaxation enhancement via the inner sphere is given by Equation 10.13 (Toth, Helm, and Merbach 2002).

$$\frac{1}{T_{1m}} < \frac{1}{\tau_m} < \frac{1}{\tau_R}, \frac{1}{T_{ie}} \tag{10.13}$$

10.5.1.2 Second-Sphere Relaxivity

Firm experimental evidence of the contribution second-sphere water molecules make to the observed relaxivity has been difficult to obtain and traditionally this contribution has been ignored or rolled into outer-sphere contributions (Lauffer 1987, Botta 2000). However, more recently experimental evidence of second-sphere contributions has been obtained and in correctly designed materials this contribution can be significant (Botta 2000, Jacques et al. 2010). This contribution becomes significant only when the residence time τ'_m is long relative to the correlation time for diffusion of water around the complex, τ_d, *given by* Equation 10.14.

$$\tau_d = \frac{a^2}{D_m + D_{H_2O}} \tag{10.14}$$

where a is the distance of closest approach of the water molecules to the metal ion and D_m and D_{H_2O} are the diffusion coefficients for the complex and water, respectively. In this case the second-sphere relaxivity, r_1^{ss}, is given by the same mechanism as the inner-sphere relaxivity so that

$$r_1^{ss} = \frac{q'/[H_2O]}{T'_{1m} + \tau'_m} \tag{10.15}$$

where q' is the number of water molecules bound to the second sphere of the metal ion, $1/T'_{1m}$ is the relaxation rate for water molecules bound to the second sphere, and $1/\tau'_m$ is the rate of exchange of the water in the second coordination sphere with the bulk. Although the distance of the second-sphere molecules to the metal ion (d') are longer than the inner-sphere waters, thereby reducing $1/T'_{1m}$, as per Equation 10.11, the number of bound waters is potentially much larger and the rate of exchange of water with the bulk, $1/\tau'_m$, is generally much faster and hence the overall contribution can be significant. In existing commercial contrast agents the second-sphere contribution is weak and is generally ignored; however, in specially developed compounds it has been suggested that the contribution of second-sphere relaxivities can be anywhere up to 50–90% of the observed relaxivity (Botta 2000, Jacques et al. 2010).

10.5.1.3 Outer-Sphere Relaxivity

The outer-sphere contribution to relaxivity arises from the translational diffusion of water near the metal ion complex. If the water and metal ion complex are assumed to behave as hard spheres, then the outer-sphere contribution takes the form:

$$r_1^{OS} = 1.572 N_A (\mu_0 \gamma_n \gamma_e h)^2 \frac{S(S+1)}{a(D_m + D_{H_2O})} [3j_1(\omega_n) + 7j_2(\omega_e)] \tag{10.16}$$

where N_A is Avogadro's number and the spectral density $j_i(\omega)$ is given by

$$j_i(\omega) = \mathrm{Re}\left[\frac{1 + 1/4(i\omega\tau_d + \tau_d/T_{ie})^{1/2}}{1 + (i\omega\tau_d + \tau_d/T_{ie})^{1/2} + 4/9(i\omega\tau_d + \tau_d/T_{ie}) + 1/9(i\omega\tau_d + \tau_d/T_{ie})^{3/2}}\right] \tag{10.17}$$

For fields and materials of interest for MRI applications, the outer-sphere contribution to the relaxivity depends primarily on a, the distance of the closest approach to the ion, which is related to the molecular dimension and charge distribution of the complex. For commercial MRI contrast agents, the outer-sphere contribution to the relaxivity corresponds to around 40%–50% of the observed relaxivity (Peters, Huskens, and Raber 1996, Botta 2000).

10.5.2 R_2 Contrast Agents

An R_2 contrast agent works by increasing the rate of dephasing of the moments in the transverse direction. This dephasing is driven by variations in the applied magnetic fields that the protons experience. Any magnetic particle or ion will generate its own dipolar fields, which will modify the applied field as shown in Figure 10.13 and cause a variation in the Larmor frequency as given by Equation 10.2. The larger the magnitude of the dipolar field, the greater the rate of dephasing and hence the larger the relaxivity. Although it is theoretically possible to use ferromagnetic particles as contrast agents, it is extremely difficult to generate a colloidally stable suspension of ferromagnetic particles due to the large dipolar interactions between particles. In general, R_2 contrast agents are based on superparamagnetic nanoparticles, normally based on magnetite (Fe_3O_4) and maghemite (γ–Fe_2O_3). It is much easier to generate colloidally stable superparamagnetic particles as in the absence of a magnetic field the particles have no time-averaged magnetic moment and hence interparticle interactions are substantially reduced. These superparamagnetic particles have high magnetic susceptibilities and therefore generate high magnetic field gradients close to their surface and so create local magnetic field inhomogeneities.

When protons are in the vicinity of the nanoparticle, they will experience the combination of both the applied magnetic field and the field associated with the nanoparticle. The field lines at the poles are in the same direction as the applied field. Therefore, the protons at the poles will experience the sum of the applied field and the particle field and will therefore precess at a faster rate. Protons at the equator will experience the difference between the applied field and the particle field, so they will precess at a slower rate. This will result in the magnetic moments dephasing at a quicker rate, increasing both R_2 and R_2^*, ultimately leading to an increase in contrast in an MR image.

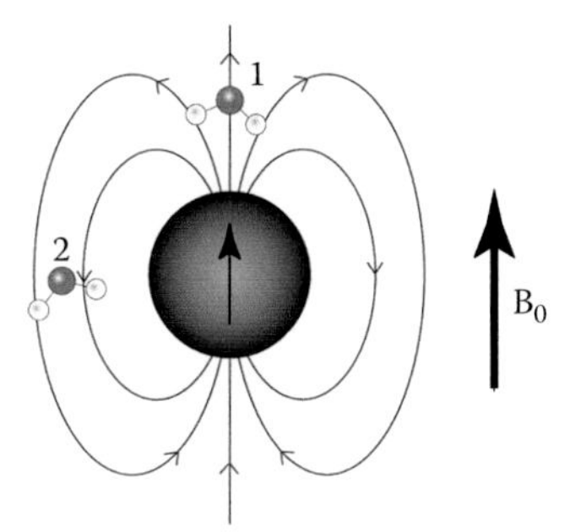

FIGURE 10.13 Schematic diagram of the dipolar field emanating from a magnetic object in an applied magnetic field, B_0. Water molecule 1 near the pole experiences a field that is greater than B_0 as the dipolar field adds to the applied field while water molecule 2 near the equator experiences a field less than B_0 as the dipolar field is in the opposite direction to the applied field.

Several investigators have modeled the effects of magnetic nanoparticles on proton transverse relaxation rates (Yablonskiy and Haacke 1994, Koenig and Kellar 1995, Jensen and Chandra 2000, Gillis, Moiny, and Brooks 2002, Matsumoto and Jasanoff 2008). These models are generally categorized into three regimes based on the characteristic time, τ_D, for a water molecule to diffuse a distance of the order of the radius of the magnetic nanoparticles given by (Roch et al. 2005):

$$\tau_D = d_p/4D_{H_2O} \tag{10.18}$$

where d_p is the diameter of the magnetic nanoparticle and D_{H_2O} is the self-diffusion coefficient of water. These regimes are the motional averaging, static dephasing, and echo limited regimes.

10.5.2.1 Static Dephasing

The simplest model is to assume that all water molecules are stationary at fixed points around the nanoparticle. This regime is known as the static dephasing regime and applies when

$$\tau_D > \frac{\pi\sqrt{3}}{2\Delta\omega} \tag{10.19}$$

where $\Delta\omega$ is the spread of Larmor frequencies at the surface of the particle given by:

$$\Delta\omega = \frac{\mu_0 \gamma_n M}{3} \tag{10.20}$$

where μ_0 is the permeability of free space, γ_n is the nuclear gyromagnetic ratio, and M is the magnetic moment per unit volume of the particle in the measurement field (Roch et al. 2005). In this regime the variations in local magnetic fields experienced by the protons are the most extreme, hence the dephasing of the proton magnetic moments is most rapid. In the static dephasing regime, the free induction decay relaxivity, $r_2{}^*{}_{sd}$ in units of s^{-1} mM of ions^{-1}, is given by:

$$r^*_{2_{sd}} = \frac{2\pi MW\Delta\omega}{\sqrt{27}\,c\rho} \tag{10.21}$$

where MW is the molecular weight of the compound, c is the number of metal ions per formula weight, and ρ is the density of the compound in g/cm^3. In this regime the $r_2{}^*$ relaxivity is independent of particle size.

If one neglects the effects of diffusion, as this model does, then r_{2sd} would be expected to be equal to $r_2{}^*{}_{sd}$. The effects of diffusion are to lower r_2 and $r_2{}^*$, as described below in the motional averaging regime. The value of r_2 is also dependent upon the echo spacing $2t_{cp}$,[4] as described below in the echo limited regime. However, this model assumes no such dependence, a situation that would be achieved as t_{cp} approaches infinity. In this ideal situation r_2 would reach its absolute upper limit, equal to $r_2{}^*{}_{sd}$. This model is proposed to be a good approximation of r_2 when (Gillis, Moiny, and Brooks 2002)

$$1/\Delta\omega < \tau_D < 2t_{cp} \tag{10.22}$$

The static dephasing model assumes the protons are effectively stationary. This assumption is reasonable if the magnetic particles are large enough that water molecules (and therefore the protons) do not have time to diffuse very far around the particle in the measurement time, which is of the order of milliseconds. Alternatively, this assumption could be valid if there is a physical means of preventing the water molecules from diffusing. It could be hypothesized that a polymer brush coating may have such an effect. This model also predicts that r_2 is dependent upon only a single parameter, the magnetization of the particles.

10.5.2.2 Motional Averaging

The motional averaging regime (Ayant et al. 1975, Gillis, Moiny, and Brooks 2002) dominates when $\tau_D \ll 1/\Delta w$. This model describes the situation where the water molecules are able to diffuse rapidly around the nanoparticle following a random path. Generally this applies to particles smaller than about 30 nm when suspended in pure water. By following this random path, the proton experiences a range of magnetic field strengths. As the precession is proportional to the magnetic field, the rate of precession will vary from faster rates when the proton is near the poles of the nanoparticle to slower rates at the equator. An individual proton will therefore have its precession averaged as a result of its motion. The dephasing is related to the rate at which the proton magnetic moments become out of phase. Hence, for an ensemble of protons, the dephasing will take longer, i.e. r_2 will be lower. In this case the transverse relaxation rate, r_{2ma}, is given by:

$$r_{2ma}^* = r_{2ma} = \frac{4MW\,(d_p\Delta\omega)^2}{45c\rho D_{H_2O}} \tag{10.23}$$

Again the $r_2{}^*$ and r_2 are equal and the relaxivity is now proportional to the square of the particle size. For smaller particles the amount of averaging of the proton precession frequency is greater compared with larger particles and hence the degree of dephasing is smaller and the relaxivities are subsequently smaller.

10.5.2.3 Echo Limiting

One other factor that must be taken into account is the effect of the echo spacing on the r_2 relaxivity. This affects systems of larger particles usually over a few hundred nanometers although this depends upon the echo time that is used. The echo limited model becomes applicable where the refocusing pulses become effective at recovering the transverse magnetization lost due to the magnetic contrast agent. If in the time between refocusing pulses, the magnetic field experienced by a proton does not change significantly, then the refocusing pulse will allow effective recovery of the transverse magnetization. If the echo time is short, there is insufficient time for the water molecules to diffuse very far. In this situation a 180° refocusing pulse will result in the recovery of a large proportion of the transverse magnetization. Conversely, a long echo time will allow for a greater amount of diffusion, which will result in the decoherence of the magnetic moments of the water molecules and an irreversible loss of transverse magnetization. In this case the refocusing pulse is said to be ineffective. If the pulse is effective at recovering the transverse magnetization then the r_2 will be lower than that of the static dephasing regime. The r_2* remains the same as that for the static dephasing case as of course r_2* is not measured using the refocusing pulse or echoes.

The echo limited regime (Gillis, Moiny, and Brooks 2002) applies to larger particles. When $\tau_D < 2t_{cp}$ the refocusing of the proton magnetic moments is ineffective and r_2 and r_2* are equivalent. When $T_D > 2t_{cp}$ partial refocusing of the transverse magnetization leads to a decrease in r_2 compared to r_2*. The analytical model proposed by Gillis, Moiny, and Brooks (2002) gives a value for transverse relaxivity in the echo limited regime, r_{2el}, given by:

$$r_{2el} = \frac{7.2MWDx^{1/3}(1.52 + \phi x)^{5/3}}{c\rho d_p^2} \tag{10.24}$$

where ϕ is the volume fraction of particles and x is given by;

In this echo limited regime, r_2* is independent of particle size whereas r_2 is inversely proportional to square of the particle diameter.

10.5.2.4 Relaxivity

The transverse relaxivities associated with the three regimes described above are shown graphically in Figure 10.14. Figure 10.14 shows that the r_2 relaxivity is expected to initially increase as the particle size increases (motional averaging regime with $\tau_D < 1/\Delta w$), reaching a maximum at a particular size related to the magnetization of the nanoparticles (start of static dephasing regime $1/\Delta w < \tau_D < 2t_{cp}$), and remain at this value until above a larger critical size ($\tau_D > 2t_{cp}$) where the r_2 relaxivity decreases (echo limited regime). While the values in Figure 10.14 are specifically for magnetite, the general shape of the plot is the same for other materials.

10.6 BIOCOMPATIBILITY AND BIODISTRIBUTION

As mentioned at the start of this chapter, a contrast agent should produce a high degree of contrast, have minimal nonspecific binding, have a circulation time sufficient to achieve its function, should be effectively eliminated from the body, and have little or no toxicity (Choi and Frangioni 2010). Overall the toxicity is determined by the combination of biological half-life of the compounds and any potential breakdown products, the chemical stability of the compound under physiological conditions, and the physiological effect of the compound and its breakdown products. For instance, free Gd^{3+} ions are toxic (Werner et al. 2008), but if Gd ions are suitably chelated, so that the biological half-life is much shorter than the rate of disassociation of the chelated compounds, then toxicity of the chelated compound reduces to near zero as the Gd ions are excreted well before any significant amount of free Gd ions is released. This is the existing rationale for the design of clinical Gd ion-based contrast agents.

Following intravenous administration, the virtually universal method for contrast agent administration, the contrast agent will come into contact with a range of serum proteins. Depending primarily on surface charge and hydrophobicity of the contrast agent, the agents may undergo a process of opsonization, where these proteins bind to the contrast agent, increasing its size and improving its recognition as a foreign object by the immune system, specifically the reticuloendothelial system (RES)[5] (Mornet et al. 2004, Longmire, Choyke, and Kobayashi 2008). Particles recognized as foreign objects then undergo rapid endocytosis or phagocytosis by monocytes and macrophages removing them from the blood stream and concentrating the particles in organs of high phagocytic activity such as the liver, spleen, lymphatic system, and bone marrow. Particles in the RES rely almost entirely on intracellular breakdown for particle removal and therefore result in large exposure times (Longmire, Choyke, and Kobayashi 2008). The RES is very effective at picking up larger particles (above 200 nm) and hence most contrast agents are

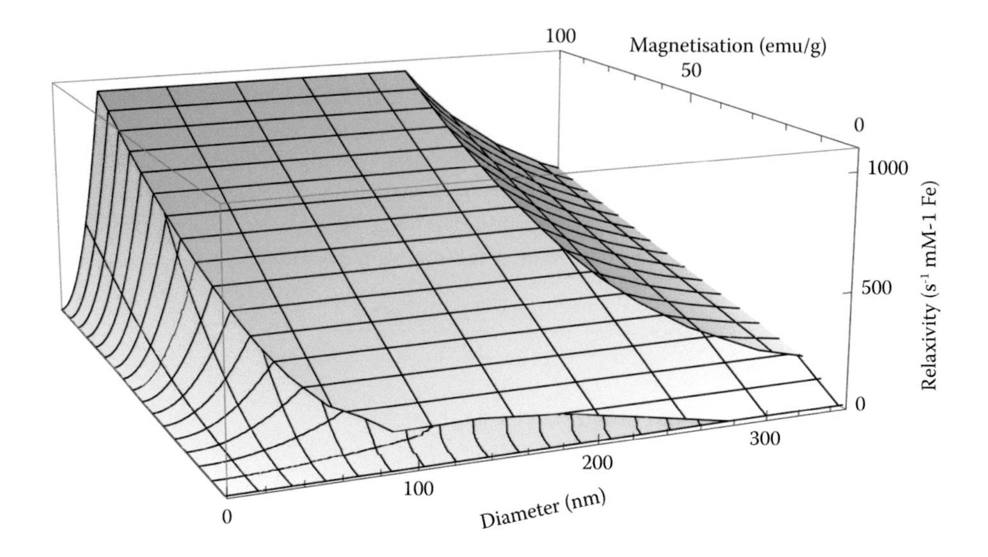

FIGURE 10.14 Predicted values of r_2 relaxivity as a function of the particle magnetization and diameter for magnetite nanoparticles. Yellow section corresponds to motional averaging regime, green section to static dephasing, and red section to echo limited regime. Note: r_2* is equal to r_2 for the motional averaging and static dephasing regimes. In the echo limited regime r_2 decays with increasing diameter while r_2* remains at the value associated with the static dephasing regime (Carroll et al. 2010).

designed with hydrodynamic sizes below this value (Moghimi, Hunter, and Murray 2005). In addition it is possible to avoid the RES via the use of suitable coatings on the particles, which reduce or eliminate opsonization. These coatings generate neutral hydrophilic outer surfaces, such as polyethylene glycol (PEG), or use zwitterionic systems such as cysteine (Mornet et al. 2004, Longmire, Choyke, and Kobayashi 2008, Choi et al. 2007).

There are two primary mechanisms for clearance of contrast agents from the body. The first is renal clearance performed by the kidneys. This is primarily a size-dependent filtration, although a number of factors such as shape, flexibility, and surface charge can also affect renal filtration. Typically particles with a hydrodynamic diameter less than 5.5–6 nm will be filtered by the kidneys and, provided they are not resorbed, will be excreted into the urine (Choi and Frangioni 2010, Longmire, Choyke, and Kobayashi 2008). The advantage of renal clearance is that the contrast agent is excreted quickly without significant cellular internalization or catabolism reducing any potential toxic effects (Longmire et al. 2011). However, the high rate of clearance may reduce the effectiveness of the agent, particularly for applications such as targeted contrast agents where long circulation times are preferred.

The second method of clearance is through the hepatobiliary system. This essential elimination of foreign substances, which do not undergo renal clearance, is through catabolization and biliary excretion via the hepatocytes in the liver (Longmire, Choyke, and Kobayashi 2008). Hepatic clearance is a more complicated process than renal clearance and while uptake of the agent from the blood may be swift, hepatic processing and biliary excretion are relatively slow, leading to relatively long whole-body half-lives.

In order to obtain long blood circulation times compatible with advanced diagnostic functions, such as targeting and molecular imaging, contrast agents need to designed to minimize opsonization and also must avoid a series of size-dependent methods of removal from the blood. As mentioned previously, particles less than 5.5 nm suffer rapid renal clearance, while particles larger than 200 nm are easily removed via the RES. In addition, one of the functions of the liver is to capture and eliminate particles of the scale of 10–20 nm (in order to remove viruses)

(Longmire, Choyke, and Kobayashi 2008). This means that the optimum hydrodynamic size for a long-circulation half-life particle is around 30–150 nm.

In addition to these considerations, issues of ionicity, osmolarity, and viscosity of the contrast agent solution need to be considered as these factors affect short- to medium-term side effects of the administered contrast agent (Lin and Brown 2007).

10.7 EXISTING MRI CONTRAST AGENTS

Clinical MRI contrast agents are presently available in two main categories. The first consist of chelated paramagnetic ions (Gd^{3+} or Mn^{2+}), which are used as positive contrast R_1 agents. The second are based on iron oxide nanoparticles, which produce negative contrast in R_2 weighted images.

10.7.1 R_1 Agents

The majority of contrast-assisted MRI scans are conducted with R_1 contrast agents based on chelated Gd^{3+} ions. There are nine different formulations, shown in Table 10.1, that are approved for clinical applications (Hasebroock and Serkova 2009). All of them are classified as analogous variations of chelated Gd complexes coordinated to either linear or cyclic polyaminocarboxylate ligands in either an ionic or nonionic form.

The majority of complexes are non-specific extracellular agents that rapidly distribute themselves between the plasma and interstitial spaces. They are commonly used for detection of tumors, inflammation, and vascular lesions. Two of the agents, MultiHance and Primovist, are useful for improving the visualization of the hepatic system. Vasovist is a blood pool agent used for MR angiography. Both Vasovist and MultiHance have improved r_1 relaxivity in vivo due to weak binding to serum proteins, which reduces the rotational rate of the bound compounds (Werner et al. 2008).

Although Gd-chelates are the most common choice for MRI, recent clinical reports have shown that there is an uncommon but significant major complication associated with their use in patients with poor or compromised renal function (Hasebroock and Serkova 2009). In these patients intravenous administration of Gd-chelates has been linked to the development of nephrogenic systemic fibrosis (NSF), a condition characterized by the formation and thickening of connective tissue in the skin joints and other organs developing over a period of days to weeks. In about 5% of patients, there is a rapid progression to a severe form of the disease and there is presently no known cure. In 2010 the growing evidence of the link between Gd-based MRI contrast agents and NSF led the FDA in the United States to require additional labeling on all Gd-based contrast agents (2010).

An examination of the reported incidences suggests that the majority of cases are associated with the use of compounds that have low thermodynamic stabilities and short half-lives, particularly Omniscan and Magnevist, with much smaller numbers reported for other compounds with much higher stabilities (Hasebroock and Serkova 2009, Idee et al. 2009, 2014). The European Medicine Agency made recommendations to minimize the risk of NSF from gadolinium-based contrast agents including a ranking of the relative risk of the present commercial agents (Thomsen 2011). Because these and similar recommendations have been adopted, the incidence of NSF has dropped significantly.

In addition to the Gd-based agents there was also one Mn^{2+}-based agent approved for clinical use, Teslascan (Mn-DPDP). Teslascan was used primarily for hepatobiliary imaging, but was withdrawn from market in 2003 (US) and 2012 (Europe). [reference to add – Manganese Contrast and Teslascan, Questions and answers in MRI, 2020 AD Elster, ELSTER LLC, http://mri-q.com/mn-agents-teslascan.html AND Teslascan (mangafodipir) Withdrawal of the marketing authorisation in the European Union, 18 July 2012, European Medicine Agency https://www.ema.europa.eu/en/documents/public-statement/public-statement-teslascan-withdrawal-marketing-authorisation-european-union_en.pdf.

TABLE 10.1
Approved Gd-Based MRI Contrast Agents

Brand Name	Name	Acronym	Form	NSF Risk[69]	Application
Magnevist	Gadopentetate dimeglumine	Gd-DTPA	Linear ionic	High	Extracellular
Omniscan	Gadodiamide	Gd-DTPA-BMA	Linear nonionic	High	Extracellular
OptiMARK	Gadoversetamide	Gd-DTPA-BMEA	Linear nonionic	High	Extracellular
MultiHance	Gadobenate dimeglumine	Gd-BOPTA	Linear ionic	Intermediate	Extracellular/ hepatobiliary
Vasovist (Ablavar)	Gadofosveset trisodium	GD-DTPA (MS325)	Linear ionic	Intermediate	Blood pool
Primovist (Eovist)	Gadoxetate disodium	Gd-EOB-DTPA	Linear ionic	Intermediate	Hepatobiliary
ProHance	Gadoteridol	Gd-HP-DO3A	Cyclic nonionic	Low	Extracellular
Gadovist	Gadobutrol	Gd-BT-DO3A	Cyclic nonionic	Low	Extracellular
Dotarem	Gadoterate meglumine	Gd-DOTA	Cyclic ionic	Low	Extracellular

10.7.2 R_2 Agents

A number of iron oxide nanoparticle-based preparations have been approved as MRI contrast agents. Table 10.2 shows a list of these and a few preparations that have proceeded to clinical trials.

Feridex was the most commonly used compound and received FDA approval in 1996. This contrast agent was an intravenous agent used for evaluation of liver lesions. Feridex consisted of 6–24 nm nanoparticles of iron oxide in clusters coated with dextran with a hydrodynamic size of 120–180 nm (Chen 2010). Feridex is rapidly cleared from the blood stream by the RES and accumulates in the spleen, liver, and lymph nodes, and was FDA approved for increasing contrast between liver lesions and normal liver tissue. It is worth noting that although for Ferridex the incidence of adverse reactions during clinical trials was not unusually large at around 9% (most commonly back pain and vasodilation), there was a relatively large percentage, 2.5%, of patients for which the reaction, in terms of pain, was so severe that the infusion was interrupted or discontinued (2008).

Gastromark is an oral contrast agent used for bowel and gastrointestinal MRI. The material is composed of clusters of 10 nm iron oxide particles coated with siloxane to generate a non-biodegradable and insoluble material with a hydrodynamic diameter of around 300 nm.

Resovist is another intravenous agent used for evaluation of liver lesions. It consists of clusters of approximately 5 nm iron oxide particles coated in carbodextran with a hydrodynamic diameter of around 60 nm (Lodhia et al. 2010). Because Resovist could be administered as a bolus (rapid) injection, it provided quicker examinations using dynamic imaging techniques as compared to Feridex, which required infusions over a longer time period.

Combidex consisted of 15–30 nm clusters of 3–7 nm iron oxide particles coated with dextran. Combidex was cleared from the blood via the RES and used to differentiate cancerous from non-cancerous lymph nodes.

TABLE 10.2
Commercial R_2-Based Contrast Agents

Brand Name	Company	Status	Coating	Generic Name	Application
Gastromark (Lumirem)	AMAG Pharmaceuticals	Approved for marketing in 1996, sold by Covidien & Guerbet	Silicon	Ferumoxsil	Gastrointestinal
Feridex (Endorem)	AMAG Pharmaceuticals	Approved for marketing in 1996, production discontinued in 2008	Dextran	Ferumoxide	Liver lesions
Resovist	Bayer Healthcare	Approved for use in 2001, production discontinued in 2009	Carboxy-dextran	Ferocarbotran	Liver lesions
Combidex (Sinerem)	AMAG Pharmaceuticals	Withdrawn from approval application in 2005–2007	Dextran	Ferumoxtran	Lymphatic MR angiography, RES-directed liver
Clariscan	Amersham Health	Development discontinued in 2007	PEG	Feruglose	Cardiovascular, blood pool

Clariscan was developed primarily for blood pool imaging and consisted of 4–7 nm iron oxide particles coated with a carbohydrate-polyethylene glycol coating with a hydrodynamic diameter of around 20 nm (Corot et al. 2006).

The majority of these preparations are no longer commercially available and production has been stopped, with the only R_2 contrast agent still on the market being Gastromark. The underlying reason for the removal of the majority of these R_2 contrast agents from sale is unclear. The reason given by the manufacturers was simple due to a "business decision" (2009).

For Resovist the business reason was related to the competition between two contrast agents, the other being a Gd-based agent—Primovist, made by the same manufacturer (Bayer) that had similar end-use applications in detecting liver lesions. Recent publications show that the performance of the liver-specific Gd-based contrast agents was at least as good or better than the iron oxide–based systems, see Figure 10.15 (Kim et al. 2010, Okada et al. 2010, Muhi et al. 2011). There is some evidence of slight to moderate improvements in detection using dual-contrast imaging techniques where both Gd and iron oxide contrast agents are used; however, this improvement in detection must be weighed against the added complexity, cost, and risk associated with the administration of two contrast agents and imaging the patient twice (Hanna et al. 2008).

In the case of Combidex, the contrast agent had completed phase 3 clinical trials in both the United States and Europe. However, in 2005 the FDA requested that rather than granting approval as a general contrast agent for metastatic cancer, clinical trial data be limited to specific cancer types. In 2007, the European distributor for Combidex (called Sinerem in Europe) withdrew is application for marketing approval, following information that the European Medicines Agency committee examining the application considered that the main study had failed to statistically demonstrate the efficacy of the product. Following this decision, the agent was used for a number of years in Europe for detection of metastasis in prostate cancer, but is no longer in production.

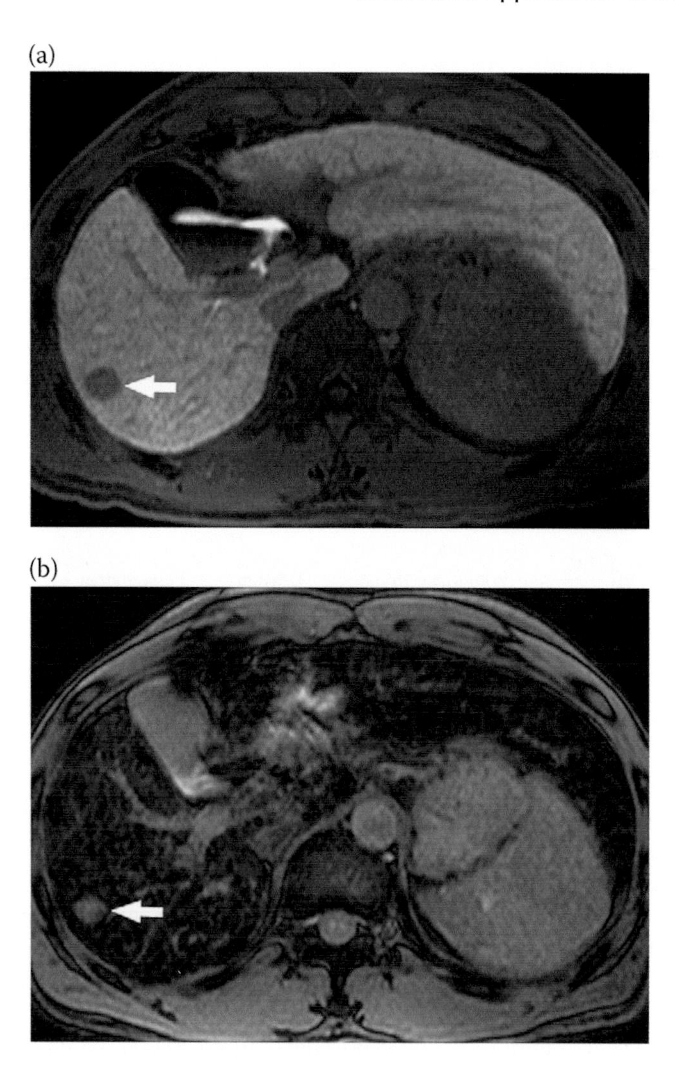

FIGURE 10.15 MR images of a 51-year-old man with a 1.6 cm hepatocellular carcinoma (HCC). **(a)** A gadoxetic acid–enhanced image showing HCC as hypointense region, **(b)** A ferucarbotran-enhanced image showing HCC as hyperintense region (Kim et al. 2010).

10.8 ADVANCED CONTRAST AGENTS

10.8.1 R_1 AGENTS

Most commercial Gd-based contrast agents have an r_1 of between 2 and 10 s^{-1} mM Gd^{-1}. However the theoretical limit for relaxivity for a Gd ion-based system is anywhere up to a few hundred s^{-1} mM Gd^{-1}, depending on the number of available water binding sites on the Gd ion (Peters, Huskens, and Raber 1996). This value of r_1 depends critically on a number of interaction time constants, which are often field-dependent, as discussed earlier in the chapter.

For Gd ion-based contrast agents, there are a number of strategies to improve the relaxivity including optimizing the rotational correlation time or increasing the number of inner sphere water molecules (Caravan et al. 2009). Many of these efforts focus on the modification of ligand structure

(Werner et al. 2008). Significant improvements in relaxivity have also been obtained by generating weak non-covalent binding between endogenous proteins and small-molecule chelated gadolinium. These compounds have the advantage of having a stable excretable material with longer intravascular retention times and significant improvements in relaxivity (Zhang, Nair, and McMurry 2005). Using these techniques, relaxivities of up to 84 s^{-1} mM Gd^{-1} have been achieved (Caravan et al. 2009).

Using these new Gd ion agents for applications such as molecular imaging can however be difficult as the binding of the agent to a target may affect molecular motion causing a loss in relaxivity. However, it is possible to design the attachment of the functional groups for targeting so that they minimize the effect on molecular motion of the Gd ion and hence maintain the majority of the relaxivity (Caravan et al. 2009).

Recently a number of authors have investigated alternatives to paramagnetic ions as R_1 contrast agents. The first of these is based on nanoparticles of inorganic compounds (Bridot et al. 2007, Park et al. 2009, Na, Song, and Hyeon 2009, Choi et al. 2010). The best results are generally based on gadolinium compounds, although compounds based on other lanthanide and transition metal compounds have been investigated. The relaxivities achieved are highest for particles with very small diameters, around 1–1.5 nm, and are at the higher range of the existing commercial contrast agents around 10 s^{-1} mM Gd^{-1} (Park et al. 2009, Choi et al. 2010). These nanoparticles have two potential advantages over more advanced Gd ion-based agents. The first is that it is possible to make these compounds fluorescent so that they provide an inherent dual-contrast mechanism (Bridot et al. 2007). The second is that although the contrast per Gd ion may be relatively low, the contrast per particle is significantly higher. This may be a significant advantage in molecular imaging and/or cellular tracking if the number of potential binding sites is relatively small (Na, Song, and Hyeon 2009).

Another alternative to Gd ion-based contrast agents are those based on systems such as full-erenes with encapsulated metal ions (Kato et al. 2003, Anderson, Lee, and Frank 2006). These systems particularly those with Gd encapsulated ions can be prepared with relatively high r_1 relaxivities of around 70 s^{-1} mM Gd^{-1}.

The final alternative to the Gd ion contrast agents is based on the concept of chemical exchange saturation transfer (CEST) (Ward, Aletras, and Balaban 2000). These contrast agents work in a way completely different from standard contrast agents, relying on a process of magnetization transfer. A detailed discussion of the mechanism of CEST and potential agents is beyond the scope of this chapter, and interested readers are recommended to read one of several excellent reviews (Sherry and Woods 2008, Hancu et al. 2010, Viswanathan et al. 2010). In simple terms the contrast agent generates a frequency shift in the Larmor frequency of water bound to the agent so that the magnetization of these water-based compounds can be manipulated separately to the bulk water. Provided that there is a reasonable rate of exchange between the bound water and the bulk water, the CEST agents can be used to create a decrease in the R_1 signal intensity. This signal is generally visualized as a difference between images with and without RF activation of the CEST agent. Originally CEST agents were based on diamagnetic compounds; however, it is possible to get much larger frequency shifts using paramagnetic agents (ParaCEST), which then allows much larger contrast. Most of these ParaCEST agents are based on lanthanide compounds due to their large frequency shifts. Note, Gd does not generate a frequency shift due to the isotropic electron orbitals (Hancu et al. 2010). Because the ParaCEST agents are based on lanthanides, which are all chemically similar, they have similar issues regarding the need for chemical stability of chelating agents, detection sensitivity, and retention time in vivo (Viswanathan et al. 2010).

With all of these different lanthanide-based agents, there is a need to maintain biocompatibility. If the lanthanides are not fully cleared from the body, then there is a chance that accumulation may result in nephrogenic systemic fibrosis (NSF). Hence, particularly for targeted systems that are interacting with endogenous proteins and compounds, the contrast agent needs to break down into a stable, easily excretable compound in a reasonable period of time.

10.8.2 R_2 Agents

As shown in Figure 10.14 the highest relaxivity for nanoparticulate R_2 contrast agents is obtained for particles in the static dephasing regime. For iron oxide particles this regime is attained when the particles are between about 25 nm and 250 nm in diameter. The relaxivity in this regime depends predominately on the magnetization of the particles and can, theoretically, reach almost 1000 s^{-1} mM Fe^{-1}. However large solid particles of this size are likely to be ferromagnetic and hence there will be a strong magnetic interparticle force between the particles and the samples are unlikely to be colloidally stable. In fact experimentally it is found that it is difficult to generate stable colloidal suspensions of magnetite particles much larger than about 15–20 nm in diameter, even though such particles are supposedly still superparamagnetic. One strategy to potentially overcome this limitation is the use of controlled clustering of smaller nanoparticles to form colloidally stable clusters. If the clusters are sufficiently large, they will fall within the static dephasing regime and can generate large relaxivities (Ai et al. 2005, Yang et al. 2007). However, in order to achieve the highest relaxivities, it is important to maintain a high density of nanoparticles within the clusters as the average magnetization of the cluster is directly related to the volume fraction of magnetic material (Vuong, Gillis, and Gossuin 2011).

In order to generate effective contrast agents, it is essential to control the surface properties through the use of surface coatings most commonly based on monomer or polymer surfactants. These surface coatings have several tasks. Firstly, they stabilize the particles in suspension in biological fluids reducing or controlling the degree of aggregation of the particles. Secondly, they control the interaction with the renal and reticuloendothelial system so as to generate a circulation time sufficient to allow the particles to perform their function as a diagnostic and or therapeutic agent. In addition, the coatings can be used to help generate more advanced diagnostic and therapeutic agents by providing a backbone for functionalization of the particles for specific biochemical targeting and by creating a controllable storage and release mechanism for the delivery of drugs and other therapeutic agents.

The binding of surfactants to iron oxide has traditionally been via carboxylates. However, although it is possible to generate stable systems in water, the surfactants tend to be displaced in solutions that are closer to biological reality such as phosphate-buffered saline. Such displacement may reduce colloidal stability and make the particles susceptible to clearance via the RES (Moghimi, Hunter, and Murray 2001). Recently greater stability in biological fluids has been attained using phosphonates and derivatives of dopamine (Goff et al. 2009, Amstad et al. 2009).

In order to improve colloidal stability, one can use either charged surfactants that stabilize the particles via electrostatic forces, or use steric forces via surfactant coatings that are dense and/or use long chains. In general both positively and negatively charged particles interact with cellular structures resulting in a reduced blood half-life. So, for applications requiring long circulation times, hydrophilic surfactants with a near neutral charge such as PEG are recommended (Sunderland et al. 2006, Berry 2009). Work on the effect of chain length on circulation times for PEG-coated particles suggests that denser coatings with smaller chains are more effective than using longer chains (Sunderland et al. 2006).

Some applications, such as cellular tracking or imaging of the RES, require efficient uptake of contrast agent by the cells. For these applications, a recent work suggests that positively charged surfactants generate enhanced non-specific cellular uptake in comparison to negatively charged or neutrally charged surfactants (Song et al. 2005, Villanueva et al. 2009). Enhanced uptake can also be generated via receptor-mediated endocytosis using peptides and antibodies (Zhao et al. 2002, Ahrens et al. 2003).

10.9 SUMMARY

There are a series of imaging techniques that make use of magnetic particles to image or improve the quality of images of living systems. In most cases these techniques make use of either ferromagnetic or superparamagnetic nanoparticles or microspheres. However, in the case of MRI, even the weak magnetic response of some paramagnetic ions can be used to provide enhanced contrast during imaging. One advantage of such techniques is that there is very little naturally occurring magnetic material in human and biological tissue and hence background signals are generally low.

Of the available techniques, MRI is the most widespread in both research and clinical environments. In MRI a combination of static magnetic fields and RF pulses are used to manipulate the small magnetic moments of the protons associated with water in the body. The signals from these protons will be modified locally by the presence of any magnetic entities nearby; hence the addition of such entities can be used to improve the local contrast.

These magnetic entities can take one of two principal forms. The first are R_1 contrast agents, predominately paramagnetic ions, which result in an increase in the longitudinal relaxation rate and an increase in signal intensity of R_1 weighted images. Most of the present commercial MRI contrast agents are of this form and are almost all based on suitably chelated gadolinium ions. The second form of contrast agent is based primarily on superparamagnetic nanoparticles of iron oxide. These systems produce a change in the transverse relaxation rate and a decrease in the signal intensity of R_2 weighted images.

There is significant room for improving the design of both R_1 and R_2 based contrast agents as commercial contrast agents are still well below the theoretical levels of contrast that can be achieved in such systems. In addition, there is a continual demand for improvements in both the passive and active targeting of such systems, incorporation of additional contrast mechanisms, and the inclusion of therapeutic modalities.

NOTES

1 Lissajous Trajectory is a space filling pattern produced by the combination of orthogonal sine waves.
2 A particle's moment is the magnetization multiplied by the particle volume.
3 These equations are commonly formatted in terms of relaxation times, which are simply the inverse of the relaxation rate.
4 The echo spacing is the time between echos in a spin echo measurement as shown in Figure 10.6.
5 Also known as the mononuclear phagocyte system (MPS).

REFERENCES

2008. Feridex offical FDA information, side effects and uses. accessed 7/12/2011. http://www.drugs.com/pro/feridex.html.

2009. AMAG Pharmaceuticals Inc. Q4 2008 Earnings Call Transcript. accessed 7/12/2011. http://seekingalpha.com/article/123206-amag-pharmaceuticals-inc-q4–2008-earnings-call-transcript.

2010. FDA Drug Safety Communication: New warnings for using gadolinium-based contrast agents in patients with kidney dysfunction. FDA, accessed 01/10/2011. http://www.fda.gov/drugs/drugsafety/ucm223966.htm.

Ahrens, E. T., M. Feili-Hariri, H. Xu, G. Genove, and P. A. Morel. 2003. Receptor-mediated endocytosis of iron-oxide particles provides efficient labeling of dendritic cells for in vivo MR imaging. *Magnetic Resonance in Medicine* 49 (6), 1006–1013.

Ai, H., C. Flask, B. Weinberg, X. Shuai, M. D. Pagel, D. Farrell, J. Duerk, and J. M. Gao. 2005. Magnetite-loaded polymeric micelles as ultrasensitive magnetic-resonance probes. *Advanced Materials* 17 (16), 1949–1952.

Aime, S., S. G. Crich, E. Gianolio, G. B. Giovenzana, L. Tei, and E. Terreno. 2006. High sensitivity lanthanide(III) based probes for MR-medical imaging. *Coordination Chemistry Reviews* 250 (11-12), 1562–1579.

Amstad, E., T. Gillich, I. Bilecka, M. Textor, and E. Reimhult. 2009. Ultrastable iron oxide nanoparticle colloidal suspensions using dispersants with catechol-derived anchor groups. *Nano Letters* 9 (12), 4042–4048.

Anderson, S. A., K. K. Lee, and J. A. Frank. 2006. Gadolinium-fullerenol as a paramagnetic contrast agent for cellular imaging. *Investigative Radiology* 41 (3), 332–338.

Ayant, Y., E. Belorizky, J. Alizon, and J. Gallice. 1975. Calculation of spectral densities for relaxation resulting from random molecular translational modulation of magnetic dipolar coupling in liquids. *Journal de Physique* 36 (10), 991–1004.

Baek, M. J., J. Y. Park, W. L. Xu, K. Kattel, H. G. Kim, E. J. Lee, A. K. Patel, J. J. Lee, Y. M. Chang, T. J. Kim, J. E. Bae, K. S. Chae, and G. H. Lee. 2010. Water-soluble MnO nanocolloid for a molecular T(1) MR imaging: A facile one-pot synthesis, in vivo T(1) MR images, and account for relaxivities. *ACS Applied Materials & Interfaces* 2 (10), 2949–2955.

Berry, C. C. 2009. Progress in functionalization of magnetic nanoparticles for applications in biomedicine. *Journal of Physics D-Applied Physics* 42 (22), 224003.

Botta, M. 2000. Second coordination sphere water molecules and relaxivity of gadolinium(III) complexes: Implications for MRI contrast agents. *European Journal of Inorganic Chemistry* 2000 (3), 399–407.

Brahler, M., R. Georgieva, N. Buske, A. Muller, S. Muller, J. Pinkernelle, U. Teichgraber, A. Voigt, and H. Baumler. 2006. Magnetite-loaded carrier erythrocytes as contrast agents for magnetic resonance imaging. *Nano Letters* 6 (11), 2505–2509.

Bridot, J. L., A. C. Faure, S. Laurent, C. Riviere, C. Billotey, B. Hiba, M. Janier, V. Josserand, J. L. Coll, L. Vander Elst, R. Muller, S. Roux, P. Perriat, and O. Tillement. 2007. Hybrid gadolinium oxide nanoparticles: Multimodal contrast agents for in vivo imaging. *Journal of the American Chemical Society* 129 (16), 5076–5084.

Brown, M. A., and R. C. Semelka. 2010. *MRI Basic Principles and Applications*. 4th ed. Hoboken, NJ: Wiley-Blackwell.

Bushberg, J. T., J. A. Seibert, E. M. Leidholdt, and J. M. Boone. 2002. *The Essential Physics of Medical Imaging*. 2nd ed. Sydney: Lippincott Williams & Wilkins.

Buxton, R. B. 2003. *Introduction to Functional Magnetic Resonance Imaging: Principles and Techniques*. Melbourne: Cambridge University Press.

Cai, W. B., and X. Y. Chen. 2007. Nanoplatforms for targeted molecular imaging in living subjects. *Small* 3 (11), 1840–1854.

Caravan, P., J. J. Ellison, T. J. McMurry, and R. B. Lauffer. 1999. Gadolinium(III) chelates as MRI contrast agents: Structure, dynamics, and applications. *Chemical Reviews* 99 (9), 2293–2352.

Caravan, P., C. T. Farrar, L. Frullano, and R. Uppal. 2009. Influence of molecular parameters and increasing magnetic field strength on relaxivity of gadolinium- and manganese-based T-1 contrast agents. *Contrast Media & Molecular Imaging* 4 (2), 89–100.

Carroll, M. R. J., R. C. Woodward, M. J. House, W. Y. Teoh, R. Amal, T. L. Hanley, and T. G. St Pierre. 2010. Experimental validation of proton transverse relaxivity models for superparamagnetic nanoparticle MRI contrast agents. *Nanotechnology* 21 (3), 035103.

Chambon, C., O. Clement, A. Leblanche, E. Schoumanclaeys, and G. Frija. 1993. Superparamagnetic iron-oxides as positive MR contrast agents—In vitro and in vivo evidence. *Magnetic Resonance Imaging* 11 (4), 509–519. doi:10.1016/0730-725X(93)90470-X.

Chen, S. 2010. Polymer-coated iron oxide nanoparticles for medical imaging. PhD, Materials Science and Engineering, Massachusetts Institute of Technology.

Choi, E. S., J. Y. Park, K. Kattel, W. L. Xu, G. H. Lee, M. J. Baek, J. H. Kim, Y. M. Chang, and T. J. Kim. 2010. Longitudinal water proton relaxivities of ultrasmall 3d and 4f transition metal oxide nanoparticles. *Journal of the Korean Physical Society* 56 (5), 1532–1536.

Choi, H. S., and J. V. Frangioni. 2010. Nanoparticles for biomedical imaging: Fundamentals of clinical translation. *Molecular Imaging* 9 (6), 291–310.

Choi, H. S., W. Liu, P. Misra, E. Tanaka, J. P. Zimmer, B. I. Ipe, M. G. Bawendi, and J. V. Frangioni. 2007. Renal clearance of quantum dots. *Nature Biotechnology* 25 (10), 1165–1170.

Chung, S. H., A. Hoffmann, K. Guslienko, S. D. Bader, C. Liu, B. Kay, L. Makowski, and L. Chen. 2005. Biological sensing with magnetic nanoparticles using Brownian relaxation (invited). *Journal of Applied Physics* 97 (10), 10R101.

2020 "Contrast Media/Contrast Agent Market by Type (Iodinated, Gadolinium, Barium, and Microbubble), Procedure (X-Ray/CT, MRI, and Ultrasound), Application (Radiology, Interventional Radiology, and Interventional Cardiology), Region – Global Forecast to 2024" Markets and Markets, https://www.marketsandmarkets.com/Market-Reports/contrast-media-market-911.html, 2020.

Coons, A. H. 1961. The beginnings of immunofluorescence. *Journal of Immunology* 87:499–503.

Corot, C., P. Robert, J. M. Idee, and M. Port. 2006. Recent advances in iron oxide nanocrystal technology for medical imaging. *Advanced Drug Delivery Reviews* 58 (14), 1471–1504.

Debbage, P., and W. Jaschke. 2008. Molecular imaging with nanoparticles: Giant roles for dwarf actors. *Histochemistry and Cell Biology* 130 (5), 845–875.

Elias, D. R., D. L. J. Thorek, A. K. Chen, J. Czupryna, and A. Tsourkas. 2008. In vivo imaging of cancer biomarkers using activatable molecular probes. *Cancer Biomarkers* 4 (6), 287–305.

Fercher, A. F., W. Drexler, C. K. Hitzenberger, and T. Lasser. 2003. Optical coherence tomography—Principles and applications. *Reports on Progress in Physics* 66 (2), 239–303.

Gibby, W. A. 2005. Basic principles of magnetic resonance imaging. *Neurosurgery Clinics of North America* 16 (1), 1–64.

Gillis, P., F. Moiny, and R. A. Brooks. 2002. On T-2-shortening by strongly magnetized spheres: A partial refocusing model. *Magnetic Resonance in Medicine* 47 (2), 257–263.

Gleich, B., and R. Weizenecker. 2005. Tomographic imaging using the nonlinear response of magnetic particles. *Nature* 435 (7046), 1214–1217.

GlobalData. 2009. Contrast Media: A Market Snapshot. Accessed 27/01/2010. http://www.researchandmarkets.com/reportinfo.asp?cat_id=12&report_id=1195353&p=1.

Goff, J. D., P. P. Huffstetler, W. C. Miles, N. Pothayee, C. M. Reinholz, S. Ball, R. M. Davis, and J. S. Riffle. 2009. Novel phosphonate-functional poly(ethylene oxide)-magnetite nanoparticles form stable colloidal dispersions in phosphate-buffered saline. *Chemistry of Materials* 21 (20), 4784–4795.

Hancu, I., W. T. Dixon, M. Woods, E. Vinogradov, A. D. Sherry, and R. E. Lenkinski. 2010. CEST and PARACEST MR contrast agents. *Acta Radiologica* 51 (8), 910–923.

Hanna, R. F., N. Kased, S. W. Kwan, A. C. Gamst, A. C. Santosa, T. Hassanein, and C. B. Sirlin. 2008. Double-contrast MRI for accurate staging of hepatocellular carcinoma in patients with cirrhosis. *American Journal of Roentgenology* 190 (1), 47–57.

Hasebroock, K. M., and N. J. Serkova. 2009. Toxicity of MRI and CT contrast agents. *Expert Opinion on Drug Metabolism & Toxicology* 5 (4), 403–416.

Helm, L. 2006. Relaxivity in paramagnetic systems: Theory and mechanisms. *Progress in Nuclear Magnetic Resonance Spectroscopy* 49 (1), 45–64.

Hendrick, R. E., and E. M. Haacke. 1993. Basic physics of MR contrast agents and maximization of image-contrast. *JMRI-Journal of Magnetic Resonance Imaging* 3 (1), 137–148.

Herranz, M., and I. Sanchez-Garcia. 2007. Medical chemistry to spy cancer stem cells from outside the body. *Mini-Reviews in Medicinal Chemistry* 7 (8), 781–791.

Idee, J. M., M. Port, C. Robic, C. Medina, M. Sabatou, and C. Corot. 2009. Role of thermodynamic and kinetic parameters in gadolinium chelate stability. *Journal of Magnetic Resonance Imaging* 30 (6), 1249–1258. doi:10.1002/jmri.21967.

Idee, J. M., N. Fretellier, C. Robic, and C. Corot. 2014. The role of gadolinium chelates in the mechanism of nephrogenic systemic fibrosis: A critical update. *Critical Reviews in Toxicology* 44 (10), 895–913. doi:10.3109/10408444.2014.955568.

Jacques, V., S. Dumas, W. C. Sun, J. S. Troughton, M. T. Greenfield, and P. Caravan. 2010. High-relaxivity magnetic resonance imaging contrast agents part 2 optimization of inner- and second-sphere relaxivity. *Investigative Radiology* 45 (10), 613–624.

Jennings, L. E., and N. J. Long. 2009. Two is better than one-probes for dual-modality molecular imaging. *Chemical Communications* (24), 3511–3524.

Jensen, J. H., and R. Chandra. 2000. NMR relaxation in tissues with weak magnetic inhomogeneities. *Magnetic Resonance in Medicine* 44 (1), 144–156.

Kato, H., Y. Kanazawa, M. Okumura, A. Taninaka, T. Yokawa, and H. Shinohara. 2003. Lanthanoid endohedral metallofullerenols for MRI contrast agents. *Journal of the American Chemical Society* 125 (14), 4391–4397.

Kim, Y. K., C. S. Kim, Y. M. Han, G. Park, S. B. Hwang, and H. C. Yu. 2010. Comparison of gadoxetic acid-enhanced MRI and superparamagnetic iron oxide-enhanced MRI for the detection of hepatocellular carcinoma. *Clinical Radiology* 65 (5), 358–365.

Koenig, S. H., and K. E. Kellar. 1995. Theory of 1/T-1 and 1/T-2 NMRD profiles of solutions of magnetic nanoparticles. *Magnetic Resonance in Medicine* 34 (2), 227–233.

Kuperman, V. 2000. Magnetic Resonance Imaging: Physical Principles and Applications. In *Academic Press Electromagnetism Series*, edited by I. Mayergoyz. Sydney: Academic Press.

Lauffer, R. B. 1987. Paramagnetic metal-complexes as water proton relaxation agents for NMR imaging - Theory and design. *Chemical Reviews* 87 (5), 901–927.

Lee, J. H., and A. P. Koretsky. 2004. Manganese enhanced magnetic resonance imaging. *Current Pharmaceutical Biotechnology* 5 (6), 529–537.

Levitt, M. H. 2008. *Spin Dynamics: Basics of Nuclear Magnetic Resonance*. 2nd ed. Chichester, England: John Wiley & Sons.

Li, Z., P. W. Yi, Q. Sun, H. Lei, H. L. Zhao, Z. H. Zhu, S. C. Smith, M. B. Lan, and G. Q. Lu. 2012. Ultrasmall water-soluble and biocompatible magnetic iron oxide nanoparticles as positive and negative dual contrast agents. *Advanced Functional Materials* 22 (11), 2387–2393. doi:10.1002/adfm.201103123.

Liang, Z.-P., and P. C. Lauterbur. 2000. *Principles of Magnetic Resonance Imaging: A Signal Processing Perspective*. New York: Wiley-IEEE Press.

Lin, S. P., and J. J. Brown. 2007. MR contrast agents: Physical and pharmacologic basics. *Journal of Magnetic Resonance Imaging* 25 (5), 884–899.

Lodhia, J., G. Mandarano, N. J. Ferris, P. Eu, and S.F. Cowell. 2010. Development and use of iron oxide nanoparticles (Part 1): Synthesis of iron oxide nanoparticles for MRI. *Biomedical Imaging and Invtervention Journal* 6 (2), e12.

Longmire, M., P. L. Choyke, and H. Kobayashi. 2008. Clearance properties of nano-sized particles and molecules as imaging agents: Considerations and caveats. *Nanomedicine* 3 (5), 703–717.

Longmire, M. R., M. Ogawa, P. L. Choyke, and H. Kobayashi. 2011. Biologically optimized nanosized molecules and particles: More than just size. *Bioconjugate Chemistry* 22 (6), 993–1000.

Lopez, A., L. Gutierrez, and F. J. Lazaro. 2007. The role of dipolar interaction in the quantitative determination of particulate magnetic carriers in biological tissues. *Physics in Medicine and Biology* 52 (16), 5043–5056.

Mansson, S., and A. Bjornerud. 2001. Physical principles of medical imaging by nuclear magnetic resonance. In: A. E. Merbach and Toth, E. (eds). *The Chemistry of Contrast Agents in Magnetic Resonance Imaging* 1–43. Sydney: John Wiley & Sons Ltd.

Matsumoto, Y., and A. Jasanoff. 2008. T-2 relaxation induced by clusters of superparamagnetic nanoparticles: Monte Carlo simulations. *Magnetic Resonance Imaging* 26 (7), 994–998.

Mehrmohammadi, M., J. Oh, S. R. Aglyamov, A. B. Karpiouk, and S. Y. Emelianov. 2009. Pulsed magneto-acoustic imaging. In *EMBC: 2009 Annual International Conference of the IEEE Engineering in Medicine and Biology Society*, edited by Zhe-Pei Liang, 1-20:4771–4774. New York: IEEE.

Moghimi, S. M., A. C. Hunter, and J. C. Murray. 2001. Long-circulating and target-specific nanoparticles: Theory to practice. *Pharmacological Reviews* 53 (2), 283–318.

Moghimi, S. M., A. C. Hunter, and J. C. Murray. 2005. Nanomedicine: Current status and future prospects. *Faseb Journal* 19 (3), 311–330.

Moller, W., S. Takenaka, M. Rust, W. Stahlhofen, and J. Heyder. 1997. Probing mechanical properties of living cells by magnetopneumography. *Journal of Aerosol Medicine-Deposition Clearance and Effects in the Lung* 10 (3), 173–186.

Mornet, S., S. Vasseur, F. Grasset, and E. Duguet. 2004. Magnetic nanoparticle design for medical diagnosis and therapy. *Journal of Materials Chemistry* 14 (14), 2161–2175.

Muhi, A., T. Ichikawa, U. Motosugi, H. Sou, H. Nakajima, K. Sano, M. Sano, S. Kato, T. Kitamura, Z. Fatima, K. Fukushima, H. Iino, Y. Mori, H. Fujii, and T. Araki. 2011. Diagnosis of colorectal hepatic metastases: Comparison of contrast-enhanced CT, contrast-enhanced US, superparamagnetic iron oxide-enhanced MRI, and gadoxetic acid-enhanced MRI. *Journal of Magnetic Resonance Imaging* 34 (2), 326–335.

Na, H. B., I. C. Song, and T. Hyeon. 2009. Inorganic nanoparticles for MRI contrast agents. *Advanced Materials* 21 (21), 2133–2148.

Oh, J., M. D. Feldman, J. Kim, C. Condit, S. Emelianov, and T. E. Milner. 2006. Detection of magnetic nanoparticles in tissue using magneto-motive ultrasound. *Nanotechnology* 17 (16), 4183–4190.

Okada, M., Y. Imai, T. Kim, S. Kogita, M. Takamura, S. Kumano, H. Onishi, M. Hori, K. Fukuda, N. Hayashi, K. Wakasa, M. Sakamoto, and T. Murakami. 2010. Comparison of enhancement patterns of histologically confirmed hepatocellular carcinoma between gadoxetate- and ferucarbotran-enhanced magnetic resonance imaging. *Journal of Magnetic Resonance Imaging* 32 (4), 903–913.

Oldenburg, A. L., F. J. J. Toublan, K. S. Suslick, A. Wei, and S. A. Boppart. 2005. Magnetomotive contrast for in vivo optical coherence tomography. *Optics Express* 13 (17), 6597–6614.

Oldenburg, A. L., V. Crecea, S. A. Rinne, and S. A. Boppart. 2008. Phase-resolved magnetomotive OCT for imaging nanomolar concentrations of magnetic nanoparticles in tissues. *Optics Express* 16 (15), 11525–11539.

Park, J. Y., M. J. Baek, E. S. Choi, S. Woo, J. H. Kim, T. J. Kim, J. C. Jung, K. S. Chae, Y. Chang, and G. H. Lee. 2009. Paramagnetic ultrasmall gadolinium oxide nanoparticles as advanced T-1 MR1 contrast agent: Account for large longitudinal relaxivity, optimal particle diameter, and in vivo T-1 MR images. *ACS Nano* 3 (11), 3663–3669.

Peters, J. A., J. Huskens, and D. J. Raber. 1996. Lanthanide induced shifts and relaxation rate enhancements. *Progress in Nuclear Magnetic Resonance Spectroscopy* 28 283–350.

Plewes, D. B., and W. Kucharczyk. 2012. Physics of MRI: A primer. *Journal of Magnetic Resonance Imaging* 35 (5), 1038–1054.

Roch, A., Y. Gossuin, R. N. Muller, and P. Gillis. 2005. Superparamagnetic colloid suspensions: Water magnetic relaxation and clustering. *Journal of Magnetism and Magnetic Materials* 293 (1), 532–539.

Rogers, W. J., C. H. Meyer, and C. M. Kramer. 2006. Technology insight: In vivo cell tracking by use of MRI. *Nature Clinical Practice Cardiovascular Medicine* 3 (10), 554–562.

Sharma, P., S. Brown, G. Walter, S. Santra, and B. Moudgil. 2006. Nanoparticles for bioimaging. *Advances in Colloid and Interface Science* 123, 471–485.

Sherry, A. D., and M. Woods. 2008. Chemical exchange saturation transfer contrast agents for magnetic resonance imaging. In *Annual Review of Biomedical Engineering*, 391–411. Palo Alto: Annual Reviews.

Shubayev, V. I., T. R. Pisanic, and S. H. Jin. 2009. Magnetic nanoparticles for theragnostics. *Advanced Drug Delivery Reviews* 61 (6), 467–477.

Song, H. T., J. S. Choi, Y. M. Huh, S. Kim, Y. W. Jun, J. S. Suh, and J. Cheon. 2005. Surface modulation of magnetic nanocrystals in the development of highly efficient magnetic resonance probes for intracellular labeling. *Journal of the American Chemical Society* 127 (28), 9992–9993.

St Pierre, T. G., P. R. Clark, W. Chua-Anusorn, A. J. Fleming, G. P. Jeffrey, J. K. Olynyk, P. Pootrakul, E. Robins, and R. Lindeman. 2005. Noninvasive measurement and imaging of liver iron concentrations using proton magnetic resonance. *Blood* 105 (2), 855–861.

Sunderland, C. J., M. Steiert, J. E. Talmadge, A. M. Derfus, and S. E. Barry. 2006. Targeted nanoparticles for detecting and treating cancer. *Drug Development Research* 67 (1), 70–93.

Thomsen, H. S. 2011. Contrast media safety—An update. *European Journal of Radiology* 80 (1), 77–82.

Toth, E., L. Helm, and A. E. Merbach. 2002. Relaxivity of MRI contrast agents. *Contrast Agents I* 221, 61–101.

Villanueva, A., M. Canete, A. G. Roca, M. Calero, S. Veintemillas-Verdaguer, C. J. Serna, M. D. Morales, and R. Miranda. 2009. The influence of surface functionalization on the enhanced internalization of magnetic nanoparticles in cancer cells. *Nanotechnology* 20 (11), 115103.

Viswanathan, S., Z. Kovacs, K. N. Green, S. J. Ratnakar, and A. D. Sherry. 2010. Alternatives to gadolinium-based metal chelates for magnetic resonance imaging. *Chemical Reviews* 110 (5), 2960–3018.

Vuong, Q. L., P. Gillis, and Y. Gossuin. 2011. Monte Carlo simulation and theory of proton NMR transverse relaxation induced by aggregation of magnetic particles used as MRI contrast agents. *Journal of Magnetic Resonance* 212 (1), 139–148.

Wagner, V., A. Dullaart, A. K. Bock, and A. Zweck. 2006. The emerging nanomedicine landscape. *Nature Biotechnology* 24 (10), 1211–1217.

Wang, G. N., X. J. Zhang, A. Skallberg, Y. X. Liu, Z. J. Hu, X. F. Mei, and K. Uvdal. 2014. One-step synthesis of water-dispersible ultra-small Fe_3O_4 nanoparticles as contrast agents for T-1 and T-2 magnetic resonance imaging. *Nanoscale* 6 (5), 2953–2963. doi:10.1039/c3nr05550g.

Wang, N., J. D. Tytell, and D. E. Ingber. 2009. Mechanotransduction at a distance: Mechanically coupling the extracellular matrix with the nucleus. *Nature Reviews Molecular Cell Biology* 10 (1), 75–82.

Ward, K. M., A. H. Aletras, and R. S. Balaban. 2000. A new class of contrast agents for MRI based on proton chemical exchange dependent saturation transfer (CEST). *Journal of Magnetic Resonance* 143 (1), 79–87.

Waters, E. A., and S. A. Wickline. 2008. Contrast agents for MRI. *Basic Research in Cardiology* 103 (2), 114–121.

Weizenecker, J., B. Gleich, J. Rahmer, H. Dahnke, and J. Borgert. 2009. Three-dimensional real-time in vivo magnetic particle imaging. *Physics in Medicine and Biology* 54 (5), L1–L10.

Werner, E. J., A. Datta, C. J. Jocher, and K. N. Raymond. 2008. High-relaxivity MRI contrast agents: Where coordination chemistry meets medical imaging. *Angewandte Chemie-International Edition* 47 (45), 8568–8580.

Wood, M. L., and P. A. Hardy. 1993. Proton relaxation enhancement. *JMRI-Journal of Magnetic Resonance Imaging* 3 (1), 149–156.

Yablonskiy, D. A., and E. M. Haacke. 1994. Theory of NMR signal behavior in magnetically inhomogeneous tissues—The static dephasing regime. *Magnetic Resonance in Medicine* 32 (6), 749–763.

Yang, J., C. H. Lee, H. J. Ko, J. S. Suh, H. G. Yoon, K. Lee, Y. M. Huh, and S. Haam. 2007. Multifunctional magneto-polymeric nanohybrids for targeted detection and synergistic therapeutic effects on breast cancer. *Angewandte Chemie-International Edition* 46 (46), 8836–8839.

Zhang, Z. D., S. A. Nair, and T. J. McMurry. 2005. Gadolinium meets medicinal chemistry: MRI contrast agent development. *Current Medicinal Chemistry* 12 (7), 751–778.

Zhao, M., M. F. Kircher, L. Josephson, and R. Weissleder. 2002. Differential conjugation of tat peptide to superparamagnetic nanoparticles and its effect on cellular uptake. *Bioconjugate Chemistry* 13 (4), 840–844.

11 Energy Dissipation by Magnetic Nanoparticles: Basic Principles for Biomedical Applications

Mythreyi Unni and Carlos Rinaldi

CONTENTS

11.1 EXAMPLES OF BIOMEDICAL APPLICATIONS OF MAGNETIC NANOPARTICLE HEATING

There are a variety of biomedical applications of MNPs that rely on the particle's ability to transform magnetic field energy into thermal energy. These include: (1) heating cancer tissues to the hyperthermia range (43–47 °C) in what is called magnetic fluid hyperthermia (Jordan et al. 2006, Zhao et al. 2012); (2) triggering the release of a drug payload by inducing a change in a polymer (Zhang, Srivastava, and Misra 2007, Rahimi et al. 2010), liposome (Chen, Bose, and Bothun 2010, Amstad et al. 2011, Kulshrestha et al. 2012), micelles (Kim et al. 2013b, Zou and Yuan 2015), or a thermally labile bond (Riedinger et al. 2013) in what can be called magnetically triggered drug release; and (3) remote actuation of membrane receptors (Mannix et al. 2008, Zhang et al. 2014). In all of these cases a time varying magnetic field, most often an alternating magnetic field (AMF) of appropriate amplitude and frequency, is applied to a collection of MNPs. In response to this AMF, the dipoles within the particles or the particles themselves oscillate and the intrinsic resistance to such oscillation leads to transformation of some of the magnetic field energy (which is doing work on the particles) into thermal energy. An important feature of this approach is that the body does not appreciably attenuate the magnitude of the AMF; hence, it is possible to manipulate particles that are located in deep tissues. The only limitation in terms of AMF application is the avoidance of non-specific heating due to eddy currents in tissue, which practically limits the amplitude and frequency of an AMF that can be applied.

11.1.1 Magnetic Fluid Hyperthermia

Magnetic fluid hyperthermia (MFH) has a long history (Roussakow 2013), starting with the work of Gilchrist (Gilchrist et al. 1957), who proposed the use of magnetic "thermoseeds" loaded into cancer tissues and cells to induce localized hyperthermia. Hyperthermia itself has been used by humans as a therapy for at least 5000 years (Breasted 1930, Zagar et al. 2010). In the case of cancer therapy, hyperthermia is attractive (Zhao et al. 2012, Wang et al. 2015, Lepage et al. 2007) because the rapidly dividing cancer cells are more susceptible to damage caused by heat (i.e. protein denaturation, membrane permeabilization, DNA damage) than normal cells (Hildebrandt et al. 2002, He et al. 2004, Mantso et al. 2016). As such, many clinical trials in the United States and in Europe have explored the application of hyperthermia to treat cancer, often in combination with other treatments such as chemotherapy and radiotherapy (Hurwitz and Stauffer 2014, Wust et al. 2002).

Hyperthermia can be classified according to the region of application into whole-body, regional, and local hyperthermia. The ultimate goal is to apply it only in the cancer tissue to avoid any side effects associated with elevated temperatures. One of the greatest challenges in effectively applying hyperthermia clinically is therefore achieving good spatial control of the temperature distribution. This need motivated Jordan and collaborators to apply heating by MNPs to the problem and created the very active field of MFH (Jordan et al. 1999). In MFH, the objective is to deliver MNPs to cancer tumors and cells, either by direct injection or systemic delivery, have the particles accumulate in the extracellular space or within the cells, and then raise the tissue temperature to the hyperthermia range by application of an AMF. Experiments using cell cultures have demonstrated that MFH can be highly effective at killing cancer cells (Jordan et al. 1996, 1999, Shinkai et al. 1996, Yan et al. 2005, Prasad et al. 2007, Creixell et al. 2011, Alvarez-Berrios et al. 2014), sometimes more effective than hyperthermia using external means of heating (Rodriguez-Luccioni et al. 2011). Pre-clinical experiments with animal models also indicate that MFH after direct injection of MNPs to a tumor (Jordan et al. 2006, Johannsen et al. 2007, Maier-Hauff et al. 2007, van Landeghem et al. 2009, Dennis et al. 2009, Moroz et al. 2001) and after systemic delivery of the MNPs (Moroz et al. 2001, DeNardo et al. 2007, Dennis et al. 2009) can be effective at reducing tumor volume and extending survival. Interestingly, MFH appears to activate an anti-cancer immune response, with experiments in rats demonstrating regression of a secondary tumor after a primary tumor was treated by MFH using magnetic liposomes as shown in Figure 11.1 (Ito et al. 2003a, b). Finally, MFH by direct delivery of amine-coated MNPs is currently approved for treating glioblastoma patients in Europe (Maier-Hauff et al. 2007, 2011) and a whole-body AMF applicator is commercially available (Figure 11.2) (Jordan et al. 2001). Still, even with this much progress, the field of MFH has yet to achieve its full potential (Kozissnik et al. 2013). Significant challenges still remain in the systemic delivery and targeting of MNPs to cancer tumors and cells, particularly in cases of metastatic disease. Furthermore, raising the tissue temperature to the desired hyperthermia range is difficult, particularly in small tumors and cell masses where prohibitively high concentrations of MNPs would be required (Dutz and Hergt 2014, Chiu-Lam and Rinaldi 2016). Also, monitoring the tumor temperature and temperature distribution non-invasively is challenging, posing problems with dose management during therapy and forcing the field to rely on predictive models of the temperature distribution (Cetas 1984, Lebihan, Delannoy, and Levin 1989). Current efforts make use of magnetic particle spectroscopy to monitor temperature of the

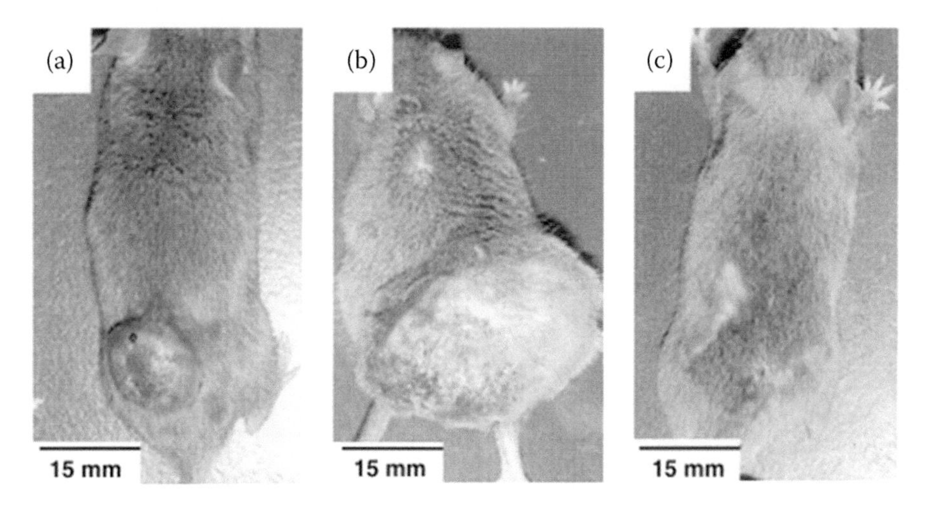

FIGURE 11.1 Photographs of representative mice treated with frequent repeated hyperthermia (RH). **(a)** Before AMF irradiation, **(b)** Control mouse on day 45, **(c)** Cured mouse after frequent RH treatment on day 45 (Ito et al. 2003a).

microenvironment surrounding the nanoparticles, where the relaxation time estimated from nanoparticle magnetization induced on the particles by an AMF at various applied frequencies is used to calculate the temperature (Reeves and Weaver 2014, Perreard et al. 2014). Finally, the eventual fate of the MNPs used in MFH after treatment is unclear, particularly for MNPs that have been coated with synthetic polymers that are not susceptible to biodegradation.

11.1.2 Magnetically Triggered Drug Release

In magnetically triggered drug release, the objective is to actuate release of a therapeutic cargo after accumulation of MNPs in the targeted tissue. In this way, dose distribution can be controlled such that side effects can be minimized. Additionally, one can envision external control of treatment schedule after delivering a sufficient amount of particles by applying AMFs at different time points. Ideally, MNPs for magnetically triggered drug release should display negligible release of the cargo in the absence of an AMF and controlled rates of release under the application of an AMF.

There are various examples of how to design MNPs for magnetically triggered release, including release of single-stranded DNA from MNPs functionalized with complementary strands (von Maltzahn et al. 2011, Ruiz-Hernandez, Baeza, and Vallet-Regi 2011), release due to contraction of a thermoresponsive polymer such as poly(*N*-isopropyl acrylamide) (Figure 11.3) (Gelbrich, Feyen, and Schmidt 2006, Polo-Corrales and Rinaldi 2012, Derfus et al. 2007), release due to melting of the hydrophobic core of block copolymer micelles or due to enhanced diffusion of the drug cargo from the micelle core (Bennett et al. 2011, Zhang, Guo, et al. 2015, Kim et al. 2013a), release due to melting of a liposome's membrane (Figure 11.4) (Amstad et al. 2011, Chen, Bose, and Bothun 2010), and release due to breaking of thermally labile bonds between the MNP and a drug cargo (Figure 11.5) (Zhang and Misra 2007, Riedinger et al. 2013). Having such a wide range of possible MNP designs and mechanisms of release holds great promise for biomedical application of magnetically triggered release, but this promise is far from being realized. The majority of reports focus on demonstrating triggered release under idealized laboratory conditions, such as in organic solvents or in de-ionized water. Testing of efficacy in vitro and in vivo is limited, and the potential effects of the complexity of biological milieu (i.e. high ionic strength, presence of proteins and other macromolecules) remain unexplored. This is particularly relevant to the use of thermoresponsive polymers such as pNIPAM, whose thermoresponsive properties can be affected by ionic species.

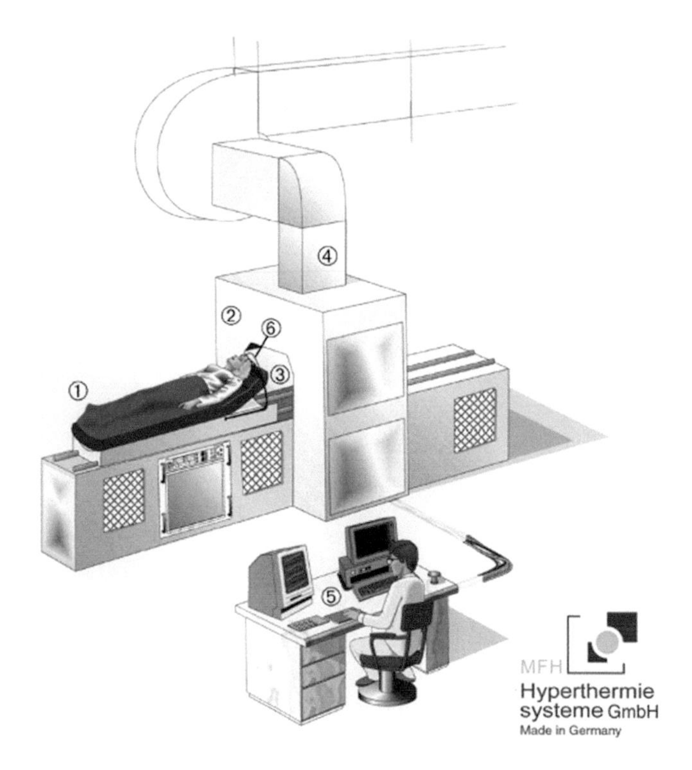

FIGURE 11.2 Sketch of the first prototype MFH therapy system. The oscillating magnetic field axis is perpendicular to the axial direction of the patient couch **(1)**. The therapy system is for universal application, i.e. suitable for MFH within, in principle, any body region. It is a ferrite-core applicator **(2)** operating at a frequency of 100 kHz with an adjustable vertical aperture of 30–50 cm **(3)**. The field strength is adjustable from 0 to 15 kA/m. The system is air cooled **(4)**. Aperture, field strength, thermometry, and further system parameters are monitored and adjusted manually by the physician at the control unit **(5)**. The temperature is measured invasively with Fluoroptic temperature probes within the tumor and at reference points outside the patient **(6)** (Jordan et al. 2001).

Furthermore, many of the proposed MNP designs display sustained release in the absence of an AMF, with only an increase in release rate once the AMF is applied. Although this may be acceptable for some applications, it is certainly not ideal and is in fact undesirable when a highly toxic drug is being delivered. One example of a design that overcomes these limitations is Ruhle et al.'s mesoporous silica particles, which were designed to have pores blocked by a thermally reversible polymer attached via retro Diels–Alder reaction (Ruhle et al. 2016). The passive release of the cargo contained within the mesoporous silica was found to increase or decrease through the controlled opening and closing of the pores via cycloreversion of the Diels–Alder bonds under an AMF. Finally, although the combination of magnetically triggered release and MFH is exciting given the potential synergy between many anti-cancer drugs and heat, it is not clear if magnetically triggered release will be effective for delivery of disease-modifying drugs (that is, under situations where the goal is not to kill a cell but rather to repair it). That is, it remains to be determined if MNPs can be designed for magnetically triggered drug release in such a way that the heat dissipated by the MNPs in the AMF to trigger the release does not have a negative impact on cell viability.

11.1.3 Magnetothermal Actuation of Cell Receptors

The application of magnetic micro- and nanoparticles for the remote actuation of mechanoresponsive cell receptors has been an active field of research for many decades (Dobson 2008). Ingber et al. activated intracellular calcium signaling in single mast cells under an AMF by

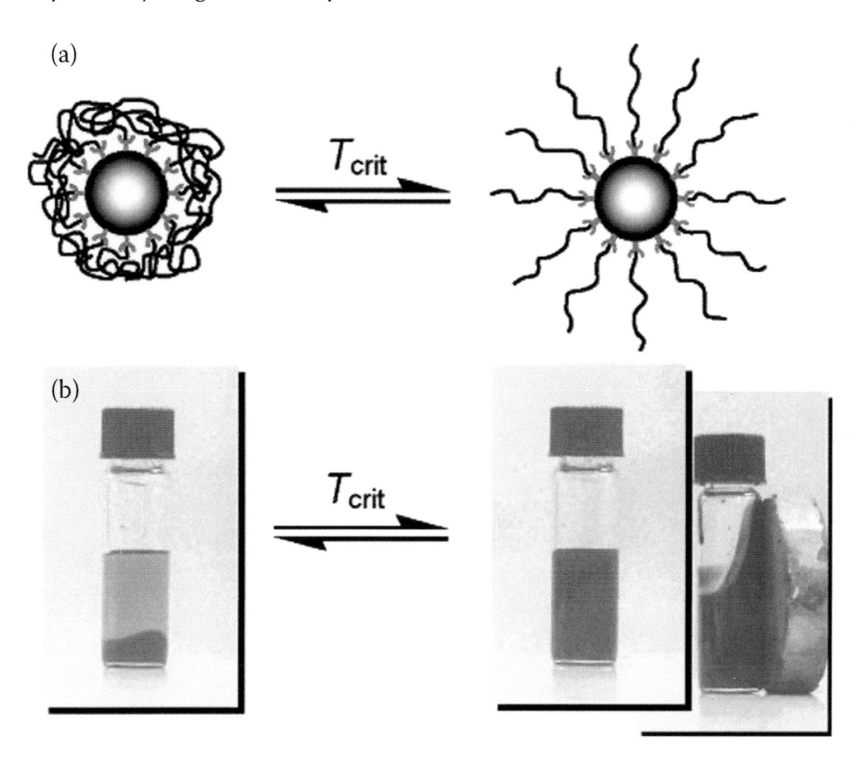

FIGURE 11.3 **(a)** Schematic of behavior of the polymer brush shell in thermoreversible magnetic fluids, **(b)** Photographs of Fe_3O_4@PMEMA43 in DMF. The particles precipitate below T_{crit} (at 20°C, left-hand). A particle dispersion is formed above T_{crit} (at 50°C, right-hand) that reacts collectively under the influence of a permanent magnet (Gelbrich, Feyen, and Schmidt 2006).

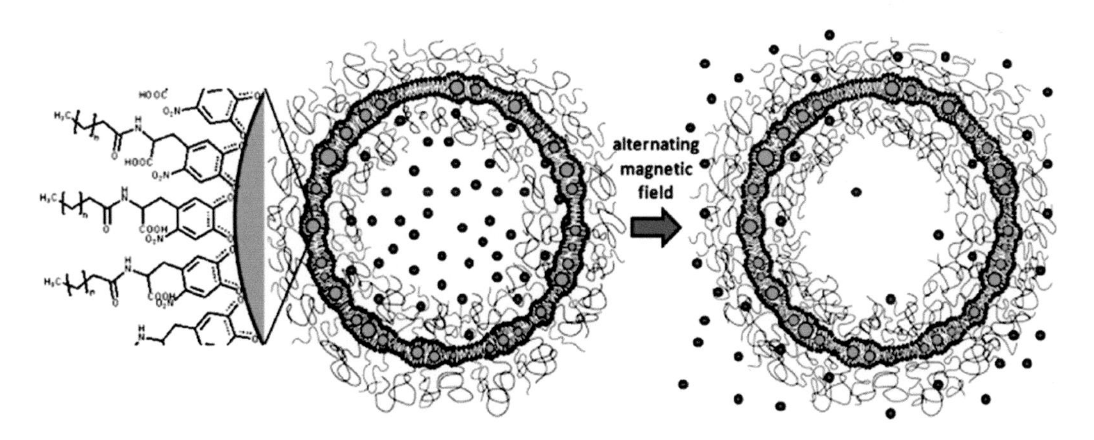

FIGURE 11.4 Schematic of liposomes containing iron oxide nanoparticles (IONPs) in their bilayer. NitroDOPA-palmityl stabilized IONPs are embedded in liposome membranes consisting of PEGylated and unmodified lipid and the release of MNPs in the presence of AMF (Amstad et al. 2011).

dinitrophenyl modified 30 nm superparamagnetic MNPs that bind to FciRi receptors (Mannix et al. 2008). Huang et al. (2010) demonstrated that thermal energy delivered by MNPs targeting cell surface receptors can also be harnessed to actuate a biological response (see Figure 11.6). This was demonstrated by activating TRPV1 receptors in the brain of the worm *Caenorhabditis elegans*,

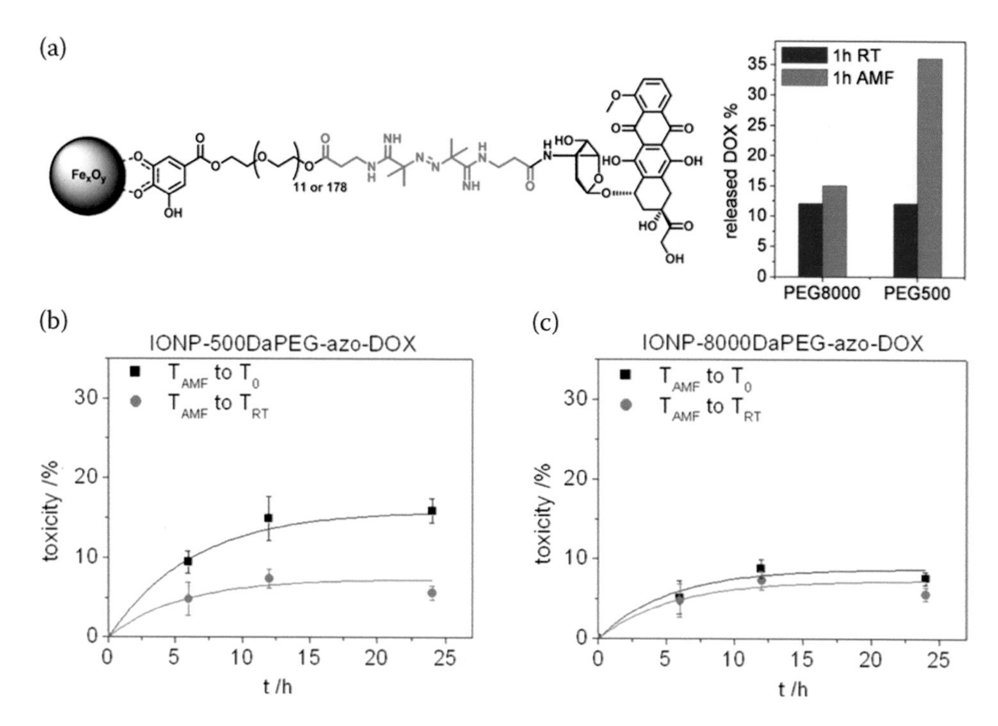

FIGURE 11.5 **(a)** Sketch of functionalized iron oxide nanoparticles (IONPs) bearing DOX, which is covalently linked through VA057 azo to the PEG spacers of two different Mw (500 and 8000 Da) together with the release profiles for samples exposed for one hour at room temperature and under AMF (334.5 kHz, 17 mT), respectively, **(b)** Cytotoxicity within 24 hours of DOX released from IONPs-500 Da PEG after AFM treatment compared to a sample incubated one hour at room temperature T_{RT} and to a blank kept for one hour at room temperature (T_0), **(c)** Cytotoxicity of DOX released from IONPs-8000 Da PEG after AFM treatment compared to a sample incubated one hour at room temperature T_{RT} and to a blank (T_0) (Riedinger et al. 2013).

resulting in a change in the direction of motion of the worm that was actuated by application of an AMF. Anikeeva's group (Loynachan et al. 2015) has demonstrated magnetothermal disruption of amyloid-β protein incubated in hippocampal neurons by specific targeting of leucine-proline-phenylalanine-phenylalanine-aspartic acid (LPFFD) modified 22 nm MNPs (see Figure 11.7). This mechanism of cellular actuation by magnetically mediated energy delivery is exciting and deserves further development to actuate other receptors relevant for biomedical applications.

Evidently, thermal energy delivery by MNPs in AMF has great potential applications in biomedicine. The common feature to all of these applications is the transformation of magnetic field energy to thermal energy effected by the MNPs. In all of these applications, a fundamental understanding of the relationship between MNP properties and magnetic field conditions with the mechanism and extent of thermal energy release is needed to design MNPs for the intended application. The rest of this chapter deals with our current understanding of the mechanisms and extent of thermal energy release by single-domain MNPs that are relevant to understanding the performance of energy-dissipating MNPs in biological systems.

11.2 RESPONSE OF SINGLE-DOMAIN MAGNETIC NANOPARTICLES TO OSCILLATING MAGNETIC FIELDS

When a collection of MNPs is subjected to a magnetic field that changes direction, their magnetic dipoles will experience a torque that will drive their alignment with the magnetic field.

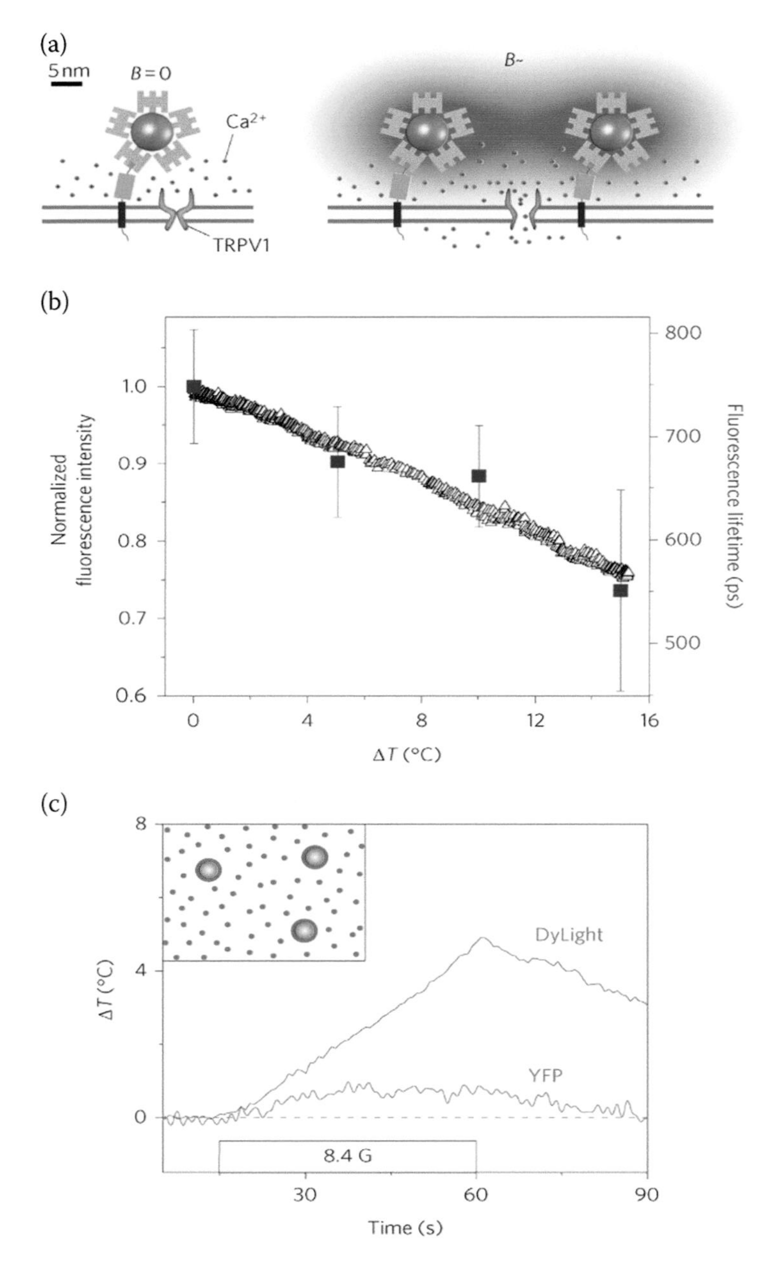

FIGURE 11.6 **(a)** Schematic showing local heating of streptavidin–DyLight549 (orange)-coated superparamagnetic nanoparticles (grey) in an AMF and heat (red)-induced opening of TRPV1, **(b)** Temperature dependence of the fluorescence intensity and lifetime of streptavidin–DyLight549 ($\Delta F/F$ = –Tempera–em measured in an externally heated nanoparticle dispersion), **(c)** Applying an AMF to a nanoparticle dispersion increased the nanoparticle surface temperature (DyLight trace, change in temperature measured by DyLight549 fluorescence) while only moderately changing the solution temperature (YFP trace, change in temperature measured by YFP fluorescence). Inset shows a schematic of the nanoparticle dispersion (free dots represent YFP; rings around the larger circles indicate the streptavidin–DyLight549 coating around the nanoparticles [grey]) (Huang et al. 2010)(see online version for color).

Alternatively, when the magnetic field does not change direction but only changes in magnitude, the MNPs and their dipoles are subject to thermal fluctuations that drive their dipoles out of alignment with the field. In that case, the equilibrium distribution of dipole orientations, and the rate at which this equilibrium distribution is reached, depend on the relative magnitudes of thermal

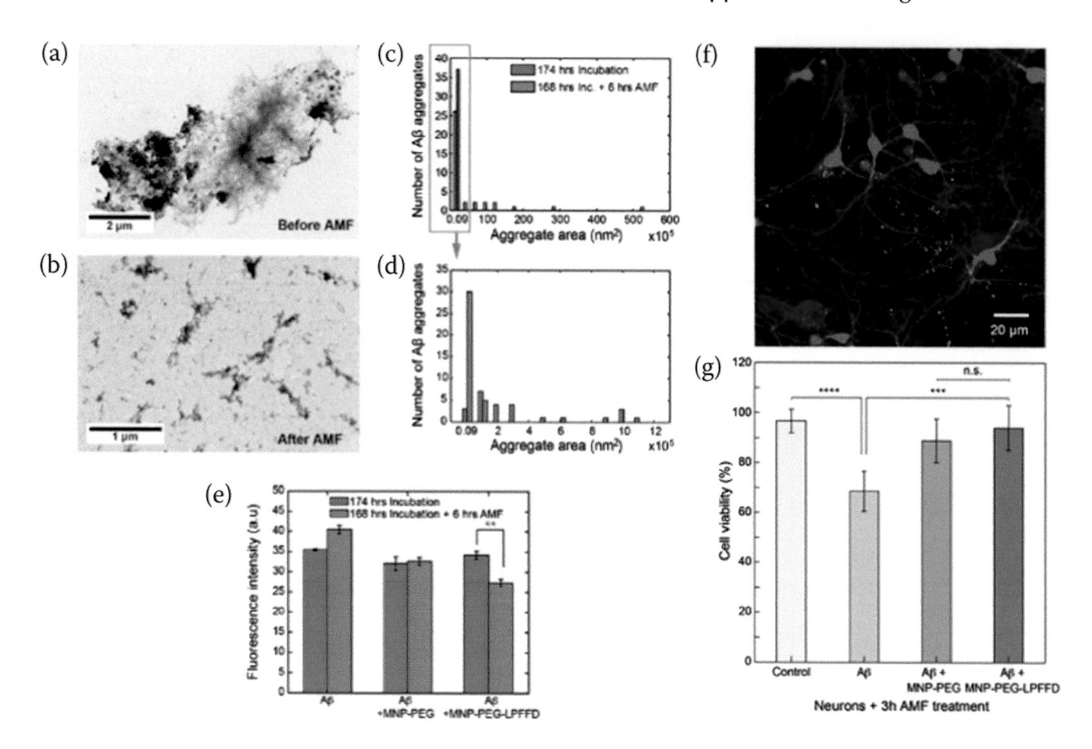

FIGURE 11.7 (*Left*) Representative TEM images of Aβ (168 hours) decorated with MNP-PEG-LPFFD, **(a)** Untreated and **(b)** Treated with six hours of AMF, **(c)** Histogram of the areas of 37 randomly selected aggregates (n = 8 images) from AMF-treated (orange) and untreated samples (green), **(d)** Areas binned into smaller intervals to examine distribution of AMF-treated samples, **(e)** Fluorescence intensity of AMF-treated and untreated samples (gain = 177). A decrease in intensity is observed for Aβ targeted with MNP-PEG-LPFFD following AMF, corresponding to the disruption of the β-sheet structure. Number of trials n = 3, student's t-test $p = 0.003$. (*Right*) **(f)** Confocal image of hippocampal neurons incubated with Aβ aggregates. Neurons were transfected with mCherry and Aβ was stained with thioflavin T, **(g)** Cell viability assessed by MTT assay of hippocampal neurons after three hours AMF treatment (control), and after three hours AMF treatment in presence of Aβ, Aβ+MNPs-PEG or Aβ+MNPs-PEG-LPFFD (Loynachan et al. 2015)(see online version for color).

energy and magnetic torque on the dipoles. As shown in Figure 11.8, depending on their properties and the properties of the surrounding medium, single-domain MNPs can respond to time-varying magnetic fields by two basic mechanisms: (1) physical particle rotation in what is called Brownian relaxation, and (2) internal dipole rotation in what is called Néel relaxation.

When particles relax by the Brownian mechanism, their dipoles are thought to be locked along a given crystal direction; therefore, the torque on the magnetic dipole induces rotation of the particle as a whole and against the hydrodynamic drag due to the surrounding fluid. In contrast, when particles relax by the Néel mechanism, their dipoles are thought to either freely rotate between crystal directions, or, more accurately, switch between crystal directions that correspond to energy minima and are often called "easy axes." Of course, in reality both mechanisms occur simultaneously albeit with different rates, quantified by their characteristic relaxation times, and the dominant mechanism will be the one with the fastest rate (i.e. shortest relaxation time). In the simple case of non-interacting single-domain MNPs, these relaxation times depend on particle properties such as core size, hydrodynamic diameter, magnetocrystalline anisotropy, and on the properties of the surrounding fluid, such as viscosity. For these simple cases there are theories available to relate the properties of the MNPs and medium to the characteristic relaxation times. These theories are described below.

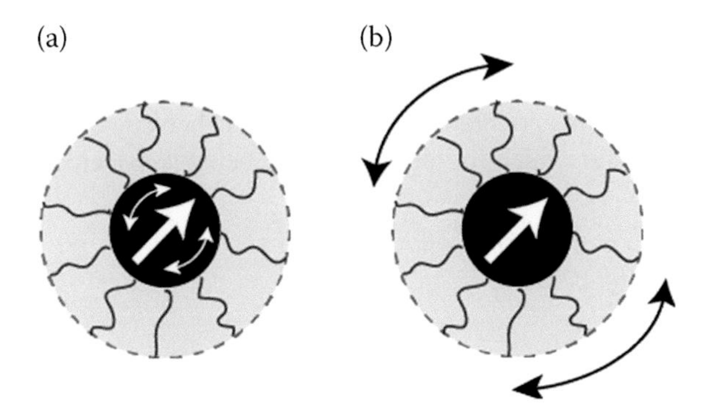

FIGURE 11.8 Illustration of mechanisms by which particles relax in the presence of an alternating magnetic field. **(a)** Internal dipole rotation: Neel mechanism, **(b)** Physical particle rotation: Brownian mechanism.

11.2.1 Brownian Relaxation by Single-Domain Magnetic Nanoparticles

The Brownian relaxation mechanism takes its name from the fact that the characteristic relaxation time is obtained from the rotational diffusivity of the magnetic particles, which itself is a consequence of the so-called Brownian motion induced by collisions between the particles and the molecules of the solvent. This phenomenon of stochastic motion due to molecular collisions is attributed to Robert Brown, even though he was not the first to report the phenomenon. However, Robert Brown was the first to experimentally rule out a variety of theories aimed at describing prior observations and to demonstrate that the phenomenon was intrinsic to the particles and bath (Mazo 2002). The theoretical description of Brownian motion had to wait until the work of Einstein (1906) and Perrin (1909, 1908). Einstein introduced a relationship between the thermal energy, hydrodynamic drag on a particle, and the translational/rotational diffusivity that is now referred to as the Stokes–Einstein relationship shown in Equation 11.1:

$$D_t = \frac{k_B T}{6\pi\eta R} \tag{11.1}$$

where D_t is the translational diffusivity, k_B is Boltzmann constant, T is the absolute temperature, η is the viscosity of the medium, and R is the molecular radius. An analogous relation (Equation 11.2) can be obtained for rotational motion:

$$D_r = \frac{k_B T}{8\pi\eta R^3} \tag{11.2}$$

where D_r is the rotational diffusivity.

Now we demonstrate how the rotational diffusivity given by the Stokes–Einstein relationship is related to the characteristic Brownian relaxation time. To do so we consider a collection of non-interacting MNPs whose magnetic dipoles are locked in a particular crystal axis suspended in a liquid. We will consider a uniaxial oscillating magnetic field such that the orientation of the magnetic dipoles of the collection of MNPs will be described by an orientation distribution $p(\theta)$, whose integral over orientation space is unity and for which the fraction of dipoles with orientations between θ and $\theta + d\theta$ is given by $f(\theta)d\theta$. The average dipole moment $\langle m_z \rangle$ along the z-direction is given by

$$\langle m_z \rangle = \int_0^{\pi} p(\Theta) m \sin^2 \Theta d\Theta \tag{11.3}$$

The effects of thermal energy, tending to randomize the dipole orientations, magnetic torque, tending to align the dipoles with the field, and the hydrodynamic drag exerted on the particle by the fluid govern the dynamics of this orientation distribution. Mathematically this can be represented by the Smoluchowski equation, which for the simple case being considered here can be written as

$$\frac{\partial p}{\partial t} + \nabla_s \cdot \left(-D_R \nabla_s p + \frac{\mu_0 m}{8\pi\eta R^3} \mathbf{H} \cdot \left(\underline{\underline{\mathbf{I}}} - \hat{\mathbf{m}}\hat{\mathbf{m}} \right) p \right) = 0 \tag{11.4}$$

Or in spherical coordinates,

$$\frac{1}{D_R}\frac{\partial p}{\partial t} = \frac{1}{\sin\theta}\frac{\partial}{\partial\theta}\left(\sin\theta\frac{\partial p}{\partial\theta}\right) + \frac{\mu_0 mH}{k_B T}\frac{1}{\sin\theta}\frac{\partial}{\partial\theta}(\sin^2\theta p) \tag{11.5}$$

where we assume $H = H(t)i_z$

For the case of a constant magnetic field, the solution of the Smoluchowski equation is the Boltzmann distribution given by

$$p(\theta) = \frac{\alpha \exp(\alpha\cos\theta)}{2\sinh\alpha} \tag{11.6}$$

where $\alpha = \frac{\mu_0 mH}{k_B T}$ is the Langevin parameter.

To solve Equation 11.5 for the case of an oscillating magnetic field, we assume the magnitude of the magnetic field is small, such that the Langevin parameter is small and the Boltzmann distribution can be expanded using a Taylor series truncated after the second term.

$$p(\theta) \approx \frac{1}{2}[1 + \alpha\cos\theta + \theta(\alpha^2)] \tag{11.7}$$

Furthermore, we assume that the time-dependence of the orientation distribution enters only in the second term of the expansion of the Boltzmann distribution, such that we may write

$$p(\theta, t) \approx \frac{1}{2}[1 + \alpha g(t)\cos\theta] \tag{11.8}$$

Equation 11.8 is substituted into Equation 11.5, and the following equation for $g(t)$ is obtained:

$$\frac{dg(t)}{dt} = \frac{2}{D_R}[\cos(\omega t) - g(t)] \tag{11.9}$$

where $H(t) = H\cos\omega t$

The fact that an ordinary differential equation is obtained for $g(t)$ is indicative that the assumed form of the orientation distribution is correct in the limit of small applied magnetic fields. Equation 11.9 is solved to obtain an expression for $g(t)$.

$$g(t) = \frac{1}{1 + \omega^2\tau^2}[\omega\tau\sin(\omega t) + \cos(\omega t)] \tag{11.10}$$

This solution and the assumed form for the orientation distribution are used in Equation 11.3 to obtain an expression for the magnetization of the suspension of MNPs with volume fraction ϕ

$$M = \chi' H_0\cos(\omega t) + \chi'' H_0\sin(\omega t) \tag{11.11}$$

where the in-phase and out-of-phase susceptibilities have been introduced:

$$\hat{\chi} = \chi' - \chi'' \tag{11.12}$$

$$\chi' = \frac{\chi_0}{1 + \Omega^2\tau^2} \tag{11.13}$$

$$\chi'' = \frac{\chi_{0^{\Omega\tau}}}{1 + \Omega^2\tau^2} \tag{11.14}$$

$$\chi_0 = \frac{\mu_0 \phi M_d^2 d^3}{18 k_B T} \tag{11.15}$$

and where the characteristic Brownian relaxation time is given by:

$$\tau_B = \frac{1}{2D_R} = \frac{3\eta V_h}{k_B T} \tag{11.16}$$

This expression illustrates the relationship between the characteristic Brownian relaxation time and the rotational diffusivity of the particle, thereby explaining the name given for this mechanism of magnetic relaxation.

We have chosen to illustrate the derivation of the expression for the Brownian relaxation time on the basis of the orientation distribution for a suspension of particles in an oscillating magnetic field because of its relevance to our subsequent discussion of energy dissipation in AMFs. The derivation of the Smoluchowski equation for rotation of a collection of dipoles and its use to obtain expressions for the equilibrium magnetization (given by the so-called Langevin function) and for relaxation of a collection of magnetic dipoles from an applied constant magnetic field are given in greater detail by Mazo (2002).

Equations 11.11 and 11.12 show that the net magnetization of a collection of dipoles in an oscillating magnetic field can be described by a complex magnetic susceptibility $\hat{\chi}$

$$\hat{\chi} = \chi' - i\chi'' \tag{11.17}$$

and in fact, Equation 11.12 for the real (in-phase) and imaginary (out-of-phase) components of the complex (or dynamic) susceptibility was obtained by Debye as a model to describe the frequency dependence of the dielectric permittivity (Debye 1929). Hence Equation 11.12 is often referred to as the Debye model for the complex (or dynamic) magnetic susceptibility. Figure 11.9 illustrates the predicted frequency dependence of the components of the complex susceptibility. Notably, the in-phase and out-of-phase susceptibilities are seen to cross at the peak of the out-of-phase susceptibility, where the condition $\Omega\tau = 1$ is satisfied. Hence, the Brownian relaxation time could be estimated from measurements of complex magnetic susceptibility spectra. This can be done using so-called AC magnetic susceptometers, a technique pioneered for MNP suspensions by Fannin (2002). Measurements of the complex magnetic susceptibility are increasingly important in understanding energy dissipation by MNPs, and additionally have applications in MNP-based sensors (Park et al. 2011, Koh and Josephson 2009, Fornara et al. 2008) and in the use of MNPs as nanoscale probes of the mechanical properties of complex fluids (Calero-DdelC, Santiago-Quinonez, and Rinaldi 2011, Barrera et al. 2010).

Finally, we note that the expressions for the Brownian relaxation time and for the complex magnetic susceptibility given here are limited to collections of non-interacting, single-domain, monodisperse spherical MNPs in a Newtonian fluid and in an applied magnetic field of small enough magnitude such that the Langevin parameter is much less than unity. These are severely limiting assumptions, as MNPs used in biomedical applications always have some degree of

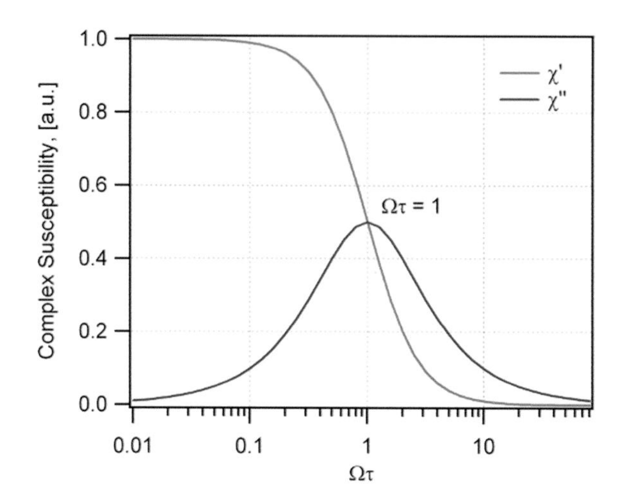

FIGURE 11.9 Debye plot for the AC susceptibility–frequency dependence of susceptibility.

polydispersity in size, sometimes have non-spherical shape, often interact magnetically, are often suspended in non-Newtonian fluids, and are subjected to magnetic fields of moderate to high amplitudes. Although the effect of polydispersity can be easily accounted for (Martens et al. 2013, Perrin 1909), some of the other effects remain to be described.

11.2.2 Néel Relaxation by Single-Domain Magnetic Nanoparticles

The mechanism of magnetic relaxation where the dipoles rotate internally, independent of physical particle rotation, is attributed to Louis Néel due to his introduction of the concept of a "magnetic after effect" or "magnetic viscosity" (Néel 1949) to account for the fact that there is a finite time required for the magnetization of a collection of dipoles to completely randomize after an applied field is removed. Although Néel obtained a simple model for the characteristic relaxation time, W.F. Brown Jr. obtained a more rigorous theory to describe the rate of so-called Néel relaxation in single-domain MNPs (Brown 1963). The theory by Brown is an extension of the theory for Brownian motion, briefly described above, and in fact involves the solution of the Fokker–Planck equation for the orientation distribution of the magnetic dipoles subject to magnetic torques and thermal fluctuations, with the addition of the effect of the magnetocrystalline anisotropy energy as a way to account for the existence of preferred crystal axes for magnetization. The corresponding expression for the relaxation time τ_N is

$$\tau_N = \frac{\sqrt{\pi}}{2}\tau_0 \frac{\exp(KV/k_B T)}{(KV/k_B T)^{1/2}} \tag{11.18}$$

where τ_0 is the characteristic time, K is the uniaxial magnetocrystalline anisotropy constant, V is the magnetic volume of the particle, k_B is the Boltzmann constant, and T is the absolute temperature. A similar form has been used by several authors to describe the Néel relaxation time of MNPs (Ferguson et al. 2013, Fortin, Gazeau, and Wilhelm 2008). However, in the field of MNP suspensions (also known as ferrofluids) and magnetic fluid hyperthermia, a simplified form (Equation 11.19) is often used:

$$\tau_N = \tau_0 \frac{\exp(KV)}{(k_B T)} \tag{11.19}$$

where τ_0 is a characteristic time, K is the particle uniaxial magnetocrystalline anisotropy constant, D_m is the diameter of the magnetic core, k_B is the Boltzmann constant, and T is the absolute temperature.

Equations 11.18 and 11.19 would not generally result in the same values of the Néel relaxation time as a function of temperature but, depending on the value of τ_0, the two can be made to agree in a limited temperature range.

Now one could proceed much as with the case of the Brownian relaxation time and solve the Fokker–Planck equations derived by W. F. Brown to describe the orientation distribution of the dipoles in particles with Néel relaxation for the case of an oscillating magnetic field. However, this would be quite a complex calculation. Instead here we apply a phenomenological approach that is often used in the field of ferrofluids. To do so, we will solve the so-called magnetization relaxation equation, which describes the time dependence of the net magnetization of a suspension of magnetic dipoles with a single characteristic relaxation time. The phenomenological equation obtained by Shliomis (1972) has been shown to be accurate for the case of low-amplitude magnetic fields (H), and has the form

$$\frac{\partial \mathbf{M}}{\partial \tau} = \frac{1}{\tau}(\mathbf{M} - \chi_0 \mathbf{H}) \tag{11.20}$$

where M is the magnetization of the suspension, τ is the relaxation time, and χ_0 is the initial susceptibility.

This equation has been widely used to describe the flow of ferrofluids and has been demonstrated to be accurate for small amplitude of the magnetic field and for frequencies that are smaller than the inverse magnetic relaxation time (i.e. $\Omega\tau \ll 1$). We consider the case where the collection of dipoles is subjected to an oscillating magnetic field of the form $H = H_0 \cos(\omega t)$. The solution to Equation 11.20 in that case is

$$M(t) = \chi' \cos(\omega t) + \chi'' \sin(\omega t) \tag{11.21}$$

where we see that, just like in the case of Brownian relaxation, the magnetization is described by an in-phase and an out-of-phase component of a complex susceptibility given by Debye's model.

11.2.3 Comparison of Brownian and Néel Relaxation Times

As noted earlier, typically both mechanisms of magnetic relaxation are possible for a given collection of MNPs. For a single given nanoparticle, the dipole will tend to align with a changing magnetic field by both internal rotation and physical rotation, and the dominant mechanism will be the faster of the two. Detailed models for this situation are lacking, with some work relying on computer simulations to investigate the behavior of ferrofluids with both mechanisms of relaxation (Berkov 1998). Instead, one often uses the above arguments to justify the assumption that in the case of particles with both modes of magnetic relaxation, the dynamics are described by an effective relaxation time τ_{eff} given by

$$\tau_{eff} = \frac{\tau_B \tau_N}{\tau_B + \tau_N} \tag{11.22}$$

Because of the difference in their functional dependence with diameter, it is expected that there will be a critical size for which the dominant mechanism of relaxation changes from one to another. This is illustrated in Figure 11.10. The Néel relaxation time was calculated for various values of the magnetocrystalline anisotropy constant representative of bulk magnetite (11 kJ/m^3) (Birks 1950), some reports for nanoscale iron oxides (110 kJ/m^3) (del Castillo and Rinaldi 2010), and for cobalt ferrite (190 kJ/m^3) (Shenker 1957). On the other hand, the Brownian relaxation times were calculated for a representative coating shell thickness of 10 nm and for values of the viscosity representative of water (0.001 Pa s), blood plasma (0.003–0.004 Pa s) (Tefferi 2003), extracellular matrix (0.12–0.26 Pa s) (Kuimova 2012), and cell cytoplasm (0.01–0.03 Pa s)

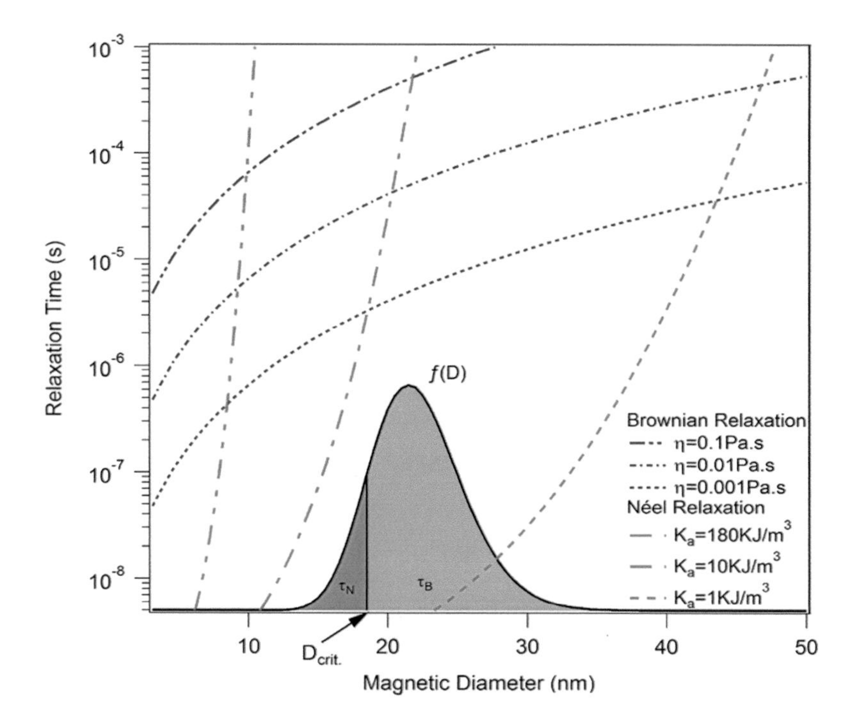

FIGURE 11.10 Comparison of Brownian and Néel relaxation times as a function of particle size for fixed shell thickness of 2 nm and in carrier fluid with different viscosities (Maldonado-Camargo et al. 2016).

(Kuimova 2012). From Figure 11.10, it is seen that generally the Néel relaxation time will dominate for smaller particles whereas the Brownian relaxation time will dominate for larger particles. Furthermore, it is seen that the critical diameter for transition from one dominant mechanism to another is principally a function of the magnetocrystalline anisotropy of the particles, with a weaker dependence on the viscosity of the surrounding medium. This is an important point when one takes into consideration the wide range of magnetocrystalline anisotropies that have been reported in the literature (Lehlooh, Mahmood, and Williams 2002, Fortin et al. 2007, Peddis et al. 2008, Vallejo-Fernandez and O'Grady 2013) and that depend on the state of aggregation of the particles, their morphology, crystallinity, and many other factors.

The above discussion focused on the possibility of the two relaxation mechanisms being active for a single given MNP and illustrated the current view that the dominant mechanism will be the faster of the two. Another important factor to take into consideration when studying suspensions of MNPs is the effect of size distribution on the relaxation time and on the mechanism of magnetic relaxation. Depending on the method used to synthesize particles, one can obtain either wide size distributions (the typical case for so-called co-precipitation methods [Massart 1981]) or narrow size distributions (the typical case for so-called thermal decomposition methods [Sun and Zeng 2002, Park et al. 2004, Vreeland et al. 2015]). Figure 11.10 illustrates a typical size distribution for nanoparticles obtained by co-precipitation, along with the diameter dependence of the Brownian and Néel relaxation times for particles with magnetocrystalline anisotropies representative of either bulk magnetite or bulk cobalt ferrite. It is seen that for both values of the magnetocrystalline anisotropy, if the particles have a size distribution representative of a co-precipitation synthesis there will be particles with dominant Brownian and dominant Néel relaxation in the suspension. It is important to emphasize that the calculations made to obtain Figure 11.10 are only as valid as the underlying assumptions for physical parameters such as magnetocrystalline anisotropy and the fact that we are neglecting the effects of particle–particle interactions. However, the calculations serve to illustrate the importance of thorough characterization of the mechanisms of relaxation active for a given batch of MNPs, as well as their characteristic relaxation time.

11.3 ENERGY DISSIPATION BY SINGLE-DOMAIN MAGNETIC NANOPARTICLES IN OSCILLATING MAGNETIC FIELDS

In this section we obtain a simple model for the rate of energy dissipation of a suspension of non-interacting single-domain MNPs with a single characteristic magnetic relaxation time. The derivation is similar to the approach by Rosensweig (2002), who considered the adiabatic temperature rise for a ferrofluid.

We start by considering the change in internal energy for a ferrofluid. According to the first law of thermodynamics, this change is equal to the sum of work done on the ferrofluid and the heat transferred to the ferrofluid by its surroundings:

$$dU = \partial Q + \partial W \tag{11.26}$$

We will consider the case where an AMF is applied to a suspension of MNPs and where the particles transfer heat to their surroundings in a way such that their internal energy remains constant (that is, the temperature of the suspension remains constant). In this case, the heat transferred by the nanoparticles to their surroundings will be equal to the work done on the particles by the AMF.

$$\partial Q = -\partial W \tag{11.27}$$

Next we take into account that the AMF is periodic; hence, the work done by the field on the MNP suspension can be calculated from Equation 11.28 (Rosensweig 2002, Soto-Aquino and Rinaldi 2015):

$$W = \mu_0 \int_0^{2p} H \frac{dM}{dt} dt \tag{11.28}$$

where $\mu_0 = 4\pi \cdot 10^{-7} (\text{TmA}^{-1})$ is the permeability of free space, M (A/m) is the magnetization, $2p$ is the period of the cycle, and H (A/m) is the magnetic field intensity. With this, the heat transferred by the nanoparticles to their surroundings becomes

$$Q = \mu_0 \int_0^{2p} H \frac{dM}{dt} dt \tag{11.29}$$

Next we define an average rate of heat transfer per cycle by dividing through by the period

$$\langle Q \rangle = \frac{1}{2p} \mu_0 \int_0^{2p} H \frac{dM}{dt} dt \tag{11.30}$$

Now we take into account that the applied magnetic field and the magnetization are collinear and use integration by parts to obtain

$$\langle Q \rangle = \frac{1}{2p} \mu_0 \int_0^{2p} M \frac{dH}{dt} dt \tag{11.31}$$

Next we assume that the time dependence of the AMF is given by the expression

$$H = H_0 \cos(\omega t) \tag{11.32}$$

To obtain

$$\langle Q \rangle = \frac{\mu_0 H_0 \omega}{2p} \int_0^{2p} M \sin(\omega t) dt \tag{11.33}$$

To proceed we would need to obtain a solution for the magnetization as a function of time under an applied sinusoidal magnetic field. We obtained such a solution in Section 11.2.3 when we solved the phenomenological relaxation equation of Shliomis (Shliomis 1972) for the magnetization of a suspension of non-interacting MNPs in an oscillating magnetic field. We therefore now must restrict the analysis to situations where that solution is valid, in which case the magnetization is given by Equation 11.21. Substituting this result in Equation 11.33 where we obtain for the rate of energy dissipation:

$$\langle Q \rangle = \frac{\mu_0 H_0^2 \omega \chi''}{2} \tag{11.34}$$

Equation 11.34 is the expression obtained by Rosensweig (2002) in his seminal paper on the subject. From the derivation above, it is clear that this simplified expression is subject to the following assumptions: (1) non-interacting spherical monodisperse single-domain MNPs, such that their magnetic relaxation is described by the Brownian or Néel mechanisms; (2) linearly polarized sinusoidal magnetic field, such that the magnetization and magnetic field are collinear; (3) small amplitude of the magnetic field, such that the Langevin parameter is much less than unity; (4) field frequencies smaller than the inverse magnetic relaxation time (i.e. $\Omega\tau \ll 1$) such that the magnetic relaxation equation of Shliomis (1972) is applicable; and (5) particles with a single magnetic relaxation time, such that the Debye model accurately describes their response to the AMF.

The assumptions underlying Equation 11.34 are seldom applicable in practical situations for a variety of reasons. MNPs used in biomedical applications often consist of small aggregates of interacting primary magnetic particles. Magnetic interactions can result in partial blocking of the magnetic dipoles in the aggregate and in an effective magnetic anisotropy, higher than that of the bulk material (del Castillo and Rinaldi 2010). Often, the particles are not spherical and therefore their response to time-varying magnetic fields is influenced by shape anisotropy, which can also give rise to a high effective magnetic anisotropy. In other cases, the particles are no longer single-domain, and domain wall motion and hysteresis may influence their energy dissipation rate (Li et al. 2010, Sun et al. 2012). Another point of deviation is the use of non-sinusoidal AMFs, and the use of different waveforms for the AMF has been shown to have an influence on the rate of energy dissipation (Lancarotte and Penteado 2001, Venkatachalam et al. 2002). Furthermore, applied field amplitudes as high as 50 kA/m are often used. For a 12 nm diameter iron oxide nanoparticle this corresponds to a Langevin parameter of (11.6), hence the assumption of small amplitude of the magnetic field is not accurate. On the other hand, although the frequencies used in energy dissipation experiments are low with respect to most iron oxide nanoparticle Néel relaxation times, this may not be the case for some cobalt ferrite and other thermally blocked nanoparticles; therefore, the phenomenological magnetization relaxation equation of Shliomis ceases to provide a valid description of the magnetization of the suspension in response to the AMF. Finally, as noted in -Section 11.2.3, depending on their size, MNPs may respond to time-varying magnetic fields by either Brownian or Néel relaxation or by a combination of both, when the diameter is close to the threshold diameter for transition from one mechanism to another. Hence, polydispersity in magnetic diameter and simultaneous relaxation by the two mechanisms can have a profound influence in their energy dissipation rate. While polydispersity can be easily accounted for in the model by

Rosensweig (2002), the effect of simultaneous relaxation by the two mechanisms remains unexplored.

Regardless of its limitations, the model obtained by Rosensweig has been of great value for the advancement of the field, mainly due to its simplicity. The key predictions of this model are: (1) the rate of energy dissipation is proportional to the concentration of the particles; (2) the rate of energy dissipation should increase approximately linearly with applied field frequency if the field frequency is smaller than the inverse relaxation time (i.e. $\Omega\tau \ll 1$); (3) the rate of energy dissipation should increase quadratically with the applied field amplitude if the amplitude is small, such that the Langevin parameter is much less than unity; (4) the rate of energy dissipation is maximal with respect to frequency at a frequency that is equal to the inverse magnetic relaxation time (i.e. $\Omega\tau = 1$); (5) the rate of energy dissipation can be enhanced by using materials with high domain magnetization; and (6) an optimum core diameter exists where the rate of energy dissipation is maximal for a given applied field amplitude and frequency. These predictions have guided and motivated many recent developments in obtaining MNPs with high energy dissipation rates, such as the use of iron nanoparticles (Fortin et al. 2007), the development of core-shell nanoparticles (Zhang, Castellanos-Rubio, et al. 2015), and controlled aggregates (Martinez-Boubeta et al. 2013, Lartigue et al. 2012) with tuned relaxation times for optimal heat dissipation at the frequencies used in practice, and optimization of the particle diameter to raise the energy dissipation rate (Khandhar et al. 2012).

11.3.1 Measurement of Energy Dissipation by Magnetic Nanoparticles—Specific Absorption Rate and Intrinsic Loss Power

Before we go further and consider the effect of mechanism of energy relaxation and of size distribution on energy dissipation rates, we now consider how energy dissipation by MNPs is measured and how it is typically reported. This will allow easier comparison of the subsequent calculations with values from the literature and critical evaluation of the accuracy of and relationship between the values reported for the rate of energy dissipation.

In the laboratory, energy dissipation rate is typically measured by monitoring the rate at which temperature rises in a suspension of MNPs under an applied AMF as shown in Figure 11.11. To do so, a thermal camera or a non-magnetic/non-conducting thermal probe is used to measure the temperature of the sample, which is placed in the internal space of a coil used to generate the AMF. For the measurement to be representative of a given applied field amplitude and frequency, it is important that the sample be subjected to as uniform a magnetic field as possible. Practically this means that, say for a cylindrical sample in a cylindrical coil, the height of the sample must be much less than the height of the coil, ideally half or less. This will therefore require the use of coils with multiple turns constructed in such a way that the field is as uniform as possible. These aspects of coil design and sample size are further discussed in the literature (Bordelon et al. 2012, Huang et al. 2012).

Another important practical aspect in measuring the rate of energy dissipation of MNP suspensions is heat transfer within the sample and to/from the surroundings. Various authors have discussed these aspects in detail (Wildeboer, Southern, and Pankhurst 2014), with some favoring measurements under completely adiabatic conditions (Natividad, Castro, and Mediano 2008, Urtizberea et al. 2010). The emphasis on adiabatic conditions can perhaps be traced back to Rosensweig's derivation of the energy dissipation rate, where he assumes an adiabatic system. However, our calculation in Section 11.3 showed that the same expression can be obtained for a system in thermal equilibrium with a heat sink, such that the internal energy of the system is held constant. Adiabatic setups for measuring rate of energy dissipation have the advantage of providing very accurate measurements, as the influence of heat transfer to/from the surroundings has been eliminated (Natividad, Castro, and Mediano 2008). However, practically it is difficult to

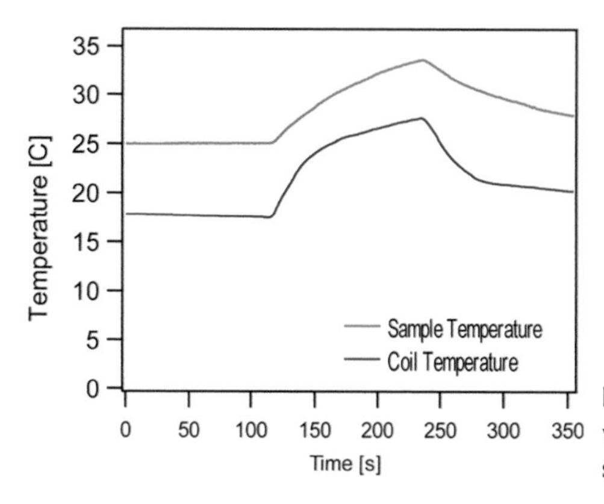

FIGURE 11.11 Temperature–time curve for a system with constant energy dissipation and fixed heat loss to the surroundings.

design and build adiabatic calorimeters compatible with applied AMFs suitable for energy dissipation rate measurements in MNP suspensions.

Rather than using adiabatic systems, most groups use systems where the rate of heat loss to the surroundings is reduced, such that its effect is negligible compared to the rate of energy dissipation by the nanoparticles. Here we analyze such a situation from the point of view of a lumped parameter heat transfer model (Deen 1998). Our use of a lumped parameter model implicitly assumes that the temperature profile in the sample is uniform at all times, an assumption that is expected to be valid for small samples such that the thermal diffusion time is smaller than the characteristic time for temperature rise.

We consider a sample with an initial temperature τ_N in thermal equilibrium with its surroundings at temperature T_a. At time t_0, the sample begins to dissipate thermal energy at a constant uniform volumetric rate P. We will assume the rate of thermal energy dissipation is not a function of the sample temperature. This thermal energy dissipation results in an increase in the sample temperature and in heat being transferred to the surroundings. We will quantify the rate of heat transfer to the surroundings using a global heat transfer coefficient h and the effective surface area for heat transfer A. Under these assumptions, the evolution of the temperature of the sample is described by the following first-order ordinary differential equation

$$m_s \hat{C}_p \frac{dT(t)}{dt} = Q - Ah[T(t) - T_a] \tag{11.35}$$

where $T(t)$ is temperature at time t, ρ is the density of the sample, and C_p is the heat capacity. The solution of this equation is

$$T = T_Q + \frac{Q}{Ah} + \left(T_0 - T_a - \frac{Q}{Ah}\right)\exp\left(-\frac{Ah}{m_s \hat{C}_p}t\right) \tag{11.36}$$

This solution predicts that the temperature will initially rise linearly with time and will eventually reach an equilibrium temperature T_e given by

$$T_e = T_a + \frac{Q}{Ah} \tag{11.37}$$

Note that the equilibrium temperature T_e depends on the contributions of energy dissipation by the particles and heat transfer to the surroundings. Practically what this means is that the maximum temperature rise observed in an experiment measuring energy dissipation rate is not a measure of the intrinsic heating properties of the particles, but rather a measure of how well-insulated the sample is, relative to the rate of energy dissipation. One could use this to estimate the heating capacity of the nanoparticles but in practice this is difficult because it requires independent estimation of the product of effective surface area and heat transfer coefficient.

A much more straightforward way to quantify the heating capacity of the nanoparticles is to focus on their initial rate of energy dissipation. To do so we take advantage of the fact that the sample is initially in thermal equilibrium with its surroundings, therefore the net rate of heat transfer to/from the surroundings is zero, i.e. $T_0=T_a$. In that case Equation 11.35 reduces to

$$m_s \hat{C}_p \frac{dT(t)}{dT} = Q \tag{11.38}$$

Solving for the rate of thermal energy dissipation yields

$$Q \approx m_s \hat{C}_p \frac{\Delta T}{\Delta t} \tag{11.39}$$

Under the assumptions outlined above, this volumetric rate of energy dissipation should be the same as that given by the model by Rosensweig, Equation 11.34, which indicates that the volumetric rate of energy dissipation is linearly proportional to the concentration of the particles. To facilitate comparison between measurements, the rate of energy dissipation is typically reported relative to the mass of magnetic particles in the sample, in units of W/g particles or W/g iron, and is typically called the specific absorption rate (SAR) of the particles. In the case of units of W/g particles, the SAR is related to the volumetric rate of energy dissipation (Q) by

$$Q = (SAR)\rho_p \tag{11.40}$$

and combining Equation 11.38 with Equation 11.39, the experimental SAR can be determined from

$$SAR_E = m_s \frac{\hat{C}_p}{m_{Fe}} \frac{\Delta T}{\Delta t} \tag{11.41}$$

As discussed above, Rosensweig's model for the volumetric energy dissipation rate has many simplifying assumptions, some of which are not accurate in experiments where MNPs dissipate heat under an applied AMF. However, it is informative to see what Rosensweig's model predicts for the SAR. If we assume that Rosensweig's model is applicable, then the predicted SAR would be

$$SAR_R = \frac{\pi\mu_0^2}{36k_B\rho} * \frac{H^2\Omega^2\tau D^3 M_d^2}{T(1 + (\Omega\tau)^2)} \tag{11.42}$$

Clearly, if the assumptions of Rosensweig's model do not apply, then the experimental and predicted SAR will differ quantitatively. However, we expect that qualitatively many of the predictions of Rosensweig's model should apply. The most important difference between the predictions in Section 11.3 and those of Equation 11.42 is that the SAR should not be a function of the concentration of particles used in its measurement. From the point of view of Rosensweig's model, this is a consequence of the assumption of particles that do not interact and the fact that the SAR is a measure relative to the mass of the particles. Hence, experimental measurements of SAR that vary with particle concentration could be indicative of significant particle–particle interactions, such as in chain formation. However, the experimentally determined SAR could also appear to vary with concentration if the particle concentration is so low that heat transfer to the surroundings is comparable to the rate of energy dissipation, even at short times.

Another important prediction of Equation 11.42 is that SAR is expected to vary quadratically with the amplitude of the AMF and, for small field frequencies such that $\Omega\tau \ll 1$, linearly with the field frequency. Hence, simply comparing SAR values for different particles measured under different magnetic field conditions is not a reliable way to compare the intrinsic heating capacity of the nanoparticles. Put another way, the SAR is a measure of the heating capacity of a suspension of MNPs under specific values of the AMF amplitude and frequency. This fact alone probably accounts for the large variation in SAR values reported in the literature, as almost every laboratory uses different field amplitudes and frequencies, determined by the details of the design of their coils and the source used to generate current through the coil. This limitation of the SAR as a reliable metric of the energy dissipation rate motivated Kallumadil et al. (2009) to introduce the concept of intrinsic loss power (ILP):

$$ILP_E = \frac{SAR_E}{H^2 f} \tag{11.43}$$

It is informative to consider the predictions of Rosensweig's model for the ILP, which are given by the equation

$$ILP_R = \frac{\pi^2 \mu_0^2}{18 k_B \rho} * \frac{f\tau D^3 M_d^2}{T\,[1 + (\Omega\tau)^2]} \tag{11.44}$$

According to Rosensweig's model, the ILP should be independent of the amplitude of the AMF and should depend on the applied field frequency only for cases where $\Omega\tau \sim 1$ or $\Omega\tau \gg 1$. Equation 11.44 illustrates that although using the ILP is certainly a step in the direction of standardizing measurements of energy dissipation rates of MNPs, it also suffers from its own limitations. These limitations stem from the use of the functional prediction of energy dissipation proportional to the square of the magnetic field amplitude and proportional to the field frequency, found in Rosensweig's model for small applied field frequencies. Therefore, under conditions where Rosensweig's model is accurate, the ILP will be a reliable standardized measure of the particle's energy dissipation capacity, but if Rosensweig's model does not apply, then comparing ILP will no longer be an accurate method of comparing the heating capacity. One example of this would be cases where the experimentally measured energy dissipation rate does not vary quadratically with applied field amplitude. For example, the field dependence of the energy dissipation rate has been observed to be linear (Carrey, Mehdaoui, and Respaud 2011, Usov and Grebenshchikov 2009) and has been reported to have regions of quadratic and linear dependence, followed by saturation (Mamiya and Jeyadevan 2011, Mehdaoui et al. 2011). The transition from quadratic to linear dependence can perhaps be explained by magnetic saturation of the sample, which happens when the Langevin parameter is of the order of or greater than unity. In this case, comparing ILP values for particles under conditions where the Langevin parameter is not much less than unity would be an inaccurate comparison of the particles' heating capacities. Another situation where ILP would be an inaccurate measure to compare heating capacity would be for particles with inverse relaxation times that are of the order of the applied magnetic field frequency. For such cases the rate of energy dissipation according to Rosensweig's model is no longer linearly dependent on the applied field frequency; hence, the ILP does not accurately account for the effect of frequency in the measurements. Based on the above arguments, the comparisons between particles in terms of heating capacity based on ILP should be made with caution and ideally comparisons of the heating performance of MNPs should be performed under identical magnetic field conditions and particle concentrations.

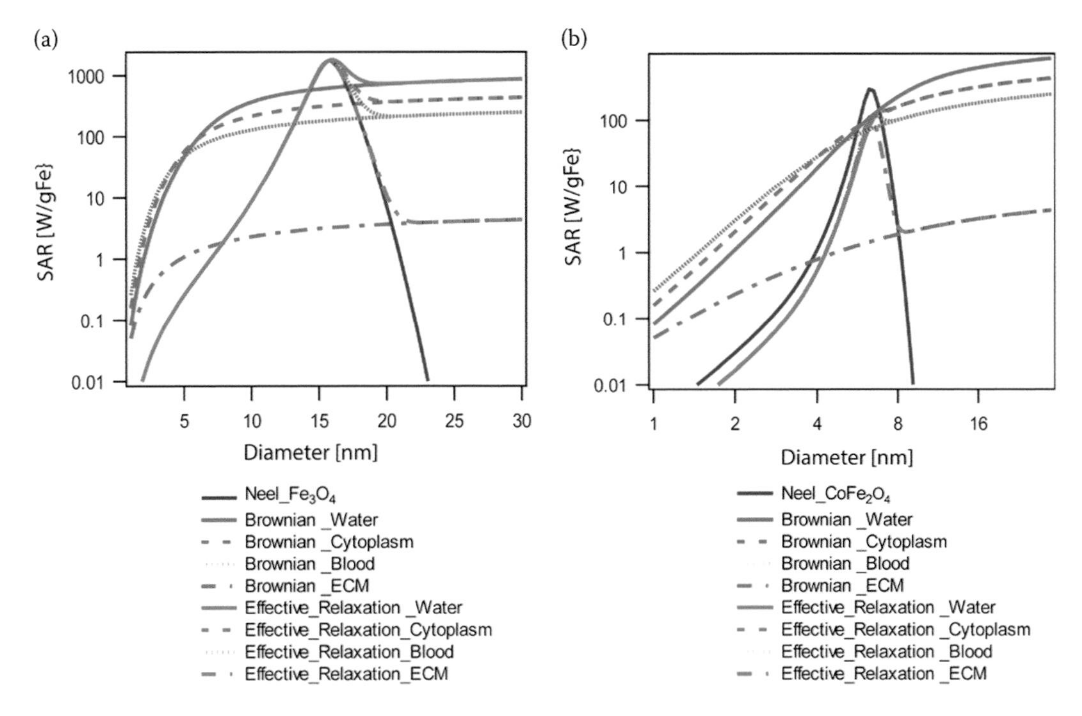

FIGURE 11.12 Comparison of Brownian and Néel relaxation energy dissipation rates in water (η = 0.001 Pas), blood (η = 0.0035 Pas), cytoplasm (η = 0.002 Pas), ECM (η = 0.2 Pas), and fixed matrix for **(a)** iron oxide (K = 13 kJ/m^3), **(b)** Cobalt ferrite nanoparticles (200 kJ/m^3).

11.3.2 Comparison of Brownian and Néel Relaxation Models in Terms of Energy Dissipation

As discussed in Section 11.2, the response of a particle to an AMF may either be by dipole rotation within the particle or by rotation of the particle as a whole. The dominant mechanism for power dissipation is assumed to be the one corresponding to the shorter of the relaxation times. There exists a critical size below which the particles dissipate heat due to Néel relaxation and above which heating due to Brownian relaxation dominates. This critical size is dependent on the magnetocrystalline anisotropy of the nanoparticles and the viscosity of the medium. Figure 11.12 provides a comparison of the rates of energy dissipation by particles that relax predominantly through the Brownian or Néel mechanisms, for both iron oxide and cobalt ferrite particles subjected to AMFs of 20 kA/m and 300 kHz in media as a function of viscosity for a particle shell thickness of 2 nm.

The energy dissipated for iron oxide nanoparticles is highest for particles with about 16 nm magnetic core diameter, irrespective of the medium viscosity, as the Néel relaxation mechanism dominates the heat dissipation rate, as seen in Figure 11.12a. Above this critical diameter, the energy dissipated by the Néel mechanism decreases while the energy dissipated by the Brownian mechanism increases and becomes dominant at large diameters. For cobalt ferrite nanoparticles, the highest energy dissipated according to these calculations corresponds to a diameter of ~6.25 nm and is predominantly through the Néel mechanism. The anisotropy constant is an order of magnitude higher for cobalt ferrite than for magnetite, resulting in transition from predominant Néel relaxation to predominant Brownian relaxation at a much lower critical diameter. As is shown in Figure 11.12, for both types of nanoparticles the viscosity of the medium has a significant effect on the rate of energy dissipation when the predominant mechanism of relaxation is the Brownian mechanism.

11.3.3 Effect of Particle Size Distribution on Energy Dissipation by Magnetic Nanoparticles

MNPs, both commercially available and synthesized in academic labs, always have a size distribution associated with them. This is often described using the lognormal particle size distribution

$$n_N\left(D_p\right) = \frac{1}{\sqrt{2\pi}D_p \ln \sigma_g} \exp\left(-\frac{\ln^2\left(D_p/D_{pg}\right)}{2\ln^2 \sigma_g}\right) \tag{11.45}$$

where D_{pg} is the number median diameter and $\ln \sigma_g$ geometric deviation.

The polydispersity of nanoparticles in a sample has a negative influence on the rate of energy dissipation, as shown by Rosensweig (2002) (Figure 11.13). Much effort has been devoted to work with samples that have a very narrow size distribution. Fortin et al. (2007) showed that higher energy dissipation rates can be obtained with samples with a lower standard deviation for a given field amplitude and frequency. Furthermore, increasing polydispersity of the nanoparticles could result in the simultaneous contribution of Néel and Brownian relaxation mechanisms, making it difficult to predict the energy dissipation.

11.4 PRACTICAL LIMITS OF ALTERNATING MAGNETIC FIELDS FOR IN VIVO APPLICATIONS

Optimizing the magnetic properties of the particles, maximizing the concentration of particles, and working at high amplitudes and frequencies of the AMF are some of the major routes by which one can maximize energy dissipation at the site of a tumor (Deatsch and Evans 2014). Unfortunately there are practical limitations on the field amplitude and frequency that can be applied to a human patient. This is because eddy currents induced by the AMF can lead to ohmic heating in tissues. Eddy currents are strongly dependent on the diameter of the object and this non-specific heat at high field strengths could result in heating of healthy tissue (Atkinson, Brezovich, and Chakraborty 1984). The International Commission on Non-Ionizing Radiation Protection (ICNIRP) has set a whole-body average tissue SAR limit of 0.4 W/kg for an occupational exposure of 30 minutes and a frequency range of 100 kHz to 300 GHz. Studies by Atkinson et al. on AMF exposure in tissues (Atkinson, Brezovich, and Chakraborty 1984) found that fields of amplitudes greater than 35.8 A/m at 13.56 MHz (for a 15 cm radius) lead to discomfort. Blistering was reported at frequencies of 13.56 MHz by Oledson et al. (1984). Based on these observations, a value of 4.85×10^8 A/(ms) for the product of AMF amplitude and frequency has been suggested as an upper limit by many authors. However, Kozissnik et al. (2013) point out that the frequency dependence of the electrical conductivity of skin and tissues should be taken into account, along with the diameter of the region to which AMF is subjected, when estimating a maximum field-frequency product for MFH. The ICNIRP guidelines and FDA guidelines for MRI, when translated into estimates of maximum allowable magnetic field amplitude and frequency product for heat dissipation to avoid damage to skin correspond to a field amplitude and frequency product of 8.1×10^7A/(ms) and 2.55×10^8 A/(ms), respectively, at 13.56 MHz. This implies that higher amplitudes of magnetic field and frequency product can be practically used. Thus much scope for understanding the effects of AMF on healthy tissue and determining the limits of exposure to AMF exists.

11.5 SUMMARY

The use of MNPs as a source of heat generation in various therapeutic applications has seen much progress over the past two decades. From initial application solely to deliver heat to cancer cells, the

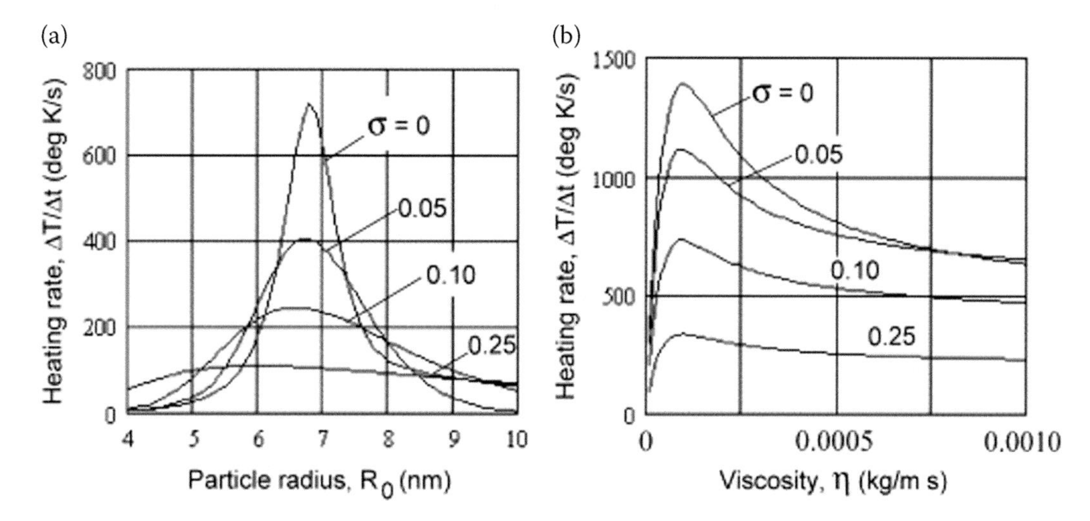

FIGURE 11.13 Temperature rise rate for 7 nm particles with $\varphi = 0.071$ that relax by **(a)** Néel mechanism ($f = 300$ kHz) and **(b)** Brownian ($B_0 = 0.06$ T, $f = 900$ kHz) mechanism (Rosensweig 2002).

field has expanded to approaches where energy dissipated by MNPs is used to control activation of cell surface receptors and to trigger drug release, illustrating the enormous potential of the phenomenon of energy dissipation by MNPs. This chapter aimed to introduce the reader to the basic models for energy dissipation by single-domain MNPs that respond to time-varying magnetic fields through the Néel and Brownian relaxation mechanisms. The limitations of these simple models have been clearly pointed out. Extension of these models to more realistic conditions, including the effects of particle–particle interactions, combined relaxation mechanisms, and non-linear magnetization, is an active area of research.

REFERENCES

http://www.icnirp.org/cms/upload/publications/ICNIRPemfgdl.pdf.

Alvarez-Berrios, M. P., A. Castillo, C. Rinaldi, and M. Torres-Lugo. 2014. Magnetic fluid hyperthermia enhances cytotoxicity of bortezomib in sensitive and resistant cancer cell lines. *International Journal of Nanomedicine* 9:145–153. doi:10.2147/ijn.s51435.

Amstad, E., J. Kohlbrecher, E. Mueller, T. Schweizer, M. Textor, and E. Reimhult. 2011. Triggered release from liposomes through magnetic actuation of iron oxide nanoparticle containing membranes. *Nano Letters* 11 (4):1664–1670. doi:10.1021/nl2001499.

Atkinson, W. J., I. A. Brezovich, and D. P. Chakraborty. 1984. Usable frequencies in hyperthermia with thermal seeds. *IEEE Transactions on Biomedical Engineering* 31 (1):70–75. doi:10.1109/tbme.1984.325372.

Barrera, C., V. Florian-Algarin, A. Acevedo, and C. Rinaldi. 2010. Monitoring gelation using magnetic nanoparticles. *Soft Matter* 6 (15):3662–3668. doi:10.1039/c003284k.

Bennett, J. B., A. L. Glover, D. E. Nikles, J. A. Nikles, and C. S. Brazel. 2011. Magnetothermally-triggered drug delivery using temperature-responsive polymeric micelles. PMSE Preprints, 105 953–954.

Berkov, D. V. 1998. Evaluation of the energy barrier distribution in many-particle systems using the path integral approach. *Journal of Physics-Condensed Matter* 10 (5):L89–L95. doi:10.1088/0953-8984/10/5/002.

Birks, J. B. 1950. The properties of ferromagnetic compounds at centimetre wavelengths. *Proceedings of the Physical Society of London Section B* 63 (362):65–74. doi:10.1088/0370-1301/63/2/301.

Bordelon, D. E., R. C. Goldstein, V. S. Nemkov, A. Kumar, J. K. Jackowski, T. L. DeWeese, and R. Ivkov. 2012. Modified solenoid coil that efficiently produces high amplitude AC magnetic fields with enhanced uniformity for biomedical applications. *IEEE Transactions on Magnetics* 48 (1):47–52. doi:10.1109/tmag.2011.2162527.

Breasted, J. H. 1930. *The Edwin Smith Surgical Papyrus*. Chicago: University of Chicago Oriental Institute Publications.

Brown, W. F. 1963. Thermal fluctuations of a single-domain particle. *Physical Review* 130 (5):1677. doi:10.1103/PhysRev.130.1677.

Calero-DdelC, V. L., D. I. Santiago-Quinonez, and C. Rinaldi. 2011. Quantitative nanoscale viscosity measurements using magnetic nanoparticles and SQUID AC susceptibility measurements. *Soft Matter* 7 (9):4497–4503. doi:10.1039/c0sm00902d.

Carrey, J., B. Mehdaoui, and M. Respaud. 2011. Simple models for dynamic hysteresis loop calculations of magnetic single-domain nanoparticles: Application to magnetic hyperthermia optimization. *Journal of Applied Physics* 109 (8). doi:10.1063/1.3551582.

Cetas, T. C. 1984. Will thermometric tomography become practical for hyperthermia treatment monitoring. *Cancer Research* 44 (10):4805–4808.

Chen, Y., A. Bose, and G. D. Bothun. 2010. Controlled release from bilayer-decorated magnetoliposomes via electromagnetic heating. *ACS Nano* 4 (6):3215–3221. doi:10.1021/nn100274v.

Chiu-Lam, A., and C. Rinaldi. 2016. Nanoscale thermal phenomena in the vicinity of magnetic nanoparticles in alternating magnetic fields. *Advanced Functional Materials* 26 (22):3933–3941.

Creixell, M., A. C. Bohorquez, M. Torres-Lugo, and C. Rinaldi. 2011. EGFR-targeted magnetic nanoparticle heaters kill cancer cells without a perceptible temperature rise. *ACS Nano* 5 (9):7124–7129. doi:10.1021/nn201822b.

DeNardo, S. J., G. L. DeNardo, A. Natarajan, L. A. Miers, A. R. Foreman, C. Gruettner, G. N. Adamson, and R. Ivkov. 2007. Thermal dosimetry predictive of efficacy of In-111-ChL6 nanoparticle AMF-induced thermoablative therapy for human breast cancer in mice. *Journal of Nuclear Medicine* 48 (3):437–444.

Deatsch, A. E., and B. A. Evans. 2014. Heating efficiency in magnetic nanoparticle hyperthermia. *Journal of Magnetism and Magnetic Materials* 354:163–172. doi:10.1016/j.jmmm.2013.11.006.

Debye, P. 1929. *Polar Molecules*. New York: Chemical Catalog Co., Inc.

Deen, W. M. 1998. *Analysis of Transport Phenomena*. Cary, NC: Oxford University Press.

Dennis, C. L., A. J. Jackson, J. A. Borchers, P. J. Hoopes, R. Strawbridge, A. R. Foreman, J. van Lierop, C. Gruttner, and R. Ivkov. 2009. Nearly complete regression of tumors via collective behavior of magnetic nanoparticles in hyperthermia. *Nanotechnology* 20 (39). doi:10.1088/0957-4484/20/39/395103.

Derfus, A. M., G. von Maltzahn, T. J. Harris, T. Duza, K. S. Vecchio, E. Ruoslahti, and S. N. Bhatia. 2007. Remotely triggered release from magnetic nanoparticles. *Advanced Materials* 19 (22):3932–3936. doi:10.1002/adma.200700091.

Dobson, J. 2008. Remote control of cellular behaviour with magnetic nanoparticles. *Nature Nanotechnology* 3 (3):139–143. doi:10.1038/nnano.2008.39.

Dutz, S., and R. Hergt. 2014. Magnetic particle hyperthermia-a promising tumour therapy? *Nanotechnology* 25 (45). doi:10.1088/0957-4484/25/45/452001.

Einstein, A. 1906. The theory of the Brownian motion. *Annalen Der Physik* 19 (2):371–381.

Fannin, P. C. 2002. Magnetic spectroscopy as an aide in understanding magnetic fluids. *Journal of Magnetism and Magnetic Materials* 252 (1-3):59–64. doi:10.1016/s0304-8853(02)00600-5.

Ferguson, R. M., A. P. Khandhar, C. Jonasson, J. Blomgren, C. Johansson, and K. M. Krishnan. 2013. Size-dependent relaxation properties of monodisperse magnetite nanoparticles measured over seven decades of frequency by AC susceptometry. *IEEE Transactions on Magnetics* 49 (7):3441–3444. doi:10.1109/tmag.2013.2239621.

Fornara, A., P. Johansson, K. Petersson, S. Gustafsson, J. Qin, E. Olsson, D. Ilver, A. Krozer, M. Muhammed, and C. Johansson. 2008. Tailored magnetic nanoparticles for direct and sensitive detection of biomolecules in biological samples. *Nano Letters* 8 (10):3423–3428. doi:10.1021/nl8022498.

Fortin, J. P., C. Wilhelm, J. Servais, C. Menager, J. C. Bacri, and F. Gazeau. 2007. Size-sorted anionic iron oxide nanomagnets as colloidal mediators for magnetic hyperthermia. *Journal of the American Chemical Society* 129 (9):2628–2635. doi:10.1021/ja067457e.

Fortin, J. P., F. Gazeau, and C. Wilhelm. 2008. Intracellular heating of living cells through Neel relaxation of magnetic nanoparticles. *European Biophysics Journal with Biophysics Letters* 37 (2):223–228. doi:10.1007/s00249-007-0197-4.

Gelbrich, T., M. Feyen, and A. M. Schmidt. 2006. Magnetic thermoresponsive core-shell nanoparticles. *Macromolecules* 39 (9):3469–3472. doi:10.1021/ma060006u.

Gilchrist, R. K., R. Medal, W. D. Shorey, R. C. Hanselman, J. C. Parrott, and C. B. Taylor. 1957. Selective inductive heating of lymph nodes. *Annals of Surgery* 146 (4):596–606.

He, X. M., W. F. Wolkers, J. H. Crowe, D. J. Swanlund, and J. C. Bischof. 2004. In situ thermal denaturation of proteins in dunning AT-1 prostate cancer cells: Implication for hyperthermic cell injury. *Annals of Biomedical Engineering* 32 (10):1384–1398. doi:10.1114/B:ABME.0000042226.97347.de.

Hildebrandt, B., P. Wust, O. Ahlers, A. Dieing, G. Sreenivasa, T. Kerner, R. Felix, and H. Riess. 2002. The cellular and molecular basis of hyperthermia. *Critical Reviews in Oncology Hematology* 43 (1):33–56. doi:10.1016/s1040-8428(01)00179-2.

Huang, H., S. Delikanli, H. Zeng, D. M. Ferkey, and A. Pralle. 2010. Remote control of ion channels and neurons through magnetic-field heating of nanoparticles. *Nature Nanotechnology* 5 (8):602–606. doi:10.1038/nnano.2010.125.

Huang, S., S. Y. Wang, A. Gupta, D. A. Borca-Tasciuc, and S. J. Salon. 2012. On the measurement technique for specific absorption rate of nanoparticles in an alternating electromagnetic field. *Measurement Science and Technology* 23 (3). doi:10.1088/0957-0233/23/3/035701.

Hurwitz, M., and P. Stauffer. 2014. Hyperthermia, radiation and chemotherapy: The role of heat in multidisciplinary cancer care. *Seminars in Oncology* 41 (6):714–729. doi:10.1053/j.seminoncol.2014.09.014.

Ito, A., K. Tanaka, H. Honda, S. Abe, H. Yamaguchi, and T. Kobayashi. 2003a. Complete regression of mouse mammary carcinoma with a size greater than 15 mm by frequent repeated hyperthermia using magnetite nanoparticles. *Journal of Bioscience and Bioengineering* 96 (4):364–369. doi:10.1016/s1389-1723(03)90138-1.

Ito, A., K. Tanaka, K. Kondo, M. Shinkai, H. Honda, K. Matsumoto, T. Saida, and T. Kobayashi. 2003b. Tumor regression by combined immunotherapy and hyperthermia using magnetic nanoparticles in an experimental subcutaneous murine melanoma. *Cancer Science* 94 (3):308–313. doi:10.1111/j.1349-7006.2003.tb01438.x.

Johannsen, M., U. Gneueckow, B. Thiesen, K. Taymoorian, C. H. Cho, N. Waldofner, R. Scholz, A. Jordan, S. A. Loening, and P. Wust. 2007. Thermotherapy of prostate cancer using magnetic nanoparticles: Feasibility, imaging, and three-dimensional temperature distribution. *European Urology* 52 (6):1653–1662. doi:10.1016/j.eururo.2006.11.023.

Jordan, A., P. Wust, R. Scholz, B. Tesche, H. Fahling, T. Mitrovics, T. Vogl, J. CervosNavarro, and R. Felix. 1996. Cellular uptake of magnetic fluid particles and their effects on human adenocarcinoma cells exposed to AC magnetic fields in vitro. *International Journal of Hyperthermia* 12 (6):705–722. doi:10.3109/02656739609027678.

Jordan, A., R. Scholz, P. Wust, H. Fahling, and R. Felix. 1999. Magnetic fluid hyperthermia (MFH): Cancer treatment with AC magnetic field induced excitation of biocompatible superparamagnetic nanoparticles. *Journal of Magnetism and Magnetic Materials* 201:413–419. doi:10.1016/s0304-8853(99)00088-8.

Jordan, A., R. Scholz, K. Maier-Hauff, M. Johannsen, P. Wust, J. Nadobny, H. Schirra, H. Schmidt, S. Deger, S. Loening, W. Lanksch, and R. Felix. 2001. Presentation of a new magnetic field therapy system for the treatment of human solid tumors with magnetic fluid hyperthermia. *Journal of Magnetism and Magnetic Materials* 225 (1–2):118–126. doi:10.1016/s0304-8853(00)01239-7.

Jordan, A., R. Scholz, K. Maier-Hauff, F. K. H. van Landeghem, N. Waldoefner, U. Teichgraeber, J. Pinkernelle, H. Bruhn, F. Neumann, B. Thiesen, A. von Deimling, and R. Felix. 2006. The effect of thermotherapy using magnetic nanoparticles on rat malignant glioma. *Journal of Neuro-Oncology* 78 (1):7–14. doi:10.1007/s11060-005-9059-z.

Kallumadil, M., M. Tada, T. Nakagawa, M. Abe, P. Southern, and Q. A. Pankhurst. 2009. Suitability of commercial colloids for magnetic hyperthermia. *Journal of Magnetism and Magnetic Materials* 321 (10):1509–1513. doi:10.1016/j.jmmm.2009.02.075.

Khandhar, A. P., R. M. Ferguson, J. A. Simon, and K. M. Krishnan. 2012. Tailored magnetic nanoparticles for optimizing magnetic fluid hyperthermia. *Journal of Biomedical Materials Research Part A* 100 A (3):728–737. doi:10.1002/jbm.a.34011.

Kim, D.-H., E. A. Vitol, J. Liu, S. Balasubramanian, D. J. Gosztola, E. E. Cohen, V. Novosad, and E. A. Rozhkova. 2013b. Stimuli-responsive magnetic nanomicelles as multifunctional heat and cargo delivery vehicles. *Langmuir* 29 (24):7425–7432. doi:10.1021/la3044158.

Kim, D. H., E. A. Vitol, J. Liu, S. Balasubramanian, D. J. Gosztola, E. E. Cohen, V. Novosad, and E. A. Rozhkova. 2013a. Stimuli-responsive magnetic nanomicelles as multifunctional heat and cargo delivery vehicles. *Langmuir* 29 (24):7425–7432. doi:10.1021/la3044158.

Koh, I., and L. Josephson. 2009. Magnetic nanoparticle sensors. *Sensors* 9 (10):8130–8145. doi:10.3390/s91008130.

Kozissnik, B., A. C. Bohorquez, J. Dobson, and C. Rinaldi. 2013. Magnetic fluid hyperthermia: Advances, challenges, and opportunity. *International Journal of Hyperthermia* 29 (8):706–714. doi:10.3109/02656736.2013.837200.

Kuimova, M. K. 2012. Mapping viscosity in cells using molecular rotors. *Physical Chemistry Chemical Physics* 14 (37):12671–12686. doi:10.1039/c2cp41674c.

Kulshrestha, P., M. Gogoi, D. Bahadur, and R. Banerjee. 2012. In vitro application of paclitaxel loaded magnetoliposomes for combined chemotherapy and hyperthermia. *Colloids and Surfaces B-Biointerfaces* 96:1–7. doi 10.1016/j.colsurfb.2012.02.029.

Lancarotte, M. S., and A. D. Penteado. 2001. Estimation of core losses under sinusoidal or non-sinusoidal induction by analysis of magnetization rate. *IEEE Transactions on Energy Conversion* 16 (2):174–179. doi:10.1109/60.921469.

Lartigue, L., P. Hugounenq, D. Alloyeau, S. P. Clarke, M. Levy, J. C. Bacri, R. Bazzi, D. F. Brougham, C. Wilhelm, and F. Gazeau. 2012. Cooperative organization in iron oxide multi-core nanoparticles potentiates their efficiency as heating mediators and MRI contrast agents. *ACS Nano* 6 (12):10935–10949. doi:10.1021/nn304477s.

Lebihan, D., J. Delannoy, and R. L. Levin. 1989. Temperature mapping with MR imaging of molecular-diffusion—Application to hyperthermia. *Radiology* 171 (3):853–857.

Lehlooh, A. F., S. H. Mahmood, and J. M. Williams. 2002. On the particle size dependence of the magnetic anisotropy energy constant. *Physica B-Condensed Matter* 321 (1–4):159–162. doi:10.1016/s0921-4526(02)00843-8.

Lepage, M., J. Jiang, J. Babin, B. Qi, L. Tremblay, and Y. Zhao. 2007. MRI observation of the light-induced release of a contrast agent from photo-controllable polymer micelles. *Physics in Medicine and Biology* 52 (10):N249–N255. doi:10.1088/0031-9155/52/10/n04.

Li, Z. X., M. Kawashita, N. Araki, M. Mitsumori, M. Hiraoka, and M. Doi. 2010. Magnetite nanoparticles with high heating efficiencies for application in the hyperthermia of cancer. *Materials Science & Engineering C-Materials for Biological Applications* 30 (7):990–996. doi:10.1016/j.msec.2010.04.016.

Loynachan, C. N., G. Romero, M. G. Christiansen, R. Chen, R. Ellison, T. T. O'Malley, U. P. Froriep, D. M. Walsh, and P. Anikeeva. 2015. Targeted magnetic nanoparticles for remote magnetothermal disruption of amyloid-beta aggregates. *Advanced Healthcare Materials* 4 (14):2100–2109. doi:10.1002/adhm.201500487.

Maier-Hauff, K., R. Rothe, R. Scholz, U. Gneveckow, P. Wust, B. Thiesen, A. Feussner, A. von Deimling, N. Waldoefner, R. Felix, and A. Jordan. 2007. Intracranial thermotherapy using magnetic nanoparticles combined with external beam radiotherapy: Results of a feasibility study on patients with glioblastoma multiforme. *Journal of Neuro-Oncology* 81 (1):53–60. doi:10.1007/s11060-006-9195-0.

Maier-Hauff, K., F. Ulrich, D. Nestler, H. Niehoff, P. Wust, B. Thiesen, H. Orawa, V. Budach, and A. Jordan. 2011. Efficacy and safety of intratumoral thermotherapy using magnetic iron-oxide nanoparticles combined with external beam radiotherapy on patients with recurrent glioblastoma multiforme. *Journal of Neuro-Oncology* 103 (2):317–324. doi:10.1007/s11060-010-0389-0.

Maldonado-Camargo, L., I. Torres-Diaz, A. Chiu-Lam, M. Hernandez, and C. Rinaldi. 2016. Estimating the contribution of Brownian and Neel relaxation in a magnetic fluid through dynamic magnetic susceptibility measurements. *Journal of Magnetism and Magnetic Materials* 412:223–233. doi:10.1016/j.jmmm.2016.03.087.

Mamiya, H., and B. Jeyadevan. 2011. Hyperthermic effects of dissipative structures of magnetic nanoparticles in large alternating magnetic fields. *Scientific Reports* 1. doi:10.1038/srep00157.

Mannix, R. J., S. Kumar, F. Cassiola, M. Montoya-Zavala, E. Feinstein, M. Prentiss, and D. E. Ingber. 2008. Nanomagnetic actuation of receptor-mediated signal transduction. *Nature Nanotechnology* 3 (1):36–40. doi:10.1038/nnano.2007.418.

Mantso, T., G. Goussetis, R. Franco, S. Botaitis, A. Pappa, and M. Panayiotidis. 2016. Effects of hyperthermia as a mitigation strategy in DNA damage-based cancer therapies. *Seminars in Cancer Biology* 37–38:96–105. doi:10.1016/j.semcancer.2016.03.004.

Martens, M. A., R. J. Deissler, Y. Wu, L. Bauer, Z. Yao, R. Brown, and M. Griswold. 2013. Modeling the Brownian relaxation of nanoparticle ferrofluids: Comparison with experiment. *Medical Physics* 40 (2). doi:10.1118/1.4773869.

Martinez-Boubeta, C., K. Simeonidis, A. Makridis, M. Angelakeris, O. Iglesias, P. Guardia, A. Cabot, L. Yedra, S. Estrade, F. Peiro, Z. Saghi, P. A. Midgley, I. Conde-Leboran, D. Serantes, and D. Baldomir. 2013. Learning from nature to improve the heat generation of iron-oxide nanoparticles for magnetic hyperthermia applications. *Scientific Reports* 3. doi:10.1038/srep01652.

Massart, R. 1981. Preparation of aqueous magnetic liquids in alkaline and acidic media. *IEEE Transactions on Magnetics* 17 (2):1247–1248. doi:10.1109/tmag.1981.1061188.

Mazo, R. M. 2002. *Brownian Motion: Fluctuations, Dynamics, and Applications*. Oxford University Press on Demand.

Mehdaoui, B., A. Meffre, J. Carrey, S. Lachaize, L.-M. Lacroix, M. Gougeon, B. Chaudret, and M. Respaud. 2011. Optimal size of nanoparticles for magnetic hyperthermia: A combined theoretical and experimental study. *Advanced Functional Materials* 21 (23):4573–4581. doi:10.1002/adfm.201101243.

Moroz, P., S. K. Jones, J. Winter, and B. N. Gray. 2001. Targeting liver tumors with hyperthermia: Ferromagnetic embolization in a rabbit liver tumor model. *Journal of Surgical Oncology* 78 (1):22–29. doi:10.1002/jso.1118.

Natividad, E., M. Castro, and A. Mediano. 2008. Accurate measurement of the specific absorption rate using a suitable adiabatic magnetothermal setup. *Applied Physics Letters* 92 (9). doi:10.1063/1.2891084.

Néel, L. 1949. Théorie du traînage magnétique des ferromagnétiques en grains fins avec application aux terres cuites. Annals of Geophysics (C.N.R.S.).5: pp. 99–136.

Oleson, J. R. 1984. A review of magnetic induction methods for hyperthermia treatment of cancer. *IEEE Transactions on Biomedical Engineering* 31 (1):91–97. doi:10.1109/tbme.1984.325374.

Park, J., K. J. An, Y. S. Hwang, J. G. Park, H. J. Noh, J. Y. Kim, J. H. Park, N. M. Hwang, and T. Hyeon. 2004. Ultra-large-scale syntheses of monodisperse nanocrystals. *Nature Materials* 3 (12):891–895. doi:10.1038/nmat1251.

Park, K., T. Harrah, E. B. Goldberg, R. P. Guertin, and S. Sonkusale. 2011. Multiplexed sensing based on Brownian relaxation of magnetic nanoparticles using a compact AC susceptometer. *Nanotechnology* 22 (8). doi:10.1088/0957-4484/22/8/085501.

Peddis, D., M. V. Mansilla, S. Morup, C. Cannas, A. Musinu, G. Piccaluga, F. D'Orazio, F. Lucari, and D. Fiorani. 2008. Spin-canting and magnetic anisotropy in ultrasmall $CoFe_2O_4$ nanoparticles. *Journal of Physical Chemistry B* 112 (29):8507–8513. doi:10.1021/jp8016634.

Perreard, I. M., D. B. Reeves, X. Zhang, E. Kuehlert, E. R. Forauer, and J. B. Weaver. 2014. Temperature of the magnetic nanoparticle microenvironment: Estimation from relaxation times. *Physics in Medicine and Biology* 59 (5). doi:10.1088/0031-9155/59/5/1109.

Perrin, J. 1908. Stokes law and Brownian motion. *Comptes Rendus Hebdomadaires Des Seances De L Academie Des Sciences* 147:475–476.

Perrin, J. 1909. Brownian motion and molecular reality. *Annales De Chimie Et De Physique* 18:5–114.

Polo-Corrales, L., and C. Rinaldi. 2012. Monitoring iron oxide nanoparticle surface temperature in an alternating magnetic field using thermoresponsive fluorescent polymers. *Journal of Applied Physics* 111 (7). doi:10.1063/1.3680532.

Prasad, N. K., K. Rathinasamy, D. Panda, and D. Bahadur. 2007. Mechanism of cell death induced by magnetic hyperthermia with nanoparticles of gamma-Mn_xFe_2-xO_3 synthesized by a single step process. *Journal of Materials Chemistry* 17 (48):5042–5051. doi:10.1039/b708156a.

Rahimi, M., A. Wadajkar, K. Subramanian, M. Yousef, W. N. Cui, J. T. Hsieh, and K. T. Nguyen. 2010. In vitro evaluation of novel polymer-coated magnetic nanoparticles for controlled drug delivery. *Nanomedicine-Nanotechnology Biology and Medicine* 6 (5):672–680. doi:10.1016/j.nano.2010.01.012.

Reeves, D. B., and J. B. Weaver. 2014. Magnetic nanoparticle sensing: Decoupling the magnetization from the excitation field. *Journal of Physics D-Applied Physics* 47 (4). doi:10.1088/0022-3727/47/4/045002.

Riedinger, A., P. Guardia, A. Curcio, M. A. Garcia, R. Cingolani, L. Manna, and T. Pellegrino. 2013. Subnanometer local temperature probing and remotely controlled drug release based on azo-functionalized iron oxide nanoparticles. *Nano Letters* 13 (6):2399–2406. doi:10.1021/nl400188q.

Rodriguez-Luccioni, H. L., M. Latorre-Esteves, J. Mendez-Vega, O. Soto, A. R. Rodriguez, C. Rinaldi, and M. Torres-Lugo. 2011. Enhanced reduction in cell viability by hyperthermia induced by magnetic nanoparticles. *International Journal of Nanomedicine* 6:373–380. doi:10.2147/ijn.s14613.

Rosensweig, R. E. 1985. *Ferrohydrodynamics, Cambridge Monographs on Mechanics and Applied Mathematics*. Cambridge; New York: Cambridge University Press.

Rosensweig, R. E. 2002. Heating magnetic fluid with alternating magnetic field. *Journal of Magnetism and Magnetic Materials* 252 (1–3):370–374. doi:10.1016/s0304-8853(02)00706-0.

Roussakow, S. 2013. *The History of Hyperthermia Rise and Decline*. Galenic Research Institute. Moscow, Russia: Conference Papers in Medicine.

Ruhle, B., S. Datz, C. Argyo, T. Bein, and J. I. Zink. 2016. A molecular nanocap activated by superparamagnetic heating for externally stimulated cargo release. *Chemical Communications* 52 (9):1843–1846. doi:10.1039/c5cc08636a.

Ruiz-Hernandez, E., A. Baeza, and M. Vallet-Regi. 2011. Smart drug delivery through DNA/magnetic nanoparticle gates. *ACS Nano* 5 (2):1259–1266. doi:10.1021/nn1029229.

Shenker, H. 1957. Magnetic anisotropy of cobalt ferrite ($CO_{1.01}FE_{2.00}O_{3.62}$) and nickel cobalt ferrite ($NI_{0.72}FE_{0.20}CO_{0.08}FE_2O_4$). *Physical Review* 107 (5):1246–1249. doi:10.1103/PhysRev.107.1246.

Shinkai, M., M. Yanase, H. Honda, T. Wakabayashi, J. Yoshida, and T. Kobayashi. 1996. Intracellular hyperthermia for cancer using magnetite cationic liposomes: In vitro study. *Japanese Journal of Cancer Research* 87 (11):1179–1183.

Shliomis, M. I. 1972. Effective viscosity of magnetic suspensions. *Soviet Physics Jetp-Ussr* 34 (6):1291–1294.

Soto-Aquino, D., and C. Rinaldi. 2015. Nonlinear energy dissipation of magnetic nanoparticles in oscillating magnetic fields. *Journal of Magnetism and Magnetic Materials* 393:46–55. doi:10.1016/j.jmmm.2015.05.009.

Sun, S. H., and H. Zeng. 2002. Size-controlled synthesis of magnetite nanoparticles. *Journal of the American Chemical Society* 124 (28):8204–8205. doi:10.1021/ja026501x.

Sun, X. L., N. F. Huls, A. Sigdel, and S. H. Sun. 2012. Tuning exchange bias in core/shell FeO/Fe_3O_4 nanoparticles. *Nano Letters* 12 (1):246–251. doi:10.1021/nl2034514.

Tefferi, A. 2003. A contemporary approach to the diagnosis and management of polycythemia vera. *Current Hematology Reports* 2:3.

Urtizberea, A., E. Natividad, A. Arizaga, M. Castro, and A. Mediano. 2010. Specific absorption rates and magnetic properties of ferrofluids with interaction effects at low concentrations. *Journal of Physical Chemistry C* 114 (11):4916–4922. doi:10.1021/jp912076f.

Usov, N. A., and Y. B. Grebenshchikov. 2009. Hysteresis loops of an assembly of superparamagnetic nanoparticles with uniaxial anisotropy. *Journal of Applied Physics* 106 (2). doi:10.1063/1.3173280.

Vallejo-Fernandez, G., and K. O'Grady. 2013. Effect of the distribution of anisotropy constants on hysteresis losses for magnetic hyperthermia applications. *Applied Physics Letters* 103 (14). doi:10.1063/1.4824649.

Venkatachalam, K., C. R. Sullivan, T. Abdallah, and H. E. Tacca. 2002. Accurate prediction of ferrite core loss with nonsinusoidal waveforms using only Steinmetz parameters. IEEE Workshop on Computers in Power Electronics.

Vreeland, E. C., J. Watt, G. B. Schober, B. G. Hance, M. J. Austin, A. D. Price, B. D. Fellows, T. C. Monson, N. S. Hudak, L. Maldonado-Camargo, A. C. Bohorquez, C. Rinaldi, and D. L. Huber. 2015. Enhanced nanoparticle size control by extending LaMer's mechanism. *Chemistry of Materials* 27 (17):6059–6066. doi:10.1021/acs.chemmater.5b02510.

Wang, Y., Q. Zhao, N. Han, L. Bai, J. Li, J. Liu, E. Che, L. Hu, Q. Zhang, T. Jiang, and S. Wang. 2015. Mesoporous silica nanoparticles in drug delivery and biomedical applications. *Nanomedicine-Nanotechnology Biology and Medicine* 11 (2):313–327. doi:10.1016/j.nano.2014.09.014.

Wildeboer, R. R., P. Southern, and Q. A. Pankhurst. 2014. On the reliable measurement of specific absorption rates and intrinsic loss parameters in magnetic hyperthermia materials. *Journal of Physics D-Applied Physics* 47 (49). doi:10.1088/0022-3727/47/49/495003.

Wust, P., B. Hildebrandt, G. Sreenivasa, B. Rau, J. Gellermann, H. Riess, R. Felix, and P. M. Schlag. 2002. Hyperthermia in combined treatment of cancer. *Lancet Oncology* 3 (8):487–497. doi:10.1016/s1470-2045(02)00818-5.

Yan, S. Y., D. S. Zhang, N. Gu, J. Zheng, A. W. Ding, Z. Y. Wang, B. L. Xing, M. Ma, and Y. Zhang. 2005. Therapeutic effect of Fe_2O_3 nanoparticles combined with magnetic fluid hyperthermia on cultured liver cancer cells and xenograft liver cancers. *Journal of Nanoscience and Nanotechnology* 5 (8):1185–1192. doi:10.1166/jnn.2005.219.

Zagar, T. M., J. R. Oleson, Z. Vujaskovic, M. W. Dewhirst, O. I. Craciunescu, K. L. Blackwell, L. R. Prosnitz, and E. L. Jones. 2010. Hyperthermia combined with radiation therapy for superficial breast cancer and chest wall recurrence: A review of the randomised data. *International Journal of Hyperthermia* 26 (7):612–617. doi:10.3109/02656736.2010.487194.

Zhang, E. M., M. F. Kircher, M. Koch, L. Eliasson, S. N. Goldberg, and E. Renstrom. 2014. Dynamic magnetic fields remote-control apoptosis via nanoparticle rotation. *ACS Nano* 8 (4):3192–3201. doi:10.1021/nn406302j.

Zhang, J., and R. D. K. Misra. 2007. Magnetic drug-targeting carrier encapsulated with thermosensitive smart polymer: Core-shell nanoparticle carrier and drug release response. *Acta Biomaterialia* 3 (6):838–850. doi:10.1016/j.actbio.2007.05.011.

Zhang, J. L., R. S. Srivastava, and R. D. K. Misra. 2007. Core-shell magnetite nanoparticles surface encapsulated with smart stimuli-responsive polymer: Synthesis, characterization, and LCST of viable drug-targeting delivery system. *Langmuir* 23 (11):6342–6351. doi:10.1021/la0636199.

Zhang, Q., I. Castellanos-Rubio, R. Munshi, I. Orue, B. Pelaz, K. I. Gries, W. J. Parak, P. del Pino, and A. Pralle. 2015. Model driven optimization of magnetic anisotropy of exchange-coupled core-shell ferrite nanoparticles for maximal hysteretic loss. *Chemistry of Materials* 27 (21):7380–7387. doi:10.1021/acs.chemmater.5b03261.

Zhang, X. M., K. Guo, L. H. Li, Z. B. Sheng, and B. J. Li. 2015. Multi-stimuli-responsive magnetic assemblies as tunable releasing carriers. *Journal of Materials Chemistry B* 3 (29):6026–6031. doi:10.1039/c5tb00845j.

Zhao, Q., L. N. Wang, R. Cheng, L. D. Mao, R. D. Arnold, E. W. Howerth, Z. G. Chen, and S. Platt. 2012. Magnetic nanoparticle-based hyperthermia for head and neck cancer in mouse models. *Theranostics* 2 (1):113–121. doi:10.7150/thno.3854.

Zou, H., and W. Z. Yuan. 2015. Temperature- and redox-responsive magnetic complex micelles for controlled drug release. *Journal of Materials Chemistry B* 3 (2):260–269. doi:10.1039/c4tb01518e.

del Castillo, V., and C. Rinaldi. 2010. Effect of sample concentration on the determination of the anisotropy constant of magnetic nanoparticles. *IEEE Transactions on Magnetics* 46 (3):852–859. doi:10.1109/tmag.2009.2032240.

van Landeghem, F. K. H., K. Maier-Hauff, A. Jordan, K. T. Hoffmann, U. Gneveckow, R. Scholz, B. Thiesen, W. Bruck, and A. von Deimling. 2009. Post-mortem studies in glioblastoma patients treated with thermotherapy using magnetic nanoparticles. *Biomaterials* 30 (1):52–57. doi:10.1016/j.biomaterials.2008.09.044.

von Maltzahn, G., J. H. Park, K. Y. Lin, N. Singh, C. Schwoppe, R. Mesters, W. E. Berdel, E. Ruoslahti, M. J. Sailor, and S. N. Bhatia. 2011. Nanoparticles that communicate in vivo to amplify tumour targeting. *Nature Materials* 10 (7):545–552. doi:10.1038/nmat3049.

12 Toxicology of Magnetic Nanoparticles

Stephen J. Klaine, Jordan T. Burbage, Paul W. Millhouse, Unaiza Uzair, and Jeffrey N. Anker

CONTENTS

12.1 INTRODUCTION

Magnetic particles are useful for many biomedical applications including contrast agents for magnetic resonance imaging (MRI), magnetically controlled drug delivery, magnetically directed energy delivery and hyperthermia, and magnetic gene transfection. For all these applications, it is critical to understand how the required therapeutic dose compares with the dose needed to elicit adverse effects. Another consideration is the ultimate fate and potential unintended consequences of the nanoparticles after the intervention is completed. To address these questions, researchers have investigated the interactions between particles and biological systems both in whole organisms (in vivo) and in cell cultures (in vitro). It is generally understood that interactions between nanoparticles and biological systems depend upon particle size, shape, and surface chemistry. The goal of biomedical research with nanoparticles is to design drug carriers and sensors that are improvements over existing technologies, or that achieve other therapeutic or diagnostic goals, while minimizing any unintended consequences associated with their use. This chapter provides an introduction to toxicology, discusses the potential adverse effects of magnetic nanoparticles (MNPs) on mammalian systems, examines the potential unintended consequences of MNPs on the environment, and provides some perspective on future research needs. Section 12.2 provides a general introduction to toxicology, Section 12.3 reviews the toxicity of iron oxide nanoparticles (IONPs) in cell culture and small animal studies, and Section 12.4 discusses doses used in human patient populations especially for MRI and iron replacement applications.

12.2 INTRODUCTION TO TOXICOLOGY

Toxicology is the branch of science that sets safe limits on the use of chemicals, drugs, food additives, and other substances that humans may consume, purchase, interact with, or that might enter the ecosystem. Toxicity is a relative property reflecting a material's potential to harm a living organism and is a function of the concentration of the material and the duration and method of exposure (e.g. oral, trans-dermal, inhalation, or intravenous injection). For a given exposure route and time, organism response as a function of quantity (either internal administered dose or external exposure concentration) can be illustrated graphically using results of a bioassay with a test organism or biological system (e.g. cell culture) (Figure 12.1). These dose-response relationships generally have sigmoidal shapes that saturate at high concentrations and are used predominantly to compare the response of different organisms to a given substance or compare the response of an organism to different substances. This relationship is further manipulated by transforming the *y*-axis to a probability or probit scale to linearize the data (Figure 12.1). From this linear relationship it is possible to estimate response metrics with confidence intervals (discussed below) (Curtis Klaassen 2013). Toxicological studies on engineered nanoparticles used in therapeutic and diagnostic applications usually begin with in vitro cellular assays and progress to studies involving live animals (in vivo). It is important to remember that for a substance to exert a toxic effect, it must reach the active site at a sufficient concentration for a sufficient amount of time. The disposition of nanoparticles within an organism governs these considerations. Patil et al. (2018) discuss a variety of in vitro and in vivo superparamagnetic iron oxide nanoparticles (SPION) toxicity studies done through various means (cellular, genetic, immune, etc.). They also summarized the results of some in vitro cytotoxicity studies of IONPs on different cell types in a table (Patil et al. 2018). Similarly, Patil et al. (2015) review the available methods and techniques (both spectrophotometric and imaging) used to quantify and evaluate the toxicity of IONPs both in vitro and in vivo (Patil et al. 2015).

12.2.1 Nanoparticle Disposition

Nanoparticles used in biomedical application are typically injected into the patient intravenously. These particles may be used for drug delivery (therapeutics) or to enhance imaging (diagnostics). Less often, nanoparticles are administered orally, trans-dermally, or inhaled. However, it is important to consider these other routes as they are more likely ways in which people may be unintentionally exposed to the substances. Once administration or exposure has occurred, these particles are subject to the normal pharmacodynamics that control the disposition of any substance in an organism (Figure 12.2). Movement through membranes (e.g. absorption, endocytosis), distribution throughout the body, metabolism by enzymes, storage in adipocytes and physiological compartments, and excretion are processes that ultimately dictate the disposition of particles within an organism. Nanoparticle disposition influences its efficacy as a therapeutic or diagnostic agent. If a nanoparticle is rapidly excreted through the kidneys, for instance, it is less likely to be effective for drug delivery to the liver or to enhance imaging of the heart. Further, disposition influences potential toxicity of the nanoparticle by controlling the concentration of the particles at a site of action. A nanoparticle that accumulates in the liver and that is also hepatotoxic is more hazardous than one that accumulates in the bones and has a site of toxic action in the liver. Finally, excretion of nanoparticles from the patient results in additional considerations from a waste management perspective. These concerns are often overlooked by healthcare professionals although they may have significant unintended consequences on the sewage system and the environment. These considerations are no different from those for anti-cancer drugs that have the potential to both disrupt wastewater treatment systems by adversely affecting microorganisms and affect wild plants and animals upon entering the environment through effluent discharge (Xie 2012, Nassour et al. 2020).

12.2.2 In Vitro Toxicity Testing

In vitro toxicity testing characterizes the response of cultured cells, usually mammalian, to contaminants. In vitro testing has several advantages over in vivo research, including reduced use of live

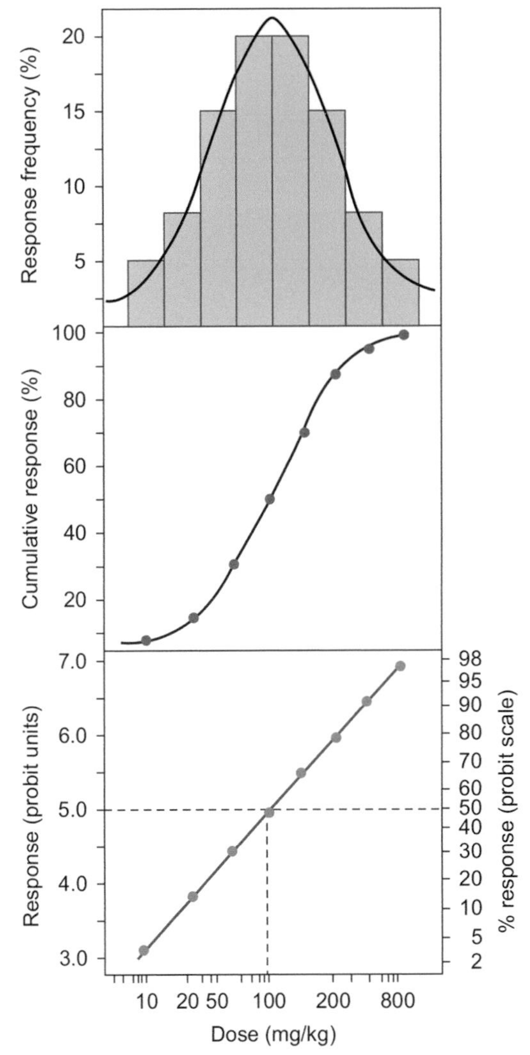

FIGURE 12.1 Diagram of quantal dose–response relationship. The abscissa is a log dosage of the chemical. In the top panel the ordinate is response frequency, in the middle panel the ordinate is percent response, and in the bottom panel the response is in probit units (see text), figure and caption used with permission from (Curtis Klaassen 2013).

animals, greater potential to elucidate mechanisms of action, reduced cost, and decreased experimental time. The rationale behind using cell culture data to predict in vivo results stems from the basal cell toxicity work of Ekwall (1983), who proposed that for most chemicals, toxicity is a result of aberrations in cellular functions (Joshi et al. 2013). Several authors have reported a good correlation between the results of in vitro and in vivo toxicity tests. Ekwall (1999) was able to explain 77% of the variability in an in vivo data set by using the results of three in vitro studies on the same chemical poisons. Evans et al. (2001) reported that in vitro IC_{35} (inhibitory concentration) cytotoxicity values using Chinese hamster ovary cells explained 85% of the variation in the most toxic LD_{50} values for the same set of diverse chemicals. However, the predictive value of in vitro testing may be significantly reduced for nanoparticles (Warheit, Sayes, & Reed, 2009, Joshi et al. 2013). This illustrates the most significant disadvantage of cellular culture testing with nanoparticles: it circumvents the uptake processes that must occur in an organism for the particle to move from the environment into the body through ingestion, inhalation, or trans-dermal absorption. Chemicals used in the studies by Ekwall (1999) and Evans 2001 were readily bioavailable and easily moved into the organism. This is not often true for nanoparticles. Sayes, Reed, and Warheit (2007) used different sizes of ZnO nanoparticles and instilled these materials into the lungs of rats. Lung membranes, especially in the alveolar (gas-

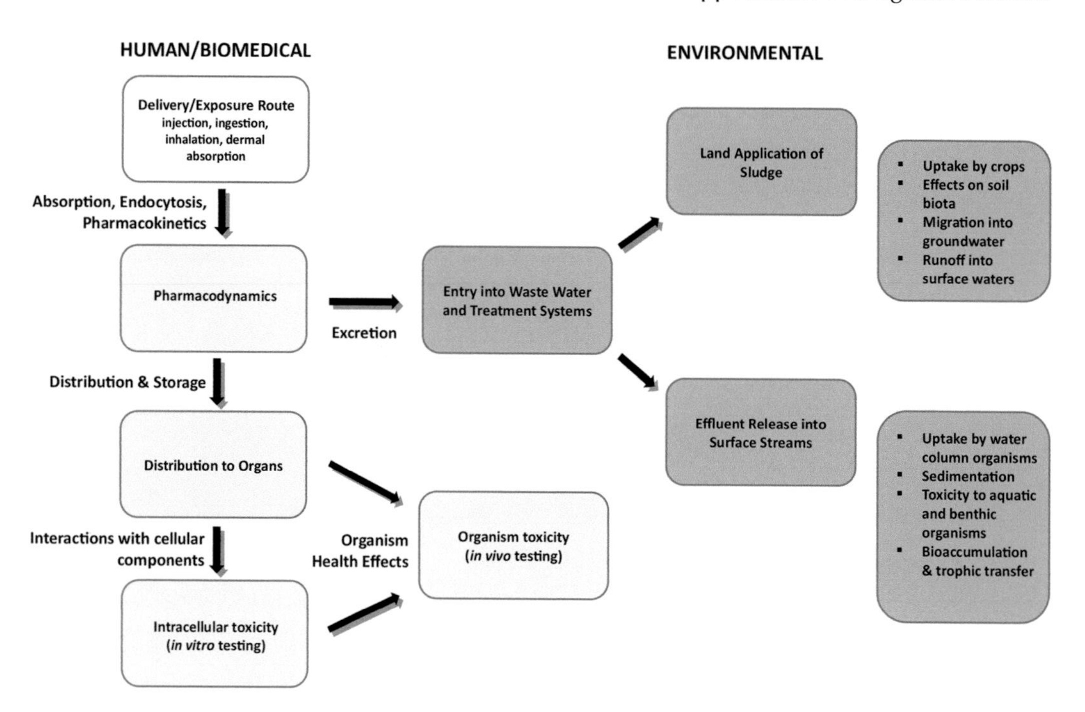

FIGURE 12.2 Nanoparticle disposition showing human and environmental compartments, movement within the human organism, and movement within the ecosystem after excretion. Note that toxic effects can be observed at the molecular/cellular and whole-organism level in humans and in multiple organisms within the environment.

exchange) region represent the thinnest membranes in living organisms. There was poor correlation between these results and those from in vitro studies. While uptake through lungs and intestines has been demonstrated in some organisms with a limited number of nanoparticles, most particles stay in the airways or the intestinal tract. Hence, Warheit *et al.* (2009) reported transient organismal responses of metal fume fever (acute self-limiting flu-like illness resulting from inhalation of airborne metal dust or fumes, commonly zinc oxide) (Wardhana and Datau 2014) that did not correlate well with the cytotoxic responses that occurred using rat lung epithelial cells and alveolar macrophages. If the nanoparticle is not injected intravenously into the organism, in vitro tests are unlikely to predict whole-organism responses. Even in the event of injected nanoparticle exposure, there is still the potential for an inflammatory response in the blood stream leading to capillary occlusion in various parts of the body, especially the lungs. The authors summarized that comparisons of in vitro and in vivo toxicity measurements demonstrated little convergence from fine- or nano-sized ZnO particle exposures. They concluded that the current cell culture methodologies did not accurately predict in vivo exposure hazards (Warheit *et al.* 2009). However, in vitro studies often involve orders of magnitude higher doses or faster administration, thus overwhelming physiologic cellular defenses (Nel et al. 2013). A multi-perspective consortium recently agreed that in vitro may thus be better suited to predict if or when adverse effects might occur, i.e. to highly active substances (Nel et al. 2013).

By far the biggest advantages of cellular bioassays are screening for potential toxic mechanisms and using these results to prioritize in vivo studies. In fact, the National Research Council (2007) recommended that new materials be evaluated for adverse effects using a tiered testing program that begins with high-throughput in vitro screening and ends in prioritized whole-animal studies. The goal is to develop adverse outcome pathways (AOPs) that begin with an initiating molecular event and end with an apical outcome with population-level consequences (National Research Council 2007).

Endpoints used in in vitro studies range from cell viability to assessments of specific mechanisms of toxic action. These include the MTT 3-(4,5-dimethylthiazol-2-yl)-2,5-diphenyltetrazolium bromide assay for cell metabolic activity and cytotoxicity, apoptosis (programmed cell death) assay, assessment of reactive oxygen species (ROS) production, and genetic toxicity (e.g. DNA damage, comet assay) (Oberdörster 2010). Nel (2013) reviewed many of these endpoints and their relationship with the mechanistic injury pathways in whole organisms (Figure 12.3). ROS, such as oxygen ions and peroxides, react with many molecules including DNA, RNA, proteins, and lipids, and can cause DNA mutations, damage to cell structures, and cell death (also known as oxidative stress). Antioxidants and repair mechanisms help protect against this.

12.2.3 In Vivo Toxicity Testing

In vivo toxicity bioassays test the effects of a substance on whole organisms. These tests are more expensive and may be long and tedious. However, they provide a more accurate assessment of the toxic effects of a substance. For human health assessment, mammalian models are generally used, whereas for ecotoxicological assessment, numerous organisms representing different trophic levels are involved.

Acute-toxicity bioassays expose organisms to a test material for a short period of time, generally hours to days. These bioassays using whole organisms are usually the first step in the characterization of the hazard associated with material exposure. The metric usually reported for an acute-toxicity bioassay is the **L**ethal **C**oncentration that kills 50% of the organisms in a given period of time (i.e. 48 hours LC_{50}). For therapeutic and diagnostic materials, another metric is the **M**aximum **T**olerated **D**ose (MTD). The MTD is the highest administered dose that does not cause unacceptable side effects; in animal models, veterinarians will monitor appetite and body weight, physical appearance, clinical signs, and behavior.

Results of these tests may then be used to design and perform chronic bioassays. In chronic bioassays, organisms are exposed to materials over a long period of time, usually weeks, months, or years. The determination of test duration is based on knowledge of the test organism and typically lasts over 80% of its expected life cycle. While mortality is assessed, material exposure concentrations are usually set to facilitate the quantification of sub-lethal endpoints such as growth and reproduction. As such, the metrics typically reported for results of a chronic toxicity bioassay are an EC_{50} value (the **E**ffective **C**oncentration that reduces reproduction or growth by 50%), a **N**o **O**bservable **E**ffect **C**oncentration (NOEC), or a **L**owest **O**bservable **E**ffect **C**oncentration (LOEC) (Figure 12.1).

A plethora of standardized methods exist for the characterization of human safety and the ecotoxicity of chemical substances. These standardized tests have been reviewed by Crane, Handy, Garrod, and Owen (2008) and are supported by both international (e.g. Organization for Economic Co-operation and Development [OECD]) and national (e.g. American Society of Testing and Materials [ASTM] and U.S. Environmental Protection Agency [EPA]) organizations. These standard assays cover mammalian models, soil and sediment dwelling organisms, marine and freshwater organisms, and various taxa including bacteria, plants, invertebrates, and vertebrates. The aspects common to all of these bioassays include: (1) the test is widely accepted, conducted by the scientific community, economical, and easy to conduct; (2) the bioassay has defined protocols, is repeatable, and generates similar results in different laboratories; (3) the endpoints should be easily measured, quantitatively related to exposure concentrations and duration, amenable to statistical analysis, and useful in ecological risk assessment; and (4) results of the bioassay should have some field relevance for similar species.

Choice of test species is often based on several criteria including (1) adequate background on the organism that facilitates ease of culture in the laboratory and interpretation of endpoints; (2) the organism should be widely available, abundant, and indigenous when possible; and (3) the organism should be sensitive to a wide range of toxic substances and endpoints should reflect ecological fitness. Most mammalian toxicity evaluations use rodents as models for human effects.

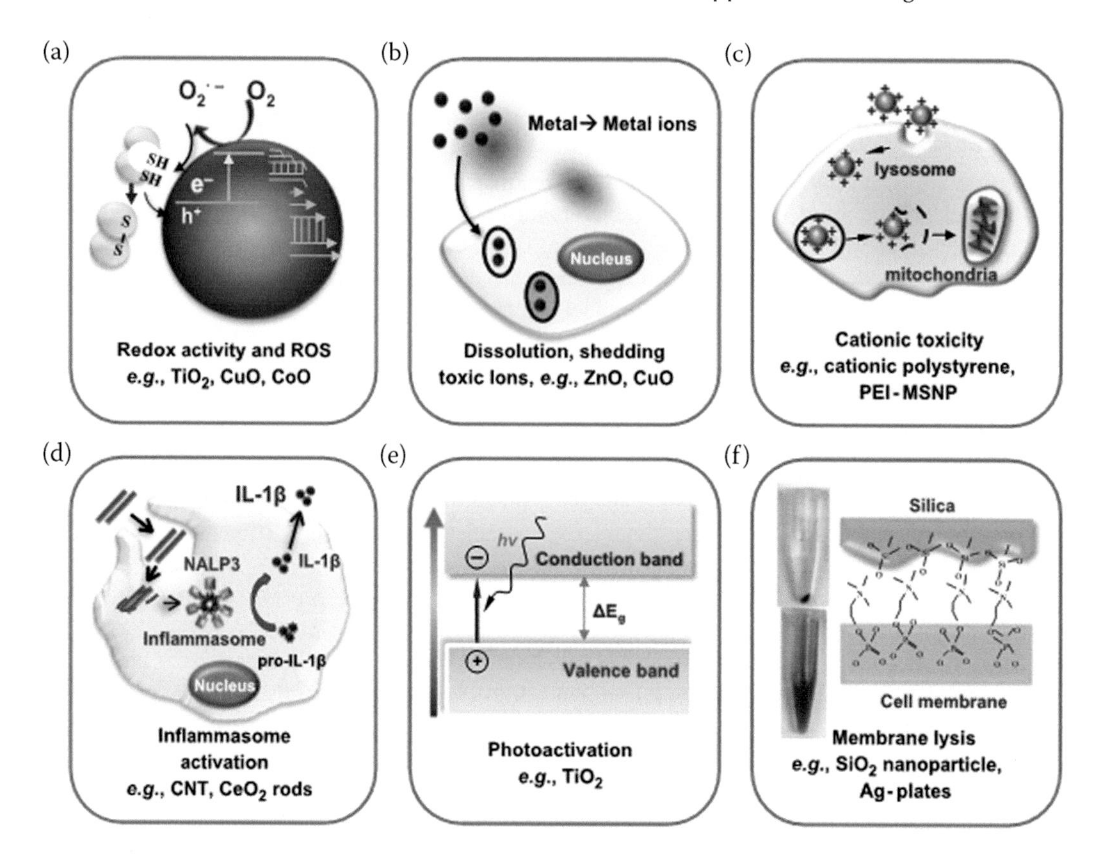

FIGURE 12.3 Pathways of toxicity (POTs) that can be used for in vitro screening, but also have a role in the pathophysiology of disease. POTs are carefully chosen to allow in vitro screening to be predictive of in vivo outcome, thereby enabling large numbers of engineered nanomaterials (ENMs) to be simultaneously assayed, ranked, and then used for focused in vivo investigation. **(a)** induction of oxidative stress by redox active ENMs capable of ROS generation by the material itself or as a consequence of interactions at the nano-bio interface, **(b)** Material dissolution or shedding of toxic metal ions by metal oxide nanoparticles, **(c)** Cationic injury to the surface membrane or the lysosome, **(d)** Activation of the NRLP3 inflammasome by long aspect ratio materials, which also produce fibrosis, **(e)** Photoactivation and generation of electron hole pairs leading to ROS production, **(f)** Membrane lysis by reactive surface chemistry of a variety of materials, figure and caption from (Nel 2013).

For ecotoxicity evaluations, there is a wide range of sensitivities among species and these can vary among toxicants. Hence, it is always useful to quantify the toxicological effects of a substance through several bioassays employing species from different trophic levels (e.g. algae, invertebrate, and vertebrate).

These tests, however, were established with more traditional chemical toxicants in mind, not necessarily nanomaterials. Consequently, some debate has ensued about the utility of existing bioassay methods for nanoparticles. The consensus is that the existing methods and framework for hazard assessment (standard test organisms, mortality, growth, and reproduction endpoints) are generally fit for this purpose, although the specific details within each group of tests may require modification or optimization to work well with nanomaterials (Crane et al. 2008). While some recommendations for incorporating nanomaterials into these bioassays have been made, international consensus has yet to be reached. This is probably due to our relative inexperience assessing the hazards of nanomaterials. The OECD has begun the long process of validating the assessment of nanomaterials with standardized tests (OECD 2010). Further, it has made recommendations regarding nanomaterial sample preparation and dosimetry (OECD 2012). In addition, academic researchers are constantly refining methodologies as the scientific community gains experience in working with ENMs.

Assessing nanoparticle toxicity requires considerations above and beyond those traditionally used for other solutes. Because these materials are not in true solution, quantitative analysis of the toxicant exposure requires more than just assessing the mass concentration. Further difficulties include the fact that during the bioassay some nanomaterials may dissolve, releasing potentially toxic ionic species, or agglomerate or aggregate changing relative particle size. These changes in exposure must be measured if organism response is to be objectively explained with respect to the nanomaterial. Given these observations, confirming that the test organisms have been exposed to the nanomaterial is particularly challenging in the case of ENMs. Further, quantifying the consistency of that exposure can be the Achilles heel of many nanotoxicology research efforts.

In mammalian tests, intravenous injection ensures that the dose is known. This is not true with feeding studies that assume that all particles are taken up from the gastrointestinal tract or inhalation studies that assume that instilled or inhaled particles are taken in through gas-exchange tissues. It is now believed that most nanoparticles do not cross in large quantities into the organism from the gut, lungs, and particularly through the skin.

For ecotoxicity studies, injection is not usually an option nor is it a realistic representation of expected exposure. Hence, exposure must be provided in the media, usually water or soil, and this often results in a scenario that is difficult to characterize and that often changes during the bioassay. In aquatic studies, one way to decrease inconsistency in exposure is to use a flow-through bioassay, which continuously introduces fresh toxicant solution. This approach comes with increased cost, complexity, maintenance, and potential for mishap. The study of nanomaterials in aquatic flow-through bioassay, especially for fish, would require large quantities of potentially expensive nanomaterials and generate a wastewater disposal problem. The alternative is a static-renewal bioassay where the test organisms are transferred to new exposure media regularly. The frequency of transfer usually depends on the fate of the toxicant in the bioassay media and the kinetics of organism waste accumulation. The prime purpose of changing the media is to ensure that the exposure is maintained and that the general quality of the media meets the test organism's requirements (Handy et al. 2012).

In order to be a valid bioassay, aquatic tests are often required to maintain at least 80% of the nominal test concentration during the exposure. This is an important validation criterion, and many regulatory agencies may reject the data if this is not achieved. This criterion may not be achievable for many ENM studies. The experimenter also may be faced with the technical challenge of maintaining several parameters in colloid chemistry simultaneously, including primary particle size and shape, as well as the dispersion state and therefore energy additions to the test system, just to achieve a notional 80% of the target concentration.

In acute bioassays, replacing the test media is an option, but even this may need to be done more frequently compared with traditional chemicals. Federici, Shaw, and Handy (2007) reported that milligram levels of TiO_2 could be maintained at a mass concentration greater than 80% nominal in freshwater for only a few hours. Consequently, water changes may be needed every eight hours. This frequency of solution changes, however, may not be a practical or financially viable proposition for a commercial contract research laboratory. Further, the additional handling of the bioassay organism adds another level of stress that may confound the results of the bioassay. The decision on exactly when, or if, to change the solution should be driven by the characteristics of the test system and should not be some arbitrary decision (e.g. water changes every 24 hours just because it is logistically convenient) (Handy et al. 2012).

Another unresolved issue in both human and environmental health testing is the nanomaterial measurement metric. In addition to simply monitoring, the mass concentration, particle size distribution, particle number, and particle surface area in the test media may need to be monitored to aid data interpretation or simply to give the researcher the option of plotting the data against a dose metric other than concentration (Auffan et al. 2009, Pace, Lesher, & Ranville, 2010, Klaine et al. 2012).

Given the previous discussion, quantifying and characterizing nanomaterial exposure in toxicity bioassays is essential for conducting a valid test, explaining test results, facilitating test duplication, and comparing results with those of other bioassays with different organisms or nanomaterials. While the quantification of mass concentration may be relatively straightforward, little guidance has been given to human toxicologists or ecotoxicologists for further characterization of particle suspensions (Klaine 2009). One proposed approach for a minimum set of parameters for nanomaterial characterization in toxicological studies included (1) appearance—particle size/size distribution, agglomeration state/aggregation, and shape; (2) composition—overall composition (including chemical and crystal structure); and (3) factors affecting interactions of a material with its surroundings—surface area, surface chemistry (including reactivity and hydrophobicity), and surface charge (Minimum Information for Nanomaterial Characterization (MINChar) Initiative 2008).

The small size and large surface area-to-volume ratio of MNPs have resulted in their manipulation to carry a wide variety of drugs to specific regions in the body (e.g. tumors). MNPs are used for the targeted delivery of cancer-targeting antibodies and radionuclides in antiangiogenic and tumor chemotherapy. They are also widely used as contrast agents in MRI. The particles often have many components, including magnetic materials for imaging, heating, and magnetophoresis, a matrix to encapsulate drugs and fluorescent sensors, and a surface coating with cloaking and molecular targeting moieties to improve colloidal stability/circulation time while targeting specific cells.

Nanoparticle toxicity depends upon nanoparticle composition, biodistribution within organs, and local concentration in cells. The biodistribution depends strongly on particle size, shape, and surface chemistry (especially charge/zeta potential and molecular targeting moieties). Distribution of magnetic particles can also be altered using external magnetic fields.

12.3 TOXICITY OF MAGNETIC NANOPARTICLES

In this section, we will first discuss the composition of MNPs and secondly how different properties of these MNPs affect their toxicity in cell culture and small animal studies.

12.3.1 Magnetic Nanoparticle Composition

MNPs have a magnetic component as well as a surface coating, and optionally a matrix loaded with drugs and dyes. Generally, the matrix and coatings are selected to be non-toxic at administered concentrations, and any drug toxicity depends strongly on the drug. Therefore, we will focus here on the magnetic component. Ferromagnetic and ferrimagnetic materials with strong magnetic moments for separation include Fe, Co, Ni, CrO_2, Gd (at < ~20°C), and a variety of alloys including FeCo and FePt. In strong magnetic fields, paramagnetic materials such as Gd and Mn exhibit small but significant magnetophoretic mobility in water. These paramagnetic materials are important in MR imaging and detailed reviews are presented elsewhere (Caravan, Ellison, McMurry, & Lauffer, 1999, Lu et al. 2009, Na et al. 2007, Park et al. 2009). We will focus here on ferromagnetic materials, which have additional applications to drug delivery, separations, hyperthermia, magnetic relaxation assays, and cellular interactions due to their large magnetic moments and magnetophoretic mobility.

Of the ferromagnetic materials, the iron oxide particles magnetite (Fe_3O_4) and its oxidized form maghemite (γ-Fe_2O_3) are by far the most commonly used for in vivo applications due to their stability, ease of synthesis, superparamagnetic properties at small sizes, and relatively low toxicity. Other high-magnetic-moment materials such as nickel and cobalt may be useful in cases where magnetic moment is crucial, such as single-cell tracking in MRI, although Co and Ni are toxic and easily oxidized. Iron oxide is already incorporated in FDA-approved formulations and may be used as bare particles. Particles containing cobalt and nickel in their formulation are often coated with gold or graphene to minimize ion release in these applications.

Most MNPs are composed of an iron oxide core, either magnetite or maghemite. The core may consist of one or more nanoparticles, typically smaller than 20 nm in diameter, so that thermal fluctuations eliminate the remanent magnetic moment on the time scales of the measurements. This loss of average magnetic moment effectively prevents magnetically induced aggregation in the absence of the field (such aggregation could lead to occlusion of fine capillaries, which can be as small as 8 μm in diameter). As such, these are referred to as superparamagnetic iron oxide nanoparticles (SPIONs) in the literature. Iron oxide is metabolized in the hepato-renal system, and the body may then incorporate the iron into the synthesis of hemoglobin molecules (Weissleder et al. 1989).

The goal of drug targeting is to treat malignancy without systemic toxicity. For optimal dosage control, MNPs are often injected loco-regionally for targeted delivery to the site of cancer (Alexiou et al. 2006). Unfortunately, uncoated MNPs are rapidly cleared from the body by the reticuloendothelial system, limiting the effectiveness of the treatment. To increase their effective circulatory half-life and stability, particle surfaces can be functionalized via surface chemistry modifications (Stolnik, Illum, & Davis, 1995). However, increasing the systemic circulation time and the chemical modifications can potentially have harmful side effects.

Surface chemistry must be designed for biocompatibility (non-toxicity) and may be designed to function with a specific cellular receptor (Alexiou et al. 2006). During therapy, this surface layer is what the patient's body recognizes. These surface chemical groups, not the iron oxide core, become the organism-particle interface and dictate any specific biological responses. Hence, nanoparticles are commonly surface-modified with unreactive, branched polymers. In a biological medium (e.g. blood), a corona of proteins will associate to the surface of the particle based on its surface chemistry.

A common theme found in the literature is that MNP toxicity is a function of particle size and surface properties (e.g. charge, functional groups, adsorbed proteins). Size effects are usually relatable to cell type and/or mechanism of uptake (e.g. pore size and endocytosis). Bare IONPs have shown some toxic effects that were thought to be linked to generating reactive oxygen species (Karlsson 2010). Particles coated with polymers such as polyethylene glycol (PEG) are considered relatively non-toxic due to their limited cellular uptake (Goodman, McCusker, Yilmaz, & Rotello, 2004). In general, MNPs with neutral surface chemistry are the most biocompatible, while cationic surfaces are usually more toxic than anionic surfaces (Gupta and Gupta 2005). Although for some applications, such as imaging inflammation, positively charged surfaces in water (negative after immersion in plasma) can increase the uptake by macrophages (Neuberger, Schöpf, Hofmann, Hofmann, & Von Rechenberg, 2005). Other factors contributing to cytotoxicity include pH and solubility. The redox-active iron oxide surface of SPIONs tends to adsorb biological molecules including proteins, amino acids, vitamins, and ions (Mahmoudi et al. 2009a). In order to achieve consistent results, researchers could pre-saturate their SPIONs in a biological medium pertinent to the test environment, for example Dulbecco's Modified Eagle's medium (DMEM). The majority of the in vitro studies included in this review use DMEM.

12.3.2 Cell Culture and Small Animal Studies of Magnetic Nanoparticles

In vitro assays with predetermined endpoints have been used in most attempts to quantify adverse effects of MNPs. These assays are usually restricted to a certain cell type and thus avoid systemic mechanisms associated with particle uptake, metabolism, distribution, and excretion. Cytotoxicity is a common endpoint for many in vitro bioassays. Other endpoints include inhibition of cell-to-plate adherence, stoppage of growth, and membrane damage. Cell viability can be assessed in vitro by examining the ability of each cell to reduce the dye MTT, a yellow tetrazole, to the insoluble formazan, which is purple in color. These and other endpoints have been used to assess in vitro toxicity of SPIONs.

12.3.2.1 Toxicity Studies in Different Cell Lines

The above mentioned in vitro assays can be carried out on a variety of cell lines, including macrophages, cancer lines of various types, fibroblasts, stem cells, and so on. These cell lines may be purchased through the ATCC (American Type Culture Collection) or primary cells can be obtained directly from animals. In addition to composition, the structure of the nanoparticle may also play a role. Dönmez Güngüneş, Şeker, Elçin, and Elçin. (2017) studied the cytotoxicity of IONPs, fullerenes (C_{60}), and single-walled carbon nanotubes (SWCNTs) on different cell types and concluded that the toxicity is highly dependent on the concentration and exposure of the nanoparticles and is also specific to cell type (Dönmez Güngüneş et al. 2017). Some other studies carried out in different cell lines are reviewed below.

12.3.2.1.1 Macrophage Cell Assays

Macrophages are specialized white blood cells that engulf and then digest cellular debris, pathogens, and particulate contaminants in a process called phagocytosis. As such they play an essential role in particle clearance and toxicity as they are the primary mechanism for removal of microparticles from the lungs and circulation. Macrophages also generate reactive-oxygen species (ROS) and inflammatory cytokines in order to deactivate pathogens and degrade particles; cytokines in turn recruit more macrophages by causing differentiation of the monocytes (another type of white blood cells). If the particles are not cleared, chronic inflammation can ensue and the resulting long-term ROS exposure can lead to DNA damage associated with cancer, arthritis, and other diseases. The cascade of effects arising from a macrophage engulfing a particle and unable to digest the material has been well-documented with exposure to asbestos. This has been reviewed and even compared to potential consequences of exposure to carbon nanotubes (Donaldson et al. 2013). This is one reason that much work has been done on nanoparticle–macrophage interactions. For example, tumor-associated peritoneal phagocytes were shown to ingest starch-coated IONPs (average diameter of 100 nm) and deliver them to peritoneal tumors, and it was further shown that this process damages tumor cells (Toraya-Brown et al. 2013). Surprisingly, similar results were seen with dextran- and mannan-coated iron oxide particles. Although the phagocyte uptake was accomplished in vitro and the cells were injected into mice, this research demonstrates the feasibility of this type of therapy. No adverse effects in non-target tissues were reported.

12.3.2.1.2 Fibroblast Cell Assays

Fibroblasts are the most common cells of mammalian connective tissue and are concerned mainly with tissue maintenance. They synthesize collagen and the extracellular matrix (ECM), the structural framework (stroma) for animal tissues, and play a critical role in wound healing. A series of studies have found that internalized dextran-SPIONs (24–72 hours, 50 μg/mL exposure) induce varying levels of cytotoxicity (including cell death and inhibited proliferation) in human dermal fibroblasts by altering tubulin and other cytoskeletal proteins in several ways (Berry, Wells, Charles, & Curtis, 2003, Berry, Wells, Charles, Aitchison, & Curtis, 2004, Gupta and Gupta 2005). This effect was also observed with uncoated SPIONs as well. The pattern of cytoskeletal disruption was altered and the toxic effects eliminated when the same particles were coated with the polysaccharide pullulan. Commonly used as a water-soluble food additive, pullulan was chosen for its non-ionic, non-immunogenic, non-antigenic, and non-toxic properties (Yuen 1974). Transmission electron microscopy (TEM) and immunofluorescence assays revealed a difference in endocytosis based on the presence or absence of pullulan (Gupta and Gupta 2005). The authors suggested that the mechanism of internalization depended upon surface chemistry. However, this could have been a size effect because pullulan-modified particles had mean diameters ranging from 13.6 nm cores to 42.0 nm. Likewise, endocytosis was prevented by the application of lactoferrin and ceruloplasmin coatings to the 13.6 nm SPIONs dosed across a range of 0–1 mg/mL; the protein-coated SPIONs merely adsorbed to the cell membrane (Gupta and Curtis 2004). Fibroblasts exposed to 50 μg/mL of 15 nm albumin-coated SPIONs for 24–72 hours were viable and proliferating, unlike their dextran-coated equivalents, as visualized by bromodeoxyuridine (BrdU) staining, which indicated DNA replication (Berry et al. 2003). In a related group of studies, a 24-hour MTT assay was employed to assess SPION cytotoxicity. At 250 μg/mL, bare particles decreased fibroblast viability by 25–50%; on the other hand, fibroblasts exposed to 1 mg/mL PEG-coated particles remained 99% viable (Gupta and Curtis 2004). The authors posit irreversible endocytosis as the means of uptake. Additionally, SPIONs surface-modified with insulin were not endocytosed. Rather, they adsorbed to the cell membrane, likely to insulin receptors expressed on the surface, resulting in increased cell proliferation (Gupta, Berry, Gupta, & Curtis, 2003, Gupta and Wells 2004). Mahmoudi et al. (2009a) carried out multiple MTT-based cytotoxicity studies on mouse fibroblast (L929) cell morphology. Although 17 nm polyvinyl alcohol (PVA)-SPIONs instigated detachment by 4 hours, cell viability remained acceptable through 48 hours at 0.2, 1, 5, and 20 mM concentrations. Flow cytometry did not detect apoptosis. Furthermore, the study approved the SPIONs as biocompatible after magnetically guiding the ferrofluid through a stimulatory blood vessel (Mahmoudi et al. 2009a). However, by 72 hours, bare 82 nm SPIONs had distorted L929 cell morphology at a concentration of 800 mM. The same bare SPIONs, when pre-soaked in DMEM, did not significantly affect cell shape because the reactive SPION surface was already saturated with DMEM proteins. In addition, polyethylene glycol-*co*-fumarate (PEGF)-coated SPIONs did not harm the cells (Mahmoudi et al. 2009a). In two other MTT studies, 12 nm SPIONs caused little toxicity to various fibroblasts (Brunner et al. 2006, Lee et al. 2007).

12.3.2.1.3 Mesenchymal Stem Cell Assays

Mesenchymal stem cells (MSCs) are multipotent stem cells that can differentiate into a variety of cell types, including osteoblasts (bone-forming cells), chondrocytes (cartilage cells), and adipocytes (fat cells). MSCs are traditionally found in the bone marrow but can also be isolated from other tissues including cord blood, peripheral blood, the fallopian tube, and fetal liver and lung tissue. MTT and comet assay results suggest that SPIONs do not adversely affect apoptotic frequency in human mesenchymal stem cells (hMSCs) (Arbab et al. 2003, Omidkhoda, Mozdarani, Movasaghpoor, & Fatholah, 2007). Ferumoxides (colloidal iron oxide) are used as an MRI contrast agent, particularly for the liver. In a study by Arbab et al. (2003), ferumoxides functionalized with (+)-poly-L-lysine were captured in the endosomes of non-dividing hMSCs. Iron was still detected in cells seven weeks later although it did not appear to alter growth rate or viability as compared to untreated controls (Arbab et al. 2003). Rat MSCs were not adversely affected after 48 hours by 25 or 50 μg/mL of 1-hydroxyethylidene-1.1-bisphosphonic acid (HEDP)-coated SPIONs; however, viability and in vitro differentiation potential of rMSCs diminished to 70% at 100 μg/mL (measured by MTS) (Delcroix et al. 2009).

12.3.2.1.4 Cancerous Cell Assays

MNP studies were also performed on various cell lines from different cancerous organs. Mitochondria were altered in human ovarian (A2780) and breast (MCF-7) cancer cells incubated for 1–3 days in culture medium with both dextrane- and phosphatidylcholine/cholesterol-coated SPIONs at 100 μg/mL. Oxygen uptake diminished and the mitochondrial ultrastructure was altered (Yurchenko et al. 2010). A 72 hours MTS assay of 5.2 nm and 7.6 nm bare SPIONs, from 0.1 μM to 100 μM, resulted in no observable toxicity to either normal or cancerous human breast cells (SKBR3, H184B5F5/M10; T47D, MB157) (Huang et al. 2008b). Human melanoma (SK-MEL-37) cells were interacted with laurate-, citrate-, and DMSA-coated MNPs, resulting in respective half maximal inhibitory concentrations (IC_{50}) of 254, 433, and 2260 μg-iron/mL. The hydrodynamic diameter (D_h) and zeta potential (ζ) also correlated with cytotoxicity (de Freitas et al. 2008).

12.3.2.2 Toxicity Studies Regarding Effect of Particle Properties

Several studies have reported cell toxicity as a function of particle surface chemistry. Among the variables considered are charge (positive, negative, neutral), brush length and presence of various chemical groups, as well as packing density, radius of curvature, and compositions.

12.3.2.2.1 Studies with Different Surface Coatings

MNPs with various coatings manifest different properties and effects, depending upon the type and thickness of the coating. MNPs with 440–890 nm diameter were dispersed in ethanol and coated with low- or high-molecular-weight polylactides (PLA; 2000 g/mol and 152,000 g/mol, respectively) or polylactide/glycolide (PLA/GA; 1:1 ratio; 13,000 g/mol) (Müller, Maaβen, Weyhers, Specht, & Lucks, 1996). Cell viability and particle internalization were tested on suspensions of human granulocytes using the MTT assay (colorimetric assay for mitochondrial reduction) and chemiluminescence (CL), respectively. MTT assays done on microtiter plates revealed the granulocyte EC_{50} at MNP concentrations of 0.38% high-weight PLA, 0.30% low-weight PLA, and 0.15% PLA/GA. The low-weight PLA and the PLA/GA coatings yielded MNPs with the greatest toxicity. Results of the CL assay suggested that cytotoxicity was a function of MNP uptake into granulocytes. In comparison, control solid lipid nanoparticles (SLNs) were less cytotoxic than the polyestered MNPs by tenfold. The adsorbed lipid-poloxamer layers led to longer circulation time in vivo by reducing phagocytic uptake of SLNs (Müller, Maassen, Schwarz, & Mehnert, 1997).

Häfeli et al. (2009) investigated the in vitro toxicity of MNPs coated with a PEG triblock copolymer. Toxicity of the PEG-MNPs was inversely correlated to the PEG tail length and was consistently observed in three human cell lines (prostate cancer cell lines [PC3, C4-2], human umbilical vein endothelial cells [HUVECs], and human retinal pigment epithelial cells [HRPEs]). Biocompatibility for in vivo applications was discerned at a >2 kDa PEG tail length (Häfeli et al. 2009).

Prabhu and others (2015) reported that no significant toxicity was associated with systematic exposure (1/10th dose of LD_{50}) of SPIONs functionalized with PEG. To determine the acute toxicity and LD_{50}, animals were given 350–750 mg/kg body weight of PEGylated SPIONs. Animals injected with a dose of >500 mg/kg died within 24 hours, which was speculated to be due to the primary effects on the cardiovascular system, liver, and an overload of ROS (Prabhu, Mutalik, Rai, Udupa, & Rao, 2015).

Toso et al. (2008) reported no significant differences in cell viability between untreated human islet of Langerhans cells (dendritic cells/antigen-presenting immune cells of the skin and mucosa) and those exposed for 48 hours to 60 nm dextran-SPIONs at 280 μg/mL. These results were successfully extended to in vivo studies to better assess the feasibility of SPION-labeling for use in the clinical MRI monitoring of islet transplants (Toso et al. 2008). In another study, no significant alterations to cell viability were observed in Schwann cells exposed to ≤4 mg/mL of 86 nm dextran-SPIONs for 48 hours (Dunning et al. 2004).

In a study by Chorny et al. (2007), PLA-NPs were loaded with oleate-coated iron oxide nanocrystals and stabilized with a bovine serum albumin (BSA) surface coat. Calcein green staining and alamarBlue® (Thermo Fischer Scientific, Waltham, MA) assays were employed to evaluate the viability of cells exposed to these ~290 nm MNPs. Twenty-four hour BAEC viability at the highest MNP dose (90 μg/mL or 0.3 ng internalized per cell) was 83 ± 3% compared to untreated cells. In an identical experiment, cell survival was 92 ± 2% for a 58 μg/mL (or 0.2 ng internalized per cell) dose (Polyak et al. 2008). The high survival/low toxicity observed in these two studies was attributed to the biomolecular coatings on the MNPs. Chorny et al. (2007) described particles transfecting into cells with plasmid DNA as the outermost layer, while Polyak et al. coated their particles with BSA (Polyak et al. 2008). These bioavailable layers had no adverse effect to cells, unlike the redox-active iron oxide core and charged coatings beneath, because they are biocompatible and likely homologous to biomolecules already in the cells.

In a study that concluded A549 uptake of MNPs was mediated by energy-dependent endocytosis, the IC_{50} of 50 nm silica-encapsulated SPIONs was established at 4 mg/mL (Kim et al. 2006). In another study, H441 lung cell viability remained >80% through 48 hour incubation with 90 μg/mL PEI-SPIONs (63 ± 36 nm) (Mykhaylyk, Antequera, Vlaskou, & Plank, 2007).

Malvindi et al. (2014) analyzed the effects of surface passivation of silica-iron oxide nanoparticles (Fe_3O_4/SiO_2). They used bare, amine-, and sulfonate-silane modified IONPs. The authors deduced that the surface modification limited the toxicity by causing an increase in resistance to lysosomal acidity that in turn lowers the release of iron ions from the nanoparticles, which trigger the production of ROS (Malvindi et al. 2014).

Remya, Syama, Sabareeswaran, and Mohanan (2016) studied the in vivo toxicokinetics and biodistribution of dextran-stabilized IONPs in rats, mice, and guinea pigs through various routes of exposure including oral, dermal, and intraperitoneal ranging from a period of 7 days up to 12 months. They suggest the dextran-IONPs to be non-irritant, non-toxic, and biocompatible with no reports of altered physiological processes, behavior changes, or visible pathological lesions in the tested animals (Remya et al. 2016).

In summary, the thickness and type of the coating used determines the extent of toxicity of the MNPs; biomolecular coatings usually have minimal adverse effects.

12.3.2.2.2 Studies with Uncoated SPIONs

In a study carried out by Perumal et al. (2018), the authors investigated the underlying molecular mechanisms behind neurotoxicity and demonstrated that neuronal cell cycle regulation, DNA damage, repair, and apoptosis were linked to neurotoxicity of bare IONPs. There is an established association between the accumulation of iron and both aging and neurodegenerative disorders such as Parkinson's and Alzheimer's disease (Hagemeier, Geurts, & Zivadinov, 2012). High levels of iron can lead to increased oxidative stress. The authors argue that increased exposure to bare IONPs increases ROS generation, promotes neuronal cell cycle re-entry, causes DNA damage (as the nanoparticles can enter the nuclei), and induces neuronal apoptosis. It is noted that the animals were dosed at 25 and 50 mg/kg of bare IONPs (<50 nm) orally for 30 days (Manickam, Dhakshinamoorthy, & Perumal, 2018).

A six hour treatment of 10 μg/mL bare SPIONs (proprietary size) resulted in repressed astrocyte attachment ability and marked mitochondrial dysfunction. Three-week-old human astrocytes showed a marked increase of 12.8% more MTS production as compared to the untreated control ($p < 0.01$) (Au et al. 2007).

MTT studies in rat hepatic cells (BRL 3 A) exposed to varying concentrations of uncoated SPIONs for 24 hours revealed no mitochondrial-related cytotoxicity up to 100 μg/mL. However, leakage of lactate dehydrogenase (LDH) was induced indicating compromised membrane integrity. A significant effect was seen at 250 μg/mL that was equivalent to the EC_{50}. No significant difference was seen between 30 nm and 47 nm SPIONs (Hussain, Hess, Gearhart, Geiss, & Schlager, 2005).

Bare SPIONs (20–30 nm) at 40 and 80 μg/L caused a significantly increased number of DNA lesions via oxidative stress in human lung epithelial (A549) cells after just 4 hours (Comet assay; $p < 0.05$) (Karlsson 2010). However, other SPION-lung cell studies found differing results using the MTT assay. One-day cell mortality increased in a dose-dependent manner across 20–40 nm SPION concentrations from 0.01 to 100 mg/mL (Horie et al. 2009).

In summary, bare SPIONs are generally considered to be more toxic than coated SPIONs; however, the extent of toxicity of the MNPs also depends upon the cell line being tested and is dose-dependent. Bare SPIONs can induce mitochondrial dysfunction, oxidative stress, and compromised membrane integrity.

12.3.2.2.3 Effect of Particle Surface Charge

Tran, Nguyen, Fox, and Tran (2018) studied the toxicity of chitosan- and PVA-stabilized IONPs. The authors suggest that the bare IONPs have a slight negative surface charge, which causes the otherwise neutral PVA-coated IONPs to have a negative zeta potential of –1.29 mV. The chitosan-coated IONPs had a positive surface charge and were found to be more toxic than PVA (cell viability of 60% vs 80% for PVA at 256 ppm of Fe) (Tran et al. 2018).

Huang et al. (2008a) targeted human hepatocytes (HepG2) via the asialoglycoprotein receptor (ASGPR) expressed on the surface of hepatic cells. Galactose-terminated, amino-functionalized SPIONs (ASPIONs) with an average particle size of 61 nm were produced and later modified by lactosylation (Gal-ASPION; 278 nm) or *N*-acetylation (NAc-ASPION; 302 nm). Despite definite receptor-mediated association of Gal-ASPIONs to ASGPR-expressing liver cells, a low 5d-LD_{50} of 1500 μg/mL was quantified by MTT. Neutral NAc-ASPIONs did not associate with liver cells; neither did Gal-ASPIONs associate with non-ASPGR-expressing 143B liver cells. This study emphasized the crucial factor surface engineering plays in nanoparticle targeting and fate. Nanoparticle surface charge (measured by zeta potential [ζ]) was correlated to cytotoxicity. According to their study, nanoparticles of neutral and negative ζ did not have cytotoxic results, but cytotoxicity was severe with a highly positive ζ (Huang et al. 2008a).

In another study, monkey kidney cells (Cos-7) were incubated for 4 hours with 9 nm (core) HO-$(CH_3)_4N^+$-SPIONs. A Prussian blue staining assay found the particles abundantly present in the cytoplasm, yet no cytotoxicity was observed within a 0.92–23.05 mM range (iron cations concentration), as gauged by MTT assay. Hemolysis was only detected at a significant level (0.5 g/dL) at a 0.1 M NP iron concentration. As such, these SPIONs were deemed kidney-biocompatible (Cheng et al. 2005).

In a study by Wilhelm et al. (2003), 35 nm anionic SPIONs (functionalized by DMSA) had a high affinity to adsorb to the cell membrane of HeLa (human ovarian tumor) cells, and were subsequently internalized by the cells with 30× higher efficiency than commonly used dextran-coated SPION control. Incubating the particles with bovine serum albumin in order to coat the anionic nanoparticles' surface with albumin (causing them to increase to 46 nm in size) significantly reduced the non-specific interactions with the cell membrane and cellular uptake. Comparable results were produced in mouse macrophages (RAW 264.7) as well (Wilhelm et al. 2003).

In a dose-dependent manner, rat pheochromocytoma (PC12) cells exposed to 0.15–15 mM of anionic SPIONs (5–12 nm, dimercaptosuccinic acid [DMSA] surface) lost their viability and ability to execute their normal biological response (extend neurites in the presence of nerve growth factor) (Pisanic, Blackwell, Shubayev, Fiñones, & Jin, 2007).

HeLa cells (human cervical cancer cell line) were incubated for 24 hours with 4.5–10 nm (<200 nm D_h) SPIONs functionalized with four differently charged carbohydrates: dextran (Ø), DMSA (–), heparin (–), and aminodextran (+). Neutrally charged dextran-SPIONs accumulated outside the cell membrane but were not internalized. Anionic SPIONs were internalized only at the highest concentration of 0.5 mg/mL. No toxicity was observed from either of the anionic SPIONs, but heparin-SPIONs caused mitotic spindle disarrangement. Internalized SPIONs were collected into endosomes located around (but never inside) the nucleus where they were easily and consistently visualized by Prussian blue staining and, for the first time, by bright light microscopy (Villanueva et al. 2009). Cationic aminodextran-SPIONs first aggregated around the cell membrane before being efficiently internalized into the cells in a concentration-dependent manner (likely by endocytosis). An earlier study explained MNP uptake in two stages: (1) adsorption of MNPs to the cell membrane, and (2) endocytosis of MNPs into magnetic vesicles, or endosomes (Rivière et al. 2007). There within the endosomes the SPIONs remained visible, without inciting toxicity, even 48 hours after removal from nanoparticle-dosed media. It was actually observed that internalized aminodextran-SPIONs partitioned into daughter cells with cycles of cell division (Villanueva et al. 2009). Another HeLa cell study found intracellular iron disappeared within 5–8 divisions (Arbab et al. 2003). The enhanced internalization and biocompatibility of aminodextran-SPIONs (relative to the other SPIONs) is promising for in vivo cell targeting and tracking via MRI (Villanueva et al. 2009).

Petri-Fink, Chastellain, Juillerat-Jeanneret, Ferrari, and Hofmann (2005) functionalized PVA-SPIONs (9 nm core) with various ratios of amine (+), carboxylic acid (–), and thiol (*S*) groups. Amino-PVA SPIONs associated strongly with the cell membranes of multiple human melanoma cell lines (Me191, Me237, Me275, and Me300) and were non-toxic. The authors paired the favorable uptake of cationic liposomes used for transfection with the preferential uptake of positively charged SPIONs. They also observed an energy-dependent, saturable mechanism of internalization, suspecting an endosomal route of MNP uptake. Mahmoudi et al. (2009a) found 17 nm PVA-SPIONs were biocompatible with K562 human leukemia cells as well as L929 mouse fibroblasts already mentioned.

The studies reviewed above yield contradictory results regarding surface charge. This is likely because interactions between the nanoparticles and cells are affected by adsorption of proteins in the medium. Anionic particles are generally considered to have high affinity to adsorb to the cell membrane and thus are internalized at a greater rate.

12.3.2.2.4 Effect of Particle Size

Size of the nanoparticles can determine the uptake and endocytosis of the particles and hence influence toxicity. A two-week study of human monocytes exposed to dextran-coated SPIONs (5 nm core, 30 nm D_h) resulted in only mild toxicity at the highest dose, 10 mg/mL, but no toxicity at 1 mg/mL. This was discerned by MTT and nitroblue tetrazolium (NBT) assays (Müller et al. 2008). However, an exposure to human macrophages of larger dextran-coated SPIONs (120–150 nm, 100 μg/mL) resulted in 80% mortality at day seven as measured by BrdU staining and 3-(4,5-dimethylthiazol-2-yl)-5-(3-carboxymethoxyphenyl)-2-(4-sulfophenyl)-2H-tetrazolium (MTS) assays (Pawelczyk et al. 2008).

Chorny et al. (2007) were successful in magnetically steering MNPs to transfect plasmid DNA into rat aortic smooth muscle cells (A10). Viability and growth inhibition of A10 cells were evaluated in response to differing doses of various sized PLA-based, magnetite-loaded MNPs. Small (≈185 nm), medium (≈240 nm), and large (≈375 nm) MNPs ($_S$MNP, $_M$MNP, and $_L$MNP, respectively) were surface-modified for DNA binding with an ion-coupled pairing of oleate and polyethylenimine (PEI), yielding a + 40 mV ζ. Though inefficient, cellular uptake of fluorescently tagged MNPs was dose-dependent across the range of 10–50 μg/mL, with $_S$MNPs and $_M$MNPs showing saturation midway at 30 μg/mL but $_L$MNPs remaining quasi-linear. Large MNPs localized to the perinuclear zone most efficiently, avoiding lysosomes. Notably, the magnetic field improved the uptake of MNPs versus non-magnetized $_L$NPs. The alamarBlue® viability assay, a redox-sensitive test similar to the MTT assay and often more sensitive (Hamid, Rotshteyn, Rabadi, Parikh, & Bullock, 2004), was exacted in triplicate 72 hours post-treatment. Very low cytotoxicity occurred, even at 50 μg/mL, where cell viability was 87 ± 2% for all MNP sizes. The same study was carried out on bovine aortic endothelial cells (BAECs) as on A10 rat smooth muscle cells outlined already. Similar trends were observed in BAECs as in A10s, including an even better cell survival rate of 96 ± 2% at the highest dose (50 μg/mL) of all PEI-oleate MNP sizes (~185, ~240, and ~375 nm) (Chorny et al. 2007).

Yang et al. (2015) evaluated the effect of carboxyl-coated IONPs in vivo (mice). They used 10, 20, 30, and 40 nm nanoparticles dosed at 20 mg/kg body weight. No acute toxicity was reported but 10 nm and 20 nm nanoparticles altered gene expression due to oxidative stress and other processes. They also observed size-dependent biodistribution with the smallest nanoparticles cleared predominantly through the liver and the larger nanoparticles cleared mainly by spleen (Yang et al. 2015).

Table 12.1 summarizes the research results discussed in this section.

From the available research results discussed so far, the following conclusions can be drawn:

> **For the nanoparticles, the number of particles present per mg and the surface area are more important than the total mass of the particles.** Because of small size and large surface-to-volume ratios, nanoparticles have more active surface area for interaction. These

interactions can be favorable or unfavorable. Particle size also affects their probability of being endocytosed. Smaller nanoparticles can more easily infiltrate into the cell and can affect functions of different organelles, especially the mitochondria where they have been found to induce oxidative stress as seen with redox-active bare iron oxide particles.

The surface chemistry of these particles modulates molecular and cellular interaction and thus plays an important role in determining their toxicity. In general, bare SPIONs are more toxic than particles modified with biocompatible surface coatings. This is in part due to the redox activity of iron oxide and in part because the coatings reduce nanoparticle aggregation and interactions with cells.

The most biocompatible coatings are those of naturally occurring materials. Bioavailable layers, both synthetic (e.g. long-chain PEG polymer) and naturally occurring such as albumin, pullulan, lactoferrin, and ceruloplasmin, had little or no adverse effect to cells unlike the redox-active iron oxide core because they are biocompatible and likely homologous to biomolecules already in the cells.

Nanoparticle surface charge (measured by zeta potential [ζ]) can be correlated to cytotoxicity. It was observed that mostly nanoparticles of neutral and negative ζ did not have significant cytotoxic results, but cytotoxicity could be severe with a highly positive ζ (Gupta and Gupta 2005).

The toxic effects of MNPs are dose-dependent. The optimal dose can vary for different types of particles depending upon the particle size, material, and surface chemistry.

Most applications require a dose of less than 1 mg/mL. Most in vitro studies so far have shown that the tolerable doses are much greater than 1 mg/mL.

Although in vivo applications can involve more interactions than in vitro, **there is generally a good correlation between in vitro and in vivo studies** (see Section 12.2.2). Because in vitro testing circumvents the uptake process, it is unlikely to predict whole-organism responses. Nevertheless, the results of in vitro studies can serve as guidelines for designing in vivo experiments and elucidating toxicity mechanisms.

12.4 PRE-CLINICAL AND CLINICAL STUDIES

There are many regulatory bodies worldwide that control market distribution and set safe limits of use for various products to be used for human and animal consumption in order to minimize the toxic effects. They also require satisfactory data from pre-clinical and clinical trials prior to their market availability to ensure that these products are safe to use. The Food and Drug Administration (FDA) is the major regulatory body in the United States responsible for protecting the public health by assuring the safety, efficacy, and security of human and veterinary drugs, biological products, medical devices, food supply, cosmetics, and products that emit radiation. Similarly, the European Medicines Agency (EMA) is a European Union agency for the evaluation of medicinal products.

Several MNP formulations have been approved by the FDA for use as MRI contrast agents. The contrast of proton MRI is dependent on the density of nuclear (proton) spins, the relaxation times of the nuclear magnetization (T1, longitudinal and T2, transverse), the magnetic environment of the tissues, and the blood flow to the tissues. Inadequate contrast between normal and infected tissues requires the development of contrast agents (see Chapter 10). Most of the contrast agents affect the T1 and T2 relaxation of the surrounding nuclei, mainly the protons of water. In addition to MRI, MNPs are also being applied to other medications such as drug delivery (Chapter 9), sensors (Chapter 8), and hyperthermia (Chapter 11).

Generally, SPIONs range in size from 4 nm to 12 nm iron core and 10–400 nm hydrodynamic diameter. Wang, Hussain, and Krestin (2001) reviewed that SPIONs are administered in well-tolerated doses of 0.5–8 mg of Fe per kg body weight for MRI contrast agents, up to a total of 157 mg Fe for oral contrast agents, and 200–1000 mg Fe (up to a maximum of 20 mg Fe per kg body weight) for iron replacement therapy. SPION agents are categorized as large SPION

TABLE 12.1
Toxicity Assays for Iron Oxide Nanoparticles in Cell Culture and Animal Studies

Particle Description	Application/Assay Type	Cell Type	Metric (LC_{50}, EC_{50}, etc.)/Effects/Findings	Source(s)
		Effect of Particle Surface Chemistry		
MNPs coated with a polyethylenoxide (PEO) triblock copolymer 0.75–15 kDa tail lengths 8.8±2.7 nm core	MTT assay up to 5 mg/mL, 48 hr	Three human cell lines: prostate cancer cell lines (PC3, C4-2), umbilical vein endothelial cells (HUVECs), and human retinal pigment epithelial cells (HRPEs)	—Biocompatibility for in vivo applications was discerned at a >2 kDa PEG tail length —Pure polymers were more toxic than the MNPs —Pure polymer (0.75–15 kDa) IC_{50}: 0.9 mg/mL PEO-MNPs (15 kDa) IC_{50}: >5 mg/mL PEO-MNPs (2 kDa) IC_{50}: 2 mg/mL	Häfeli et al. (2009)
Starch-, dextran-, and mannan-coated iron oxide nanoparticles (average diameter = 100 nm)	*In vivo:* 280 µg of iron *In vitro:* 56 µg of iron (not a cytotoxicity study)	—NPs injected intraperitoneally into tumor bearing mice (tumor associated peritoneal phagocytes delivering NPs to peritoneal tumors) *In vitro:* Murine ovarian cancer cell line	*In vivo:*—Tumor cells damaged by hyperthermia —No adverse effects in non-target tissues were reported —No significant difference in IONP accumulation in the tumor between the three different coatings *In vitro:* —A dramatic difference due to the coating in IONP uptake by tumor cells (Cell uptake: starch >mannan> dextran)	Toraya-Brown et al. (2013)
MNPs (440–890 nm diameter) coated with low- or high-molecular-weight polylactides (PLA; 2000 g/mol and 152,000 g/mol) or polylactide/glycolide (PLA/GA; 1:1 ratio; 13,000 g/mol)	Particle internalization tested by: —MTT assay for mitochondrial reduction —chemiluminescence (CL) assay	Human granulocytes	*MTT assay:* EC_{50}:3.8 mg/mL (0.38%) high-PLA, 3.0 mg/mL (0.30%) low-PLA, 1.5 mg/mL (0.15%) PLA/GA low-weight PLA and the PLA/GA coated MNPs showed greatest toxicity *CL assay:*	Müller et al. (1997)

(*Continued*)

TABLE 12.1 (Continued)

Particle Description	Application/Assay Type	Cell Type	Metric (LC_{50}, EC_{50}, etc.)/Effects/Findings	Source(s)
			cytotoxicity was a function of MNP uptake into granulocytes —Control solid lipid nanoparticles (SLN) were less cytotoxic than the polyestered MNPs by 10-fold due to reduced phagocytic uptake and longer circulation time in vivo	
Pullulan-coated SPIONs (42±2.5 nm diameter, 13.6 nm core) 1.71×10^{17} particles/g MNP	0–2 mg/mL, 24 hr immunofluorescence assays, TEM, MTT assay	Human dermal fibroblasts (hTERT-BJ1)	No toxic effects, difference in endocytosis between coated and uncoated SPIONs, (pullulan is non-ionic, non-immunogenic, non-antigenic and non-toxic) 92% viable relative to control at concentration as high as 2.0 mg/mL IC_{50}: >2 mg/mL	Gupta and Gupta (2005)
Lactoferrin- and ceruloplasmin-coated SPIONs (13.6 nm core)	0–1000 μg/mL, 24 hr MTT assay	Human dermal fibroblasts	The protein-coated SPIONs merely adsorbed to the cell membrane No endocytosis IC_{50}: >>1 mg/mL	Gupta and Curtis (2004)
Dextran and albumin-coated SPIONs (10 nm)	24 hr, 0.05 mg/mL BrdU staining and fluorescence microscopy	Human dermal fibroblasts (hTERT-BJ1)	Fibroblasts were viable and proliferating after 24–72 hr exposure to 50 μg/mL of albumin-SPIONs Bare and dextran SPIONs caused some cell death IC_{50}: 0.05 mg/mL for dextran-SPIONs IC_{50}: >0.05 mg/mL for albumin-SPIONS	Berry et al. (2003, 2004)
	1 mg/mL exposure, 24 hr MTT assay		99% viability at 1000 μg/mL after 24 hr	

(Continued)

TABLE 12.1 (Continued)

Particle Description	Application/Assay Type	Cell Type	Metric (LC_{50}, EC_{50}, etc.)/Effects/Findings	Source(s)
PEG-coated particles 40–50 nm		Human dermal fibroblasts (hTERT-BJ1)		Gupta and Wells (2004)
Insulin-coated SPIONs Size: not mentioned, <20 nm	Up to 1 mg/mL, 24 hr MTT assay	Human dermal fibroblasts (hTERT-BJ1)	No endocytosis, adsorbed to cell membrane, likely to insulin receptors resulting in increased cell proliferation IC_{50}: >>1 mg/mL for insulin-SPIONs IC_{50}: >250 μg/mL for uncoated SPIONS	Gupta et al. (2003)
Polyethylene-glycol-*co*-fumarate (PEGF)-coated SPIONs 82±12 nm, (ζ 3.24)	44.676 mg/mL (800 mM Fe), 72 hr	Mouse fibroblast (L929) cells	No harm to cells after 72 hr exposure to 4.468 mg/mL iron	Mahmoudiet al. (2009b)
PLA-NPs loaded with oleate-coated iron oxide nanocrystals and stabilizing with BSA surface coat 290 nm	24 hr, up to 90 μg/mL exposure Calcien green staining and alamarBlue assays	Bovine aortic endothelial cells (BAECs)	—83±3% viability at 90 μg/mL or 0.3 ng internalized per cell after 24 hr —92±2% viability at 58 μg/mL or 0.2 ng internalized per cell after 24 hr	Jain et al. (2008)
Dextrane- and phosphatidylcholine/ cholesterol-coated SPIONs	100 μg/mL, 1–3 days	Human ovarian (A2780) and breast (MCF-7) cancer cells	Oxygen uptake diminished and the mitochondrial ultrastructure altered at 100 μg/mL after 1–3 days	Yurchenko et al. (2010)
Laurate-coated MNPs	MTT assay and TEM	Human melanoma (SK-MEL-37) cells	IC_{50}: 254 μg-iron/mL D_h and ζ correlated with cytotoxicity	de Freitas et al. (2008)
Citrate-coated MNPs	MTT assay and TEM	Human melanoma (SK-MEL-37) cells	IC_{50} :433 μg-iron/mL D_h and ζ correlated with cytotoxicity	de Freitas et al. (2008)
DMSA-coated MNPs	MTT assay and TEM	Human melanoma (SK-MEL-37) cells	IC_{50} : 2.3 mg-iron/mL D_h and ζ correlated with cytotoxicity	de Freitas et al. (2008)
50 nm silica-encapsulated SPIONs	TEM study	Human lung epithelial (A549) cells	IC_{50}: 4.0 mg/mL uptake of MNPs was mediated by energy-dependent endocytosis	Kim et al. (2006)
PEI-SPIONs (63±36 nm)	48 hr incubation with 90 μg/mL	H441 lung cells	Viability remained >80% after 48 hr incubation with 90 μg/mL PEI-SPIONs	Mykhaylyk et al. (2007)

(*Continued*)

TABLE 12.1 (Continued)

Particle Description	Application/Assay Type	Cell Type	Metric (LC_{50}, EC_{50}, etc.)/Effects/Findings	Source(s)
Dextran-coated SPIONs (<25 nm core, 40–160 nm D_h)	*In vivo:* 25–3000 mg/kg body wt., 7 days to 12 months, clinical evaluations, histopathological analysis of organs postmortem *In vitro:* 100–800 µg/mL, 24 hr, MTT assay	*In vivo* (mice, rats, guinea pigs): Oral, dermal, and intraperitoneal exposure, evaluations at molecular, cellular, genetic, and immune levels. *In vitro:* L929 fibroblast cells	*In vivo:* No visible pathological lesions, allergic reactions, altered physiological processes or behavior changes observed *In vitro:* Cell viability >80% at 800 µg/mL after 24 hr	Remya et al. (2016)
Silica-SPIONs (12 nm core, 7 nm silica shell, 32 nm D_h, −24 mV ζ): -amine modified silica-SPIONs-sulfonate-silane modified silica SPIONs	0.1–5 nM NPs conc. 48–96 hr exposure WST-8 assay (cell viability) LDH assay (membrane integrity) DCF assay (ROS generation) Comet assay (genotoxicity)	A549 and HeLa cells	Surface-modified silica-SPIONs did not show signs of toxicity. Cytotoxicity correlated with NP uptake. Unmodified silica-SPIONs release twice the amount of Fe ions than the surface-passivated NPs.	Malvindi et al. (2014)
PEG-coated SPIONs (16 nm, −30.49 mV ζ) and bare SPIONS (11 nm, −2.65 mV ζ)	*In vivo:* 350–750 mg/kg body wt. intravenous injections, up to 14 days, serum biochemical parameters, histopathology	*In vivo* (mice): Acute intravenous toxicity, genotoxicity, biodistribution studies	LD_{50} : 508 ± 41 mg/kg body wt. for PEG-SPIONs PEGylation reduces toxicity of bare SPIONs	Prabhu et al. (2015)
		Bare SPIONs		
Bare SPIONs	24 hr, 250 µg/mL MTT assay	Human dermal fibroblasts (hTERT-BJ1)	Fibroblast viability decreased by 25–50% LC_{50}: 250 µg/mL	Gupta and Wells (2004)
Bare SPIONs (proprietary size by NanoSonics)	6 hr, 10 µg/mL LDH and MTS assay	Three-week-old human astrocytes	Repressed young astrocyte attachment ability and altered mitochondrial functioning	Au et al. (2007)
Bare SPIONs (13.6 nm core) 1.71×10^{17} particles/g MNP	0–2 mg/mL, 24 hr immunofluorescence assays, TEM, MTT assay	Human dermal fibroblasts (hTERT-BJ1)	Toxic even at 0.05 mg/mL Disrupted cytoskeleton 60% loss of cell viability at 2 mg/mL IC_{50}: <2 mg/mL	Gupta and Gupta (2005)

(*Continued*)

TABLE 12.1 (Continued)

Particle Description	Application/Assay Type	Cell Type	Metric (LC_{50}, EC_{50}, etc.)/Effects/Findings	Source(s)
Bare SPIONs (30 and 47 nm)	24 hr, up to 250 μg/mL MTT and LDH assays	Rat hepatic cells (BRL 3A)	—No mitochondrial-related cytotoxicity up to 100 μg/mL —Leakage of lactate dehydrogenase (LDH) was induced indicating compromised membrane integrity at 100–250 μg/mL —EC_{50}: >250 μg/mL —No significant difference seen between 30 and 47 nm SPIONs	Hussain et al. (2005)
Bare SPIONs 82 nm	44.676 mg/mL (800 mM), 72 hr, MTT assay	Mouse fibroblast (L929) cells	Distorted cell morphology after 72 hr exposure to 44,676 μg/mL of Fe	Mahmoudi et al. (2009b)
Bare SPIONs 82 nm (pre-soaked in DMEM) (−21.4 mV ζ)	44.676 mg/mL (800 mM), 72 hr, MTT assay	Mouse fibroblast (L929) cells	No effect on cell shape as the reactive SPION surface was already saturated with DMEM proteins after 72 hr exposure to 44,676 μg/mL of Fe	Mahmoudiet al. (2009b)
Bare SPIONs calculated diameter: 12 nm Hydrodynamic diameter from XRD: 50 nm specific surface area: 93 m^2/g	0–30 ppm NPs (μg/g≈μg/mL) 3 days —MTT assay —DNA Hoechst assay for cell proliferation	Human mesothelioma MSTO-211H and rodent 3T3 fibroblast cells	—Cell-type specific response—Slower proliferating 3T3 cells slightly affected by addition of up to 30 ppm (≈30 μg/mL) iron oxide —Faster growing MSTO cells drastically reduced upon exposure to as little as 3.75 ppm (≈3.75 μg/mL) iron oxide —IC_{50} (MSTO): 3.75 μg/mL —IC_{50} (3T3): >30 μg/mL	Brunner et al. (2006)
Surfactant-coated (oleic acid and ethylene glycol) SPIONs 8.7–12 nm core	1 μg/mL, 24 hr incubation MTT assay	Human skin fibroblast HS 68 cells	—No significant difference in cell viability compared to the control group —IC_{50}: >1 μg/mL	Lee et al. (2007)
Bare SPIONs 5.2 and 7.6 nm	0.1, 1, 10, 100 μg/mL 72 hr MTS assay	Normal and cancerous human breast cells (SKBR3, H184B5F5/M10; T47D, MB157)	No observable toxicity IC_{50}: >100 μg/mL	Huang et al. (2008b)

(Continued)

TABLE 12.1 (Continued)

Particle Description	Application/Assay Type	Cell Type	Metric (LC_{50}, EC_{50}, etc.)/Effects/Findings	Source(s)
Bare SPIONs 20–30 nm	40 and 80 μg/mL, 4 hr Comet assay	Human lung epithelial (A549) cells	Caused a significantly increased number of DNA lesions via oxidative stress	Karlsson (2010)
Ultrafine NiO particles 20 nm (S.A. 50–80 m^2/g) 100 nm (S.A. >650 m^2/g)	0.1–10 mg/mL, 24 hr LDH and MTT assay	Human lung carcinoma A549 cells and human keratinocyte HaCaT cells	One-day cell mortality increased in a dose-dependent manner from a dose of 100–100,000 μg/mL due to release of Ni^{2+} by the UFPs	Horie et al. (2009)
		Effect of Particle Surface Charge		
Anionic SPIONs (5–12 nm, dimercaptosuccinic acid [DMSA] surface)	8.38–837.68 μg/mL (0.15–15 mM) exposure	Rat pheochromocytoma (PC12) cells	—Dose-dependent diminishing cell viability and ability to execute their normal biological response —Viability ≥60% at 0.15 and 1.5 mM after 48 hr —Viability <20% at 15 mM after 48 hr	Pisanic et al. (2007)
61 nm (−43.9 mV ζ) amino-functionalized SPIONs modified by: —lactosylation (Gal-ASPION; 278 nm) —*N*-acetylation (NAc-ASPION; 302 nm) neutral	MTT assay	—ASGPR expressing human hepatocytes (HepG2) —Non-ASPGR-expressing 143B liver cells	—LD_{50} of 1500 μg/mL for Gal-ASPION-278 —Receptor-mediated association and subsequent endocytosis of Gal-ASPIONs to ASGPR-expressing liver cells —Neutral NAc-SPIO-302 are not taken up by HepG2 cells —Both neutral NAc-ASPIONs and Gal-ASPIONs did not associate with non-ASPGR-expressing 143B liver cells —NPs of neutral and negative ζ did not have cytotoxic results, but cytotoxicity was severe with a highly positive ζ	Huang et al. (2008a)
9 nm (core) HO-$(CH_3)_4N^+$-SPIONs	4 hr, 51.4–1287 μg/mL (0.92–23.05 mM) range (iron cations concentration)	Monkey kidney cells (Cos-7)	—No cytotoxicity to Cos-7 —Hemolysis (0.5 g/dL) of 1 mL human whole blood at a 0.1 M (5580 μg/mL) NP iron concentration	Cheng et al. (2005)

(Continued)

TABLE 12.1 (Continued)

Particle Description	Application/Assay Type	Cell Type	Metric (LC_{50}, EC_{50}, etc.)/Effects/Findings	Source(s)
	Prussian blue staining and MTT assay			
PLA-based, magnetite-loaded MNPs with PEI-oleate surface modification (+40 mV ζ) 185, 240 and 375 nm	Alamar Blue 72 hr 10–50 μg/mL dosage	Rat aortic smooth muscle cells (A10)	—Dose-dependent cellular uptake improved by the magnetic field —Very low cytotoxicity even at 50 μg/mL —Cell viability 87±2% for all MNP sizes	Chorny et al. (2007)
PEI-oleate MNPs (+40 mV ζ) 185, 240, and 375 nm	alamarBlue 72 hr 10–50 μg/mL dosage	Bovine aortic endothelial cells (BAECs)	96±2% cell survival rate at 50 μg/mL of all PEI-oleate MNP sizes	Chorny et al. (2007)
—35 nm anionic SPIONs (functionalized by DMSA) —Dextran-MNPs (Ferumoxtran) for comparison	15 min to 12 hr at 4 and 37°C 0.1–20 mM Fe Magnetophoresis and Electron Spin Resonance Assays for kinetics of cellular uptake	HeLa (human ovarian tumor) cells and mouse macrophages (RAW 264.7)	—3× more efficient internalization of bare anionic maghemite nanoparticles by cells than commonly used dextran-coated SPION control —AMNP uptake saturation at 15 mM Fe for HeLa cells and 1.5 mM for macrophages —Dextran-coated particles show non-saturable internalization —Tweaking the particle coating with BSA (now 46 nm size) significantly reduced the non-specific interactions with the cell membrane and cellular uptake	Rivière et al. (2007)
4.5–10 nm (<200 nm D_h) SPIONs functionalized with four differently charged carbohydrates: dextran (Ø), DMSA (−), heparin (−), and aminodextran (+)	Up to 500 μg/mL 24 hr incubation Prussian blue staining and bright light microscopy	HeLa cells	—Neutrally charged dextran-SPIONs accumulated outside cell membrane but not internalized —Anionic SPIONs internalized only at 500 μg/mL —No toxicity observed from either of the anionic SPIONs —Cationic aminodextran-SPIONs efficiently	Villanueva et al. (2009)

(*Continued*)

TABLE 12.1 (Continued)

Particle Description	Application/Assay Type	Cell Type	Metric (LC_{50}, EC_{50}, etc.)/Effects/Findings	Source(s)
			internalized into the cells in a concentration-dependent manner and partitioned into daughter cells with cycles of cell division	
PVA-SPIONs (9 nm core) functionalized with various ratios of amine (+), carboxylic acid (–), and thiol (*S*) groups	*MTT assay* *exposure:* 1–10×10^{12} NPs/mL (equivalent to 12–145 µg iron/mL) for 2 hr	Multiple human melanoma cell lines (Me191, Me237, Me275, and Me300)	—Only amino-PVA SPIONs interacted strongly with the cell membranes but were non-toxic —IC_{50}: 145 µg/mL	Petri-Fink et al. (2005)
Ferumoxides, functionalized with (+)-poly-L-lysine (size not mentioned)	—25 µg/mL ferumoxides overnight (>12 hr) incubation —MTT assay for cell viability —ROS (fluorescent probe CM-H_2DCFDA) assay —Flow cytometry (for apoptosis) —MRI	Human mesenchymal stem cells and HeLa cells	—IC_{50}: >25 µg/mL —Iron was still detected in non-dividing mesenchymal cells after 7 weeks but did not alter growth rate or viability —In rapidly dividing cells, intracellular iron disappeared by five to eight divisions —No increase in rates of apoptosis or ROS formation —Very good labeling efficiency with very low concentration of iron oxide (25 µg/mL)	Arbab et al. (2003)
1-hydroxyethylidene-1.1-bisphosphonic acid (HEDP)-coated SPIONs 20 nm (4 nm core)	—24–48 hr, 25, 50, and 100 µg/mL —MTS assay —In vivo MRI	Rat MSCs	—No adverse effect at 25 or 50 µg/mL —Viability and in vitro differentiation potential diminished to 70% at 100 µg/mL —IC_{50}≈50 µg/mL	Delcroix et al. (2009)
SPIONs stabilized with: —Chitosan, +41 ± 4.5 mV ζ (5–10 nm core, 63 ± 26 nm D_h) —PVA, −1.3 ± 0.7 mV ζ (10–15 nm core, 70 ± 4 nm D_h)	—0.1–256 ppm Fe conc. —24 hr incubation —MTS assay	Mouse fibroblast cells (3T3)	—Dose-dependent toxicity observed for both types of NPs —Chitosan-SPIONs EC_{50}: 32 ppm, cell viability 60% at 256 ppm —PVA-SPIONS EC_{50}: 64 ppm, cell viability 80% at 256 ppm	Tran et al. (2018)
		Effect of Particle Size		
		Human monocytes		Müller et al. (2008)

(Continued)

TABLE 12.1 (Continued)

Particle Description	Application/Assay Type	Cell Type	Metric (LC_{50}, EC_{50}, etc.)/Effects/Findings	Source(s)
Dextran-SPIONs (5 nm core, 30 nm D_h)	1–10 mg/mL nitrobluete trazolium (NBT) and MTT assays		—Mild toxicity at 10 mg/mL (10,000 µg/mL) —No toxicity at 1 mg/mL (1000 µg/mL)	
PVA-SPIONs 17 nm	MTT-based cytotoxicity study on cell morphology 11.17, 55.85, and 279.23 µg/mL (0.2, 1, 5, and 20 mM), 48 hr	Mouse fibroblast (L929) cell	—No apoptosis detected by flow cytometry —cell viability remained acceptable through 48 hr at 0.2, 1, 5, and 20 mM concentrations —Biocompatible with K562 human leukemia cells as well as L929 mouse fibroblasts	Mahmoudiet al. (2009b)
Dextran-SPIONs 60 nm	48 hr, 280 µg/mL exposure	Human islet of Langerhans cells	No significant difference in cell viability	Toso et al. (2008)
Dextran-SPIONs 86 nm	48 hr, 4000 µg/mL exposure	Schwann cells (from 2-day-old Fischer rat pups)	No significant difference in cell viability exposed to ≤4000 µg/mL for 48 hr	Dunning et al. (2004)
Dextran-SPIONs 120–150 nm	100 µg/mL BrdU stain and MTS assay	Human macrophages	80% mortality at day 7	Pawelczyk et al. (2008)
Carboxyl-SPIONs 10, 20, 30, 40 nm	20 mg/kg body wt. tail vein injections 1–7 days	*In vivo* (mice): biodistribution, histopathology, hematology and PCR analysis	—No acute toxicity observed —Size-dependent biodistribution and transportation: 10 nm highest accumulation in liver, 40 nm highest accumulation in spleen —No histopathological abnormalities or lesions —10–20 nm SPIONs altered gene expression of susceptible genes (pcsk9 and Hmox1)	Yang et al. (2015)

(LSPIONs) SPION agents, standard SPION (SSPIONs) agents, ultrasmall SPION (USPIONs) agents, and monocrystalline iron oxide nanoparticle (MION) agents depending on their sizes with the coating material (about 20 nm to 3500 nm in diameters). Large iron oxide particles are selected for bowel contrast (AMI-121 [i.e. Lumirem and Gastromark] and ferristene [i.e. Abdoscan] [mean diameter no less than 300 nm]), and for liver/spleen imaging (AMI-25 [i.e. Endorem and Feridex IV, diameter 80–150 nm]; SHU 555 A [i.e. Resovist, mean diameter 60 nm]). Smaller iron oxide particles are chosen for lymph node and bone marrow imaging (AMI-227 [i.e. Sinerem and Combidex, diameter 20–40 nm]), perfusion imaging and MR angiography (NC100150 [i.e. Clariscan, mean diameter 20 nm]). The monocrystalline iron oxide nanoparticles are under research for receptor-directed MRI and magnetically labeled cell probe MRI. All iron oxide particles for IV injection are biodegradable and oral SPIONs are coated with insoluble material (Wang *et al.* 2001).

Among the approved SPIONs are Feridex IV (ferumoxides injectable solution, a.k.a. Endorem® [EU]), Resovist® (a.k.a. Cliavist), Sinerem (a.k.a. Combidex), Lumirem (a.k.a. Gastromark), Clariscan, and Feraheme. Most of these formulations have been discontinued, either due to safety concerns (e.g. aggregation of blocked magnetic particles in the MRI's magnetic field has the potential to cause embolism) or competing Gd-based formulations, with the exception of oral MRI contrast agents and Feraheme (Wang 2011).

Ferumoxides (dextran-coated) have been tested in clinical trials as negative contrast agents that decrease signal on T2 images, and have been used in liver, spleen, and myocardial perfusion MRI (Leung 2004a). Ferumoxide-enhanced MRI considerably increases tumor and liver contrast and improves tumor lesion recognition on T2-weighted images (Bartolozzi, Lencioni, Donati, & Cioni, 1999). A 95% sensitivity for metastatic tumor detection was reported by using ferumoxide-enhanced MRI (Senéterre et al. 1996). In a large multi-center trial, additional liver tumors were identified on 27% of ferumoxide-enhanced MR images compared with non-enhanced series (Ros et al. 1995). Feridex and Resovist were both approved specifically for MRI of the liver. The difference is that Resovist can be administered as a rapid bolus (and thus can be used with both dynamic and delayed imaging), whereas Feridex should be administered as a slow infusion and is used solely in delayed-phase imaging. Also, the safety profile appears to be more favorable for Resovist (Wang 2011).

The oral contrast agent ferumoxsil (GastroMARK®) is effective in gastrointestinal MRI for demarcation of normal and pathologic structures in the bowel. Ferumoxsil 10 nm iron oxide core coated with a non-biodegradable and insoluble matrix (siloxane) and suspended in viscosity-increasing agents such as starch and cellulose is used in oral large SPIO preparations as a negative contrast agent that decreases signal on T2 images (Leung 2004b). In an MRI study of AMI-121 in 15 volunteers, loss of T2-weighted enhancement signal intensity was observed in the proximal and distal small bowel. Enhanced images showed improved marking of the head and tail of the pancreas, anterior regions of the kidneys, and para-aortic region (Hahn et al. 1990). AMI-121 MRI was helpful in discerning normal bowel from solid lesions and in identifying slight gastrointestinal tract mass change. In a study of 30 consecutive patients suspected of having gastrointestinal pathology, oral contrast-enhanced computed tomography was more sensitive than AMI-121 MRI in detecting abdominal pathology. Conversely, AMI-121 MRI was more specific than computed tomography (Johnson, Stoupis, Torres, Rosenberg, & Ros, 1996).

SPIONs are also administered intravenously in the management of anemia in patients with chronic kidney disease (CKD) to provide supplemental iron. Currently available IV iron preparations include low-molecular-weight iron dextran (CosmoFer), sodium ferric gluconate (Ferrlecit), iron sucrose (Venofer), iron carboxymaltose (Ferinject/Injectafer), ferumoxytol (Feraheme), and iron isomaltoside 1000 (Monofer). Many of these formulations do not require a test dose and can be administered as a large bolus injection (Macdougall 2009). Ferric carboxymaltose, licensed in Europe in 2007, is a dextran-free iron compound with low immunogenic potential and limited labile iron toxicity. When administered as a slow intravenous push, quantities as high as 750 mg of iron may be administered at the rate of approximately 100 mg (2 mL) per minute for patients weighing 50 kg (110 lb) or more (Luitpold Pharma. 2013). Ferumoxytol was developed in the United States initially as an MRI contrast agent and has been licensed by the FDA (AMAG Pharma. 2009) for use in CKD. It consists of a highly bioavailable IONP with minimal immunological reactivity and minimal, detectable, and releasable iron. It can be administered in doses of 510 mg over 17 seconds (Spinowitz et al. 2008). Because of their small size, low molecular weight, and semi-synthetic carbohydrate coating, ferumoxytol nanoparticles (hydrodynamic diameter of 30 nm) are attractive for other potential biomedical applications such as magnetic fluid hyperthermia (MFH), cell tracking, loading optimization, nanomagnetic cellular actuation, and nanomagnetic drug and gene delivery. The carboxylated polymer coating affords functionalization of the particle surface and the small size increases accumulation in highly vascularized tumors via the enhanced permeability and retention effect as they escape the body's reticuloendothelial surveillance system (RES). Bullivant et al. (2013) showed that the significant and rapid heating of ferumoxytol nanoparticles above ambient temperature led to the potential to induce cell apoptosis and necrosis if delivered to solid tumors at appropriate concentrations (see Chapter 11) (Bullivant et al. 2013).

The small size and hydrophilic coating of ferumoxtran-10 (Combidex [U.S.], Sinerem [EU]) results in a longer intravascular circulation and allows the particles to escape rapid accumulation in the RES. These particles are phagocytosed by macrophages and accumulate in the lymphatic system over 24–36 hours, thus post-contrast imaging is usually obtained 24 hours after administration of the contrast agent (Wang 2011). Although Harisinghani et al. (2003) presented extraordinary results of the way in which ferumoxtran-10 could exhibit the presence of positive lymph nodes in patients with prostate cancer, a recent multi-center study showed a 24.1% false positive rate, leading to unnecessary surgical intervention. Heesakkers et al. (2009) evaluated the use of ferumoxtran-10 and MRI to identify lymph node metastases occurring outside the normal area of pelvic lymph node dissection in 296 patients with prostate cancer. All patients had intermediate to high risk for nodal metastases. Clinical development was discontinued due to insufficient data to differentiate metastatic from non-metastatic lymph nodes although the ferumoxtran-10 safety profile appeared favorable as a contrast agent (Heesakkers et al. 2009).

The final utility of the iron oxide particles in cardiovascular applications is still not established and further work is required. The distinctive intravascular characteristics of SPIO contrast agents in non-pathologic vessels and the uptake in macrophages afford applications not possible with the Gd-chelates that are in common use (Bjørnerud and Johansson 2004). The Gd-based MNPs for MRI include gadodiamide (Omniscan), gadobenic acid (MultiHance), gadopentetic acid (Magnevist), gadoteridol (ProHance), gadofosveset (Vasovist, Ablavar), gadoversetamide (OptiMARK), gadoxetic acid (Eovist [US], Primovist in other parts of the world), Gadoteric acid (Dortarem [Guerbet 1989] and Artirem [Guerbet 2002]). Guerbet is also developing another Gd-based MRI contrast agent, Vistarem.

MRI has also been successfully used for detection of MSCs labeled with superparamagnetic iron oxide particles. Although preliminary studies reported the labeling of MSCs with SPIO to be safe without affecting cell biology, recent studies have found effects of SPIO labeling on MSC metabolism and functional biology (Schäfer et al. 2010). Although SPIO labeling does not appear to affect cell viability or differentiation (Bos et al. 2004, Schäfer et al. 2009), there may be an impact on iron metabolism, colony formation capacity, and migration ability (Schäfer et al. 2007, Schäfer *et al.* 2009, Schäfer *et al.* 2010, Pawelczyk, Arbab, Pandit, Hu, & Frank., 2006). SPIO stem cell labeling is not an FDA-approved indication as of the date of this publication (Wang 2011).

The composition and preparation method of the contrast particles may also play a role. Gu, Fang, Sailor, and Park (2012) compared iron oxide nanocrystals prepared via thermal decomposition of organometallic precursors with those prepared using the aqueous precipitation method. They suggest that the nanocrystals synthesized by thermal decomposition show an increase in MR signal but also have a longer in vivo residence time that may cause long-term toxicity effects (Gu et al. 2012).

In addition to MRI, magnetic microparticles have also been used in human patients to study the uptake and clearance of particles in the lungs. The work dates back to Valberg and Brain (1988) who discovered that iron oxide particles inhaled by workers during welding could be magnetized by an external magnetic field and detected non-invasively through tissue. The work was expanded by Möller et al. (2001), who had patients inhale small quantities (1 mg) of iron oxide microparticles and measured the magnetization, relaxation rate, and clearance by magnetopneumography for a subsequent 300 days. They found that patients with idiopathic pulmonary fibrosis and sarcoidosis exhibited reduced clearance rates (half-lives of 275 ± 109 days and 756 ± 345 days, respectively) when compared to healthy individuals. Clearance half-lives for young, healthy non-smokers and smokers were 124 ± 66 and 220 ± 74 days respectively while those for older subjects were 162 ± 120 for non-smokers and 459 ± 334 for smokers, which further demonstrated that cigarette smoking reduced alveolar clearance (Möller et al. 2001). The work was complemented by cell culture studies using cell magnetometry, which show that intracellular transport relies upon cytoskeletal structure and is affected by ROS-generating nanoparticles that attack the cytoskeleton (Möller et al., 2000, Möller et al., 2002, Nemoto and Moeller 2000). This all demonstrates the utility of magnetic particles for non-invasively investigating the health of patients' lungs, augmenting the findings from far more common MRI.

Magnetic particle imaging (MPI) is an emerging medical imaging modality that uses non-radioactive, "kidney-safe" SPIO tracers. Keselman et al. (2017) used MPI to track the short-term biodistribution and long-term clearance of two SPIO nanoparticles: Ferucarbotran and LS-008. The latter is a long circulating tracer cleared mainly through spleen with a half-life of 18.2 days as compared to ferucarbotran, which is cleared predominantly through the liver with a half-life of about 5.6 days. Through in vivo assays they demonstrated that three-dimensional MPI was able to quantitatively assess short-term biodistribution and long-term tracking and clearance of these tracers (Keselman et al. 2017).

Krishnan et al. (2015) reviewed the various studies of iron oxide nanoparticles focusing on their toxicity, biodistribution, and pharmacokinetics, and summarized post-injection blood half-lives of different types of IONPs. They suggest that the in vivo performance of the IONPs depend on their characteristics such as size, shape, morphology, surface coating, and charge (Arami, Khandhar, Liggitt, & Krishnan, 2015).

In conclusion, magnetic particles hold promising biomedical applications and are currently being safely administered as oral MRI SPIO formulations and as supplemental IV iron solutions. The iron oxide particle size of the SPIO MRI contrast agents varies widely, which affects the physicochemical and pharmacokinetic properties, and thus their clinical application. In vitro studies have suggested tolerable doses of iron MNPs in ranges much greater than 1 mg/mL (1 g/L ≈ 1 g/kg) in most cases. So far, iron formulations have been safely administered in concentrations much less than 1 g/kg body weight (0.5–8 mg/kg Fe as contrast agent and up to 20 mg/kg Fe as supplemental iron) as shown in Table 12.2. As a supplement, higher doses can be given to iron-deficient patients; however, these MNP formulations are employed as a source of iron after degradation and are not intended for use in strong magnetic fields as MRI contrast agents. Improper handling of the MNPs in a magnetic field before administration can lead to adverse effects such as chaining of the MNPs, leading to embolism.

12.5 SUMMARY

Much research has focused on the use of SPIONs for imaging and drug delivery. Generally well-tolerated, assessment of these particles for unintended consequences suggests that surface chemistry and size may play a role in the adverse effects on biological systems. More importantly, the toxic effects of the MNPs are dose-dependent and the optimal dose can vary for different types of particles depending on their specific properties. In general, most applications require a dose below 1 mg/mL whereas doses significantly greater than 1 mg/mL are generally well-tolerated in in vitro studies. Most authors who have reported on the toxicity of MNPs for in vivo studies have employed test amounts smaller than the toxicity range of MNPs, 0.5–8 mg/kg body weight for in vivo MRI studies, up to 25 μg/mL for magnetic transfection studies and up to 1 mg/mL for different in vitro cell studies. FDA-approved formulations are also used at levels well below the toxic doses.

TABLE 12.2
Iron Oxide Pharmaceutical Preparations with Their Availability Status and Recommended Dosage

Trade Name	Manufacturer	Indication	Status	Coating	Core Diameter (nm)	Particle hydrodynamic Diameter (nm)	Dose	Source(s)
Ferumoxides/AMI-25 (Standard SPION)								
Feridex IV Endorem	AMAG Pharma Guerbet	Liver/spleen imaging, myocardiaPerfusion imaging	Discontinued	Dextran	~ 5	120–180	0.56 mg Fe/kg Or 15 µmol Fe/kg	Leung (2004a), Bartolozzi et al. (1999), Senéterre et al. (1996), Ros et al. (1995), Weissleder et al. (1989)
Ferumoxtran-10/AMI-227 (Ultra-Small SPION)								
Combidex Sinerem	AMAG Pharma Guerbet	Lymph node and bone marrow imaging	Discontinued	Low mol. wt. dextran	4–6	20–40	2.6 mg Fe/kg diluted with 100 mL normal saline	Wang (2011), Drugs.com (2005), Michel et al. (2001)
Feruglose/NC100150 (Ultra-Small SPION)								
Clariscan (PEG-fero)	(Nycomed) Amersham Health (now GE Healthcare)	MR angiograph agent	Discontinued	PEG	4–7	~20	0.5–8 mg Fe/kg body wt., up to 2–5 mg Fe/kg	Wang (2011), Johansson (2006)
(Oral Contrast Agents)								
Ferumoxsil/AMI-121 (Large SPION)								
Lumirem GastroMARK	Guerbet 1993 Advanced Magnetics 1996	Gastrointestinal imaging/ bowel contrast	Available	Poly-[*N*-(2-aminoethyl)-3-aminopropyl] Siloxane	~10	~400	600 mL (105 mg Fe) at a rate of 300 mL over 15 minutes. Max. oral dose is 900 mL (157.5 mg Fe)	Leung (2004b), Hahn et al. (1990), Johnson et al. (1996), Drugs.com (2010)

(*Continued*)

TABLE 12.2 (Continued)

Trade Name	Manufacturer	Indication	Status	Coating	Core Diameter (nm)	Particle hydrodynamic Diameter (nm)	Dose	Source(s)
Ferristene (Oral/Rectal Large SPION)								
Abdoscan/ Dynosphere M-035	Nycomed Imaging/BCM Pharma	Bowel contrast	Discontinued	Insoluble	~10	~400	200–400 mL (23.4 mg Fe/200 mL)	Michel et al. (2001), Jacobsen et al. (1996)
Ferucarbotran/SHU 555A (Standard SPION)								
Resovist/Cliavist	Bayer Pharma (discontinued) Fujifilm Ri Pharma Co. (Japan)	Liver/spleen imaging	Available in Japan only	Carboxy-dextran	~4.5	45–60	<60 kg: 0.45 mmol Fe (0.9 mL) >60 kg: 0.7 mmol Fe (1.4 mL)	Wang (2011)
LS-008	Lode Spin Labs, Seattle WA	Long circulating MPI tracer	Pre-clinical development	PMAO-PEG	26	78	2.35 mg/mL Fe	Kemp, Ferguson, Khandhar, & Krishnan, 2016), R. M. Ferguson et al. (2013), R. M. Ferguson et al. (2015)
Supplemental Iron (Ultra-Small SPION)								
Feraheme (ferumoxytol)	AMAG Pharma, USA (2009)	Iron replacement therapy	Available	Polyglucose sorbitol carboxymethylether	~5	17–31	510 mg iron/17 mL in single-use vials	Macdougall (2009), Spinowitz et al. (2008), Bullivant et al. (2013), AMAG Pharma. (2009), Jahn et al. (2011)
Rienso (ferumoxytol)	Takeda Pharma, Denmark (2012)	Iron replacement therapy	Discontinued (2015)	Polyglucose sorbitol carboxymethylether	~5	17–31	510 mg iron/vial/50 kg body wt.	Scott, Lyseng-Williamson, and McCormack (Scott, Lyseng-Williamson,

(Continued)

TABLE 12.2 (Continued)

Trade Name	Manufacturer	Indication	Status	Coating	Core Diameter (nm)	Particle hydrodynamic Diameter (nm)	Dose	Source(s)
								and McCormack. 2013), Wáng and Idée (2017)
Ferinject/ Injectafer (iron carboxy-maltose)	Vifor Pharma	Iron replacement therapy	Available	Carbohydrate (dextran-free) with akaganeite core	11.7	23.1	200–1000 mg Fe (up to a maximum of 15 mg/kg body wt.) in single dose	Macdougall (2009), Vifor France (2017), Kulnigg et al. (2008), Jahn et al. (2011)
CosmoFer (iron dextran)	Teva, Germany	Iron replacement therapy	Available	Low-molecular-weight dextran	5.6	12.2	100–200 mg Fe (2–3 times a week) or up to 20 mg/kg in single dose	Jahn et al. (2011), Pharmacosmos (2014)
Ferrlecit (sodium ferric gluconate)	Sanofi-Aventis, Germany	Iron replacement therapy	Available	Carbohydrate	4.1	8.6	10 mL of Ferrlecit (125 mg Fe) in single dose	Jahn et al. (2011)
Monofer (iron isomaltoside 1000)	Pharmacosmos, Denmark	Iron replacement therapy	Available	Carbohydrate	6.3	9.9	Up to 20 mg Fe/kg body wt.	Jahn et al. (2011), Wikström et al. (2011)
Venofer (iron sucrose)	Vifor, Germany	Iron replacement therapy	Available	Polynuclear iron (III)-hydroxide in sucrose	5.0	8.3	Up to 1000 mg Fe	Jahn et al. (2011)
For Hyperthermia								
NanoTherm	Mag Force Nanotechnologies	Hyperthermia	Approved	Aminosilane	~12	~16	0.28 mL of magnetic fluid per cm^3 of tumor volume	Maier-Hauff et al. (2011)

To date, most magnetic nanoparticle toxicity research has been piecemeal, and the field would benefit from a systematic research effort that quantifies the influence of particle size, shape, and surface chemistry. In such a study, particle characteristics could be controlled and exposures quantified. Results of such a study would be useful for developing models to predict the toxicity of particles proposed for biomedical purposes Further, this same approach could be used to examine the effects of magnetic nanoparticles on organisms that may be exposed upon release into the environment (Figure 12.2).

REFERENCES

Alexiou, C., R. J. Schmid, R. Jurgons, M. Kremer, G. Wanner, C. Bergemann, E. Huenges, T. Nawroth, W. Arnold, and F. G. Parak. 2006. Targeting cancer cells: Magnetic nanoparticles as drug carriers. *European Biophysics Journal: EBJ* 35 (5):446–450. https://doi.org/10.1007/s00249-006-0042-1.

AMAG Pharma. 2009. FERAHEME® (ferumoxytol) injection prescribing information. U.S. Food and Drug Administration. https://www.accessdata.fda.gov/drugsatfda_docs/label/2009/022180lbl.pdf.

Arami, H., A. Khandhar, D. Liggitt, and K. M. Krishnan. 2015. In vivo delivery, pharmacokinetics, biodistribution and toxicity of iron oxide nanoparticles. *Chemical Society Reviews* 44 (23):8576–8607. https://doi.org/10.1039/c5cs00541h.

Arbab, A. S., L. A. Bashaw, B. R. Miller, E. K. Jordan, B. K. Lewis, H. Kalish, and J. A. Frank. 2003. Characterization of biophysical and metabolic properties of cells labeled with superparamagnetic iron oxide nanoparticles and transfection agent for cellular MR imaging. *Radiology* 229 (3):838–846. https://doi.org/10.1148/radiol.2293021215.

Au, C., L. Mutkus, A. Dobson, J. Riffle, J. Lalli, and M. Aschner. 2007. Effects of nanoparticles on the adhesion and cell viability on astrocytes. *Biological Trace Element Research* 120 (1-3):248–256. https://doi.org/10.1007/s12011-007-0067-z.

Auffan, M., J. Rose, J.-Y. Bottero, G. V. Lowry, J.-P. Jolivet, and M. R. Wiesner. 2009. Towards a definition of inorganic nanoparticles from an environmental, health and safety perspective. *Nature Nanotechnology* 4 (10):634–641. https://doi.org/10.1038/nnano.2009.242.

Bartolozzi, C., R. Lencioni, F. Donati, and D. Cioni. 1999. Abdominal MR: Liver and pancreas. *European Radiology* 9 (8):1496–1512. https://doi.org/10.1007/s003300050876.

Berry, C. C., S. Wells, S. Charles, and A. S. G. Curtis. 2003. Dextran and albumin derivatised iron oxide nanoparticles: Influence on fibroblasts in vitro. *Biomaterials* 24 (25):4551–4557.

Berry, C. C., S. Wells, S. Charles, G. Aitchison, and A. S. G. Curtis. 2004. Cell response to dextran-derivatised iron oxide nanoparticles post internalisation. *Biomaterials* 25 (23):5405–5413. https://doi.org/10.1016/j.biomaterials.2003.12.046.

Bjørnerud, A., and L. Johansson. 2004. The utility of superparamagnetic contrast agents in MRI: Theoretical consideration and applications in the cardiovascular system. *NMR in Biomedicine* 17 (7):465–477. https://doi.org/10.1002/nbm.904.

Bos, C., Y. Delmas, A. Desmoulière, A. Solanilla, O. Hauger, C. Grosset, I. Dubus, et al. 2004. In vivo MR imaging of intravascularly injected magnetically labeled mesenchymal stem cells in rat kidney and liver. *Radiology* 233 (3):781–789. https://doi.org/10.1148/radiol.2333031714.

Brunner, T. J., P. Wick, P. Manser, P. Spohn, R. N. Grass, L. K. Limbach, A. Bruinink, and W. J. Stark. 2006. In vitro cytotoxicity of oxide nanoparticles: Comparison to asbestos, silica, and the effect of particle solubility. *Environmental Science & Technology* 40 (14):4374–4381.

Bullivant, J. P., S. Zhao, B. J. Willenberg, B. Kozissnik, C. D. Batich, and J. Dobson. 2013. Materials characterization of feraheme/ferumoxytol and preliminary evaluation of its potential for magnetic fluid hyperthermia. *International Journal of Molecular Sciences* 14 (9):17501–17510. https://doi.org/10.3390/ijms140917501.

Caravan, P., J. J. Ellison, T. J. McMurry, and R. B. Lauffer. 1999. Gadolinium(III) chelates as MRI contrast agents: Structure, dynamics, and applications. *Chemical Reviews* 99 (9):2293–2352.

Cheng, F.-Y., C.-H. Su, Y.-S. Yang, C.-S. Yeh, C.-Y. Tsai, C.-L. Wu, M.-T. Wu, and D.-B. Shieh. 2005. Characterization of aqueous dispersions of Fe(3)O(4) nanoparticles and their biomedical applications. *Biomaterials* 26 (7):729–738. https://doi.org/10.1016/j.biomaterials.2004.03.016.

Chorny, M., B. Polyak, I. S. Alferiev, K. Walsh, G. Friedman, and R. J. Levy. 2007. Magnetically driven plasmid DNA delivery with biodegradable polymeric nanoparticles. *FASEB Journal: Official Publication of the Federation of American Societies for Experimental Biology* 21 (10):2510–2519. https://doi.org/10.1096/fj.06-8070com.

Crane, M., R. D. Handy, J. Garrod, and R. Owen. 2008. Ecotoxicity test methods and environmental hazard assessment for engineered nanoparticles. *Ecotoxicology (London, England)* 17 (5):421–437. https://doi.org/10.1007/s10646-008-0215-z.

Curtis Klaassen. 2013. *Casarett & Doull's Toxicology: The Basic Science of Poisons*. 8th ed. McGraw-Hill Professional. https://doi.org/10.1036/9780071769228.

de Freitas, E. R. L., P. R. O. Soares, R. de Paula Santos, R. L. dos Santos, J. R. da Silva, E. P. Porfirio, S. N. Báo, E. C. de Oliveira Lima, P. C. Morais, and L. A. Guillo. 2008. In vitro biological activities of anionic gamma-Fe_2O_3 nanoparticles on human melanoma cells. *Journal of Nanoscience and Nanotechnology* 8 (5):2385–2391.

Delcroix, G. J.-R., M. Jacquart, L. Lemaire, L. Sindji, F. Franconi, J.-J. Le Jeune, and C. N. Montero-Menei. 2009. Mesenchymal and neural stem cells labeled with HEDP-coated SPIO nanoparticles: In vitro characterization and migration potential in rat brain. *Brain Research* 1255 (February):18–31. https://doi.org/10.1016/j.brainres.2008.12.013.

Donaldson, K., C. A. Poland, F. A. Murphy, M. MacFarlane, T. Chernova, and A. Schinwald. 2013. Pulmonary toxicity of carbon nanotubes and asbestos—Similarities and differences. *Advanced Drug Delivery Reviews* 65 (15):2078–2086. https://doi.org/10.1016/j.addr.2013.07.014.

Dunning, M. D., A. Lakatos, L. Loizou, M. Kettunen, C. ffrench-Constant, K. M. Brindle, and R. J. M. Franklin. 2004. Superparamagnetic iron oxide-labeled Schwann cells and olfactory ensheathing cells can be traced in vivo by magnetic resonance imaging and retain functional properties after transplantation into the CNS. *The Journal of Neuroscience: The Official Journal of the Society for Neuroscience* 24 (44):9799–9810. https://doi.org/10.1523/JNEUROSCI.3126-04.2004.

Drugs.com. 2005. Combidex (Ferumoxtran-10) FDA approval status. Drugs.Com. March 2005. https://www.drugs.com/history/combidex.html.

Drugs.com. 2010. Gastromark—FDA prescribing information, side effects and uses. Drugs.Com. February 2010. https://www.drugs.com/pro/gastromark.html.

Dönmez Güngüneş, Ç., Ş. Şeker, A. E. Elçin, and Y. M. Elçin. 2017. A comparative study on the in vitro cytotoxic responses of two mammalian cell types to fullerenes, carbon nanotubes and iron oxide nanoparticles. *Drug and Chemical Toxicology* 40 (2):215–227. https://doi.org/10.1080/01480545.2016.1199563.

Ekwall, B. 1983. Screening of toxic compounds in mammalian cell cultures. *Annals of the New York Academy of Sciences* 407:64–77.

Ekwall, B. 1999. Overview of the final MEIC results: II. The In vitro—In vivo evaluation, including the selection of a practical battery of cell tests for prediction of acute lethal blood concentrations in humans. *Toxicology in Vitro: An International Journal Published in Association with BIBRA* 13 (4-5):665–673.

Evans, S. M., A. Casartelli, E. Herreros, D. T. Minnick, C. Day, E. George, and C. Westmoreland. 2001. Development of a high throughput in vitro toxicity screen predictive of high acute in vivo toxic potential. *Toxicology in Vitro* 15 (4-5):579–584.

Federici, G., B. J. Shaw, and R. D. Handy. 2007. Toxicity of titanium dioxide nanoparticles to rainbow trout (*Oncorhynchus mykiss*): Gill injury, oxidative stress, and other physiological effects. *Aquatic Toxicology (Amsterdam, Netherlands)* 84 (4):415–430. https://doi.org/10.1016/j.aquatox.2007.07.009.

Ferguson, R. M., A. P. Khandhar, H. Arami, L. Hua, O. Hovorka, and K. M. Krishnan. 2013. Tailoring the magnetic and pharmacokinetic properties of iron oxide magnetic particle imaging tracers. *Biomedizinische Technik. Biomedical Engineering* 58 (6):493–507. https://doi.org/10.1515/bmt-2012-0058.

Ferguson, R. M., A. P. Khandhar, S. J. Kemp, H. Arami, E. U. Saritas, L. R. Croft, J. Konkle, et al. 2015. Magnetic particle imaging with tailored iron oxide nanoparticle tracers. *IEEE Transactions on Medical Imaging* 34 (5):1077–1084. https://doi.org/10.1109/TMI.2014.2375065.

Goodman, C. M., C. D. McCusker, T. Yilmaz, and V. M. Rotello. 2004. Toxicity of gold nanoparticles functionalized with cationic and anionic side chains. *Bioconjugate Chemistry* 15 (4):897–900. https://doi.org/10.1021/bc049951i.

Gu, L., R. H. Fang, M. J. Sailor, and J.-H. Park. 2012. In vivo clearance and toxicity of monodisperse iron oxide nanocrystals. *ACS Nano* 6 (6):4947–4954. https://doi.org/10.1021/nn300456z.

Gupta, A. K., and A. S. G. Curtis. 2004. Lactoferrin and ceruloplasmin derivatized superparamagnetic iron oxide nanoparticles for targeting cell surface receptors. *Biomaterials* 25 (15):3029–3040. https://doi.org/10.1016/j.biomaterials.2003.09.095.

Gupta, A. K., and M. Gupta. 2005. Cytotoxicity suppression and cellular uptake enhancement of surface modified magnetic nanoparticles. *Biomaterials* 26 (13):1565–1573. https://doi.org/10.1016/j.biomaterials.2004.05.022.

Gupta, A. K., and S. Wells. 2004. Surface-modified superparamagnetic nanoparticles for drug delivery: Preparation, characterization, and cytotoxicity studies. *IEEE Transactions on Nanobioscience* 3 (1):66–73.

Gupta, A. K., C. Berry, M. Gupta, and A. Curtis. 2003. Receptor-mediated targeting of magnetic nanoparticles using insulin as a surface ligand to prevent endocytosis. *IEEE Transactions on Nanobioscience* 2 (4):255–261.

Hagemeier, J., J. J. G. Geurts, and R. Zivadinov. 2012. Brain iron accumulation in aging and neurodegenerative disorders. *Expert Review of Neurotherapeutics* 12 (12):1467–1480. https://doi.org/10.1586/ern.12.128.

Hahn, P. F., D. D. Stark, J. M. Lewis, S. Saini, G. Elizondo, R. Weissleder, C. J. Fretz, and J. T. Ferrucci. 1990. First clinical trial of a new superparamagnetic iron oxide for use as an oral gastrointestinal contrast agent in MR imaging. *Radiology* 175 (3):695–700. https://doi.org/10.1148/radiology.175.3.2343116.

Hamid, R., Y. Rotshteyn, L. Rabadi, R. Parikh, and P. Bullock. 2004. Comparison of alamar blue and MTT assays for high through-put screening. *Toxicology in Vitro: An International Journal Published in Association with BIBRA* 18 (5):703–710. https://doi.org/10.1016/j.tiv.2004.03.012.

Handy, R. D., G. Cornelis, T. Fernandes, O. Tsyusko, A. Decho, T. Sabo-Attwood, C. Metcalfe, et al. 2012. Ecotoxicity test methods for engineered nanomaterials: Practical experiences and recommendations from the bench. *Environmental Toxicology and Chemistry* 31 (1):15–31. https://doi.org/10.1002/etc.706.

Harisinghani, M. G., J. Barentsz, P. F. Hahn, W. M. Deserno, S. Tabatabaei, C. H. van de Kaa, J. de la Rosette, and R. Weissleder. 2003. Noninvasive detection of clinically occult lymph-node metastases in prostate cancer. *The New England Journal of Medicine* 348 (25):2491–2499. https://doi.org/10.1056/NEJMoa022749.

Heesakkers, R. A. M., G. J. Jager, A. M. Hövels, B. de Hoop, H. C. M. van den Bosch, F. Raat, J. A. Witjes, P. F. A. Mulders, C. H. van der Kaa, and J. O. Barentsz. 2009. Prostate cancer: Detection of lymph node metastases outside the routine surgical area with ferumoxtran-10-enhanced MR imaging. *Radiology* 251 (2):408–414. https://doi.org/10.1148/radiol.2512071018.

Horie, M., K. Nishio, K. Fujita, H. Kato, A. Nakamura, S. Kinugasa, S. Endoh, et al. 2009. Ultrafine NiO particles induce cytotoxicity in vitro by cellular uptake and subsequent Ni(II) release. *Chemical Research in Toxicology* 22 (8):1415–1426. https://doi.org/10.1021/tx900171n.

Huang, G., J. Diakur, Z. Xu, and L. I. Wiebe. 2008a. Asialoglycoprotein receptor-targeted superparamagnetic iron oxide nanoparticles. *International Journal of Pharmaceutics* 360 (1-2):197–203. https://doi.org/10.1016/j.ijpharm.2008.04.029.

Huang, J. H., H. J. Parab, R.-S. Liu, T.-C. Lai, M. Hsiao, C.-H. Chen, H.-S. Sheu, J.-M. Chen, D.-P. Tsai, and Y.-K. Hwu. 2008b. Investigation of the growth mechanism of iron oxide nanoparticles via a seed-mediated method and its cytotoxicity studies. *The Journal of Physical Chemistry C* 112 (40):15684–15690.

Hussain, S. M., K. L. Hess, J. M. Gearhart, K. T. Geiss, and J. J. Schlager. 2005. In vitro toxicity of nanoparticles in BRL 3A rat liver cells. *Toxicology in Vitro: An International Journal Published in Association with BIBRA* 19 (7):975–983. https://doi.org/10.1016/j.tiv.2005.06.034.

Häfeli, U. O., J. S. Riffle, L. Harris-Shekhawat, A. Carmichael-Baranauskas, F. Mark, J. P. Dailey, and D. Bardenstein. 2009. Cell uptake and in vitro toxicity of magnetic nanoparticles suitable for drug delivery. *Molecular Pharmaceutics* 6 (5):1417–1428. https://doi.org/10.1021/mp900083m.

Jacobsen, T. F., M. Laniado, B. E. Van Beers, B. Dupas, F. P. Boudghène, E. Rummeny, T. H. Falke, P. A. Rinck, D. MacVicar, and B. Lundby. 1996. Oral magnetic particles (Ferristene) as a contrast medium in abdominal magnetic resonance imaging. *Academic Radiology* 3 (7):571–580.

Jahn, M. R., H. B. Andreasen, S. Fütterer, T. Nawroth, V. Schünemann, U. Kolb, W. Hofmeister, M. Muñoz, K. Bock, and M. Meldal. 2011. A comparative study of the physicochemical properties of iron isomaltoside 1000 (Monofer®), a new intravenous iron preparation and its clinical implications. *European Journal of Pharmaceutics and Biopharmaceutics* 78 (3):480–491.

Jain, T. K., M. K. Reddy, M. A. Morales, D. L. Leslie-Pelecky, and V. Labhasetwar. 2008. Biodistribution, clearance, and biocompatibility of iron oxide magnetic nanoparticles in rats. *Molecular Pharmaceutics* 5 (2):316–327. https://doi.org/10.1021/mp7001285.

Johansson, L. 2006. Method of magnetic resonance imaging. US7082326B2, issued July 25, 2006.

Johnson, W. K., C. Stoupis, G. M. Torres, E. B. Rosenberg, and P. R. Ros. 1996. Superparamagnetic iron oxide (SPIO) as an oral contrast agent in gastrointestinal (GI) magnetic resonance imaging (MRI): Comparison with state-of-the-art computed tomography (CT). *Magnetic Resonance Imaging* 14 (1):43–49.

Joshi, R., V. Feldmann, W. Koestner, C. Detje, S. Gottschalk, H. A. Mayer, M. G. Sauer, and J. Engelmann. 2013. Multifunctional silica nanoparticles for optical and magnetic resonance imaging. *Biological Chemistry* 394 (1):125–135. https://doi.org/10.1515/hsz-2012-0251.

Karlsson, H. L. 2010. The comet assay in nanotoxicology research. *Analytical and Bioanalytical Chemistry* 398 (2):651–666. https://doi.org/10.1007/s00216-010-3977-0.

Kemp, S. J., R. M. Ferguson, A. P. Khandhar, and K. M. Krishnan. 2016. Monodisperse magnetite nanoparticles with nearly ideal saturation magnetization. *RSC Advances* 6 (81):77452–77464.

Keselman, P., E. Y. Yu, X. Y. Zhou, P. W. Goodwill, P. Chandrasekharan, R. M. Ferguson, A. P. Khandhar, et al. 2017. Tracking short-term biodistribution and long-term clearance of SPIO tracers in magnetic particle imaging. *Physics in Medicine and Biology* 62 (9):3440–3453. https://doi.org/10.1088/1361-6560/aa5f48.

Kim, J.-S., T.-J. Yoon, K.-N. Yu, M.-S. Noh, M. Woo, B.-G. Kim, K.-H. Lee, et al. 2006. Cellular uptake of magnetic nanoparticle is mediated through energy-dependent endocytosis in A549 cells. *Journal of Veterinary Science* 7 (4):321–326.

Klaine, S. J. 2009. Considerations for research on the environmental fate and effects of nanoparticles. *Environmental Toxicology and Chemistry* 28 (9):1787–1788. https://doi.org/10.1897/09-203.1.

Klaine, S. J., A. A. Koelmans, N. Horne, S. Carley, R. D. Handy, L. Kapustka, B. Nowack, and F. von der Kammer. 2012. Paradigms to assess the environmental impact of manufactured nanomaterials. *Environmental Toxicology and Chemistry* 31 (1):3–14. https://doi.org/10.1002/etc.733.

Kulnigg, S., S. Stoinov, V. Simanenkov, L. V. Dudar, W. Karnafel, L. C. Garcia, A. M. Sambuelli, G. D'Haens, and C. Gasche. 2008. A novel intravenous iron formulation for treatment of anemia in inflammatory bowel disease: The ferric carboxymaltose (FERINJECT) randomized controlled trial. *The American Journal of Gastroenterology* 103 (5):1182–1192. https://doi.org/10.1111/j.1572-0241.2007.01744.x.

Lee, K.-J., J.-H. An, J.-S. Shin, D.-H. Kim, C. Kim, H. Ozaki, and J.-G. Koh. 2007. Protective effect of maghemite nanoparticles on ultraviolet-induced photo-damage in human skin fibroblasts. *Nanotechnology* 18 (46):465201. https://doi.org/10.1088/0957-4484/18/46/465201.

Leung, K. 2004a. Ferumoxides. In: *Molecular Imaging and Contrast Agent Database (MICAD)*. Bethesda, MD: National Center for Biotechnology Information (US). http://www.ncbi.nlm.nih.gov/books/NBK23037/.

Leung, K. 2004b. Ferumoxsil. In: *Molecular Imaging and Contrast Agent Database (MICAD)*. Bethesda, MD: National Center for Biotechnology Information (US). http://www.ncbi.nlm.nih.gov/books/NBK22994/.

Lu, J., S. Ma, J. Sun, C. Xia, C. Liu, Z. Wang, X. Zhao, et al. 2009. Manganese ferrite nanoparticle micellar nanocomposites as MRI contrast agent for liver imaging. *Biomaterials* 30 (15):2919–2928. https://doi.org/10.1016/j.biomaterials.2009.02.001.

Macdougall, I. C. 2009. Evolution of IV iron compounds over the last century. *Journal of Renal Care* 35 (Suppl 2):8–13. https://doi.org/10.1111/j.1755-6686.2009.00127.x.

Luitpold Pharma. 2013. INJECTAFER® (ferric carboxymaltose injection) prescribing information. U.S. Food and Drug Administration. https://www.accessdata.fda.gov/drugsatfda_docs/label/2013/203565s000lbl.pdf.

Mahmoudi, M., M. A. Shokrgozar, A. Simchi, M. Imani, A. S. Milani, P. Stroeve, H. Vali, U. O. Häfeli, and S. Bonakdar. 2009a. Multiphysics flow modeling and in vitro toxicity of iron oxide nanoparticles coated with poly (vinyl alcohol). *The Journal of Physical Chemistry C* 113 (6):2322–2331.

Mahmoudi, M., A. Simchi, M. Imani, A. S. Milani, and P. Stroeve. 2009b. An in vitro study of bare and poly (ethylene glycol)-*co*-fumarate-coated superparamagnetic iron oxide nanoparticles: A new toxicity identification procedure. *Nanotechnology* 20 (22):225104. https://doi.org/10.1088/0957-4484/20/22/225104.

Maier-Hauff, K., F. Ulrich, D. Nestler, H. Niehoff, P. Wust, B. Thiesen, H. Orawa, V. Budach, and A. Jordan. 2011. Efficacy and safety of intratumoral thermotherapy using magnetic iron-oxide nanoparticles combined with external beam radiotherapy on patients with recurrent glioblastoma multiforme. *Journal of Neuro-Oncology* 103 (2):317–324. https://doi.org/10.1007/s11060-010-0389-0.

Malvindi, M. A., V. De Matteis, A. Galeone, V. Brunetti, G. C. Anyfantis, A. Athanassiou, R. Cingolani, and P. P. Pompa. 2014. Toxicity assessment of silica coated iron oxide nanoparticles and biocompatibility improvement by surface engineering. *PLoS One* 9 (1):e85835. https://doi.org/10.1371/journal.pone.0085835.

Manickam, V., V. Dhakshinamoorthy, and E. Perumal. 2018. Iron oxide nanoparticles induces cell cycle-dependent neuronal apoptosis in mice. *Journal of Molecular Neuroscience: MN* 64 (3):352–362. https://doi.org/10.1007/s12031-018-1030-5.

Michel, S. C. A., B. Marincek, J. Fröhlich, D. Nanz, D. Fink, R. Caduff, and R. Kubik-Huch. 2001. Lymph node staging with a lymph node-specific contrast agent (Sinerem®, Guerbet) in patients with breast cancer. *Proceedings of the International Society for Magnetic Resonance in Medicine* 9:569.

Mykhaylyk, O., Y. S. Antequera, D. Vlaskou, and C. Plank. 2007. Generation of magnetic nonviral gene transfer agents and magnetofection in vitro. *Nature Protocols* 2 (10):2391–2411. https://doi.org/10.1038/nprot.2007.352.

Minimum Information for Nanomaterial Characterization (MINChar) Initiative. 2008. Recommended minimum physical and chemical parameters for characterizing nanomaterials on toxicology studies. *Health & Environmental Research Online (HERO)* ID 594265.

Möller, W., I. Nemoto, T. Matsuzaki, T. Hofer, and J. Heyder. 2000. Magnetic phagosome motion in J774A.1 macrophages: Influence of cytoskeletal drugs. *Biophysical Journal* 79 (2):720–730. https://doi.org/10.1016/S0006-3495(00)76330-2.

Möller, W., W. Barth, M. Kohlhäufl, K. Häussinger, W. Stahlhofen, and J. Heyder. 2001. Human alveolar long-term clearance of ferromagnetic iron oxide microparticles in healthy and diseased subjects. *Experimental Lung Research* 27 (7):547–568.

Möller, W., T. Hofer, A. Ziesenis, E. Karg, and J. Heyder. 2002. Ultrafine particles cause cytoskeletal dysfunctions in macrophages. *Toxicology and Applied Pharmacology* 182 (3):197–207.

Müller, K., J. N. Skepper, T. Y. Tang, M. J. Graves, A. J. Patterson, C. Corot, E. Lancelot, P. W. Thompson, A. P. Brown, and J. H. Gillard. 2008. Atorvastatin and uptake of ultrasmall superparamagnetic iron oxide nanoparticles (Ferumoxtran-10) in human monocyte-macrophages: Implications for magnetic resonance imaging. *Biomaterials* 29 (17):2656–2662. https://doi.org/10.1016/j.biomaterials.2008.03.006.

Müller, R. H., S. Maaβen, H. Weyhers, F. Specht, and J. S. Lucks. 1996. Cytotoxicity of magnetite-loaded polylactide, polylactide/glycolide particles and solid lipid nanoparticles. *International Journal of Pharmaceutics* 138 (1):85–94.

Müller, R. H., S. Maassen, C. Schwarz, and W. Mehnert. 1997. Solid lipid nanoparticles (SLN) as potential carrier for human use: Interaction with human granulocytes. *Journal of Controlled Release* 47 (3):261–269.

Na, H. B., J. H. Lee, K. An, Y. I. Park, M. Park, I. S. Lee, D.-H. Nam, et al. 2007. Development of a T1 contrast agent for magnetic resonance imaging using MnO nanoparticles. *Angewandte Chemie (International Ed. in English)* 46 (28):5397–5401. https://doi.org/10.1002/anie.200604775.

Nel, A. E. 2013. Implementation of alternative test strategies for the safety assessment of engineered nanomaterials. *Journal of Internal Medicine* 274 (6):561–577. https://doi.org/10.1111/joim.12109.

Nassour, C., S. J. Barton, S. Nabhani-Gebara, et al. 2007. Occurrence of anticancer drugs in the aquatic environment: a systematic review. *Environ Sci Pollut Res* 27, 1339–1347. https://doi.org/10.1007/s11356-019-07045-2.

National Research Council. 2007. *Toxicity Testing in the 21st Century: A Vision and a Strategy*. National Academies Press.

Nel, A. E., E. Nasser, H. Godwin, D. Avery, T. Bahadori, L. Bergeson, E. Beryt, et al. 2013. A multi-stakeholder perspective on the use of alternative test strategies for nanomaterial safety assessment. *ACS Nano* 7 (8):6422–6433. https://doi.org/10.1021/nn4037927.

Nemoto, I., and W. Moeller. 2000. A viscoelastic model of phagosome motion within cells based on cytomagnetometric measurements. *IEEE Transactions on Bio-Medical Engineering* 47 (2):170–182. https://doi.org/10.1109/10.821751.

Neuberger, T., B. Schöpf, H. Hofmann, M. Hofmann, and B. Von Rechenberg. 2005. Superparamagnetic nanoparticles for biomedical applications: Possibilities and limitations of a new drug delivery system. *Journal of Magnetism and Magnetic Materials* 293 (1):483–496.

Omidkhoda, A., H. Mozdarani, A. Movasaghpoor, and A. A. P. Fatholah. 2007. Study of apoptosis in labeled mesenchymal stem cells with superparamagnetic iron oxide using neutral comet assay. *Toxicology in Vitro: An International Journal Published in Association with BIBRA* 21 (6):1191–1196. https://doi.org/10.1016/j.tiv.2007.03.010.

OECD. 2010. Current Developments/Activities on the Safety of Manufactured Nanomaterials, Tour de Table at the 7th Meeting of the Working Party on Manufactured Nanomaterials. Tour de Table at the 7th Meeting of the Working Party on Manufactured Nanomaterials No. 26. Series on the Safety of Manufactured Nanomaterials. Paris, France: Organisation for Economic Co-operation and Development.

OECD. 2012. Guidance on Sample Preparation and Dosimetry for the Safety Testing of Manufactured Nanomaterials. Tour de Table at the 7th Meeting of the Working Party on Manufactured Nanomaterials No. 36. Series on the Safety of Manufactured Nanomaterials. Paris, France: Organisation for Economic Co-operation and Development.

Pace, H. E., E. K. Lesher, and J. F. Ranville. 2010. Influence of stability on the acute toxicity of CdSe/ZnS nanocrystals to daphnia magna. *Environmental Toxicology and Chemistry* 29 (6):1338–1344. https://doi.org/10.1002/etc.168.

Park, J. Y., M. J. Baek, E. S. Choi, S. Woo, J. H. Kim, T. J. Kim, J. C. Jung, K. S. Chae, Y. Chang, and G. H. Lee. 2009. Paramagnetic ultrasmall gadolinium oxide nanoparticles as advanced T1 MRI contrast agent: Account for large longitudinal relaxivity, optimal particle diameter, and in vivo T1 MR images. *ACS Nano* 3 (11):3663–3669. https://doi.org/10.1021/nn900761s.

Patil, R. M., Nanasaheb D. T., P. B. Shete, P. A. Bedge, S. Gavde, M. G. Joshi, S. A. M. Tofail, and R. A. Bohara. 2018. Comprehensive cytotoxicity studies of superparamagnetic iron oxide nanoparticles. *Biochemistry and Biophysics Reports* 13 (March):63–72. https://doi.org/10.1016/j.bbrep.2017.12.002.

Patil, U. S., S. Adireddy, A. Jaiswal, S. Mandava, B. R. Lee, and D. B. Chrisey. 2015. In vitro/in vivo toxicity evaluation and quantification of iron oxide nanoparticles. *International Journal of Molecular Sciences* 16 (10):24417–24450. https://doi.org/10.3390/ijms161024417.

Pawelczyk, E., A. S. Arbab, S. Pandit, E. Hu, and J. A. Frank. 2006. Expression of transferrin receptor and ferritin following ferumoxides-protamine sulfate labeling of cells: Implications for cellular magnetic resonance imaging. *NMR in Biomedicine* 19 (5):581–592. https://doi.org/10.1002/nbm.1038.

Pawelczyk, E., A. S. Arbab, A. Chaudhry, A. Balakumaran, P. G. Robey, and J. A. Frank. 2008. In vitro model of bromodeoxyuridine or iron oxide nanoparticle uptake by activated macrophages from labeled stem cells: Implications for cellular therapy. *Stem Cells (Dayton, Ohio)* 26 (5):1366–1375. https://doi.org/10.1634/stemcells.2007-0707.

Petri-Fink, A., M. Chastellain, L. Juillerat-Jeanneret, A. Ferrari, and H. Hofmann. 2005. Development of functionalized superparamagnetic iron oxide nanoparticles for interaction with human cancer cells. *Biomaterials* 26 (15):2685–2694. https://doi.org/10.1016/j.biomaterials.2004.07.023.

Pisanic, T. R., J. D. Blackwell, V. I. Shubayev, R. R. Fiñones, and S. Jin. 2007. Nanotoxicity of iron oxide nanoparticle internalization in growing neurons. *Biomaterials* 28 (16):2572–2581. https://doi.org/10.1016/j.biomaterials.2007.01.043.

Pharmacosmos. 2014. CosmoFer—Summary of Product Characteristics (SPC)—(EMC). Medicines.org.uk. https://www.medicines.org.uk/emc/medicine/14139.

Polyak, B., I. Fishbein, M. Chorny, I. Alferiev, D. Williams, B. Yellen, G. Friedman, and R. J. Levy. 2008. High field gradient targeting of magnetic nanoparticle-loaded endothelial cells to the surfaces of steel stents. *Proceedings of the National Academy of Sciences of the USA* 105 (2):698–703. https://doi.org/10.1073/pnas.0708338105.

Prabhu, S., S. Mutalik, S. Rai, N. Udupa, and B. S. S. Rao. 2015. PEGylation of superparamagnetic iron oxide nanoparticle for drug delivery applications with decreased toxicity: An in vivo study. *Journal of Nanoparticle Research* 17 (10):412.

Remya, N. S., S. Syama, A. Sabareeswaran, and P. V. Mohanan. 2016. Toxicity, toxicokinetics and biodistribution of dextran stabilized iron oxide nanoparticles for biomedical applications. *International Journal of Pharmaceutics* 511 (1):586–598. https://doi.org/10.1016/j.ijpharm.2016.06.119.

Rivière, C., C. Wilhelm, F. Cousin, V. Dupuis, F. Gazeau, and R. Perzynski. 2007. Internal structure of magnetic endosomes. *The European Physical Journal. E, Soft Matter* 22 (1):1–10. https://doi.org/10.1140/epje/e2007-00014-1.

Ros, P. R., P. C. Freeny, S. E. Harms, S. E. Seltzer, P. L. Davis, T. W. Chan, A. E. Stillman, L. R. Muroff, V. M. Runge, and M. A. Nissenbaum. 1995. Hepatic MR imaging with ferumoxides: A multicenter clinical trial of the safety and efficacy in the detection of focal hepatic lesions. *Radiology* 196 (2):481–488. https://doi.org/10.1148/radiology.196.2.7617864.

Sayes, C. M., K. L. Reed, and D. B. Warheit. 2007. Assessing toxicity of fine and nanoparticles: Comparing in vitro measurements to in vivo pulmonary toxicity profiles. *Toxicological Sciences: An Official Journal of the Society of Toxicology* 97 (1):163–180. https://doi.org/10.1093/toxsci/kfm018.

Schäfer, R., R. Kehlbach, J. Wiskirchen, R. Bantleon, J. Pintaske, B. R. Brehm, A. Gerber, H. Wolburg, C. D. Claussen, and H. Northoff. 2007. Transferrin receptor upregulation: In vitro labeling of rat mesenchymal stem cells with superparamagnetic iron oxide. *Radiology* 244 (2):514–523. https://doi.org/10.1148/radiol.2442060599.

Schäfer, R., R. Kehlbach, M. Müller, R. Bantleon, T. Kluba, M. Ayturan, G. Siegel, et al. 2009. Labeling of human mesenchymal stromal cells with superparamagnetic iron oxide leads to a decrease in migration capacity and colony formation ability. *Cytotherapy* 11 (1):68–78. https://doi.org/10.1080/14653240802666043.

Schäfer, R., R. Bantleon, R. Kehlbach, G. Siegel, J. Wiskirchen, H. Wolburg, T. Kluba, et al. 2010. Functional investigations on human mesenchymal stem cells exposed to magnetic fields and labeled with clinically approved iron nanoparticles. *BMC Cell Biology* 11 (April): 22. https://doi.org/10.1186/1471-2121-11-22.

Scott, L. J., K. A. Lyseng-Williamson, and P. L. McCormack. 2013. Ferumoxytol: A guide to its use in iron deficiency anaemia in adults with chronic kidney disease in the EU. *Drugs & Therapy Perspectives* 29 (8):223–227.

Senéterre, E., P. Taourel, Y. Bouvier, J. Pradel, B. Van Beers, J. P. Daures, J. Pringot, D. Mathieu, and J. M. Bruel. 1996. Detection of hepatic metastases: Ferumoxides-enhanced MR imaging versus unenhanced MR imaging and CT during arterial portography. *Radiology* 200 (3):785–792. https://doi.org/10.1148/radiology.200.3.8756932.

Spinowitz, B. S., A. T. Kausz, J. Baptista, S. D. Noble, R. Sothinathan, M. V. Bernardo, L. Brenner, and B. J. G. Pereira. 2008. Ferumoxytol for treating iron deficiency anemia in CKD. *Journal of the American Society of Nephrology: JASN* 19 (8):1599–1605. https://doi.org/10.1681/ASN.2007101156.

Stolnik, S., L. Illum, and S. S. Davis. 1995. Long circulating microparticulate drug carriers. *Advanced Drug Delivery Reviews* 16 (2):195–214.

Toraya-Brown, S., M. R. Sheen, J. R. Baird, S. Barry, E. Demidenko, M. J. Turk, P. J. Hoopes, J. R. Conejo-Garcia, and S. Fiering. 2013. Phagocytes mediate targeting of iron oxide nanoparticles to tumors for cancer therapy. *Integrative Biology: Quantitative Biosciences from Nano to Macro* 5 (1):159–171. https://doi.org/10.1039/c2ib20180a.

Toso, C., J.-P. Vallee, P. Morel, F. Ris, S. Demuylder-Mischler, M. Lepetit-Coiffe, N. Marangon, et al. 2008. Clinical magnetic resonance imaging of pancreatic islet grafts after iron nanoparticle labeling. *American Journal of Transplantation: Official Journal of the American Society of Transplantation and the American Society of Transplant Surgeons* 8 (3):701–706. https://doi.org/10.1111/j.1600-6143.2007.02120.x.

Tran, P. A., H. T. Nguyen, K. Fox, and N. Tran. 2018. In vitro cytotoxicity of iron oxide nanoparticles: Effects of chitosan and polyvinyl alcohol as stabilizing Agents. *Materials Research Express* 5 (3):035051.

Valberg, P. A., and J. D. Brain. 1988. Lung particle retention and lung macrophage function evaluated using magnetic aerosols. *Journal of Aerosol Medicine* 1 (4):331–349.

Villanueva, A., M. Cañete, A. G. Roca, M. Calero, S. Veintemillas-Verdaguer, C. J. Serna, M. del Puerto Morales, and R. Miranda. 2009. The influence of surface functionalization on the enhanced internalization of magnetic nanoparticles in cancer cells. *Nanotechnology* 20 (11):115103. https://doi.org/10.1088/0957-4484/20/11/115103.

Vifor France. 2017. Ferinject (Ferric Carboxymaltose)-Summary of Product Characteristics (SPC)-(EMC). Medicines.org.uk. https://www.medicines.org.uk/emc/medicine/24167.

Wang, Y.-X. J. 2011. Superparamagnetic iron oxide based MRI contrast agents: Current status of clinical application. *Quantitative Imaging in Medicine and Surgery* 1 (1):35–40. https://doi.org/10.3978/j.issn.2223-4292.2011.08.03.

Wang, Y.-X. J., S. M. Hussain, and G. P. Krestin. 2001. Superparamagnetic iron oxide contrast agents: Physicochemical characteristics and applications in MR imaging. *European Radiology* 11 (11):2319–2331. https://doi.org/10.1007/s003300100908.

Wardhana, and E. A. Datau. 2014. Metal fume fever among galvanized welders. *Acta Medica Indonesiana* 46 (3):256–262.

Warheit, D. B., C. M. Sayes, and K. L. Reed. 2009. Nanoscale and fine zinc oxide particles: Can in vitro assays accurately forecast lung hazards following inhalation exposures? *Environmental Science & Technology* 43 (20):7939–7945. https://doi.org/10.1021/es901453p.

Weissleder, R., D. D. Stark, B. L. Engelstad, B. R. Bacon, C. C. Compton, D. L. White, P. Jacobs, and J. Lewis. 1989. Superparamagnetic iron oxide: Pharmacokinetics and toxicity. *AJR. American Journal of Roentgenology* 152 (1):167–173. https://doi.org/10.2214/ajr.152.1.167.

Wikström, B., S. Bhandari, P. Barany, P. A. Kalra, S. Ladefoged, J. Wilske, and L. L. Thomsen. 2011. Iron isomaltoside 1000: A new intravenous iron for treating iron deficiency in chronic kidney disease. *Journal of Nephrology* 24 (5):589–596.

Wilhelm, C., C. Billotey, J. Roger, J. N. Pons, J.-C. Bacri, and F. Gazeau. 2003. Intracellular uptake of anionic superparamagnetic nanoparticles as a function of their surface coating. *Biomaterials* 24 (6):1001–1011.

Wáng, Y. X. J., and J.-M. Idée. 2017. A comprehensive literatures update of clinical researches of superparamagnetic resonance iron oxide nanoparticles for magnetic resonance imaging. *Quantitative Imaging in Medicine and Surgery* 7 (1):88–122. https://doi.org/10.21037/qims.2017.02.09.

Xie, H. 2012. Occurrence, ecotoxicology, and treatment of anticancer agents as water contaminants. *Environ Anal Toxicol.* https://doi.org/10.4172/2161-0525.S2-002.

Yang, L., H. Kuang, W. Zhang, Z. P. Aguilar, Y. Xiong, W. Lai, H. Xu, and H. Wei. 2015. Size dependent biodistribution and toxicokinetics of iron oxide magnetic nanoparticles in mice. *Nanoscale* 7 (2):625–636. https://doi.org/10.1039/c4nr05061d.

Yuen, S. 1974. Pullulan and its applications. *Process Biochemistry*. 9(9) pp. 7-9.

Yurchenko, O. V., I. N. Todor, I. K. Khayetsky, N. A. Tregubova, N. Yu Lukianova, and V. F. Chekhun. 2010. Ultrastructural and some functional changes in tumor cells treated with stabilized iron oxide nanoparticles. *Experimental Oncology* 32 (4):237–242.

Index

Note: Page numbers in *italics* denote figures and **bold** denote tables.

D

E

F

G

L

M

N

O

P

U

V

W

X

Y

Z